Animal Behavior

ANIMAL SEVENTH EDITION
BEHAVIOR

AN EVOLUTIONARY APPROACH

JOHN ALCOCK

ARIZONA STATE UNIVERSITY

SINAUER ASSOCIATES, INC.

SUNDERLAND, MASSACHUSETTS

The Cover

A bugling elk conveys honest information about his fighting capacity to rival males. Photograph by Bruce Lyon.

The Frontispiece

The threat display of the Hamadryas baboon raises many questions for evolutionary biologists. Photograph by C. K. Lorenz.

Animal Behavior: An Evolutionary Approach, Seventh Edition

© Copyright 2001 by Sinauer Associates, Inc.

Sinauer Associates, Inc., P. O. Box 407, Sunderland, Massachusetts, 01375-0407 U.S.A. Fax: 413-548-1118.

Internet: publish@sinauer.com; http://www.sinauer.com

Library of Congress Cataloging-in-Publication Data

Alcock, John, 1942-
Animal behavior / John Alcock.--7th ed.
p. cm.
Includes bibliographical references.
ISBN 0-87893-011-6
1. Animal behavior--Evolution. I. Title

QL751 .A58 2001
591.5--dc21   2001020603

10 9 8 7 6 5 4 3 2 1

To The Nature Conservancy

Contents in Brief

Contents

Preface

Another four years or so have passed since the previous edition of Animal Behavior appeared. During this time, much that is depressing has occurred on the political and environmental fronts. Let us not dwell on these events, but instead consider the positive side of the ledger, which includes the many excellent papers on animal behavior that have been written over the last few years. Indeed, the rate at which important discoveries about behavior are being made has steadily increased throughout the lifetime of my textbook. Just pick up a recent copy of any journal in the discipline and compare the research reports there with those in the same journal 15 or 20 years previously. Thus, in 1982, the journal *Animal Behaviour* published 1264 pages of reports; in 2000 the total was 2180 pages, a figure that underestimates the actual increase because the journal substantially enlarged its page size in 1998. Moreover, the quality of the work, as well as its quantity, has changed. The average recent paper on animal behavior is more sophisticated and interesting than the average paper written in the 1980s. Scientists really do build on what others have accomplished, which sets the bar higher and higher each year.

As researchers have met the challenge of expanding our already substantial knowledge of animal behavior, they have generated an embarrassment of riches for a textbook writer. As I have revised my book again, my main difficulty has been what to leave out, not what to put in. Although there have always been many more good papers than space available for my summaries, this time around the problem has seemed especially acute, given the numbers of excellent research reports coupled with my desire to limit the length of my book. Despite efforts in the past to keep the number of pages of text under control, my book had grown bigger over the years. This time I have managed to reverse the trend, which should make the book's contents more digestible, although it does mean that some good stories were omitted. Readers of my book can, however, explore any behavioral topic of their choice by turning to the original literature. In this regard, one tool—the Web of Science (http://www.webof-science.com)—is tremendously valuable because it provides such an easy and quick way to track down the work of particular researchers and to follow the trail of papers that cite a research report of interest. I cannot recommend this search mechanism too strongly.

In addition to bringing the book up to date and reducing its length somewhat, I have been able to incorporate color illustrations throughout the text, thanks to the willingness of my publisher, Sinauer Associates, to make the move from black and white to full color. This change not only adds to the aesthetic value of the book, but should also help readers grasp the point of graphs more quickly and easily. The photographs may also help students see why so many behavioral biologists love studying real animals in real environments.

The many changes that I have made to this seventh edition of my book have all been made with one primary goal in mind: to help my readers see how researchers have been able to reach satisfactory conclusions about how and why animals do the things they do. I hope that its examples of the wonderfully interesting research done on animal behavior will make the point that scientific logic offers a powerful means to gain an understanding of nature.

Acknowledgments

The author of any textbook depends on cooperation from a surprisingly large number of people. As I have rewritten the book again, I have been very fortunate in having many generous colleagues willing to give me their time, advice, photographs, and other forms of help. Mike Maxwell was especially helpful in providing information on microsatellite analyses. All the chapters of the book have incorporated changes suggested to me by the following reviewers: Alex Basolo, Eliot Brenowitz, Ken Catania, Robin Dunbar, Bruce Lyon, Jim Marden, Bob Montgomerie, Randy Nelson, Gabrielle Nevitt, Don Owings, Steve Nowicki, Kern Reeve, Gene Robinson, Tom Seeley, Nancy Segal, David Westneat, and Jeanne Zeh. Special thanks to Bob Montgomerie for reading the entire manuscript and for offering extremely helpful suggestions about how to do things better.

A host of other colleagues have provided me with permission to use illustrations that originally appeared in their papers, and some have given me the illustrations themselves. I want to single out Bruce Lyon for special thanks; his superb photographs appear in many places throughout the text as well as on the cover of the book. I have acknowledged all suppliers of photographs in the text at the appropriate figures. Acknowledgments to the publishers who have also generously granted permission to use their copyrighted material appear on pages located between the Bibliography and the Index.

My editor at Sinauer Associates, Pete Farley, has done the hard and often unglamourous work of keeping the project moving ahead. I am very grateful to him as well as to all the other Sinauerians, especially Chelsea Holabird, Joan Gemme, Chris Small, David McIntyre, and Mara Silver. Norma Roche, who copyedited this edition as well as many others, really knows how to fix mistakes. Readers of my book are lucky that they get to read the copyedited version rather than the original draft.

Although chapter reviewers, photographers, editorial staff, and presidents of the United States often change from edition to edition, some things stay the same, which provides a certain reassuring stability to my life. My wife Sue continues to cope with my many idiosyncrasies, still listening thoughtfully whenever I vigorously denounce the likes of, say, Henry Kissinger or Antonin Scalia, still willing to live in a cramped campervan for months at a time when we are in Western Australia for another round of bee research. My younger son Nick is still in town (Tempe, Arizona), and he takes time off from helping those accused of driving under the influence to help us eat dinner and play ping-pong. As in the past, he lets me win once in a while so that I can retain some small measure of self-esteem, that most important of modern commodities. My older son Joe is close enough (Albuquerque, New Mexico) that he can join us on occasion, helping me maintain the illusion of being surrounded by family. In addition, I am happy to report that none of my friends at Arizona State (among them Dave Brown, Steve Carroll, Jim Collins, Stuart Fisher, Dave Pearson, and Ron Rutowski) has yet gone to his reward, which means that we can all get together at lunch and sometimes on Friday afternoons for beer, just as we have been doing for decades. Admittedly, the amount of beer that we consume has changed, dropping from the pitcher per person of the good old days to a glass or two currently, but the topics under discussion (the numerous disadvantages of getting older and the troubles caused by defective colleagues) have remained the same, thank goodness. To my family and friends, thank you.

1 An Evolutionary Approach to Animal Behavior

*F*or hundreds of thousands of years, humans observed animals because their lives depended on a knowledge of animal behavior. Even today, the subject still has great practical significance. Information on the reproductive behavior of insect pests, for example, may ultimately lead to their control, while knowledge of the migratory routes of an endangered whale or shorebird may enable conservationists to design adequate reserves to save the animal from extinction. Moreover, an understanding of the evolutionary basis of our own behavior might help us identify why we so often damage our environments, perhaps enabling us to reduce our destructive tendencies [1251]. But even if the only beneficiaries of studies of animal behavior were the persons who conducted the research, I suspect that work in this field would continue. Learning how and why animals behave is an intrinsically fascinating business. Perhaps you can imagine what it would be like to be the first person to discover that male damselflies actually use their penis as a scrub brush to remove the sperm of rival males from their mates [1177]; or maybe you can put yourself in the shoes of the person who first showed that female Seychelles warblers could control the sex of their offspring so as to have daughters at times when it was most advantageous [641].

In the pages ahead, you will learn about these and many other remarkable discoveries. The point of this text, however, is not only to introduce you to these entertaining findings, but also to help you understand how scientists have determined that the damselfly penis serves as a competitive weapon of sorts or that the control of offspring sex ratio by the Seychelles warbler is an adaptation with a particular purpose. I believe that the process of doing science is every bit as interesting as the findings that are its end product. If I can help you understand the logic of science, as well as appreciate the wonderful diversity of animal behavior, my textbook will have done its job.

Questions about Behavior

I lived for one summer in Monteverde, a tiny community in the mountains of Costa Rica, which was founded by pacifist Quakers from the United States around the time of the Korean War. While I was there, a friend loaned me a black light, which I hung up by a white sheet on the back porch of our home. The ultraviolet rays of the lamp attracted hundreds of moths each night, and many stayed on the sheet until I could inspect them. Some mornings I found a huge bright yellow moth belonging the genus *Automeris* clinging to the sheet. In the chilly dawn, the sluggish moth did not struggle if I picked it up carefully. But if I jostled it suddenly, or poked it sharply on its thorax, the moth abruptly lifted its forewings and held them up to expose its previously concealed hindwings. The hindwings were marvelously decorated, with circular patches that looked like two eyes, which seemed to stare back at me (Figure 1).

Anyone seeing *Automeris* abruptly expose its hindwing "eyes" will have some questions about the behavior. But no matter how long the list of questions, each query can be assigned to one of two fundamentally different categories: "how questions," about the **proximate** mechanisms inside the moth that cause the behavior, or "why questions," about the **ultimate** or evolutionary reasons for the behavior [774, 859]. "How questions" about behavior ask *how* an individual manages to carry out an activity; this category of questions requires explanations about how an animal's internal mechanisms developed and how

1   *Automeris* moth from Costa Rica.   (Left) The moth in its resting position, with forewings held over the hindwings. (Right) After being jabbed in the thorax, the moth pulls its forewings forward, at which time the "eyes" on the hindwings become visible. Photographs by the author.

they then cause the animal to behave in a certain way. In contrast, "why questions" about behavior ask *why* the animal has evolved the mechanisms that underlie its actions.

How Questions about Proximate Causes

Consider the following questions about the wing-flipping reaction of an *Automeris* moth to a sharp poke:

- How do the moth's muscles make its wings move, and what controls those muscles?
- How does the moth know when it has been touched?
- Did the foods the moth ate as a caterpillar influence how it behaves as an adult?
- Did the moth inherit this behavior from its mother or father?

What these questions have in common, despite their diversity, is an interest in the operation of mechanisms *within* the moth that cause it to pull its forewings forward, revealing the amazing hindwings. The diversity of proximate questions is great enough, however, that we can subdivide them into two complementary groups, one dealing with the interactive effects of heredity and environment on the development of the mechanisms underlying wing-flipping, and the other dealing with how the fully developed physiological mechanisms actually operate when the behavior occurs.

The developmental side of the equation has to do with how the moth's heredity—its genes, its DNA—influenced the proliferation and specialization of cells that occurred as a fertilized egg gave rise to a caterpillar, which grew into an adult with a particular kind of nervous system. The operational side of the equation has to do with how neural mechanisms within the adult moth detect certain kinds of stimulation and how messages are then relayed to activate muscular reactions. Research on the developmental and physiological aspects of behavior remain to be carried out for *Automeris*, but someday we may learn about both of these proximate causes of its behavior.

Why Questions about Ultimate Causes

Even if we already knew everything there was to know about the proximate causes of wing-flipping by *Automeris* moths, we could still ask many more questions:

- What do today's moths gain, if anything, by wing-flipping?
- Has the behavior changed over evolutionary time?
- If so, what were the predecessors of today's wing-flipping response?
- If the behavior has changed, what caused the changes?

These questions all involve the evolutionary, or ultimate, reasons why an animal does something. Why does the moth suddenly lift its wings and expose its eyespots when it is molested? The British scientist David Blest suggested that the action spread because in the past wing-flipping frightened off some bird predators when they mistook the moth's eyespots for the eyes of *their* enemy, predatory owls [116].

If Blest was right and wing-flipping behavior saved the lives of moths in the past, then the evolutionary process has contributed to the persistence of the proximate mechanisms that enable today's moths to behave the way they do. Particular genes present in the bodies of contemporary *Automeris* moths have been replicated and passed on from generation to generation, perhaps because they helped the moth develop an ability that frightened away predatory birds, enabling it to live long enough to transfer its hereditary informa-

2   Proximate and ultimate causes of behavior. At the proximate level, various internal mechanisms enable an *Automeris* moth to execute its wing-flipping behavior. At the ultimate level, the moth's reaction to bird predators determines its reproductive success, as measured by how many copies of its genes reach the next generation. Reproductive differences among individuals with different proximate mechanisms determine which genes are available to influence the development of individuals in the next generation.

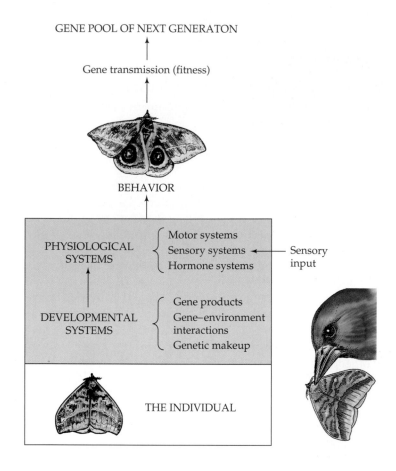

GENE POOL OF NEXT GENERATON

Gene transmission (fitness)

BEHAVIOR

PHYSIOLOGICAL SYSTEMS { Motor systems Sensory systems ← Sensory input Hormone systems

DEVELOPMENTAL SYSTEMS { Gene products Gene–environment interactions Genetic makeup

THE INDIVIDUAL

tion to some descendants. This process could help explain why *Automeris* moths living in Monteverde today receive genes from their parents that promote the development of wing-flipping behavior. The developmental plan, and therefore the behavioral abilities, of each member of the species is a result of differences among individuals in their reproductive success over evolutionary time (Figure 2).

The current function of a behavior offers insight into its possible usefulness in the past, which could help explain why the trait spread and replaced others over time. Characteristics that currently advance the reproductive chances of individuals could plausibly have had the same effect in the past, thereby affecting the course of evolution. But what traits preceded the wing-flipping behavior of modern *Automeris?* If we could go back far enough in time, we would find an ancestor of the moth that did not exhibit the behavior. Perhaps the origins of wing-flipping involved wing movements associated with taking flight, movements that have been altered during the moth's history, just as the color pattern of the hindwings has certainly changed over time [115]. A full understanding of the ultimate causes of wing-flipping requires investigation into the initial form and subsequent evolution of the behavior, as well as the processes responsible for the changes.

You should now be able to discriminate proximate (mechanistic) questions from ultimate (evolutionary) ones (Table 1). If you wanted to find out how the nervous system of *Automeris* moths controls the wing-flipping response, you would be interested in the proximate basis of behavior, as is anyone concerned with how genetic, developmental, neural, or hormonal mechanisms work within an animal's body. On the other hand, if you were interested in whether wing-flipping evolved because of past predation pressure, you would be deal-

TABLE 1 Levels of analysis in the study of animal behavior	
Proximate Causes	**Ultimate Causes**

1. Genetic–developmental mechanisms
     Effects of heredity on behavior
     Development of sensory–motor
         systems via gene–environment
         interactions
2. Sensory–motor mechanisms
     Nervous systems for the detection
         of environmental stimuli
     Hormone systems for adjusting
         responsiveness to environmental stimuli
     Skeletal–muscular systems for
         carrying out responses
	1. Historical pathways leading to a current behavioral trait
	     Events occurring over evolution
	         from the origin of the trait to the present
	2. Selective processes shaping the history of a behavioral trait
	     Past and current usefulness of the
	         behavior in promoting lifetime
	         reproductive success

Sources: Holekamp and Sherman [532], Sherman [1046], and Tinbergen [1139]

ing with an ultimate issue, as is anyone who wants the answers to questions about the reproductive value of a trait and its historical foundations (see Figure 2).

Moreover, if someone were to claim that work on the evolutionary basis of wing-flipping behavior eliminated the need to answer questions about the physiological basis of the behavior, you would (I hope) object strenuously. Proximate and ultimate hypotheses are complementary, not mutually exclusive—a concept that many people find difficult to grasp. For example, I once read that capuchin monkeys rub the oils from citrus fruits onto their fur because the chemicals may help heal skin wounds. The author then added, "Of course, the monkeys may simply enjoy the sensation," as if this explanation meant that we could ignore the medicinal benefit hypothesis. That would be a mistake. At a proximate level, monkeys may indeed derive pleasure from applying certain substances to their bodies, but this explanation does not replace the ultimate wound-healing hypothesis. If, in the past, monkeys that liked to rub citrus oils on their skin had even slightly greater reproductive success than individuals that were indifferent to oily sensations, we would better understand why all capuchins today use citrus oils in a particular fashion. The full analysis of any behavior involves answering both proximate *and* ultimate questions.

Answering Proximate and Ultimate Questions about Behavior

It is one thing to be curious about a mechanism of behavior or its evolutionary foundation and another thing to satisfy one's curiosity. Getting valid answers to biological questions requires a particular approach, called the **scientific method,** whose logic must be understood if you are to grasp why biologists accept some conclusions but not others. We shall explore this issue with examples taken from two studies done in the middle of the twentieth century by the great behavioral biologist Niko Tinbergen, one on a proximate question and the other on an ultimate one.

Beewolves and Homing Behavior
Tinbergen helped make the study of animal behavior a part of modern biology. Although Charles Darwin investigated earthworm burrowing behavior, bumblebee mating, bowerbird displays, and the facial expressions of dogs and

3 **The founders of ethology.** From left to right: Niko Tinbergen, Konrad Lorenz, and Karl von Frisch. Photographs by (left) B. Tschanz, (middle) Sybille Kalas, and (right) O. von Frisch.

humans, no scientific journals were devoted to behavioral research until the mid-1930s. At that time, the field of **ethology** originated under the guidance of Tinbergen, a native of the Netherlands, and his friend Konrad Lorenz, an Austrian. They and their colleagues investigated both proximate and ultimate questions about the behavior of gulls, jackdaws, butterflies, snow buntings, greylag geese, moth caterpillars, and many other animals in their natural environments [720, 1137]. These pioneering ethologists ultimately received the Nobel Prize in Medicine in 1973, which Tinbergen and Lorenz shared with Karl von Frisch (Figure 3), an Austrian researcher famous for his work on honey bee communication (see p. 220).

One of Tinbergen's earliest ethological studies began in 1929, more than 40 years before he was awarded the Nobel Prize, when he discovered a large number of digger wasps nesting in the sand dunes near Hulshorst, Holland. These wasps so fascinated Tinbergen that he and his fellow researchers spent weeks living in a primitive campsite and bicycling up to 70 miles a day in order to learn more about them [1137]. The species of digger wasp that caught Tinbergen's eye was *Philanthus triangulum,* the beewolf, so named because it captures and paralyzes honey bees by stinging them (Figure 4, left). Female beewolves transport captured bees to an underground nest, where they are stored in brood cells off the main tunnel. The bees are eventually eaten by the wasp's offspring when the little grub hatches out from an egg laid on a bee by the nesting female.

Some sand dunes in Hulshorst were dotted with hundreds of burrows, each marked with a low mound of yellow sand that the female beewolf had transported to the surface when excavating her nest. Tinbergen noted that when a beewolf left her burrow to go bee hunting, she covered up the opening by raking sand over it, hiding it from view, and yet when she came back a half hour or an hour later carrying a paralyzed honey bee, she darted directly to her hidden nest entrance, ignoring all the others (Figure 4, right). By giving females unique paint marks, Tinbergen verified that each wasp built and provisioned only one nest at a time.

The skill with which the marked beewolves found their hidden tunnels intrigued and puzzled Tinbergen. How could they get home so easily? The wasps provided a hint to a possible answer: when a female left her nest, particularly on her first flight of the day, she often took off slowly and looped over the nest, flying back and forth in arcs of ever-increasing length and height. After a few seconds, she abruptly turned and zipped off in a straight line (to the bee-hunting grounds, which were about a kilometer away). Tinbergen suspected that the wasps "actually took in the features of the burrow's surroundings while

4 **The beewolf wasp.** (Left) A wasp with a honey bee that it is stinging. (Right) A wasp at its nest entrance, from which it has removed the covering sand. Photographs by Erhard Strohm.

circling above" the covered nest entrance, and that by memorizing local landmarks, such as the sticks and grass clumps scattered in the sand, they were able to find the nest entrance upon their return.

Tinbergen realized that if this hypothesis—this possible explanation—was correct, he ought to be able to make it hard for a female to relocate her nest by changing the local landmarks around the burrow. So he waited until some females left to hunt for bees, then carefully swept the area around their burrow entrances, moving away tufts of grass, pebbles, and sticks that the beewolves might use to orient themselves. His test showed that he was right about the wasps' dependence on visual cues. When prey-laden females came zooming back close to their nests within the landmark-free zone, they appeared confused and hovered in midair about a meter away from their nest mound before circling out to repeat the approach again and again. Only by dropping their prey and searching more or less at random on the ground were some females eventually able to find their nest entrance.

Tinbergen's simple experiment confirmed his suspicion that beewolves formed a visual image of the area immediately around their nests, which they used to pinpoint the covered entrance when coming back with prey. But he wanted to test his hypothesis in another way, to be sure that he had it right. As he wrote in *Curious Naturalists* [1137], "The test I did next was again quite simple. If a wasp used landmarks it should be possible to do more than merely disturb her by throwing her beacons all over the place; I ought to be able to mislead her, to make her go to the wrong place, by moving the whole constellation of her landmarks over a certain distance." In other words, Tinbergen predicted the results he should get from a manipulative experiment if his landmarks hypothesis was correct. When he did the experiment, he found to his delight that it worked like a charm. By carefully moving an entire set of local landmarks around a nest 20 centimeters to the southeast, he induced the returning female to land 20 centimeters to the southeast of her real nest entrance. When he shooed the wasp away, and then shifted all the "runway beacons" back to their original locations, the beewolf circled around and came down right at her nest entrance.

Tinbergen then did yet another experiment, this time to see if he could make the wasps train themselves to landmarks that he provided. He put a ring of pine cones around some nests while the nest owners were inside their burrows. When the wasps came out, they looped back and forth over the nest area before heading off to hunt, but then carried on as usual. Two days later, Tinbergen returned to displace the circles of pine cones while the females were off hunting. He expected that if the wasps had learned the experimental landmarks, they would land within the moved circle of pine cones, rather than at their nests. The experiment worked just as Tinbergen thought it would (Figure 5) [1136].

5   Spatial learning by a beewolf wasp. Tinbergen designed a simple experiment to test his hypothesis that female beewolves (A) learned the local visual landmarks around their burrows and (B) used this information to find the nest entrance when returning with prey from distant hunting grounds. After Tinbergen [1136].

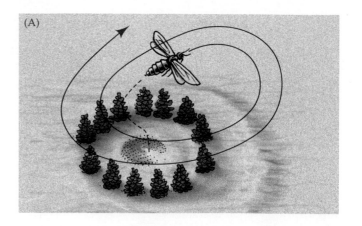

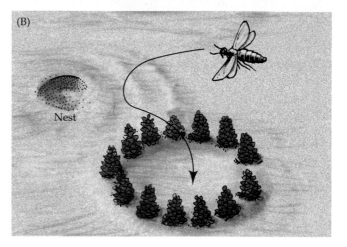

Thus, Tinbergen was able to answer a proximate question about how hunting beewolves get back to their nests. He showed that female beewolves respond to certain visual cues in the environment and that they store landmark information for use when homing. Without dissecting a single wasp, Tinbergen was able to tell us something about what goes on inside these creatures, something about their neural mechanisms—something about the proximate causes of behavior.

Tinbergen produced this result by using the scientific method. First, he asked a question: How do beewolves home to their nest entrance? Then he proposed a tentative answer, a hypothesis, about the cause of homing: the beewolves possess a visual system that gathers and stores information about local landmarks in the nesting area for use when returning to the nest entrance. To test this hypothesis, he developed logical expectations or predictions from it: for example, experimental displacement of the local landmarks around a nest should cause a homing female to shift her landing site accordingly. This prediction, when tested, proved to be correct, which gave Tinbergen confidence that his original hypothesis was indeed right. He tested the hypothesis again with the pine cone experiment, which gave him further confidence that his conclusion was correct.

Thus, we can dissect the procedure Tinbergen followed into a series of steps:

1. He started with a **causal question** about a natural phenomenon.
2. He developed a **hypothesis** to explain what he saw.
3. Based on the hypothesis, he generated **predictions** or expected results.
4. He carried out a **test** of his predictions by gathering actual results for comparison with the expected ones.

5. In so doing, he reached a **scientific conclusion** about the validity of his original explanation.

Tinbergen's acceptance of his hypothesis would have been even stronger if he had systematically tested and rejected *several* alternative hypotheses, each with its own mutually exclusive predictions. However, even by working with just one possible explanation, he made real progress in understanding beewolf homing behavior, thanks to the simple, effective logic of science.

Tinbergen's research is typical of most scientific investigations. Indeed, the scientific method is familiar enough to be used intuitively, rather than formally, by all sorts of people, not just scientists. Listen to Click and Clack, the car mechanics on National Public Radio, as they try to figure out what is causing a 1987 Volvo station wagon to stall unexpectedly as its driver, Bill from Bedford, Massachusetts, motors down the highway. Or analyze how air crash experts attempt to determine whether flight 123 came to an unhappy end because of a terrorist bomb, mechanical failure of a particular part, or fire in the luggage hold.

Whatever you call it, the scientific method works: it improves our understanding of the causes of things. By understanding what is meant by hypothesis, prediction, test, and conclusion, we can better understand how science works and why certain conclusions are justified.

Gulls and Eggshell Removal

Let's examine another example of science in action, also provided by Tinbergen but this time from a study of the *ultimate* cause of a behavior. Tinbergen observed that black-headed gulls remove broken eggshells from their nests a short time after their youngsters have hatched. Although this action might seem trivial, since it takes only a few seconds for a gull to fly off and drop the eggshell a short distance away from the nest, he was still curious about it. After all, while the gull is away from the nest, it leaves its babies unguarded, and the world of black-headed gulls is full of predators, such as other, larger gulls that like nothing better than to dine on newly hatched chicks. Therefore, Tinbergen suspected that removing eggshells must provide some considerable reproductive benefit [1140], an argument reinforced by the finding that many other birds behave the same way (Figure 6).

In essence, Tinbergen assumed that once upon a time, black-headed gulls (or the now-extinct species that eventually evolved into black-headed gulls) did *not* remove eggshells from their nests. Then the first eggshell remover appeared, perhaps as a result of a genetic mutation that had some small but particular

6   Nesting loons, like black-headed gulls, remove conspicuous empty eggshells from their nests. Photograph by Bruce Lyon.

influence on the development of the bird's nervous system. The new gene, with its distinctive developmental effect, could not have persisted unless the individual that first carried it reproduced more successfully than other birds, with their different genes and associated willingness to ignore eggshells in their nests. If the mutant bird's descendants continued to have more surviving chicks on average than the once-typical birds, the mutant gene would have made itself more and more common. But if the reproductive success of eggshell removers was lower on average than that of birds that ignored eggshells, the mutation underlying the new response would have disappeared from the species because its owners would have failed to pass it on.

In other words, Tinbergen assumed that eggshell removal evolved because of past differences among the ancestors of black-headed gulls in their behavior and reproductive success. The ultimate question he asked was, what might have caused the spread of eggshell removal in an ancestral population in which the behavior was once rare? He could not go back in time to study what happened in long-gone generations of black-headed gulls, but he could determine whether the behavior offered some reproductive advantage to the current generation. If so, he could more plausibly claim that this advantage was responsible for the current maintenance of the trait, and also possibly for the spread of the trait sometime in the past. If not, he would know that one possible explanation for the evolution of eggshell removal was probably wrong.

Tinbergen suggested that eggshell removal by today's black-headed gulls might be reproductively advantageous because it eliminated a visual cue that could give the nest away to certain predators. Black-headed gulls nest out in the open, but they appear to try to hide the nest in whatever vegetation is available. Moreover, the color of their eggs and chicks is an inconspicuous mottled brown and gray. In contrast, the white inner part of an opened gull eggshell is highly conspicuous, and it might serve as a nest-identifying beacon to carrion crows and other predators if not removed by a parent gull.

To test the hypothesis that the eggshell removal trait was passed down through generations of black-headed gulls because it foiled predators, Tinbergen first developed a prediction from the hypothesis: if eggshell removal was an antipredator device, then the presence of broken eggshells should help predators locate food. He checked this prediction with a simple experiment. He stole some intact eggs from nests in a colony of black-headed gulls and scattered them through the sand dunes that were regularly patrolled by egg-eating carrion crows. By some of the unhatched eggs he placed broken eggshells a short distance away; by others, he dropped the eggshells farther away. The intact eggs that were closest to white eggshell bits were more likely to be found and eaten by foraging crows than those that were farther away from an eggshell give-away cue (Table 2). Since this finding matched the predicted result, Tinbergen concluded that eggshell removal by nesting black-headed gulls could have evolved because birds that happened to behave this way lost fewer offspring to predators. Their surviving offspring became the ancestors of today's gulls, which possess the hereditary proximate mechanisms associated with reproductive success in the past [1140].

TABLE 2 Effect of the proximity of eggshells on egg predation by crows

Distance from eggshell to egg (cm)	Eggs taken by crows	Eggs not taken by crows	Percentage (%) taken
15	63	87	42
100	48	102	32
200	32	118	21

Source: Tinbergen [1140]

Note that although this study concerned an evolutionary or ultimate question, the procedure for answering it did not differ fundamentally from the method Tinbergen used when he studied the proximate basis of homing behavior in the beewolf wasp. He began with (1) a question: Why had black-headed gulls evolved a special response to the eggshells in their nests? He then generated (2) a speculative answer, a working hypothesis: maybe eggshell removal had spread in the past because it helped parent gulls conceal their offspring from crows and other visually hunting predators. This hypothesis led logically to (3) a prediction: he expected to see crows and other predators using the conspicuous cues offered by broken eggshells to narrow their search for unopened eggs. Tinbergen then (4) checked the prediction, using an experiment to find out what crows actually did. By matching the resulting data against the predicted outcome, he tested his hypothesis and reached (5) a scientific conclusion, namely, that his original explanation was probably right.

Darwinian Theory and Ultimate Hypotheses

When Tinbergen developed his hypothesis about the evolved function of eggshell removal, he was strongly influenced by natural selection theory, which has determined the way biologists have explored ultimate questions since 1859. In that year, Charles Darwin (Figure 7) published *On the Origin of Species* [286], with its sweeping explanation of how evolutionary change might occur within species. Darwin's great idea rests on three commonly observed features of living things:

1. **Variation:** Members of a species differ in their characteristics (Figure 8).
2. **Heredity:** Parents pass on some of their distinctive characteristics to their offspring.
3. **Differential reproduction:** Because of their distinctive inherited characteristics, some individuals within a population have more surviving offspring than others.

It was Darwin's genius to perceive that evolutionary change is *inevitable* when these three conditions occur in a species (Figure 9). If some black-headed gulls, for instance, produce more offspring than others, and if their adult offspring have inherited the trait (such as eggshell removal) that advanced their successful reproduction, then those offspring will also spread that reproduction-enhancing trait. The other side of the coin is equally clear. If some gulls leave fewer offspring than others because of their inherited characteristics, those of their off-

7 Charles Darwin, shortly after returning from his around-the-world voyage on the *Beagle*, before he wrote *On the Origin of Species*. Copyright Science Photo Library.

8 A variable species. The ladybird beetle *Harmonia axyridis* exhibits hereditary variation in its color pattern. Photographs by Mike Majerus.

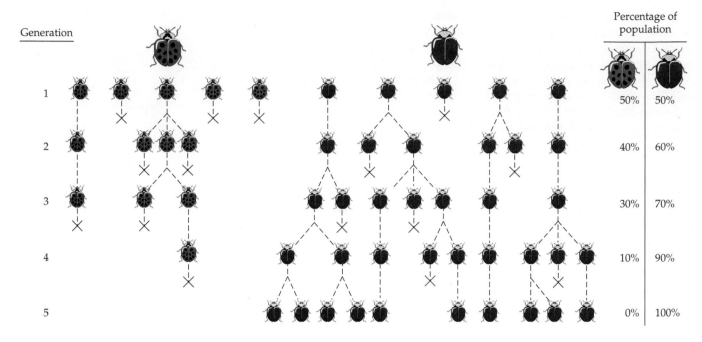

9   Natural selection. If the differences in the color patterns of ladybird beetles are hereditary, and if one type of beetle leaves more surviving offspring on average than the other, then the population will evolve, becoming more and more dominated by the reproductively successful type.

spring that do survive may inherit their disadvantage and thus may also leave relatively few surviving progeny. As a result, traits that compromise lifetime reproductive success will become progressively rarer over evolutionary time. Darwin called this process **natural selection** because he saw the elimination of traits unfavorable to reproduction and the spread of beneficial ones as the natural consequence of hereditary variation. Thus, Darwinian logic leads us to expect that evolutionary change will always be in the direction that promotes successful reproduction by individuals.

Darwin developed the theory of natural selection before critical discoveries about the nature of heredity had been made. Genes are now known to be nucleic acids that faithfully encode the information needed for the synthesis of proteins, which are critically important for all living things. Genes can be copied and transmitted to offspring. When this happens, the genes themselves are essentially reproducing, with the organism merely acting as a mortal vehicle for the process. The modern evolutionary approach to animal behavior applies Darwinian logic at the genetic level in much the same way that Darwin applied it to individuals:

1. **Genetic variation:** Genes can occur in more than one form. When alternative forms, or **alleles**, of a gene exist, the different alleles may lead to the production of slightly different forms of the same protein.
2. **Heredity:** Alleles can be transmitted from parent to offspring.
3. **Differential reproduction:** Some alleles are better than others at producing effects that cause their bearers to transmit copies of their alleles to subsequent generations.

If these three conditions apply, then alleles that help make individuals more reproductively successful will become more common in the population over evolutionary time. Other alleles that confer less reproductive success on individuals will eventually disappear. (We assume that populations cannot grow exponentially forever, so that only a limited number of copies of a gene can exist at any one time.) The logical conclusion is that selection on individuals will favor alleles that help build bodies that are unusually good at promoting the propagation of those alleles—or, as E. O. Wilson puts it, a chicken is really the way that chicken genes make more copies of themselves [1247].

Darwinian Logic and the Study of Behavior

No matter how the logic of natural selection is presented, it is a blockbuster of an idea. It means that humans and all other living things have been shaped by past selection favoring characteristics that enhanced their ability to reproduce and pass on their genes. An understanding of this point helps us identify questions worth asking while at the same time shaping the kinds of hypotheses that we will test.

Let's illustrate the utility of Darwinian theory with the case of infanticide in hanuman langurs. These graceful primates live in bands, which often consist of one large, reproductively active male and a group of smaller adult females and their offspring (Figure 10). From time to time, the resident male is pushed out of the group by a newcomer, usually after a series of violent clashes. After such a takeover, infants tend to die. Although the cause of death is often unclear, the new male is the prime suspect in many instances, and males have been seen doing the deed several times (Figure 11) [132, 553].

The phenomenon of infanticide is precisely the sort of thing likely to attract the attention of a Darwinian biologist. Why should a male langur that has just spent days in a dangerous running battle with the previous resident male turn on the offspring of the very females he has finally succeeded in joining? Having avoided incapacitating injury during his battles with the rival male, why should he now take the chance of being bitten by one of his female companions as she attempts to protect an infant (Figure 11)? In other words, how can infanticide possibly advance a male's reproductive success? This question jumps out at an evolutionary biologist precisely because infanticide by male langurs seems so unlikely to have evolved by natural selection.

Indeed, the behavior might not be an evolved trait that contributes to the male's reproductive success, but could instead be a social pathology brought on by overcrowding. Under high-density conditions, males may encounter each

10 **A band of hanuman langur females and their offspring.** Males fight to monopolize sexual access to the females in groups like this. Photograph by S. Nagendra.

11 Male infanticide in hanuman langurs. (Left) An infant-killing male langur flees from a female belonging to the band he is attempting to join. (Right) A nursing baby langur that has been paralyzed by a bite to the spine (note the open wound) by a male langur. This infant was attacked repeatedly over a period of weeks, losing an eye and finally its life at age 18 months. Photographs by (left) Volker Sommer, from Sommer [1085], and (right) Carola Borries.

other so often that fighting becomes commonplace, with the hyperaggressive males then assaulting not just rivals of the same sex, but also females and their offspring. In fact, this nonevolutionary hypothesis was the first explanation offered by some langur watchers, who knew that langurs were often fed by Indian villagers, and so perhaps had reached unnaturally high population densities.

But another student of langurs, Sarah Hrdy, believed that infanticidal behavior by males might have evolved by natural selection, despite its apparent lack of reproductive value for the male. She thought that perhaps infanticide spread as a result of sexual competition between males [553]. A killer male might increase his reproductive success if he eliminated nursing offspring sired by a rival male, leaving the mothers of those infants no other option than to mate with him and have his babies.

For the moment, let's not worry about whether Hrdy's hypothesis is right or wrong. The point is that her potential answer to the puzzle of infanticide derives from Darwinian theory. Hrdy's suggestion is that males tend to gain descendants through the selective practice of infanticide. This suggestion differs fundamentally from the social pathology hypothesis, which states that killer males gain nothing from infanticide because it is an abnormal behavior induced by unnatural conditions.

The Problem with Group Selection

Darwinian hypotheses about infanticide also differ from some other explanations for the behavior, such as the possibility that killer males evolved in order to help prevent overpopulation, which would destroy the food resources that

Hanuman langur bands need to survive. Although this hypothesis also claims that the behavior spread in the past because of certain beneficial consequences, it is *not* based on Darwinian theory. Here the beneficiary of infanticide is not the male that kills infants, but the group to which he belongs. The evolutionary mechanism for the spread of infanticide is not directly linked to differences among individuals in their reproductive success (as in Darwinian natural selection), but rather to differences among groups in their survival via a process that has been called **group selection.**

Group selection theory received its formal presentation in *Animal Dispersion in Relation to Social Behaviour* [1286], written in 1962 by V. C. Wynne-Edwards. According to Wynne-Edwards, only those groups or species that possessed population-regulating mechanisms could have survived to the present; others that lacked these mechanisms had surely become extinct through over-exploitation of the critical resources on which they depended. With groups competing unconsciously to survive, only those populations whose members reduced their reproductive output would be likely to persist. Thus, Wynne-Edwards argued that evolutionary change regularly occurs because of differences among groups in their possession of self-sacrificing individuals, which in turn affect the survival chances of the group.

This argument was challenged in 1966 by G. C. Williams in *Adaptation and Natural Selection* [1239]. Williams showed that the survival of alternative alleles was much more likely to be determined by reproductive differences among genetically different individuals than by survival differences among groups. The basis for this claim can be illustrated with reference to langurs. Imagine that in the past there really were male langurs prepared to risk their own lives by killing infants in order to reduce their population for the long-term benefit of their group. In such a case, group selection, as defined by Wynne-Edwards, would be said to favor the allele(s) for male infanticide because the group as a whole would benefit from the removal of excess infants.

However, in a population of langurs with some group-benefiting infanticidal males, Darwinian natural selection would also be at work. Suppose that there were two genetically distinct types of males: the infanticidal type, which spent its energy and sometimes shortened its life for the good of the group, and another type, which lived longer and reproduced more because it let the infanticidal males carry the burden of population reduction. Which of these two types would constitute more of the next generation? Whose hereditary material would be transmitted better to following generations? What would happen over evolutionary time to infanticidal tendencies in our hypothetical population of langurs?

The general point that Williams made is that selection acting on differences among variant individuals within a population will usually have a much stronger evolutionary effect than selection acting on differences among entire groups. For-the-good-of-the-group selection can occur, provided that groups retain their integrity for long periods and differ in their genetic constitution in ways that affect their survival chances. But if group selection favors a trait, such as reproductive self-sacrifice, while natural selection acts against it, natural selection seems likely to win, as we have just seen in our hypothetical langur example. Although research continues on forms of group selection more complex than the for-the-good-of-the-group type proposed by Wynne-Edwards [1080, 1245], almost all behavioral biologists have accepted the arguments of Williams. Researchers now carefully distinguish between group benefit and individual (or gene) benefit hypotheses. The overwhelming majority of scientists studying the evolution of animal behavior employ Darwinian theory, rather than group selection theory in any of its forms.

However, we could, if we wished, test non-Darwinian hypotheses using the logic of the scientific method that we illustrated with Tinbergen's studies. The social pathology hypothesis (which is neither Darwinian nor group selectionist) and the population regulation hypothesis (which is based on the theory of group selection) for infanticide by male langurs both happen to yield the same key prediction. If high population density really does cause abnormal behavior (the social pathology hypothesis), or if it truly threatens the survival of langur groups and so activates self-sacrificing infanticide (the population regulation hypothesis), then we would expect to see infanticide by males *only* in areas in which Hanuman langur populations are abnormally or unusually high. Contrary to this prediction, infanticide regularly occurs in troops living at moderate or even low densities in natural areas where they are not fed by people [131, 839]. This finding weakens our confidence in both the social pathology and population regulation hypotheses for langur infanticide.

Testing Alternative Hypotheses

Does our tentative rejection of these two non-Darwinian hypotheses mean that Hrdy's Darwinian explanation is correct? Clearly not. First, studies of langurs living under presumably natural conditions are rare, so the evidence against the social pathology and population regulation hypotheses is not utterly compelling. Second, Hrdy's increased reproductive opportunities hypothesis is not the only possible explanation based on Darwinian theory. For example, perhaps males commit infanticide after takeovers in order to cannibalize infants, thereby replenishing their depleted energy reserves. If so, killer males could derive benefits from their actions that could keep them alive to reproduce more than males that did not kill and consume youngsters when taking over a band. In order to feel confident that we have identified the true ultimate cause of infanticide, we will have to test alternative hypotheses in ways that help us reject the incorrect ones and retain the right explanation, if we have included it in our list.

When it comes to hypothesis testing, note that more than one explanation can lead to the same prediction, as we just saw when examining the two non-Darwinian hypotheses on infanticide. Both of those hypotheses yielded the same prediction: infanticide should be limited to high-density populations. If we had found that males killed infants only when populations had increased markedly, we would not have been able to accept one of the hypotheses and reject the other. Likewise, our two Darwinian alternatives, the increased reproductive opportunities hypothesis and the cannibalism hypothesis, both produce the prediction that infanticide by males will tend to occur soon after takeovers. In the first case, we expect males to stop killing infants by the time the first of their own offspring are born to the resident females. In the second case, we expect males to stop killing infants after they have recovered from their energetically expensive takeover. The fact that infanticide is indeed limited to the period soon after a takeover does not help us decide between these two alternatives.

If the cannibalism hypothesis is true, however, we should sometimes see male newcomers eating an infant. No such records exist, but remember that observations of males in the act of infanticide are rare. But if we could demonstrate that male langurs do not consume dead infants, we would obviously reject the adaptive cannibalism argument.

The increased reproductive opportunities hypothesis produces various predictions, including (1) attacking males should not kill their own offspring, and (2) females deprived of their young infants should mate with the very males that killed their offspring, resulting in the production of new infants sired by the killer males. These predictions have received confirmation from various sources [553, 1085]. For example, a researcher observing a langur group of several adult males living together with various females recorded 16 cases of infanticide for which DNA samples had been collected from the presumptive killer male and his victim. In every instance the killer was *not* the father of the deceased infant. Moreover, females who had lost their young promptly regained their sexual receptivity, and the new infants that resulted were fathered by the infanticidal males, as was again demonstrated conclusively through DNA testing [132]. These findings provide strong support for the increased reproductive opportunities hypothesis for infanticide.

But the more tests, the better. If natural selection has produced male Hanuman langurs that kill infants to gain more rapid sexual access to females, then we would expect to observe infanticide in many other species whose social systems resemble that of the langur. This prediction has now been confirmed through studies of various other animals in which newcomer males replace previous resident males, kill infants fathered by those males, and then mate with the females that have lost their youngsters [344]. In lions, for example, infanticide often occurs when a new group of males ousts the males from a pride containing a number of females with young cubs [943]. The incoming males hunt down cubs less than 9 months old and try to kill them (Figure 12), although, like female langurs, lionesses try (sometimes successfully) to protect their cubs. Lionesses that keep their cubs alive give birth at 2-year intervals, but females whose babies are killed resume sexual cycling at once and mate with the killers of their offspring. Since a male can expect to remain in a pride and have access to its females for just 2 years on average, the reproductive benefits of infanticide from the male's perspective are evident. Indeed, male lions probably kill a quarter of all the cubs that die in their first year in some populations [943].

The observation that lions and other animals commit infanticide under certain predictable conditions supports the increased reproductive opportunities hypothesis. If these conditions favor the evolution of male infanticide, then we can predict that infanticide should be practiced by *females* of those unusual species in which sexual access to males limits female reproductive success. This prediction has been confirmed for a giant waterbug whose males take care of egg masses (Figure 13), which are sometimes attacked by egg-stabbing females.

12 Infanticide by a male lion. The male carries a cub he has killed after displacing the adult males that once lived with the pride. Photograph by George Schaller.

13   Protection against infanticide.
This male waterbug guards a clutch of
eggs against infanticidal females that may
destroy his current clutch in order to
replace these eggs with their own.
Photograph by Bob Smith.

After the destruction of a clutch of eggs, the male associates with the infanticidal female, mates with her, and cares for a new brood of her eggs [563].

Likewise, in a water bird called the jacana, males provide exclusive care of
the eggs and young. Territory-defending females sometimes attack the chicks
of neighboring females, forcing the brooding male to abandon these offspring.
He may then mate with and accept a new clutch of eggs from the infanticidal
female. When researchers experimentally removed some territorial females,
neighboring females quickly invaded the vacated territories and, in three of four
cases, either killed the baby jacanas there or forced them to flee. Within 48 hours
the males that had lost their offspring were involved in sexual liaisons with the
infanticidal females. By committing infanticide, these females had gained caretakers for their eggs sooner than if they had waited for the males to finish
rearing their current broods [361].

Certainty and Science

You must have deduced from my summary of research on infanticide that I
think the increased reproductive opportunities hypothesis applies to langurs
and lions, waterbugs and jacanas. I do—but I could be wrong, and indeed,
some other researchers believe that langurs and some other animals do not
commit infanticide in ways that increase their reproductive success [75, 118,
276]. These disagreements remind us that *all* scientific conclusions must be considered tentative to some degree. In the past, majority opinion has changed
dramatically when a previously unconsidered hypothesis came along or new
data surfaced that destroyed an established hypothesis. When I was student
at Amherst College, my paleontology professor convinced me that the earth's
continents have always been where they are now located. However, as new
evidence came in, everyone, including me, abandoned the old view; now the
generally accepted hypothesis is that the continents "float" around the planet
on moveable plates.

The rejection of established wisdom happens all the time in science. Scientists tend to be a skeptical lot, perhaps because special rewards go to those who
can show that previously published conclusions are incorrect. Researchers constantly criticize their colleagues' ideas, in good humor or otherwise, sometimes
causing their fellow scientists to change their minds. The uncertainty about
Truth that scientists accept, at least when talking about other people's ideas,
often makes nonscientists nervous, in part because scientific results are usually
presented to the public as if they were written in stone. But anyone who has
taken a look at the history of any scientific endeavor will learn that new ideas
continually surface and old ones are regularly replaced or modified. I repeat,
complete certainty is *never* achieved in science. The strength of science stems
from the willingness of at least some scientists to consider new ideas and to test
hypotheses repeatedly.

I hope that you will keep this point in mind as we review the findings of scientists and their interpretations of evidence in the chapters ahead. We will
first examine the proximate and ultimate aspects of bird singing behavior (Chapter 2) before looking more closely at the different components of a proximate
analysis of behavior (in Chapters 3 through 6). Then we turn to ultimate questions about evolutionary history and adaptive value (in Chapters 7 through 14).
The book concludes with a chapter on the evolution of human behavior. Thanks
to the small army of behavioral researchers that have attacked these questions, there is much to say on these topics, so let's get started.

Summary

1. Basic questions about animal behavior fall into two categories. "How questions" require answers about the *proximate* causes of behavior: how do genetic–developmental and sensory–motor mechanisms cause an individual to behave? "Why questions" require answers about the *ultimate* causes of behavior: Why have certain genes and certain proximate mechanisms persisted to the present, and why has evolution followed one path instead of another?

2. Both proximate and ultimate questions can be investigated scientifically following these steps:

    1. We begin with a causal question about what causes something to happen.
    2. We devise a working hypothesis, or possible answer to the question.
    3. We predict what we expect to observe in nature if the hypothesis is true.
    4. After developing our prediction(s), we collect the appropriate data and test the hypothesis by matching the actual results against the expected ones.
    5. We reach a scientific conclusion based on the results of our test, rejecting hypotheses if their prediction(s) fail to be upheld, and tentatively accepting hypotheses whose tests are positive.

3. Charles Darwin proposed that evolutionary change occurs by natural selection. According to Darwin's theory, if a species contains genetically different individuals whose particular characteristics cause them to have different numbers of surviving offspring, the types that reproduce most successfully will become more numerous in subsequent generations.

4. V. C. Wynne-Edwards proposed a theory of group selection, arguing that evolutionary change will occur if genetically different groups vary in how well they survive because of genetic differences among them.

5. The theory that a researcher uses affects the kinds of hypotheses he or she is likely to propose and test. Users of Darwinian theory produce hypotheses on how traits might promote the survival of the genes of individuals with those traits; users of group selection theory produce hypotheses on how traits might advance the survival of the group or species to which the individual belongs.

6. Today almost all behavioral biologists use Darwinian rather than group selection theory as the foundation for their hypotheses because selection at the level of individuals should be a more powerful force for evolutionary change than selection at the level of groups. A trait favored by group selection might lead individuals to sacrifice their reproductive success for the good of the group. If other members of the group have alternative traits that better propagate their different genes, these genes will replace the ones that are being sacrificed for the benefit of the group.

7. The beauty of science lies in its logical approach to testing alternative hypotheses, whether proximate or ultimate, whether based on theory X or theory Y. Persons who use the approach can eliminate explanations that fail their tests while accepting other hypotheses that have passed their tests.

Discussion Questions

1. Why do humans eat so much candy and drink so many soft drinks? Which of the following explanations are proximate hypotheses, and which are ultimate hypotheses?

a. Candy and soft drinks contain sugar, which tastes sweet to people.

b. Sweet taste is remembered as good; the memory of pleasure leads people to eat or drink more of the same.

c. Sugar, which is present in candy and soft drinks, is an energy source that helps keep people alive.

d. Our primate ancestors depended on sugar-rich fruits; from these ancestors, we have inherited the same kind of taste perceptions that they had.

e. The genetic information in our bodies shapes the development of nerve cells that provide perceptions of sweetness and pleasure.

f. In the past, those individuals who liked sugar left more descendants than those who were indifferent to sweet-tasting foods.

g. The sensory input from taste receptors in the tongue to selected brain cells leads to a positively reinforcing sensation of sweetness.

2. Lemmings are small mouselike rodents that live in the Arctic tundra. They are known for extreme fluctuations in population size. At high population densities, large numbers leave their homes to travel long distances. In the course of their journey, many die, some by drowning as they attempt to swim across lakes and rivers. One widely circulated explanation for their behavior is that the travelers are attempting to commit suicide to relieve pressure on their population. If some die, the survivors will have something left to eat. What theory is the foundation for this hypothesis? What would G. C. Williams have to say about it? How would he use Gary Larson's cartoon (below) to make his point?

3. In Dawson's burrowing bee, males come in two sizes, large and small, with no intermediates [12]. The large males fight to mate with virgin females in places where the females are emerging from underground nests (see p. 21, photograph by the author). The small males avoid fighting, and instead patrol flowers, where they sometimes find and copulate with females that have emerged without mating. Why do small male Dawson's burrowing bees behave differently

from large ones when it comes to finding mates? The statements in the following list are either hypotheses or predictions. Identify which is which.

a. Small males should lose fights when they try to compete aggressively with larger individuals.
b. Small males search for and locate mates in areas that larger males ignore.
c. Small males use a mate-finding method that larger males are incapable of using.
d. Small males avoid direct competition with larger, unbeatable rivals.
e. Small males ought to secure at least some receptive females in the areas where they search.
f. Large males are expected to be unable to fly with the rapidity and agility of small males.

4. Secure a copy of a paper by Ethan Clotfelter [227]. What is the causal question that motivated his research? List the four hypotheses he examined, linking each with a prediction derived from it. With each prediction, present the relevant test data. Finally, what were the scientific conclusions that he drew from his work? Was his study designed to answer a proximate or an ultimate question?

5. In a spider that lives in the deserts of the Middle East, web-building females guard an egg sac until the spiderlings emerge from the fertilized eggs within; the young spiders then consume their mother. Males search for females, finding very few in their lifetime; if they encounter an egg-guarding female, they will try to remove her egg sac. Females fight back and sometimes kill (and eat) the male. But males sometimes succeed in destroying the egg sac. Such behavior reduces the potential for population growth in this species. Does this mean that male behavior is maladaptive? If not, how can you account for male behavior in this species, and what predictions follow from your explanation(s)?

Suggested Reading

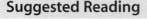

Books written by scientists that capture the sense of curiosity and excitement that biologists feel as they study the proximate basis of animal behavior include Vincent Dethier's *To Know a Fly* [313] and Kenneth Roeder's *Nerve Cells and Insect Behavior* [976]. Niko Tinbergen's *Curious Naturalists* [1137] and Konrad Lorenz's *King Solomon's Ring* [720] bridge the gap between proximate and ultimate approaches. See also Howard Evans's *Life on a Little Known Planet* [371] and Michael Ryan's *The Túngara Frog* [998].

For other books by scientists that capture the delight of field research, consider Evans's *Wasp Farm* [372], Bernd Heinrich's *In a Patch of Fireweed* [505], George Schaller's *The Year of the Gorilla* [1017], and Tinbergen's *The Herring Gull's World* [1138], as well as Jane Goodall's *In the Shadow of Man* [442], Shirley Strum's *Almost Human* [1106], Cynthia Moss's *Elephant Memories* [821], and *Journey to the Ants* by Bert Hölldobler and E. O. Wilson [537]. A realistic account of an ambitious field research project comes from Craig Packer [877]. The study of langur infanticide is described in *The Langurs of Abu* by Sarah Hrdy [554]. Bernd Heinrich's superb *Ravens in Winter* offers an unusually clear picture of how scientists test alternative hypotheses [507]. For a provocative essay on the nature of science, read [1264].

Charles Darwin had something useful to say about the logic of natural selection in *On the Origin of Species* [286], and so do Daniel Dennett [312] and Richard Dawkins [301, 304]. G. C. Williams's classic *Adaptation and Natural Selection* [1239] demolishes "for-the-good-of-the-species" arguments.

2 | Proximate and Ultimate Causes of Behavior: How and Why Birds Sing

I began bird-watching avidly at age six. In the subsequent half century plus, my "life list" has grown to well over a thousand species, with the most recent addition being the western bristlebird, a rarity found only in a few coastal heaths in southwestern Australia. I detected the bird's very loud whistled song before I saw the singer. The song, which I knew I had not heard before, sent me racing back to the car to retrieve my binoculars, and soon the bristlebird was on my life list.

Although the western bristlebird's vocalizations are unusually loud, the bird is like most others in having its own distinctive song, a fact that bird-watchers regularly use to identify birds that are out of sight. Thus, my father taught me to identify yellow warblers on the basis of their "sweet-sweet-sweeter-than-sweet," whereas the common yellowthroat (another warbler) sang "witchety-witchety-witchety" and the song sparrow produced "Madge-Madge-Madge, put on the tea kettle-ettle." Why is it that the bristlebird, the yellow warbler, the common yellowthroat, and the song sparrow sing such different songs? For that matter, why does each individual bird typically sing its own special variant of its species' distinctive song?

◀ *Studies of bird song* have relied heavily on male white-crowned sparrows, one of which is shown singing here. Photograph by Doug Nelson.

This chapter uses these questions about bird song to emphasize that every aspect of animal behavior has proximate and ultimate causes, which are different but interrelated. Thus, at the proximate level, a bird sings its own song because of the way its nervous system works, which is a function of the developmental process that produced that system with its special features. In turn, the proximate developmental and physiological mechanisms of each bird species have an evolutionary history, and therefore we can explore the ultimate causes of bird song, the better to demonstrate the complementarity of proximate and ultimate causation in animal behavior.

Different Songs: Proximate Causes

Not only do different bird species generate unique songs or calls, but even members of the same species may vocalize differently. Thus, for example, in a great many songbird species, males communicate with elaborate songs, but females do not. Take the white-crowned sparrow. In this species, males have a complex whistled vocalization that they give thousands of times each spring, whereas females produce simple call notes but rarely, if ever, a full-fledged song. Moreover, although all male white-crowns sing somewhat similarly, sharing a basic pattern that makes them all identifiable as white-crowned sparrows to bird-watchers, groups of males in the same neighborhood may all share a version of that basic pattern that differs from that found in other neighborhoods. In at least some populations of white-crowns, the local dialect has persisted with only modest changes for decades [491], creating a geographic mosaic of fairly stable song types. For example, white-crown males living in Marin, north of San Francisco Bay, sing a dialect that is easily distinguished from the type of song produced by white-crowns residing near Berkeley, about 50 miles to the south (Figure 1) [758].

Either of two proximate explanations could account for the dialects of white-crowned sparrows. On the one hand, the Marin white-crowns might all differ genetically from the Berkeley birds in a way that translates into a difference in the nerve cells that control singing behavior, and thus a difference in the songs of the two populations. Alternatively, the song differences between birds from Marin and Berkeley might have nothing to do with hereditary differences between the birds from the two places. Instead, all young male white-crowns in Marin might have learned to sing their distinctive dialect from adult males around them who sang that dialect, whereas the Berkeley males learned their song type from adults who sang a different dialect, just as a young person living in Mobile, Alabama, is likely to be exposed to and learn the English language dialect characteristic of that region as opposed to, say, the one you hear in Bangor, Maine.

Peter Marler and his colleagues wanted to know whether genetic differences or environmental differences accounted for white-crowned sparrow dialects [758]. To find out, they hatched eggs taken from white-crowned sparrow nests and hand-reared the babies in the laboratory. Some of these birds were then held in soundproof chambers without exposure to songs of any sort. Although these individuals began singing at several months of age, the product was a twittering song with only a vague similarity to a mature male's full song. The birds continued to sing as they grew older, but their songs never took on the rich character of a wild white-crowned sparrow's song [757].

Clearly something critical was missing from the hand-reared birds' *environment*, perhaps the opportunity to hear adult male white-crowns sing their songs. If so, then a young male isolated in his soundproof chamber but exposed to tapes of white-crowned sparrow song ought to be able to sing a proper dialect eventually. And that is exactly what happened when Marler and his colleagues

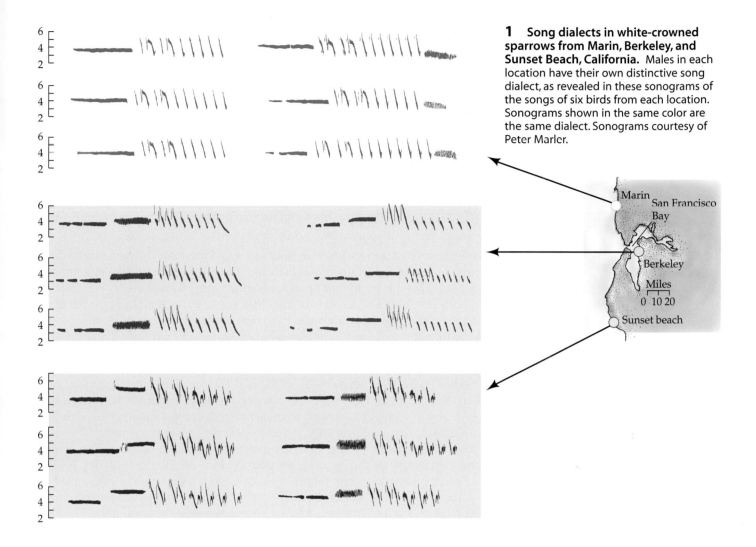

1   Song dialects in white-crowned sparrows from Marin, Berkeley, and Sunset Beach, California. Males in each location have their own distinctive song dialect, as revealed in these sonograms of the songs of six birds from each location. Sonograms shown in the same color are the same dialect. Sonograms courtesy of Peter Marler.

allowed 10- to 50-day-old youngsters to listen to tapes of white-crowned sparrow song. The isolated birds eventually reproduced the exact song they had heard on tape. If a young male had listened to a tape of a Berkeley dialect, he would mimic that song; if he had instead been treated to a steady diet of Marin song, he would sing that dialect by the time he was 200 days old.

These results offer powerful support for the environmental differences hypothesis. Young birds that grow up near Marin hear only the Marin dialect as sung by older males in their neighborhood. They store the acoustical information they acquire from their "tutors" and later match their own song output against their memories of the song, eventually coming to duplicate a particular dialect.

Additional laboratory experiments with isolated birds and taped songs led Marler to the following conclusions about song learning in the white-crowned sparrow [757]:

1. **Song recognition:** A male white-crowned sparrow is predisposed to learn his own species' song, not that of another species, especially during a critical period when he is between 10 and 50 days old. Isolated white-crowns of this age that listen to tapes of song sparrows instead of white-crowns develop aberrant songs that generally resemble those of birds that have heard no songs at all. However, if the young bird hears a tape with both

song sparrow and white-crowned sparrow songs on it, he will develop normal white-crowned sparrow song, learning the white-crown dialect from the tape while ignoring the song sparrow song [650]. The innate preference of young white-crown males for white-crowned sparrow song is so great that fledglings can acquire songs from tutor tapes that play white-crowned sparrow song *in reverse* [1219].

2. **Song practice:** If a young bird hears white-crown song during the critical period at 10 to 50 days of age, no additional acoustical experience is required until he is about 150 days of age, when he begins to produce a variable, exploratory "subsong." At this time, he must be able to hear himself sing if he is ever to develop a normal full song [649]. If he is deafened before subsong begins, he cannot match his own vocal output with the memory of his species' song acquired when he was 10 to 50 days old. As a result, development of the song halts, and the end product is highly abnormal.

3. **Song crystallization:** Once an undeafened young male has generated a large and variable repertoire of songs during the practice phase, he can select one of his own that matches a memorized tutor song. By repeatedly singing that song, the bird forms a crystallized full song, which will become his main vocalization as an adult male. In nature, therefore, most adults sing a copy of the song they heard sung by a natural tutor early in their lives.

Social Experience and Song Development

Marler interpreted the results of these taped-tutor experiments to mean that as the male white-crowned sparrow's brain develops, he acquires a neural mechanism that is highly specialized for learning his species' song. The juvenile male's brain is able to store acoustical information from singing white-crowned sparrows during an early period in his life (Figure 2) [757]. If Marler's interpretation is correct, then we can predict that male white-crowned sparrows will never be able to sing another species' song. Occasionally, however, ornithologists have heard *wild* white-crowned sparrows singing like song sparrows. Observations of this sort caused Luis Baptista to wonder whether some other factor in addition to acoustical experience might influence the song learning process. Marler's famous taped-tutor experiments were all done with birds deprived of social interactions with adults. Perhaps social stimuli can also affect what a young white-crown learns about song.

Baptista and his colleague Lewis Petrinovich tested this social learning hypothesis by placing fledgling hand-reared white-crowns in cages where they could see and hear living adult song sparrows or strawberry finches [72]. These young white-crowns learned their social tutor's song even though they could hear, but not see, other white-crowns (Figure 3). Moreover, white-crowns that were older

2   Song learning hypothesis based on laboratory experiments with white-crowned sparrows. According to this hypothesis, young white-crowns have a critical period 10 to 50 days after hatching when their neural systems can acquire information by listening to white-crown song, but not from any other species' song. Later in life, the bird matches his own subsong with his memory of the tutor's song, and eventually imitates it perfectly—unless he is deafened. Based on a diagram by Peter Marler.

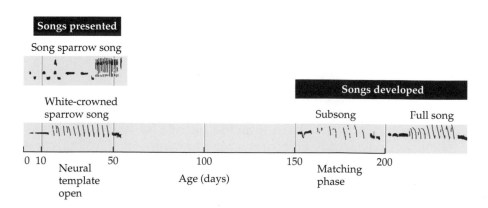

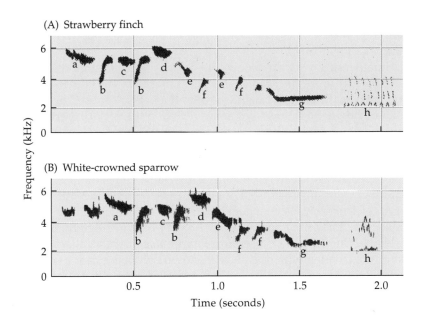

(A) Strawberry finch

(B) White-crowned sparrow

Frequency (kHz)

Time (seconds)

3 Social experience influences song development. A white-crowned sparrow that has been caged next to a strawberry finch will learn the song of its social tutor. (A) The song of a tutor strawberry finch; (B) the song of a sparrow caged nearby. The letters beneath the sonograms label the syllables of the finch song and their counterparts in the song learned by the sparrow. Sonograms courtesy of Luis Baptista.

than 50 days when they first encountered a social tutor could still learn the alien song, showing that song learning can occur past the 50-day mark under some circumstances [71]. Indeed, some wild male white-crowns modify their songs throughout their lives. By recording the songs of known individuals (using distinctive combinations of colored leg bands to identify them) and comparing their songs with those of males in neighboring territories, researchers have shown that older male white-crowns in some populations regularly incorporate song elements of their current neighbors into their vocalizations [70].

Social experience shapes singing behavior in many birds in addition to white-crowned sparrows. Captive starlings, for example, are adept at mimicking human speech, producing such phrases as "see you soon baboon" or "basic research" as well as the sounds of laughter or kissing or coughing. But they will do so *only if* they are hand-reared in a human household where they participate in the social life of their caregivers (Figure 4) [1209]. In nature, young male starlings and

4 Social effects on song learning. Kuro the starling learned to include words in his vocalizations because he had a close relationship with the family of Keigo Iizuka. Photograph by Birgitte Nielsen, courtesy of K. Iizuka.

white-crowned sparrows receive both acoustical and social stimulation from adult males of their own species, and both factors have something to do with the development of their songs.

So, in proximate terms, why do male white-crowns sing different dialects? At the developmental level, the studies of Marler, Baptista, and others have demonstrated that the differences in white-crown dialects are caused by environmental differences in the sounds that males hear and the social experiences they have with other singing males. The sounds of these special songs and the social stimuli associated with them can have their developmental effects only because the brains of young males possess networks of brain cells that are extraordinarily specialized for song learning. Thus, at the physiological level, behavioral differences in singing arise in white-crowns because they have networks of nerve cells, or neurons, that are able to record specific kinds of information from socially interactive males, almost always members of their own species. The learned information is used for a particular purpose: altering the vocal output of the listener, who must hear his own songs in order to match what he sings with his memories of the songs of territorial rivals and neighbors. This special kind of learning demands special physiological mechanisms—in other words, a neural system dedicated to song recognition and imitation.

Indeed, the neural networks underlying song learning are so specialized that they may actually encourage male white-crowned sparrows to learn the dialect of the population into which they are born. When Doug Nelson gave very young males the opportunity to listen to tapes of two different dialects, one from the sparrows' natal population and the other from a region about 200 km away, about two-thirds of the youngsters eventually imitated the natal dialect, compared with only one-third that mimicked the foreign dialect. These results demonstrate that the white-crown's brain contains elements that predispose the young male to attend to, record, and eventually copy not just any white-crowned sparrow song, but rather a particular variant of that song [835]. This result underlines the importance of *genetic* information in influencing the development of the neural basis for a *learned* behavior.

The Avian Song Control System

The white-crowned sparrow story tells us that if we want to understand the proximate foundation of bird song, we need to identify the information processing centers in the brain that control this behavior [208]. These centers have been located. They include certain **nuclei** (anatomically discrete collections of neuronal cell bodies, which are the part of the nerve cell that contains the nucleus) as well as the neural fibers that form an information highway linking one nucleus with another. In white-crowned sparrows, one such nucleus, the higher vocal center (or HVC), connects to the robust nucleus of the archistriatum (mercifully shortened to RA by anatomists), which in turn is linked with the tracheosyringeal portion of the hypoglossal nucleus (whose less successful acronym is nXIIts). This bit of brain anatomy sends neural fibers to the syrinx, the sound-producing structure of birds that is analogous to the larynx in humans. The fact that the HVC and RA can communicate with the nXIIts, which connects to the syrinx, immediately suggests that these masses of nerve cells exert control over singing behavior (Figure 5).

Various hypotheses about the precise nature of neural control of bird song have been tested. For example, if neural messages from the RA cause songs to be produced, then the experimental destruction of this center, or surgical cuts through the neural pathway leading from the RA to the nXIIts, should have devastating effects on a bird's ability to sing. Experiments designed to test these predictions have been done with a variety of songbirds [208], with the result that we can now say with confidence that the RA does indeed play a critical role in song production, while other nuclei are essential for song learning. Destruc-

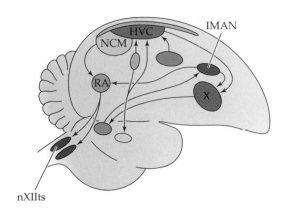

5   The song system of a typical songbird. The major components, or nuclei, involved in song production include the robust nucleus of the archistriatum (RA), the higher vocal center (HVC), the lateral portion of the magnocellular nucleus of the anterior neostriatum (IMAN), the caudomedial neostriatum (NCM), and area X (X). Motor pathways carry signals from the HVC to nXIIts to the muscles of the song-producing syrinx. Other pathways connect the nuclei, such as IMAN and X, that are involved in song learning rather than song production. After Brenowitz et al. [152].

tion of the magnocellular nucleus of the anterior neostriatum (IMAN), for example, does not strongly interfere with an adult zebra finch's ability to sing the song it had learned earlier in life, but if the operation occurs in a juvenile bird *before* it has acquired a mature song, then the bird will fail to sing a normal song in adulthood. Further evidence of the importance of this component of the control system for song learning came from checking the prediction that the IMAN should be much reduced or absent in birds that sing but do *not* learn their songs. There are many such species and, as predicted, they lack the well-defined IMANs of other birds [208].

Still other predictions follow from the hypothesis that song learning and song production are under the control of specific brain nuclei. For example, if one or another nucleus is involved in song learning or in song production, then we would expect to find changes in the activity of certain cells in that nucleus at particular stages in a bird's life. Interestingly, activity in cells of the IMAN of the zebra finch change just *before* the bird begins to sing, evidently preparing the brain to command the behavior [516]. Moreover, in those species, like the white-crowned sparrow, in which males sing but females do not, the RA should be substantially larger in males than in females, a prediction that has been shown to be correct for the sparrow and other songbirds (Figure 6) [57, 833, 847]. And in species like the white-crowned sparrow, in which singing behavior is highly seasonal, we might also expect the male song control system to undergo seasonal changes in anatomy. Just prior to the onset of singing in spring, the song system of males ought to increase in size, the better to accommodate the demands of song control or new song acquisition at this time. At the end of the breeding season, when male singing frequency declines sharply, we would expect the song nuclei to shrink. Both predictions have been confirmed via anatomical studies of bird brains. In the white-crowned sparrow, for example, the volume of the RA is higher for males during the spring breeding season than in the fall [151].

Moreover, given that some structural changes in song control system anatomy are seasonal in nature, one should be able to induce these changes by experimentally exposing captive birds to conditions that mimic those of a particular season. One such condition is the duration of the **photoperiod**, the number of hours of light in a 24-hour period. Increases in photoperiod occur as winter gives way to spring, and these increases are known to be detected by many songbirds, the males of which typically respond by stepping up the production of testosterone. When a team of biologists exposed some white-crowned sparrows to long photoperiods after keeping them in captivity under a short-day regime for 3 months, the higher vocal center (HVC) quickly became larger in the experimental birds (which also received supplemental testosterone). In fact, the birds exposed to springtime conditions had added some 50,000 new neu-

Zebra finch

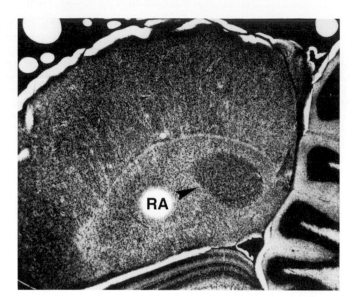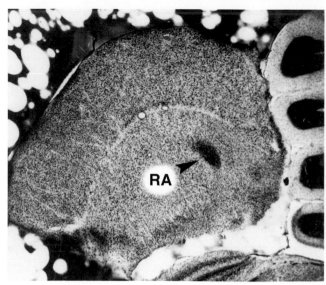

6   Differences in the size of one nucleus of the song system, the robust archistriatum (RA), in (left) the male and (right) the female zebra finch. Photographs courtesy of Art Arnold; from Nottebohm and Arnold [847].

rons to the HVC after just 1 week of the experiment [1150]. The growth of the HVC was even greater when captive male white-crowned sparrows were also provided with female company [1151]. Thus, elements in the mature song control system of male white-crowned sparrows (and other songbirds) are highly responsive to photoperiodic, hormonal, acoustical, and social stimulation.

The ability of these particular environmental inputs to have their effects arises from the special properties of functioning song control systems, which must be able to respond in very special ways. If a white-crowned sparrow's song control system is to change adaptively as opposed to randomly, certain cells within the system must in effect anticipate the receipt of certain inputs. When these key events occur, the cells must then alter their biochemistry very precisely in order to alter the bird's behavior. If, for example, what a young white-crown hears is to result in his learning a song dialect, then certain patterns of sensory signals generated by acoustical receptors in the bird's ears must be relayed to the IMAN or other song system centers where learning occurs. Chemical events in these centers that are caused by the inputs from acoustical neurons presumably alter the activity of genes in the responding cells, leading to new patterns of protein production that reshape the cells structurally or biochemically. Once they have been altered, the song system elements can do things that they could not do before the bird was exposed to the song of other males.

This theory of learning has been tested by examining cells in song system centers for changes in gene activity after a bird has heard a relevant song. You will recall that one acoustical stimulus that apparently affects the acquisition of song by white-crowned sparrows is the bird's own songs. The sparrow's ability to hear itself sing when it is 150 to 200 days old is essential for the crystallization of a normal full song from the variable subsongs it initially sings. Likewise, male zebra finches go through a period early in life in which they appear to be matching elements in their initial subsongs against stored memories of the full songs they heard others singing previously. During this process, certain neurons in the finch's anterior forebrain become more and more responsive to the bird's own song as opposed to tutor songs; by the time the finch is an adult, with a fully crystallized song, certain of its auditory neurons have

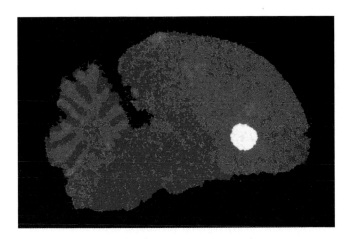

7   Gene expression in a component of the song system of the zebra finch. The white and yellow area corresponds to area X. The brightness of the site is related to the high level of activity of the *ZENK* gene, which has resulted in the production of relatively large amounts of the protein coded for by this gene. The absence of white and yellow elsewhere indicates that the *ZENK* gene was not active in this individual's robust archistriatum (RA) but was active in area X. Photograph courtesy of David Clayton.

become highly selective, responding much more strongly to this song than to any other [334]. Presumably this developmental process requires a biochemical restructuring of the properties of the neurons in question, which in turn requires genetic information that can alter the internal protein environment of these cells.

We now know that as zebra finches attempt to sing like a tutor, the activity of the *ZENK* gene rapidly increases in certain song system neurons, with corresponding increases in cellular amounts of the protein coded by this gene [790]. In other words, when a zebra finch listens to itself sing, it generates sensory feedback that activates a particular gene in certain cells. This gene's activity translates into the production of a specific protein, which presumably has something to do with subsequent alterations in the neural circuits that control the finch's song (Figure 7). This hypothesis has received support from the observation that as the zebra finch gets closer and closer to singing an accurate copy of a tutor song, *ZENK* gene activity in the RA falls [584]. As the finch gains a crystallized full song, further changes in cell architecture or biochemistry in the song control system cannot improve the ability to sing that song, and therefore the *ZENK* gene in certain key cells shuts down. Work of this sort has also been done by exposing zebra finches to songs that vary in their similarity to the tutor songs that they heard during the early song learning phase. The more similar the song played to the experimental finches, the more strongly activated were certain specific circuits in the medial caudal portion of the neostriatum (NCM), a brain region that only recently has been implicated in song learning [119]. Heightened activity in the NCM generated more production of the *ZENK* gene's protein in the activated circuits.

The Development of the Song Control System

The interaction between certain kinds of acoustical stimuli and the *ZENK* gene in special parts of the zebra finch brain is just one example of the interactive nature of development. The development of every aspect of all living things requires *both* genetic information and well-defined environmental inputs, such as specific sounds as well as the amino acids and other chemical constituents of cells. Let's illustrate the interactive nature of development by considering how a single-celled organism, a fertilized bird egg, becomes a young male white-crowned sparrow with a nervous system that can detect, identify, and record information about the sounds made by his fellow white-crowns and himself.

The nucleus of the fertilized egg contains chromosomes with a complete set of genes, which will be copied and transmitted to all the cells formed during the growth of the embryo. In addition to this genetic information, however, the egg cell also contains quantities of sugars, proteins, and fats, which are the

Avoiding a Misunderstanding

Even though the story of song system development is incomplete for zebra finches and other birds, you now know enough about behavioral development to avoid a common error, which is to divide behaviors into two groups, those that supposedly are "genetically determined" versus those that are said to be "environmentally determined." Imagine that someone tells you that the dialect sung by a white-crowned sparrow from Marin, California, is "environmentally determined" because it was learned, whereas the courtship song of a male zebra finch is "genetically determined" because it depends on the effects of the genes on the bird's sex chromosomes. How would you respond?

I hope that you would point out that in birds, song depends on the structures present in the nervous system, and these structures cannot develop without *both* genetic information and environmental inputs. Therefore, the ability to learn a dialect is not purely environmental, because the capacity for learning resides in the learning centers within the song control system of the white-crown; these nuclei and their connections with other brain components are the product of vast numbers of *gene*–environment interactions. Moreover, it also makes no sense to say that zebra finch courtship song is genetically determined, because the underlying song system is the end product of an extraordinarily large number of gene–*environment* interactions, all of which are influenced by the hormones and other environmentally supplied materials present in cells.

Now how would you respond to someone who said that the *differences* among white-crown males in their dialects are environmentally caused, whereas the *differences* between the behavior of normal adult male and normal adult female zebra finches stem from genetic differences between them? I hope that you would agree that both of these explanations are potentially legitimate. Differences in acoustical experiences can lead to differences in the gene–*environment* interactions occurring during the development of male white-crowned sparrows. Conversely, differences in the genetic makeup of male and female zebra finches lead to differences in the *gene*–environment interactions that occur during the development of the two sexes in this species. Although all the characteristics of living things are dependent on both genetic and environmental factors, the *differences* between two individuals in their attributes may be due to either one or the other component (or both). That is, a difference in either genes or environment can generate a developmental difference arising from the interplay that always takes place between the two factors within developing organisms (see Figure 20).

Different Songs: Ultimate Causes

This chapter's primary goal is to reinforce the point that behavior has both proximate and ultimate causes, which can be analyzed separately even though the two levels of analysis are intimately related. We have seen that the ability of a white-crowned sparrow to sing rests on an elaborate proximate foundation involving both developmental and neural mechanisms. Now let's bring evolution into the picture. What is the evolutionary origin of the proximate mechanisms that make it possible to learn a species-specific song? An accurate reconstruction of the evolutionary history of birds could tell us when song learning arose in these animals and what its antecedents were. Work on these matters is still in its infancy, but researchers have established that song learning occurs in species belonging to just 3 of the 23 orders of birds (Figure 10): the parrots, the hummingbirds, and the passeriform songbirds [148]. Has each group independently evolved the neural equipment that makes song learning possible?

10   The phylogeny of song learning in birds. If we assume that the long-extinct bird that gave rise to all modern species did not learn elements of its songs, but instead produced innate vocalizations, as do many modern bird groups, then song learning has evolved independently in three different lineages of modern birds. On the other hand, song learning may have originated in a common ancestor of parrots (Psittaciformes), hummingbirds (Trochiliformes), and passerines, and been retained in these three lineages while having been lost in other descendants of an ancestral song-learning species. After Brenowitz [148].

If the independent origins scenario were true, we would expect to find major differences among the song control systems of the three groups. Careful anatomical studies have revealed that the key control centers are positioned quite differently in the brains of songbirds and of parrots [150]. On the other hand, the budgerigar, a parrot, possesses the same *ZENK* gene that songbirds have. Moreover, this regulatory gene is activated in various parts of the song control system when the budgerigar hears budgie warbles or when it is vocalizing. At the genetic level, therefore, real similarities exist between songbirds and parrots, with both sharing at least one major gene involved in the development of structures that contribute to song recognition and song production [577]. As a result, we cannot say for certain whether parrots, hummingbirds, and songbirds secured their remarkable song-learning skills by inheriting the genes for their kind of song system development from a distant common ancestor, or whether they independently evolved similar abilities, making use of some of the same long-persisting genes and ancient vocal systems present in the species preceeding them.

The Adaptationist Approach

The song control systems that make learning a distinctive song possible have an evolutionary history with a beginning almost certainly followed by a long series of modifications, some of which were retained and others that were lost over evolutionary time. The current song system of any species is the prod-

11 Female red-winged blackbirds adopt the precopulatory display position (tail elevated) more often when hearing their own species' song than when listening to (A) a tape of a swamp sparrow's song or (B) a tape of a mockingbird singing their own species' song (mockingbirds mimic the songs of other species). After Orians [861] and Searcy and Brenowitz [1032].

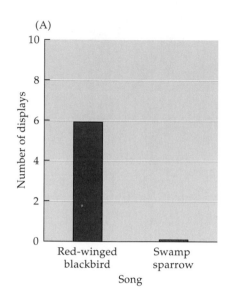

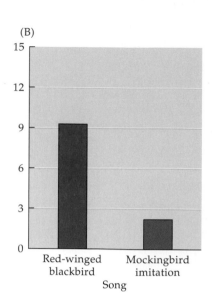

Red-winged blackbird female

uct of this historical process, which raises another complementary evolutionary question, one that focuses on why some attributes have persisted to the present. Why is it that white-crowned sparrows living today have a nervous system that enables them to acquire a distinctive vocalization different from that of any other living sparrow species? Has natural selection been responsible for the evolutionary changes that have resulted in today's proximate mechanisms that produce a species-specific song in male white-crowned sparrows?

In the past, white-crowns whose songs happened to be especially distinctive might have enjoyed a reproductive advantage because highly recognizable singers were able to communicate their species' identity more effectively to listening females. Sexually receptive female white-crowns currently respond more enthusiastically to songs of their own species than to songs of other sparrows [1088], and the same is true for other songbirds (Figure 11) [1031]. If in the past female songbirds generally reacted more positively to especially distinctive singers of their species, then natural selection driven by mate choice would result in the spread of the favored songs. We have here an example of what is called an **adaptationist hypothesis**, a possible explanation based on natural selection theory that attempts to identify the adaptive (reproductive) value of a particular attribute, in this case singing a song that announces unmistakably the species membership of the singer.

More than one adaptationist explanation exists for why different species sing different songs. Species that occupy different habitats may evolve different song properties to overcome special environmental obstacles to being heard. In forests, for example, the dense foliage of trees absorbs sound, making it harder for forest birds to broadcast their signals a long distance, especially if the song contains trilled sound frequencies above 4 kHz [208, 820]. The songs of forest-dwelling birds often do feature whistles, not trills, with frequencies below, not above, 4 kHz, whereas species that live in open woodland or savannas are more likely to use trills and notes above the 4 kHz level. This is true even for different populations of the same species occupying habitats that differ in foliage density (Figure 12) [559]. Thus, strong selection for effective propagation of acoustical messages may lead geographically separated populations of birds to produce different kinds of songs.

In addition, one can devise *nonadaptationist* explanations for the distinctive song output of white-crowned sparrows. For example, the species-specific nature of the vocalizations of white-crowned sparrows could have arisen as

Forests

Woodlands

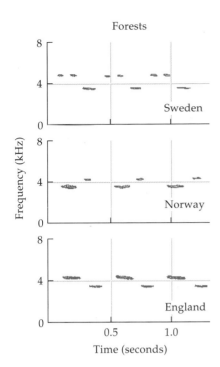

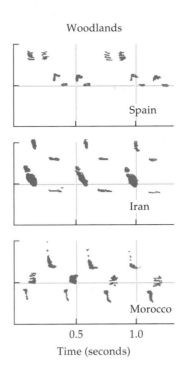

Time (seconds)

Time (seconds)

12 **Songs match habitats.** Great tits from dense forests produce pure whistles of relatively low frequency, whereas males of the same species that live in more open woodlands use more and higher sound frequencies in their more complex songs. After Hunter and Krebs [559].

Great tit

an incidental, nonadaptive result of the effects of speciation, rather than through selection for song properties that could be transmitted effectively or that facilitated species identification. This **by-product hypothesis** suggests that white-crown song was shaped during the period when white-crowned sparrows were evolving in geographic isolation from some of the species that now live with them. The gene pool of these isolated ancestral white-crowns would have been influenced by a variety of events, including random mutations and the accidental loss of genetic variation from small populations. In addition, natural selection acting nonrandomly on features other than the song per se would have also altered the gene pool of the species. As assorted genetic changes accumulated in the population ancestral to modern white-crowned sparrows, this population would have become increasingly different from other populations of sparrows, genetically speaking [775]. Some of these genetic differences might have had developmental side effects on the kinds of vocalizations the birds could produce, altering their songs as compared with their ancestor's songs and the songs of other species as well. As a result, when this population expanded its geographic range and came into contact with certain other species, they might have already been singing different songs.

We can test both adaptationist and nonadaptationist hypotheses for the unique songs of many songbirds. For example, if song distinctiveness is the product of selection for the ability to avoid confusion about species membership, then we would not expect closely related species *that live apart* to sing particularly different songs. If, however, distinctive songs are the nonadaptive effect of genetic changes occurring during speciation, then closely related species that have evolved in isolation from one another might well sing quite differently. In fact, some pairs of related species that occupy different regions do sing highly recognizable songs (Figure 13). If we could be confident that, say, black-throated gray and yellow-throated warblers did not overlap during some part of their evolutionary history, these species would provide evidence that geographic overlap among species is *not* necessary to produce species-specific songs.

The nonadaptationist by-product hypothesis also yields the prediction that the songs of species with partially overlapping ranges will not be more differ-

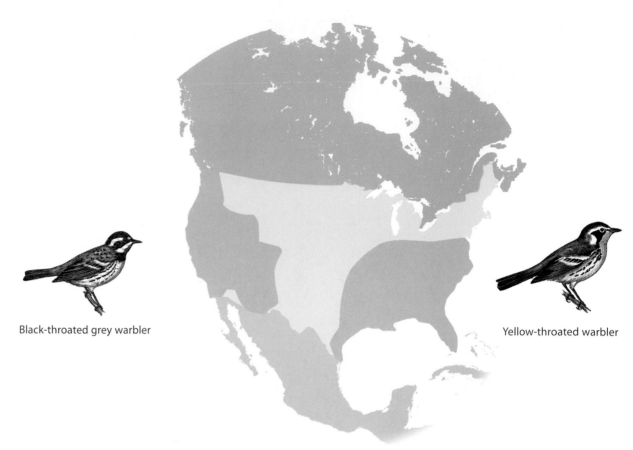

Black-throated grey warbler

Yellow-throated warbler

13   Geographic distributions and songs of the black-throated gray warbler and the yellow-throated warbler. Although the ranges of these two closely related species do not overlap, which means that they probably have not exerted selection on one another for distinctive songs, their songs are nevertheless very different.

ent in the zone of overlap than elsewhere. In contrast, the adaptationist species identity hypothesis yields the prediction that when two species of sparrow or warbler have partially overlapping ranges, the songs of the two species should diverge *more* in the zone of overlap (where confusion about species membership is possible) than in areas in which only one of the two species occurs.

Two closely related species that coexist in only part of their collective range are the blue-winged and golden-winged warblers. Frank Gill and Bertram Murray reported that in the area of overlap, the songs of the two species become less variable, and thus more distinctive [432]. As a result, the potential for female confusion about the species identity of calling males is presumably somewhat reduced in the places where the problem of recognizing singers might arise. However, the two species regularly hybridize in the zone of overlap, suggesting that the differences between their songs, which are very substantial throughout their entire ranges, have little to do with preventing hybridization.

In addition, two other closely related sparrow-like songbirds, the indigo bunting and the lazuli bunting, do *not* sing especially differently in their zone of overlap in Nebraska. Indeed, where they live together, a male indigo bunting may learn and incorporate song elements from a lazuli bunting neighbor, and vice versa [364]. And yet hybrids between the two species occur only rarely, suggesting that females have no difficulty telling the species apart despite their song similarities.

The few other bird studies on vocalization and species identification also fail to support the hypothesis that song differences are the sole, or even primary, basis for the ability of females of species A to identify males of species A [73, 208]. Thus, there is reason to doubt that natural selection for species recognition has played an important role in the evolution of species-specific bird songs. The unique features of the songs of many birds may have arisen largely or entirely during periods when an evolving species was geographically isolated from other populations of shared recent ancestry. The divergence among related lineages might have been driven by random genetic events or by selection for males able to match their songs to their habitats, as mentioned above. Or perhaps lineages came to differ because females in different regions came to have different song preferences, not because of the benefits of identifying a male of their own species but because songs convey information about the genetic quality of the singer relative to others of his kind—another adaptationist possibility that we will examine shortly. Although more work is required to sort out the various possibilities, we can at least conclude from this example that non-adaptationist hypotheses are legitimate alternatives to adaptationist ones.

Why Do Only Males Sing?

Not only are bird songs usually species-specific, but they are more often sung by males than by females. Why should this be so? Here again, one can imagine evolutionary hypotheses that explain the differences between the sexes in terms of their possible adaptive value *or* in terms of certain nonadaptive consequences of evolution. For example, it could be that females fail to sing as a nonadaptive by-product of the way their sexual development occurs. The hormones that regulate the development of sex differences could be so important to the functional development of critical female attributes, such as the female reproductive tract, that certain side effects of female hormones, such as the absence of an elaborate song system in the brain, could persist over evolutionary time. According to this hypothesis, hormonal mutations in the past that conferred singing ability on females would have so disrupted normal ovarian development, or egg production, or maternal care that they were selected against, despite the possible benefits that females might have gained from being able to sing.

Admittedly, this hypothesis is a straw man because I knew it to be incorrect before I presented it. If the development of a song control system really interferes with the development of female reproductive capability, then there should be no bird species in which females sing. And yet females of many species, especially those living in the tropics, sing elaborate and complex songs [69, 242], suggesting that there are no insuperable developmental constraints on the production of song control systems in female bird brains. I present this nonadaptationist hypothesis here merely to illustrate again what is meant by an ultimate explanation of this sort.

Adaptationist explanations exist for why male birds sing. One reproductive benefit of male song could be the attraction of females, which require information about the species membership of a singer (see above) or about the quality of a potential mate relative to other males in the population. However, singing males could also gain by communicating with rival males, in which case their songs could contain information about their fighting ability, warning other males to stay away from an occupied territory or a fertile mate.

Although these various alternatives have been tested surprisingly few times [657], they produce eminently testable predictions. The mate attraction hypothesis, for example, generates the prediction that males of monogamous species should sing loudly and often before acquiring a mate, but stop singing afterward, in contrast to polygynous species, in which males can attract several mates in sequence [350]. Males of the monogamous California towhee conform to this expectation by ceasing to sing once they have paired off with a female [761], whereas males of the polygynous house wren continue to sing after attracting a first mate [588]. And in the great tit, males whose mates have been temporarily

14   Song is a signal to females. Males of the great tit that have acquired a mate almost cease singing, but if their mate is experimentally removed, they begin to sing more frequently. Once their mate has been returned to them, however, they sing hardly at all. After Krebs et al. [654].

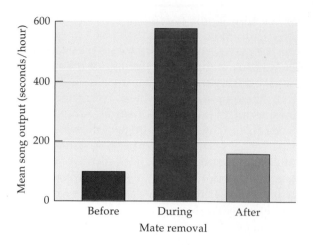

removed by experimenters resume singing at a great rate, only to reduce the rate of singing when their mate is returned to them (Figure 14).

The mate attraction hypothesis also predicts that females will approach taped song, and some limited supporting evidence has now been collected [208]. Playback studies have shown that female house wrens visit nest boxes from which recorded songs of their species are being broadcast more often than they visit "silent" nest boxes [589], although stronger evidence would have been provided if a control group of nest boxes had been advertised by tapes of another species' song.

Moreover, the only statistically significant factors influencing how rapidly male starlings pair with females are the duration of the male's songs and the size of his song repertoire (Figure 15) [822]. In a laboratory setting, female starlings prefer to sit on perches near nest boxes where they can listen to a song recording lasting about a minute as opposed to one that lasts only half as long [424]. At the proximate level, this preference has been linked to activity in a particular region of the female's brain, the ventral caudomedial neostriatum (see Figure 5), where changes in *ZENK* gene activity also occur [425].

Why might female birds have evolved the genetic and physiological systems that cause them to respond positively to relatively lengthy songs with many different phrases? Perhaps repertoire size conveys honest information about the

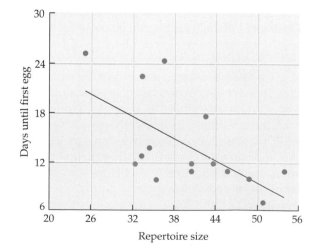

15   Female starlings prefer complex songs. The speed with which a female pairs with a male and begins to lay eggs is related to the number of song phrases in the male's repertoire. The y-axis shows the number of days from the date on which the male claimed a nest box to the date on which the first egg was laid. After Mountjoy and Lemon [822].

male's health, or age, or experience (but see [239]). In some birds, males with large repertoires are less likely to be infected by malarial parasites [809]. Males with fewer parasites may be more capable of being good parents, as has been demonstrated for the sedge warbler, another songbird in which females prefer males with a large battery of songs [117]. Male sedge warblers with the preferred song repertoires bring more food to their offspring, which grow faster, a result that almost certainly raises the reproductive success of both parents.

More generally, females might find the singers of certain kinds of learned songs attractive because the learning must have taken place early in the male's life, when genetic defects or nutritional stress could well interfere with brain development and the ability to acquire a top-quality song or a large repertoire of songs [851]. In the great reed warbler, well-fed nestlings that gain more weight have larger song repertoires as first-year adults than individuals that have been nutritionally stressed in the nest [850]. In zebra finches, the size of the male's song control nuclei is hereditary; males with a larger HVC have more complex songs than those with a smaller HVC [5, 6]. Therefore, a male's song or his song repertoire could be related to his genetic, developmental, and physiological competence, which means that his song could signal his value as a gene donor or parent. As a result, female birds that choose partners on the basis of their songs could acquire mates of high genetic or parental quality, which creates selection in favor of males able to learn a complex vocalization or a large song repertoire.

But males could also sing in order to keep rival males from entering their territories. In experiments in which resident male white-throated sparrows were removed and replaced by speakers broadcasting their songs, new males were slower to move into these territories than others from which males were removed but not replaced with taped song [376] (Figure 16). Likewise, in experiments with song sparrows in which two males were removed from their territories but only one was replaced with a speaker that broadcast his song, the empty territory was always the first to be invaded by an intruder [852]. These results support the hypothesis that song can benefit the singer by helping repel other males.

But is the value of a song to deter other males from competing for territories, or to indicate to rivals that the singing male is highly motivated and highly capable of guarding his fertile partner against intruders [806]? The mate guarding hypothesis generates a key prediction: the song rate should peak when females are *most* fertile, rather than during territory establishment and initial pair formation. Song rate and female fertility are strongly correlated in some bird species (Figure 17) [737], but not others [430, 975]. Moreover, an alternative hypothesis exists for those cases in which males do sing more when their

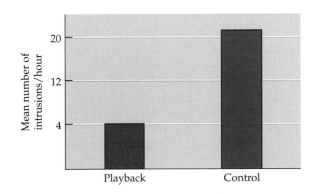

16 Does bird song repel territorial intruders? Territories from which resident male white-throated sparrows were experimentally removed attracted fewer intruders when the taped song of the removed males was broadcast from their vacant territories. After Falls [376].

17 Song and mate guarding in the great tit. The frequency with which males sing increases during the days when their partners are most fertile, perhaps as a warning to rival males that they will be attacked if they come near their mates. (Each differently colored dot represents a different pair of birds.) After Mace [737].

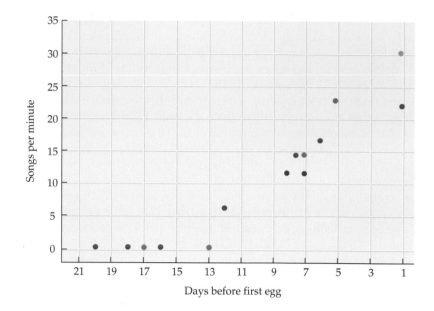

mates are most fertile: the male may be communicating primarily with his partner, attempting to induce her to copulate and thereby fertilize her eggs with his sperm [921]. This case illustrates the difficulty of determining just who the receiver of a signal really is and the importance of testing alternative hypotheses.

However, the broader point is that evolutionary biologists have been able to identify and test any number of hypotheses on the possible reproductive benefits for male birds who sing. We now know that singing actually does help males of some species attract mates, while in some other species singing keeps fellow males out of the songster's territory. Why *mated* males with well-established territories continue to sing is a topic that is sure to attract more attention in the future.

Natural Selection and Dialects

Our final evolutionary question about bird song is, why do male white-crowned sparrows, among other species, learn to sing different versions, or dialects, of their species' song? Here, as elsewhere, alternative hypotheses are available. On the one hand, perhaps *differences* in dialects are not adaptive in and of themselves, but emerge as an incidental effect of song learning mechanisms that evolved for some other reason. Once learning plays a role in song acquisition, then nongenetic historical accidents can influence what songs birds in a particular area will sing [149]. Imagine an area colonized by one or a few males, which might be juveniles with incompletely formed songs. If the population grows, young males born in the area will tend to imitate the idiosyncratic song of the older established males closest to them, forming an "island" of males singing the same distinctive song type, which becomes a dialect passed on by learning from generation to generation [680, 731]. If this scenario is true, natural selection for song distinctiveness per se may not have been part of the history of a given sparrow dialect.

However, once dialects have formed, for whatever reason, we can ask whether they have acquired adaptive value. If dialects are incidental effects of no reproductive significance, males in area B that happened to sing dialect A should suffer no disadvantage as a result. We can test this prediction by looking at the song preferences of female cowbirds from two different subspecies living in different parts of North America. Contrary to the predicted result, females of subspecies A find their own males' song type more sexually stimulating than the songs of subspecies B, judging from the frequency with

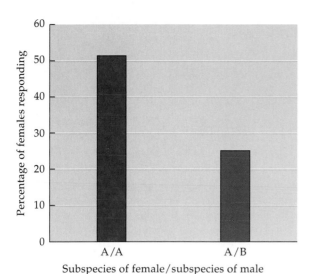

18 Song potency in cowbirds. Female cowbirds prefer the song of males of their own subspecies. When females of subspecies A hear the song of males of their own subspecies (A/A), they are much more likely to adopt the precopulatory position (see drawing) than when they hear the songs of males of another subspecies (A/B). Data from West and King [1208].

which they adopt the precopulatory position (Figure 18) in response to tapes of these songs [623, 1208]. Because females prefer their own subspecies' dialect, males living with them gain by singing that dialect. Selection is therefore maintaining dialect differences in this species. If dialects are also currently adaptive for white-crowned sparrows, then male white-crowns able to learn a dialect ought to gain higher reproductive success on average than males unable to acquire a particular dialect. Singing a specific dialect could be selectively advantageous for several reasons. As mentioned already, a dialect might have sound frequencies or patterns of sound production that permit the song message to travel farther and with less degradation in the singer's habitat. Males that sing a habitat-matching dialect might therefore communicate more effectively than if they used another dialect [208].

Cowbird female

Another adaptationist hypothesis goes like this: in a species divided into stable subpopulations, males in subpopulation X have genes that have survived natural selection in an area occupied for generations by those males' ancestors. By learning to sing the dialect associated with their place of birth, males announce their possession of traits (and underlying genes) well adapted for that area. Females born in that area gain by having a preference for males that sing the local dialect because they will endow their offspring with genetic information that promotes the development of locally adapted characteristics. Some evidence from white-crowned sparrows supports this hypothesis [58]. On the other hand, if females select mates born in their natal area, then female white-crowns should prefer males with the dialect that they heard while they were nestlings—namely, their father's dialect—and they do not, at least in one Canadian population [212]. Furthermore, as we have seen, young male white-crowns are not locked into their natal dialects, but can change them [318]; therefore, females could not rely on a male's dialect to identify his birthplace.

Still more adaptationist hypotheses about song dialects have been proposed. For example, perhaps dialects help males communicate more effectively with fellow males, rather than with females. This hypothesis is based on the presumption that when neighboring males interact regularly over long periods, they have something to gain by forging special relationships based on mutual tolerance. One way a male could announce his intention to accept the territorial borders of a neighbor would be to change his song to match that of the neighbor—as males actually do in some populations of the white-crowned sparrow [70, 88], but not all [834].

Song sparrow

The ability to match songs, or at least song elements, with rival neighbors also occurs in the aptly named song sparrow, a bird that has a repertoire of many different, distinctive song types. (White-crowned sparrows sing just one song type, with different populations having their own dialects. Song sparrows sing many song types, with less sharply defined dialects in different regions.) Young song sparrows usually learn their songs from tutors that are their neighbors in the first breeding season, with the final repertoire tending to match that of an immediate neighbor [846]. In fact, Michael Beecher and his colleagues found through playback experiments that when a male heard a tape of a neighbor's song, he tended to reply to that tape by singing a song from his own repertoire that matched one in the repertoire of that particular neighbor. Even if the subjects heard a neighbor's song that was *not* in their personal repertoire, so that they could not simply imitate it, they still picked a song to sing that matched one in that neighbor's song bank (Figure 19). Thus, song sparrows recognize their neighbors and know what songs they sing, and they use this information

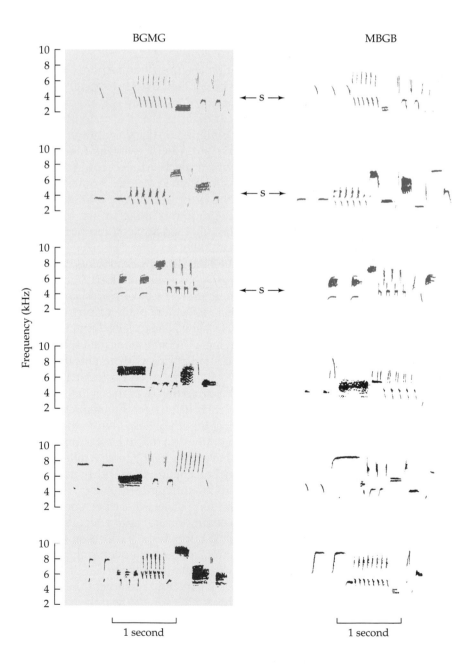

19 Repertoire matching in the song sparrow. Males BGMG and MBGB occupy neighboring territories and share three song types (the top three rows of spectrograms); six unshared song types appear on the bottom three rows. After Beecher et al. [85].

to shape their replies [85], probably conveying to a neighbor that they recognize him as an individual and will treat him more tolerantly than an intruding stranger. Such a recognition signal could save both the signaler and his neighbor time and energy.

If the ability to learn the songs, or elements of the songs, of neighbors promotes less costly (and thus adaptive) relationships among long-term neighbors, then in social settings marked by very rapid turnover, song learning of this sort should not occur. Under these ecological circumstances, males should acquire songs characteristic of their species as a whole to facilitate communication with any and all conspecifics, rather than developing a dialect characteristic of a small stable subpopulation. Donald Kroodsma and his colleagues have tested this proposition by taking advantage of the existence of two very different populations of the sedge wren. In the Great Plains, sedge wren males are highly nomadic, moving from breeding site to breeding site throughout the summer. In this population, song matching and dialects are absent; instead, the birds improvise variations on songs heard early in life as well as inventing entirely novel songs of their own, albeit all within a broad general pattern characteristic of their species [659]. In contrast, males of the same species living in Costa Rica and Brazil remain on their territories year round. In these areas, dialects and song-matching are predicted to occur, and these learning-proficient males do have dialects and do sing like their neighbors [661]. The differences in song learning that occur within this one species provide support for the general argument that learning and dialects arise when males gain by communicating with long-term neighbors.

Proximate and Ultimate Causes Are Complementary

The singing behavior of songbirds is well enough understood to illustrate the fundamental link between the proximate and ultimate causes of behavior (see Figure 2, p. 4). As we have seen, bird species generally differ in the songs they sing; moreover, in many species, males are more likely to sing than females, and individual males often vary in their songs. In each case, the differences arise because of complex interactions between the genes possessed by the species, the sexes, or individuals and the environments in which those species, sexes, or individuals develop. These proximate interactions create elaborate feedback loops in which a developmental product of some sort becomes part of the environment of the developing organism (Figure 20). Hormones, for example, are products of gene–environment interactions; once manufactured, they become part of the cellular environment where they can interact with other parts of the genome, turning some genes on and others off. The cascading effects of these highly structured events result in the production of a song control system. The way in which these systems can respond to experience is a function of the way they develop and the genes that are present in their cells, which are preprogrammed to react in particular ways to particular patterns of sensory stimulation and neural activity.

This assortment of proximate causes generates differences among the songs of birds that can have evolutionary effects. The ability of a male to announce his species membership or his other attractive qualities may influence his chances of securing a mate and reproducing, while his song qualities may also have an effect on his ability to secure and retain a territory and the female(s) within. The summed consequences of these effects translate into differences among individuals in their reproductive, and thus genetic, success. Those individuals whose proximate mechanisms have enabled them to have higher fitness than others will, by definition, contribute more copies of their genes to the next generation,

20   What causes differences among individuals? The development of an individual's traits depends on the interaction between the genetic information inherited from its parents and the nutrients, chemical substances, and experiences provided by its environment. Therefore, differences among individuals can be caused by differences in either their genes or their environments, or both. These proximate causes of individual differences have the potential to affect the evolution of a species.

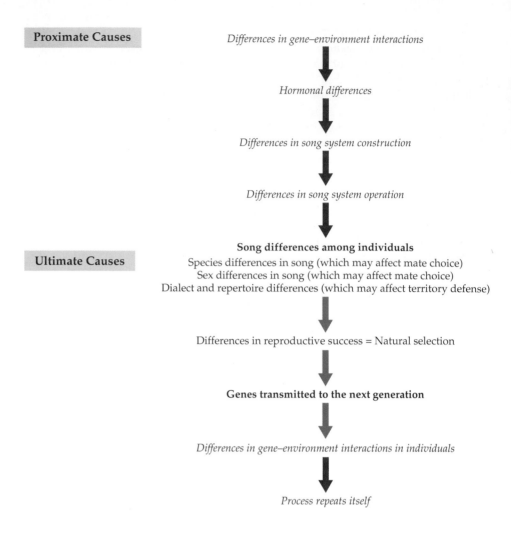

Proximate Causes

Differences in gene–environment interactions

⬇

Hormonal differences

⬇

Differences in song system construction

⬇

Differences in song system operation

⬇

Song differences among individuals
Species differences in song (which may affect mate choice)
Sex differences in song (which may affect mate choice)
Dialect and repertoire differences (which may affect territory defense)

Ultimate Causes

⬇

Differences in reproductive success = Natural selection

⬇

Genes transmitted to the next generation

⬇

Differences in gene–environment interactions in individuals

⬇

Process repeats itself

where those genes will be available to participate in interactive developmental processes within members of that generation. The hereditary attributes of these animals will in turn be measured against one another in terms of their ability to promote genetic success in a new round of selection. Thus, one cannot completely understand the proximate causes of behavior without also considering the evolutionary effects of reproductive competition, and vice versa.

Summary

1. Every behavioral trait has both proximate and ultimate causes. The proximate causes of behavior can be divided into two major components: the *underlying physiological mechanisms* that make behavior possible and the ways in which those mechanisms are *assembled during development*. The ultimate causes of behavior can also be treated as two different but interrelated issues, one concerned with the *pattern of historical changes* that occurred between the origin of a trait and its current manifestation and the other targeted on the *processes responsible for the changes* that have taken place over time.

2. Proximate and ultimate causes are complementary. The physiological mechanisms that are the basis for behavior have an evolutionary history. Physiological and developmental systems that promote the ability of individuals to reproduce and pass on their genes are far more likely to persist and be maintained than alternative systems that fail to have positive effects on reproduction.

3. Different species of birds sing different songs; in many songbirds, males sing far more often than females; and the males of any given species may also sing dialects of their species-specific song that are characteristic of a given geographic region. Each of these features of bird song can be subjected to both proximate and ultimate analyses, demonstrating the complementarity of these different levels of analysis.

4. Research on bird song also demonstrates that the ability to acquire a song characteristic of one's species or locality is the product of a complex developmental process completely dependent on *both* genetic information and environmental input, which includes the bird's acoustical and social experiences as well as the environmentally supplied chemical building blocks within its brain and body.

5. Studies of bird song at the evolutionary level have been guided by the *adaptationist approach*. Biologists who employ this approach ask why a particular attribute, such as the ability to acquire a song dialect, might advance the reproductive success of individuals more than some alternative characteristic. This question is the foundation for the development and testing of hypotheses on the role that natural selection may have played in shaping the behavioral evolution of a species.

Discussion Questions

1. Frogs and toads, as well as birds, use species-specific vocalizations as apparent mate attraction signals. The figure below contains data from a study of two closely related frog species from southeastern Australia whose ranges overlap partially [708]. These data are pertinent to what hypothesis about the evolution of species differences in mate attraction signals? What prediction can be evaluated on the basis of these data? What scientific conclusion can you derive from the test?

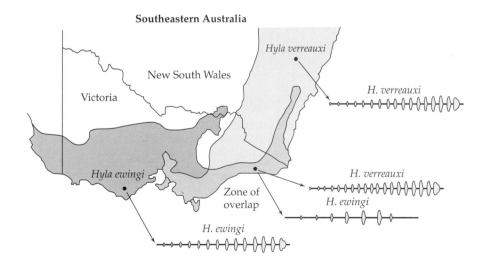

2. A natural experiment sometimes occurs in Australian woodlands when galahs (a species of parrot) lay eggs in nests that are then stolen from them by pink cockatoos (another parrot). The cockatoos then become foster parents for baby galahs. The young foster-reared galahs produce begging calls and alarm calls that are identical to those produced by galahs cared for by their genetic parents. However, the adopted galahs eventually give contact calls very much like those of their adoptive cockatoo parents, as you can see from the figure below. (The birds produce these signals to help maintain contact with others when traveling in flocks [995].) Someone claims that these observations show that begging and alarm calls are genetically determined, whereas contact calls are environmentally determined. Explain why this claim is wrong. Then defend the superficially similar statement that the alarm call differences between adopted galahs and their parents are the result of genetic differences between them. What other behavioral differences are the result of *environmental differences* between the adopted galahs and certain other individuals?

Galah

Pink cockatoo

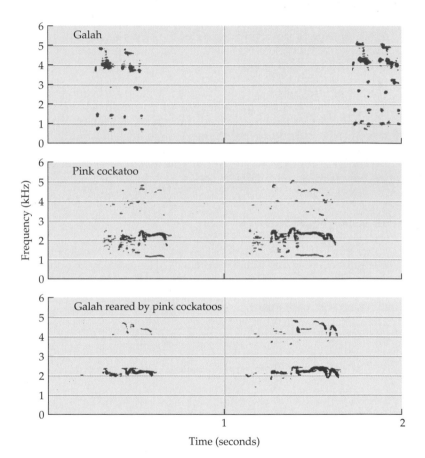

3. We observe that a white-crowned sparrow's species-specific song is more similar to that of his neighbor than to his father's dialect. Categorize the following hypotheses as "proximate" or "ultimate."

   a. Young adult white-crowned sparrow males try to match their song as closely as possible with that of their neighbors.
   b. White-crowned sparrow males possess genetic information that influences the development of a biased song learning mechanism.
   c. By singing the same song as his neighbors, the young adult male more efficiently deters invasion of his territory by those neighbors.
   d. The hormonal condition of the young adult male predisposes him to listen to and mimic the songs of his territorial neighbors.
   e. Those males with the song flexibility to mimic the songs of their neighbors have in the past attracted females more reliably than males without this ability.

4. List several well-known features of language learning in humans that are similar to song learning in birds. Do these similarities suggest certain hypotheses on the proximate bases of human language learning, especially the genetic and developmental components? Do comparisons with birds also suggest some interesting hypotheses on the adaptive value of learned language for members of our species? *After* you have attempted to answer these questions, examine information in any of the following sources [322, 333, 920] in order to test at least some of your hypotheses.

5. In the greater spear-nosed bat, unrelated females roost together in groups. The members of any given group tend to produce similar calls, but different groups have different calls. If you transfer some young bats into a new group and, at the same time, move some of their siblings into still another group, the shifted bats eventually change their calls to resemble those in their new groups [138]. What proximate hypothesis and what ultimate hypothesis on the acquisition of acoustical communication are supported by these results?

Suggested Reading

Bird Song: Biological Themes and Variations [208] is a very readable review of the proximate and ultimate causes of bird song. See also the more technical *Ecology and Evolution of Acoustic Communication in Birds* [660]. A full analysis of all forms of animal communication has been written by Jack Bradbury and Sandra Vehrencamp [143]. A number of papers provide excellent examples of the techniques used to explore the proximate basis of song learning [43, 70, 1209, 1219], while other research reports [589, 657] illustrate how one can also study the evolution of bird song.

3

The Development of Behavior: A Focus on Heredity

*I*f you had lived in Britain before 1950 and wanted to see a blackcap warbler there in December, you would have been out of luck, since all the blackcaps that bred in or traveled through Britain earlier in the year would have migrated to Africa for the winter. But starting in the 1950s, British bird-watchers began to find more and more blackcaps overwintering in their country. Why the change in warbler behavior? Did some British birds lose their hereditary ability to migrate south, and so stayed on after the breeding season? Or were these birds possessors of mutant genes that "caused" them to fly to Britain for the winter instead of heading somewhere else?

We can answer these questions thanks to geneticists who have conducted some ingenious behavioral experiments with blackcap warblers. We shall review these and other studies in order to examine the connection between genes and behavior. This topic was first raised in Chapter 2, in which the goal was to scan the entire spectrum of the causes of behavior, both proximate and ultimate. This chapter is the first of several that will go more deeply into the proximate basis of behavior, examining evidence that an

◀ **The development of migratory behavior in birds,** such as these snow geese heading north from Tule Lake, California, is influenced by their genes, but how? Photograph by Ron Sanford and Mike Agliolo

animal's genes have pervasive influences on its behavioral abilities, a proximate fact of life that has great evolutionary significance as well.

The Genetics of Behavior

Birds are able to sing, feed, fight, and migrate because they possess genetic information that facilitates the development of these abilities. In theory, therefore, a mutation, or change, in even one gene could produce a developmental change that eventually translates into a distinctive behavioral ability for those individuals that carry the mutation. However, in order to establish that genetic differences contribute to the development of one or another behavioral difference among individuals, we need to eliminate the possibility that the behavioral differences arise because of differences in these individuals' environments. (You will remember that the development of every trait requires both genetic and environmental contributions.)

If genetic differences help explain why some blackcaps spend the winter in Britain while others migrate to Africa (Figure 1), then the offspring of the "winter-in-Britain" birds ought to inherit the special behavior of their parents. One way to test this prediction is to compare the behavior of parent blackcaps with that of their sons and daughters in a controlled environment. Peter Berthold and his colleagues were up to this challenge. First, they captured some wild British blackcaps during winter and took them to Berthold's laboratory in Germany, where the warblers spent the rest of the winter indoors. Then, with the advent of spring, pairs of warblers were released into outdoor

Blackcap warbler

1 Different migratory routes of blackcap warblers. Blackcaps living in southern Germany and Scandinavia first go southwest to Spain before turning south to western Africa. Blackcaps living in eastern Europe go southeast before turning south to fly to eastern Africa. Other members of the species that breed in central Germany fly in a westerly direction to southern Britain, where they remain for the winter.

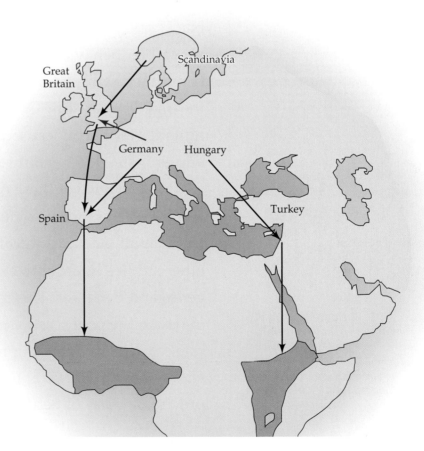

aviaries, where they bred, providing Berthold with a crop of youngsters that had never migrated [102].

Once the young birds were several months old, Berthold's team placed some in special cages that had been electronically wired to record the number of times a bird hopped from one perch to another. The electronic data revealed that when fall arrived, the young warblers became increasingly restless at night, exhibiting the kind of heightened activity characteristic of songbirds preparing to migrate. The immature blackcaps' parents also became nocturnally restless when placed in the same kind of cages in the fall. These data suggest that the British wintering population is *not* composed of birds that have lost their ability to migrate. Instead, the birds wintering in Britain could be migrants that flew to Britain from somewhere else.

But just where does the British wintering population come from? To answer this question, the researchers put some warblers in cages shaped like funnels and lined with typewriter correction paper. Whenever the bird leaped up from the base of the funnel in an attempt to take off, it landed on the paper and left scratch marks, which indicated the direction in which the bird was trying to go (Figure 2). Berthold's subjects, experienced adults and novice youngsters alike, oriented due west, jumping up in that direction over and over, judging from the footmarks left on the paper. These data showed that the adults, which had originally been captured in Britain, must have come from Belgium or central Germany, a point eventually confirmed by the discovery of some banded blackcaps from this part of Europe in Britain during the winter.

Perhaps the genetically distinct birds that pioneered the use of southern Britain as a migratory destination have improved chances of reproductive success because this region, which features mild winters, can be reached after a much shorter flight than that required to get to Africa. But many European warblers still go to Africa by way of Spain, presumably because, at the proximate level, they have inherited from their parents a different migratory mechanism than those that use Britain as a winter home. Berthold's team checked this prediction in another experiment, rearing a new crop of young blackcaps from parents captured in *southwestern,* not central, Germany. These birds were kept in identical aviaries and tested in funnel cages. As predicted by the

2 Funnel cage for recording the migratory orientation of captive birds. The bird can see the night sky through the wire mesh ceiling of the cage. As the bird jumps up onto the surface of the funnel, it leaves marks that show the direction in which it intended to fly. Photograph by Jonathan Blair.

(A) Adults from Britain

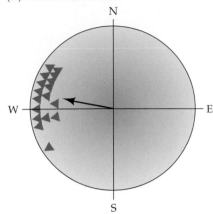

(B) F₁ offspring of British adults

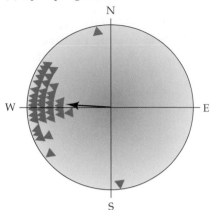

(C) Young from southern Germany

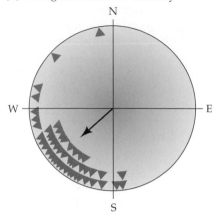

3   Hereditary differences in migratory orientation of blackcap warblers. (A) Adults captured on the British wintering grounds and held in Germany until the next fall attempted to fly west (toward Britain). (B) The offspring of those adults attempted to fly west as well. (C) The offspring of adults captured in southwestern Germany, however, oriented in a southwesterly direction that would have taken them to Spain. After Berthold et al. [101].

genetic differences hypothesis, the young birds oriented toward the southwest, rather than taking the westerly tack of the birds whose parents winter in Britain (Figure 3). This experiment clearly demonstrates that genetic differences, not environmental ones, are largely responsible for the differences in migratory behavior between the two German populations of blackcap warblers [99, 102].

Andreas Helbig has extended this story by looking at the hereditary basis for the migratory pattern of Austrian blackcaps, which travel in a southeasterly direction, not west or southwest. In the fall, Austrian blackcaps move across Turkey to Lebanon and Israel before turning due south to Ethiopia and Kenya (see Figure 1). Helbig created "hybrids" by crossbreeding captive Austrian blackcaps (which migrate toward the southeast) with blackcaps from southwestern Germany (which migrate toward the southwest). He then measured the flight orientation of the parental and hybrid birds by placing them in funnel cages. The mean orientation of the marks left on the typewriter correction paper was calculated for each bird. When the directions chosen by hybrid offspring were compared with those of their German and Austrian parents, they proved to be intermediate (Figure 4). This result supports the hypothesis that southwestern

4   A test of the genetic differences hypothesis for why migratory blackcap warblers from western and eastern Europe take different flight paths. The inner ring shows the migratory orientations of birds from southwestern Germany (tan triangles) and from Austria (white triangles). The outer ring shows the orientations of hybrid offspring whose parents differed in migratory orientation (blue triangles). The large red arrows show the mean directions taken by the three groups of birds. As predicted, the hybrids tended to orient due south, an intermediate direction compared with those of their parents. After Helbig [511].

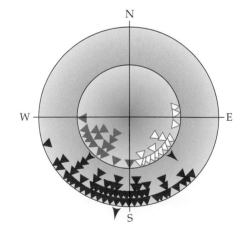

German and Austrian birds differ in their genetic makeup in ways that affect their migratory orientation. Hybrids between the two populations receive genetic information from one parent that promotes the development of one type of migration mechanism, but they also receive some different information from the other parent that has different developmental effects. The mix of parental genes has the consequence of producing a compromise device, which causes the birds to orient halfway between southeast and southwest [511].

Genetic Differences and Human Behavior

Our improved understanding of the development of blackcap warbler migratory behavior came about in part because researchers were able to compare the behavior of parents and offspring. The same approach has been highly productive when applied to our own behavior. No matter whether one is dealing with birds or human beings, we can expect the genetic makeup, or **genotype,** of parents and offspring to be similar, though not identical. For this reason, we can also expect offspring and parents to be similar in their hereditary phenotypes. (A **phenotype** is any observable trait that develops as a result of an interaction between the individual's genotype and its environment; a hereditary phenotype is a characteristic that *differs* among individuals because of genetic differences among them.)

We can specify exactly how similar genetically a child is likely to be to a parent because when eggs or sperm are formed, they contain exactly half of the genes of each parent. Since human offspring are created when an egg unites with a sperm, each child shares half its genes with each parent, and so is said to have a **coefficient of relatedness** of 0.5 with its mother (and with its father).

Other pairs of relatives are also likely to share a certain proportion of their genes in common as a result of sharing some of the same ancestors. For example, because full siblings have the same parents, they could theoretically have exactly the same genotypes (if the genotypes of the individual sperm and eggs that formed them happened to be identical, which they almost never are). Alternatively, full siblings could by chance have been formed by the union of eggs and sperm that happened to have no alleles in common (which again is close to an impossibility). Siblings therefore have coefficients of relatedness that range in a normally distributed fashion from the exceptionally rare extremes of 0 to 1 to much more common intermediate values around 0.5, such that on average, full siblings have a coefficient of relatedness of 0.5.

In contrast, half-siblings have only one parent in common and so range in relatedness from 0 to 0.5, with the average being 0.25. Therefore, we can expect half-siblings to differ, on average, more than siblings do with respect to hereditary phenotypes. In fact, half-siblings do differ on average more than full siblings in many ways, including such things as IQ scores.

By the same argument, fraternal twins should be less alike with respect to hereditary traits than identical twins. Identical, or monozygotic (MZ), twins are derived from the same fertilized egg, which divides and gives rise to two genetically identical embryos. In contrast, fraternal, or dizygotic (DZ), twins are formed from the nearly simultaneous fertilization of two different eggs by two different sperm, and as a result, DZ twins are no more similar genetically than any other pair of full siblings.

Another way to compare the degree of genetic similarity of twins is to consider the probability that DZ twins, as opposed to MZ twins, will inherit the same **allele** of a given gene from a particular parent. (An allele is a particular form of a given gene.) Imagine that the mother's genotype with respect to the A gene is A_1A_2; therefore, the chance that the A_1 allele will be present in a given egg is $\frac{1}{2}$. If behavioral development is influenced by the presence of the A_1 allele, the behavior of DZ twins should often differ (because they originated from different eggs, some of which carry the maternal A_1 allele while others have the A_2

allele). In contrast, MZ twins should behave more similarly because they originated from the same egg (and same sperm), and so possess exactly the same set of genes. In fact, fraternal twins differ from each other more than identical twins do with respect to a host of traits, including various measures of personality.

An alternative explanation exists, however, for the great behavioral similarity between identical twins. As noted earlier, behavioral development is influenced not only by genes but also by the environment, which includes the experiences an individual has while growing up. These experiential factors could be more alike for MZ than for DZ twins if parents and other people treat MZ twins very similarly because of their nearly identical appearance, and this could explain why MZ twins are so similar behaviorally (but see [517]). More generally, any similarities among related individuals could arise from similar family environments, which is why behavior geneticists have to take special steps to distinguish between the effects of shared genes versus shared environments.

One way to eliminate the effect of shared family experience is to observe the behavior of identical twins who happen to have been separated at birth and placed in different families. Unplanned experiments of this sort are rare, but they do occur. Oskar Stohr was raised as a Catholic in Nazi Germany by his grandmother while his identical twin brother, Jack Yufe, grew up on various Caribbean islands with his Jewish father. One can hardly imagine more diverse environments for the development of two human beings, and yet these men are remarkably similar in appearance (Figure 5) and behavior. Despite having lived apart for 47 years, they both "like sweet liqueurs, . . . store rubber bands on their wrists, read magazines from back to front, dip buttered toast in their coffee and have highly similar personalities" [530].

A study of more than 50 pairs of MZ twins reared apart demonstrates beyond reasonable doubt that genetic differences are responsible for a significant portion of the differences among human beings in personality, temperament, and social attitudes [133]. For example, when twins filled out elaborate questionnaires designed to provide quantitative measures of personality traits (e.g., degree of cooperativeness, sociability, and so on), the scores of MZ twins reared apart were on average more similar than the scores of DZ twins, whether those siblings had been reared apart or together. The scores of totally unrelated pairs of individuals were much less similar than those of twins of any sort.

5 **Identical twins separated at birth:**
Jack Yufe (left) and Oskar Stohr (right).
Photograph by Bob Burroughs.

From these studies, one can calculate what proportion of the variance (a statistical measure of variation among individuals) in personality traits among these groups was due to genetic differences among them. In the lingo of behavior genetics, this is the **heritability** of the phenotype in question. Heritabilities can range from 0 (none of the variance in phenotypes is caused by genetic differences) to 1 (all of the variance is due to genetic variation in the population studied). The twin studies produced a heritability of about 0.5 for personality scores; in other words, about 50 percent of the *differences* in personality scores among all the individuals tested were due to genetic differences among them. The other 50 percent of the *differences* were essentially environmental in origin.

Genetic Differences and IQ Differences

Identical twins are also more similar than fraternal twins in their IQ scores, even when the identical twins have grown up separately [134] (Table 1). By comparing the correlations between MZ twins and between DZ twins raised in various environments, Thomas Bouchard and his colleagues have calculated a heritability of about 0.7 for IQ. That is to say, about 70 percent of the differences in IQ scores among individuals *in the population they studied* stemmed from the genetic variation that existed among those individuals. Only the remaining 30 percent could be attributed to the diverse environments that the participants in the study experienced as they were growing up. Although heritability measures apply only to the population from which the data are gathered, other researchers working with different samples of twins have generated similar heritability estimates [135]. Moreover, the substantial heritability of cognitive abilities persists throughout life, as found in a study of Swedish twins who were 80 years old or older [777].

As an alternative hypothesis on the close IQ similarity between MZ twins reared apart, consider the possibility that the MZ twins lived in adoptive households that were more similar than those chosen for DZ twins that were reared apart. If that were true, the more similar home environments could have channeled the intellectual development of the separated MZ twins along similar lines, leading ultimately to their very similar IQ scores. But when Bouchard's team measured various aspects of the adoptive home environment, such as socioeconomic status, education level of both adoptive parents, intellectual orientation of the household, and the like, they found that there was little or no connection between the similarity in environments of twins reared apart and their similarity in IQ scores [136].

Yet another way to evaluate this "similar home environment" hypothesis is to examine another of its predictions, which is that unrelated persons reared together in the same household should exhibit substantial similarities in their behavioral attributes. The fact that many children have been adopted into fam-

TABLE 1 Familial correlations for IQ scores: Predicted values based on the genetic differences hypothesis and the actual correlations

Category	Predicted correlation	Actual median correlation	Number of studies
Identical (MZ) twins reared together	1.0	0.85	34
Identical (MZ) twins reared apart	1.0	0.75	5
Fraternal (DZ) twins reared together	0.5	0.58	41
Siblings reared together	0.5	0.45	69
Parent-genetic offspring	0.5	0.39	32
Parent-adoptive offspring	0.0	0.18	6

Sources: Bouchard and McGue [137]; Bouchard [134]

ilies of nonrelatives provides an opportunity to evaluate this hypothesis, which, if true, should mean that adopted children are more similar behaviorally to their foster parents than to their genetic parents. By testing the cognitive abilities of adopted children over the first sixteen years of life, Robert Plomin and his colleagues found that although adopted children showed some similarity to their adoptive parents in verbal and spatial skills when they were young, these similarities faded as the children became adolescents [928]. In other words, the more years adopted children spent with adoptive foster parents in a shared home environment, the *less* they resembled their caretakers in cognitive ability, a result totally at odds with the similar home environment hypothesis. Moreover, as adopted children grow less and less like their adoptive parents, they become increasingly similar to their genetic parents, a finding that speaks to the powerful influence of our genes in shaping our cognitive abilities (Figure 6).

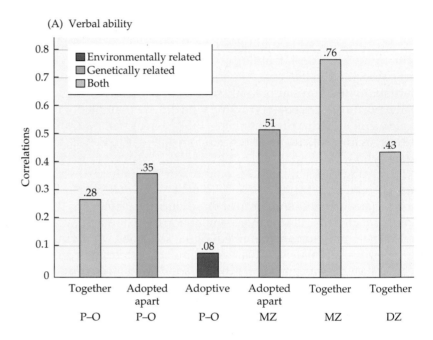

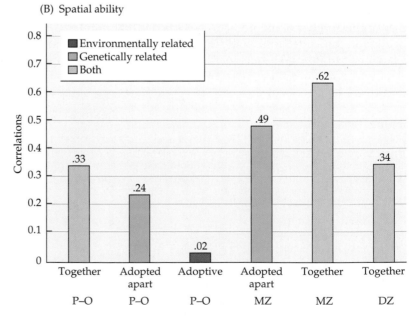

6   A test of the similar home environment hypothesis. The graphs show correlations in (A) verbal ability scores and (B) spatial ability scores for various pairs of individuals living together and apart (P–O, parents and offspring; MZ, identical twins; DZ, fraternal twins). The data are composite measures based on a number of different studies. Note especially that adopted individuals are more similar in verbal and spatial ability to their absent genetic parents than they are to their adoptive parents. Moreover, identical twins reared apart have more similar test scores on average than fraternal twins that grow up together. After Plomin et al. [926].

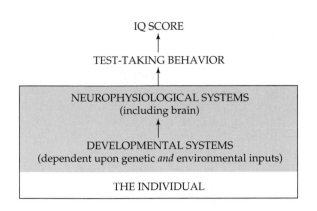

IQ SCORE

↑

TEST-TAKING BEHAVIOR

↑

NEUROPHYSIOLOGICAL SYSTEMS
(including brain)

↑

DEVELOPMENTAL SYSTEMS
(dependent upon genetic *and* environmental inputs)

THE INDIVIDUAL

7 IQ scores are not purely genetically determined. An IQ score is a phenotype dependent on neural activity, and the brain that produces this activity develops via processes in which both genes and environment play essential roles.

More evidence on this point comes from an examination of the similarity in IQ between unrelated children of the same age that are reared together in pairs from an early age, a procedure that creates "virtual" twins. The mean correlation between IQ scores for these pairs of children was 0.26 [1039], as opposed to 0.45 for full siblings, 0.58 for DZ twins, and 0.85 for MZ twins reared together [137]. These data show that the environment is clearly relevant to the development of the ability to perform well on IQ tests (otherwise, the correlation between the scores of virtual twins would be zero), but that genetic influences are obviously important as well.

Keep in mind that the high heritability of IQ scores, verbal skills, and spatial abilities does *not* mean that intelligence (or any other behavioral attribute) is "genetically determined." Let me stress that all behavioral traits, including performance on IQ tests, develop as a result of an interaction between genes and environments. Taking cognitive skills tests requires a brain, which in turn is the product of an almost incomprehensibly complex interplay between the genetic information in a fertilized egg and its environment, which includes the nongenetic materials in the egg and the egg's surroundings. These materials are utilized by the growing embryo and are absolutely essential for the construction of brain tissue. Moreover, as the brain begins to generate electrical activity and receive messages from outlying sensory receptors linked to it, the brain creates and receives experiences that are vital to its further development.

Thus, the results of the familial comparisons outlined above show just one thing: genetic *differences* among the populations of people studied to date contribute to the *differences* in the scores they achieve on IQ tests and on other tests of cognitive ability (Figure 7). But environmental differences are obviously important too, as can be seen from the average correlation between the IQs of MZ twins, which is less than 1.0. Even MZ twins develop in somewhat different environments, with different surroundings in the womb, different foods consumed, and different social interactions, all of which may lead to developmental differences between them. Even so, MZ twins are generally much more similar behaviorally than other pairs of people.

How Many Genetic Differences Are Needed to Produce a Behavioral Difference?

Comparisons among relatives do not automatically reveal what gene or genes contributed to the behavioral differences under study. But in some cases, we now know, a difference in a single gene can cause a behavioral difference between individuals. For example, a foraging difference between two types of a small roundworm stems from a difference in a single nucleotide, just one of the long chain of nucleotides that make up the DNA that comprises the *npr-1* gene. This single nucleotide difference translates into a single amino acid difference in the long chain of amino acids that constitute the protein encoded by the gene. The slight difference in this protein is enough to differentiate social roundworms, which feed together when grazing on bacteria-covered surfaces, from solitary roundworms, which keep well apart when feeding [308, 1081].

Another similar example involves two genetically different forms of *Drosophila melanogaster*, the most famous fruit fly. The larvae of one type (called "rovers") travel about four times farther when feeding on a yeast-coated petri dish during a 5-minute test than larvae of the other type (which are labeled "sitters") [307]. When adults of the two types are crossed, they produce larval offspring (the F_1 generation) that are all rovers. When these larvae mature and interbreed, they produce an F_2 generation with three times as many rovers as sitters (Figure 8). Persons familiar with Mendelian genetics will recognize that rovers must have at least one copy of the dominant allele of a gene affecting larval foraging behavior, whereas sitters must have two copies of the recessive allele. If this analysis is correct, then if one could transfer the dominant allele associated with rover behavior to an individual of the sitter genotype, the genet-

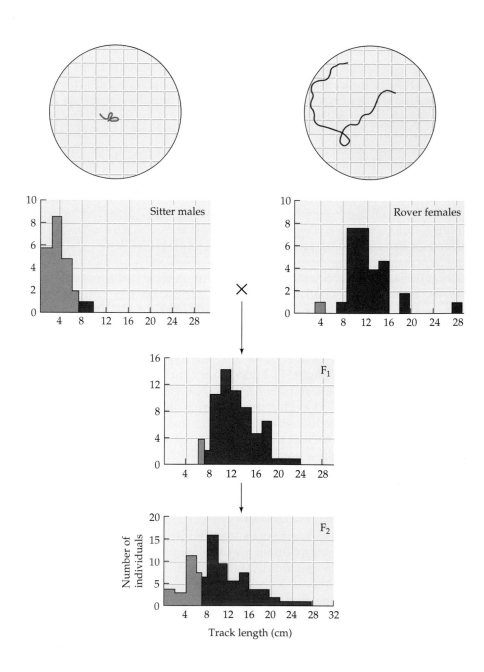

8 Genetic differences cause behavioral differences in fruit fly larvae. Representative tracks made by sitter and rover phenotypes feeding in a petri dish appear at the top of the figure. When adult male flies of the sitter strain mate with adult females of the rover strain, their larval offspring (the F_1 generation) almost all exhibit the rover phenotype (that is, they move more than 7.6 cm in 5 minutes). When flies from the F_1 generation interbreed, their offspring (the F_2 generation) are composed of rovers (purple) and sitters (tan) in the ratio of 3:1. After de Belle and Sokolowski [307].

ically altered larva should exhibit rover behavior. This experiment has been done, with positive results [1081]. So here is another case in which a difference between two behavioral types stems from a difference in the information contained in a single gene—just one of the 13,061 genes [898] located on the four chromosomes of *Drosophila melanogaster* [306].

Many other single-gene effects have been found in fruit flies, although rarely in natural populations, probably because most mutations reduce reproductive success and so have been quickly eliminated by natural selection. Finding behavioral mutants in fruit flies generally requires that they be produced experimentally by exposing laboratory populations to radiation or chemicals that cause genetic mutations. Once this has been done, the offspring of the treated flies can be scanned for behavioral oddities, and the apparent mutants tested in breeding experiments. Among the many mutant alleles formed and identified in such experiments are *stuck* (males with the mutant gene fail to dismount after the normal 20-minute copulation) and *coitus interruptus* (males with this allele disengage after just 10, not 20, minutes of copulation) [92]. Indeed, a whole series of mutations have been discovered that decrease male courtship vigor, eliminate female receptivity, alter the courtship song pattern, or cause males to court both males and females with equal intensity [483].

Do studies of this sort show that fly courtship and copulation behaviors are encoded in the base sequence of specific genes? No. When an allele makes a difference, it does so by affecting the production of a particular protein, not by specifying a behavior of some sort. Furthermore, that protein will not be manufactured unless environmentally supplied amino acids are available in the right cells at the right time. Once the protein is present in these cells, it is free to facilitate a chemical reaction, setting in motion a whole series of gene–environment interactions, one after another, each dependent on the preceding chemical event. As a result, individuals with different versions of even one protein may follow increasingly different developmental pathways that lead eventually to differences in their behavior.

For example, let's consider again the gene that influences the foraging behavior of fruit fly larvae. The techniques now available to molecular biologists have enabled researchers to precisely identify the gene in question (which goes by the label *dg2*) and to discover that it codes for a "cGMP-dependent protein kinase" [867]. This enzyme is produced only in certain cells of the larva's olfactory system and brain, suggesting that it may have a special role to play in the larva's ability to analyze its chemical environment. Since the two naturally occurring variants of the gene produce different forms of a key protein, with different efficacies in promoting a particular reaction, sitters and rovers acquire different sensory mechanisms within their nervous systems, which may provide the foundation for their behavioral differences.

Much the same sort of thing has been documented for two strains of laboratory mice that are genetically identical in every respect, except for a single gene that encodes an enzyme called α-calcium-calmodulin kinase. Because of this one genetic and enzymatic difference, members of the two strains differ in the construction of their hippocampus, a region of the vertebrate brain involved in spatial learning. These hippocampal differences between the two kinds of mice underlie differences in their performance on spatial memory tests (Figure 9) [1059].

The possibility that one gene's protein product can be critical for the development of a behavioral ability receives further support from knockout experiments, in which a known gene is inactivated by the researcher in order to determine its effects on development. For example, one can scramble the code in the *fosB* gene of laboratory mice, producing tailor-made mutants that are spectacularly indifferent to their offspring. Although normal in most other respects, mutant females ignore their newborn pups, whereas wild-type females homozygous for the active *fosB* gene invariably gather their pups together and crouch

9   A single gene affects learning ability in laboratory mice. Mice were placed in a circular water-filled arena containing one small, barely submerged platform. The mice had to find the platform in order to escape from the water. Both wild-type and mutant mice were able to learn the location of the platform. Two days later the mice were tested again. When the platform was in the same position as during the training trials, the wild-type mice found it significantly more rapidly than those with the α-calcium-calmodulin kinase mutation. When the location of the platform was set randomly, the wild-type mice did not perform any better than the mutants, showing that the difference between the two groups during the first test was due to the superior retention of learning by the wild-type mice. After Silva et al. [1059].

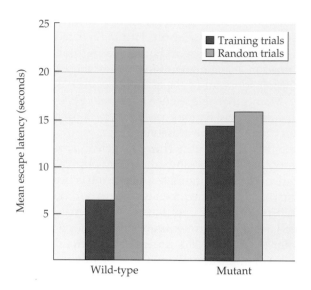

over them, keeping them warm and permitting them to nurse. Should a pup wriggle away or be displaced, the wild-type females quickly retrieve their errant offspring; the mutant females do not (Figure 10) [173].

These highly specific differences in maternal behavior have been tentatively explained in the following way. Both mutant and wild-type mice that have just given birth almost always have neural circuits in place that cause them to inspect their newborns. The resulting sensory experiences, especially olfactory stimuli from the pups, generate sensory receptor signals that travel to the female's brain. The multiple inputs that come from sniffing, touching, and hearing newborns are integrated in a particular region of the hypothalamus, called the preoptic area

10   A single gene affects maternal behavior in laboratory mice. Wild-type female mice gather their pups together and crouch over them (above), but females with inactivated *fosB* genes (below) do not exhibit these behaviors (the pups can be seen scattered in the foreground). Photographs courtesy of Michael Greenberg; from Brown et al. [173].

(POA), where volleys of neural signals serve to activate *fosB* alleles in certain cells. As a result, wild-type females quickly produce a functional FosB protein. However, these signals have no such effect on the altered *fosB* alleles of mutant females. Thus, the two types of mice differ in the effects of certain experiences on gene activation. The production of the FosB protein eventually leads to additional genetic and enzymatic changes that produce structural alterations in specific neural circuits within the POA. The modified neural machinery of wild-type females motivates them to pull their pups underneath their bodies and care for them solicitously. These essential neural changes never occur in the mutant females for lack of the gene (and allied protein) needed to initiate the series of biochemical events that would result in normal maternal behavior [173].

The link between genes and neurophysiology has also been explored with another knockout mutant mouse, this one missing the *Oxt* gene, making it impossible for the mouse's brain cells to produce oxytocin, an important hormone [381]. Male mutant mice cannot remember females with whom they have recently interacted. Each time the same female is removed from and then returned to the cage she shares with an *Oxt* mutant male, the male gives her a thorough and lengthy sniffing that is no different from his response the very first time they met (Figure 11). In contrast, if a female is placed in the cage of a normal male with a functional oxytocin gene, he remembers what she smells like, so that if she is taken from his cage but then later returned to it, he will spend less time sniffing her on this occasion compared to the first time they met. As the process is repeated, the normal male shows less and less interest in what is to him a familiar individual. Thus, the *Oxt* gene appears to have a very specific behavioral effect, which is to enhance the male's ability to remember the scent of familiar females. (The mutant males have no trouble remembering the scents of familiar foods, and they can tell the difference between the odors of lemon and chocolate just as well as companions who have functional *Oxt* genes.)

In yet another experiment, some mice had the gene *disheveled-1* knocked out. The resulting mutant mice behaved like normal mice—except that they had little desire for social contact, as reflected in a reluctance to trim the whiskers of their companions [699]. Remember that *fosB* does not code for maternal behavior, nor does *Oxt* code for social memory, nor does *disheveled-1* code for whisker trimming behavior. Instead, these genes, like all others, code for specific molecular products, which (if actually produced) can participate in a particular bio-

(A)

(B)

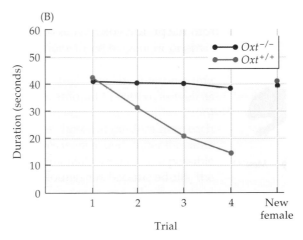

11 Social amnesia is related to the loss of a single gene. (A) Male mouse inspecting a female. (B) Knockout male mice that lack a functional *Oxt* gene carefully inspect the same female every time she is reintroduced into their cage, whereas males with the typical genotype show less and less interest in a female that they have inspected previously. (A) Photograph by Larry J. Young; (B) After Ferguson et al. [381].

differences among individuals. If, in past populations, the proximate effects of genetic variation were similar, then behavioral evolution by natural selection was inevitable. Natural selection in the past should have led to the spread of alleles and associated traits that promote the reproductive success of individuals living today. The increase in frequency of fitness-enhancing alleles typically occurs at the expense of other alleles, whose elimination reduces genetic variation in the population.

Under some circumstances, however, both genetic and behavioral variation can be maintained within a species. For example, the members of a species that live in different places under different conditions may experience different selection pressures and evolve different phenotypes. Thus, central German, southern German, and Austrian blackcap warblers differ genetically in ways that affect their migratory behavior. The members of each population fly to different places where winters are relatively mild and so the birds have a reasonable chance of surviving to return the next spring to their breeding grounds. Moreover, by going to different places, the populations avoid competing for winter food with blackcaps elsewhere, which could improve their chances of making it through the winter.

Likewise, the rover and sitter phenotypes of fruit fly larvae have probably been maintained in natural populations of *Drosophila melanogaster* because selection has favored either one or the other phenotype depending on the density of a given population [1082]. When larvae have to cope with intense competition for food, the rover phenotype (and its allele of the *dg2* gene) could be at an advantage, but in low-density populations where food is abundant, the sitters (and their different allele) may gain because they need not invest in energetically expensive travel. If this hypothesis is correct, then if you were to rear larvae generation after generation in either low-density or high-density conditions, natural selection in the laboratory ought to result in behavioral divergence between the members of the two kinds of populations. The experiment has been done (Figure 17), and its results support the proposition that variation in population density contributes to the coexistence of two hereditary foraging phenotypes in fruit fly larvae [1082].

Another example of variation among populations in genetics and behavior involves the garter snake *Thamnophis elegans*, which occupies much of western North America, including both foggy coastal California and the drier, elevated inland areas of that state [44]. The diets of snakes in the two areas, referred to hereafter as "coastal" and "inland" snakes, differ markedly. Coastal snakes feast

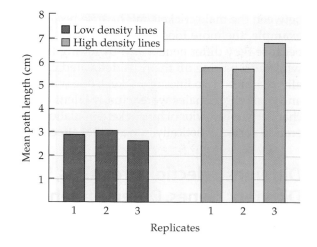

17 The results of natural selection on the foraging behavior of fruit fly larvae. Larvae reared in the laboratory for many generations under low-density conditions (purple bars) move shorter distances in the same period of time than larvae from populations reared under high-density conditions (green bars). Three different populations for each of the two conditions are presented, showing the consistent effect of density-dependent selection on larval behavior. After Sokolowski et al. [1082].

18   **A coastal Californian garter snake** about to consume a banana slug, a favorite food of snakes in this region. Photograph by Steve Arnold.

on the banana slugs that abound in humid coastal California, whereas inland snakes find other things to eat, primarily fish and frogs, in arid inland California where slugs do not roam. Banana slugs are large, presumably nutritious prey, but one can only admire the willingness of coastal snakes to consume these creatures (Figure 18). I once made the mistake of picking up a banana slug. Ten minutes later I was still at the kitchen sink scrubbing frantically, trying to remove the repulsively sticky mucus that the slug had liberally applied to my hand.

In any event, coastal and inland snakes do not behave the same way. Steve Arnold wanted to know whether genetic differences were involved, so he took pregnant female snakes from the two populations into the laboratory, where they were held under identical conditions. When the females gave birth to a litter of babies (garter snakes produce live young rather than laying eggs), each baby was placed in a separate cage, away from its littermates and its mother, to remove these possible environmental influences on its behavior. Some days later Arnold offered each baby snake a chance to eat a small chunk of freshly thawed banana slug by placing it on the floor of the young snake's cage. Most naive young coastal snakes ate all the slug hors d'oeuvres they received; most of the inland snakes did not (Figure 19). In both populations, slug-refusing snakes ignored the slug bit completely.

Arnold took another group of isolated newborn snakes that had never fed on anything and offered them a chance to respond to the *odors* of different prey items. He took advantage of the readiness of newborn snakes to flick their tongues at, and even attack, cotton swabs that have been dipped in fluids from some species of prey (Figure 20). Chemical scents are carried by the tongue to the vomeronasal organ in the roof of the snake's mouth, where the odor molecules are analyzed as part of the process of detecting prey. By counting the number of tongue flicks that hit the swab during a 1-minute trial, Arnold measured the relative responsiveness of inexperienced baby snakes to different odors.

Populations of inland and coastal snakes reacted about the same to swabs dipped in toad tadpole solution (a prey of both groups), but behaved very differently toward swabs daubed with slug scent (Figure 21). Within each group, not every snake responded identically, but almost all inland snakes ignored the slug odor, whereas almost all coastal snakes rapidly flicked their tongues at it. Because all the young snakes had been reared in the same environment, the differences in their willingness to eat slugs and to tongue-flick in reaction to slug odor must have been caused by genetic differences among them.

19   Response of newborn, naive garter snakes to slug chunks. Young snakes from coastal populations tend to have high feeding scores (e.g., a score of 10 indicates that the snake ate a slug cube on each of the 10 days of the experiment). Inland garter snakes were much less likely than coastal snakes to eat even one slug cube (which would yield a score of 1). After Arnold [44].

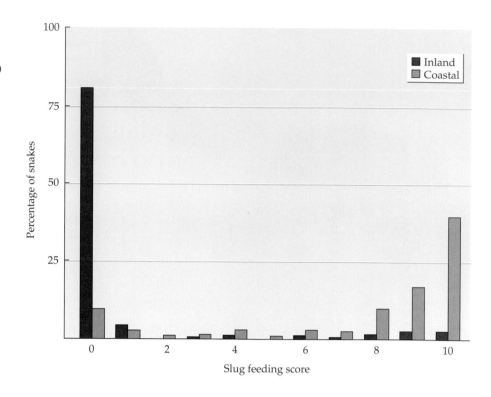

Arnold then did a heritability study of the sort we discussed earlier in the context of human twin studies. By comparing the tongue-flick scores of siblings, Arnold determined that *within* each population, only about 17 percent of the *differences* in responsiveness to slug odor stemmed from genetic differences among individuals. The low heritability of tongue flicking within either population simply means that the behavioral variation that exists *within* each population is largely caused by environmental differences.

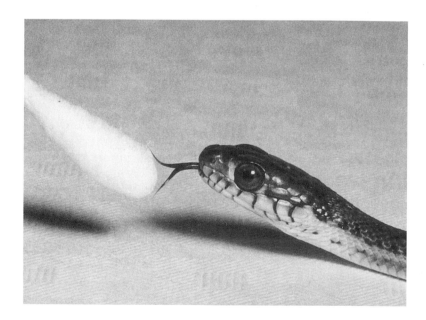

20   A tongue-flicking newborn garter snake senses odors from a cotton swab that has been dipped in slug extract. Photograph by Steve Arnold.

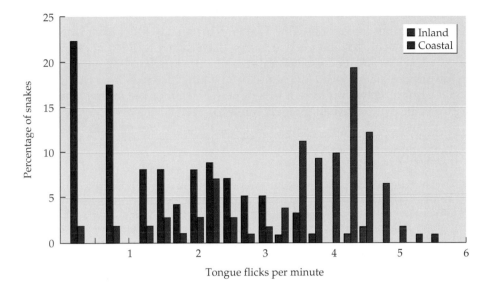

21　Odor preferences of inland and coastal garter snakes as measured by the frequency of tongue flicking in response to cotton swabs dipped in slug extract. Coastal snakes tongue-flicked much more than inland snakes. After Arnold [44].

If the feeding differences *between* the two populations arise because most coastal snakes have a different allele or alleles than most inland snakes, then crossing adults from the two populations should generate a great deal of variation in the resulting group of hybrid offspring. Arnold conducted the appropriate experiment and found the expected result, confirming again that the differences *between* populations have a strong genetic component.

Having identified a genetic basis for why coastal and inland garter snakes have different dietary preferences, Arnold turned his attention to the evolutionary basis for these differences. He proposed that among the original colonizers of the coastal habitat were a very few individuals that carried the then rare allele(s) for slug acceptance. (The garter snake almost certainly colonized coastal California more recently than inland western North America.) These slug-eating individuals were able to take advantage of an abundant food resource in their new habitat. If, as a result, their reproductive success was as little as 1 percent higher than that of their slug-rejecting fellows, the coastal population could have reached its present state of divergence from the inland population in less than 10,000 years.

It is easy to imagine why slug-accepting alleles might enjoy an advantage and spread rapidly in coastal populations, where slugs provide an abundant source of potential food. But why would slug acceptors be nearly eliminated from inland populations by natural selection? Perhaps because slug-eating snakes will also consume aquatic leeches. These blood-sucking animals are absent from coastal California but plentiful in inland lakes. Leeches can survive even after being swallowed by a garter snake, and if they attach themselves to the wall of the snake's digestive tract, they might damage their consumer. Therefore, snakes with slug-accepting alleles might survive and reproduce less well in inland California (a point that still needs to be documented). If so, the alleles that contribute to the development of the ability to attack slugs (and leeches) would be eliminated in this area.

Thus, the geographic differences in the feeding behavior of garter snakes can be explained in terms of their proximate basis: different alleles predominate in the two populations, which cause the development of differences in the chemoreceptors that detect certain molecules present in slugs and leeches. At the ultimate level, certain genes and sensory mechanisms are associated with different degrees of reproductive success in different areas. Inland snakes must contend with potentially dangerous leeches, whereas coastal snakes are exposed only to sticky but edible slugs. This difference in the snakes' environments influences the survival and reproductive success of behaviorally different phenotypes in the two areas [44]. Because natural selection has acted in different directions in the two populations, hereditary and behavioral differences now characterize snakes from the two regions, a fine demonstration of the interrelationship between the proximate and ultimate causes of animal behavior.

Summary

1. Genetic differences among individuals may result in behavioral differences among them. To demonstrate that a particular allele contributes to the development of a behavioral characteristic is not to say that the trait is "genetically determined." The statement, "There is an allele for slug acceptance," for example, is shorthand for the following: "A particular allele in an individual's genotype codes for a distinctive protein whose contribution to cell biochemistry affects the development of the physiological mechanisms needed for the ability to recognize slugs as food."

2. Researchers in behavior genetics have developed many ways to test the hypothesis that specific behavioral differences between individuals are caused by genetic differences. Supportive results have come from research based on comparisons between relatives, especially parents and offspring as well as fraternal and identical twins. Further evidence for the effect of genes—even single genes—on behavioral development has been gathered through manipulations of an individual's genotype by removing (or adding) particular genes. In addition, artificial selection experiments demonstrate that certain behavioral differences arise because individuals do not have the same genotype.

3. The finding that artificial selection can cause modern populations to evolve in the laboratory is evidence that natural selection acting in the past could have caused ancestral populations to evolve in the wild. Therefore, we can expect different populations of the same species that have been subjected to different selective pressures to exhibit different hereditary traits. For example, differences in the foods available to garter snakes living in different parts of California appear to have led to the spread of different alleles and different evolved prey preferences in the two areas. This case illustrates again the close connection between the proximate (i.e., genetic–developmental) causes of behavior and the ultimate (i.e., evolutionary) basis of behavior.

Discussion Questions

1. The cliff swallow forms breeding colonies that contain from 2 to 3,700 nests. Birds born in large colonies tend to seek out large colonies when they breed for the first time, should they disperse from their natal colony; birds whose parents nested in small groups also tend to nest in small groups if they disperse from their birthplace. Charles and Mary Brown demonstrated a heritable basis for the social differences among cliff swallows [166]. What alternative hypothesis must they have ruled out? Design an experiment in which you move nestlings from their natal nests in order to collect the kind of data that would test this alternative hypothesis (and the genetic differences hypothesis as well).

2. A few blackcap warblers live year-round in southern France, although 75 percent of the breeding population migrates from this area in winter. Perhaps the difference between the two behavioral phenotypes is environmentally induced. Make a prediction about the outcome of an artificial selection experiment in which the experimenter tries to select for both nonmigratory and migratory be-

havior. Describe the procedure and present your predicted results graphically. Check your predictions against the actual results (see [99]).

3. Analyze the science behind Carol Lynch's study of nest-building behavior in mice described in this chapter. What question was she trying to answer? What hypothesis or hypotheses did she propose? What prediction(s) and test(s) of the hypotheses did she produce? What conclusion did she reach?

4. In 1998, a team of geneticists claimed that men with two copies of the I allele of the *ACE* gene were much less able to improve their physical condition through exercise than people with a different allele (D) of this gene. According to these researchers, individuals with two copies of the I allele were very much underrepresented in groups of mountain climbers who climb to 7000 meters without oxygen [814]. Dispute the claim that the D allele is a "gene for mountain climbing." What role does the environment play in the development of the biochemical product of the *ACE* gene? Why are many genes almost certainly involved in the development of a person who enjoys mountain climbing? How, then, could a single difference in the *ACE* gene potentially result in a difference in mountain-climbing behavior between two people?

5. Two different behavioral traits sometimes coexist within the same population. Consider the cichlid fish *Perissodus microlepis*, which makes its living, believe it or not, by snatching scales from the bodies of other fish in Lake Tanganyika in Africa. Two structurally different forms of the scale-eater exist, one with its jaw twisted somewhat toward the right, the other with its jaw bent toward the left [549] (see the figure below). Scale-eaters with a jaw turned to the right always take scales from the prey's left flank, while the other form invariably goes for the prey's right flank. Right-twisted parents usually produce offspring with the same jaw shape and feeding behavior. Ditto for left-twisted cichlids. What does this tell us about the proximate causes of the behavioral variation in the population? Why is this case puzzling from an evolutionary perspective? See [549] for a solution to the puzzle.

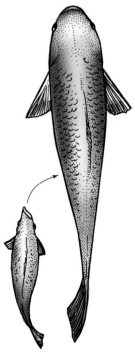 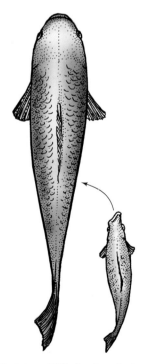

"Right-jawed" *Perissodus* attack prey from the left rear side

"Left-jawed" *Perissodus* attack prey from the right rear side

6. Is it possible that some genetic differences among people might affect how rewarding they find certain experiences, such as listening to music, or taking cocaine, or demanding attention? If so, would these differences in rewards motivate persons differently with respect to the experiences they wanted to have and, thus, the environments that they sought out? If so, does it make sense to categorize the behavioral *differences* among people as either genetic or environmental in origin?

Suggested Reading

Steve Arnold's work on the genetics, physiology, and ecology of garter snake feeding behavior is a classic [44]. The alternative hypotheses for behavioral similarities in human twins reared apart have been nicely analyzed by Thomas Bouchard [134]. A recent textbook reviews the entire field of behavior genetics [927], while Nancy Segal examines the twin studies in detail for a general audience [1038].

4 The Development of Behavior: A Focus on the Environment

The previous chapter dealt primarily with the genetic component of the gene–environment interactions that underlie all behavioral development. What about the other side of the equation, the environmental contribution to development? If you were to remove the yolk from a fertilized white-crowned sparrow egg, without harming so much as one gene in the nucleus of the egg cell, the development of the white-crowned sparrow would never get off the ground. The environmentally supplied nutrients in the yolk are essential ingredients for the construction of white-crowned sparrow brains and bodies. Likewise, if you were to rear a young male white-crowned sparrow in a laboratory where he heard only tapes of song sparrow songs, that male would never vocalize like a typical white-crowned sparrow (see Chapter 2). Indeed, sensory experiences are as critical as the material environment in the development of the behavioral repertoire of the white-crowned sparrow. Nor is there anything special about this species. Indeed, all animals require many different kinds of environmental inputs if they are to develop the ability to behave adaptively. As we shall see shortly, the prenatal hormonal environment of an embryonic mouse can influence the development of its

◄ When a male wasp is deceived into trying to mate with an orchid, the experience changes his behavior, thanks to a special learning mechanism possessed by the insect. Photograph by the author

behavior, while the experiential environment can change the course of brain development in a marsh tit or the timing of the transition between roles played by a honey bee worker in her colony. The ways in which the environment can have these kinds of proximate developmental effects is the subject of this chapter. The chapter examines evolutionary issues as well, especially why some evolved developmental mechanisms produce behavioral flexibility whereas others do not.

The Interactive Theory of Development: Hormones and Behavior

The development of all living things depends upon a complex interaction between genotype and environment. The genetic information in a fertilized white-crowned sparrow egg specifies what enzymes the embryonic sparrow can make, which in turn translates into differences in the hormones produced by endocrine cells in male and female embryos. These hormonal differences then regulate the resulting development of the song control circuitry in the brains of male and female birds. The song control system of the nestling male white-crown also has the capacity to respond to early social and acoustical experiences. All these environmental events lead to the activation or deactivation of selected genes in selected neurons, altering the biochemistry of the brain in such a way that the young male eventually learns to sing the full song of his species.

In many other animals besides songbirds, behavioral development comes under the influence of the external social environment and the internal hormonal environment. We shall illustrate the hormone–development connection with three examples, the first of which is taken from studies of laboratory mice. Among mammals in general, and mice specifically, males and females differ genetically in ways that result in the development of sperm-producing testes in males and egg-producing ovaries in females. As the testes develop in embryonic males, genes are activated that result in the production and release of testosterone, a hormone that switches the development of target cells in the brain into the masculine pathway. As a result of having a masculinized brain, the adult mouse will behave aggressively toward rival males while treating receptive females quite differently. The brains of adult female mice follow a different developmental pathway because their neurons are not exposed to testosterone early in life. As a result, adult females are not nearly so aggressive as males, nor do they attempt to copulate with estrous females, although they will respond appropriately to sexually interested males.

If it is true that the hormonal environment within cells determines the developmental fate of the mammalian brain, then the sex of a male's siblings in his mother's uterus can be predicted to affect the degree to which his embryonic brain will be masculinized. In a litter of mouse embryos, some male embryos will by chance be sandwiched between two sisters (0M embryos), while others will rub shoulders with one male and one female sibling (1M embryos); still others will happen to lie between two brothers (2M embryos). These three categories of males are exposed to different concentrations of male and female sex hormones, which are released in tiny quantities by their littermates and diffuse into their bodies [1173].

In order to test whether the slight differences in exposure to sex hormones caused by embryo position have the predicted effect on the behavior of males, Frederick vom Saal and his colleagues first delivered mouse pups via caesarean section to document their positions in the uterus. The males were then castrated, and later in life given implants of testosterone, to ensure that any differences

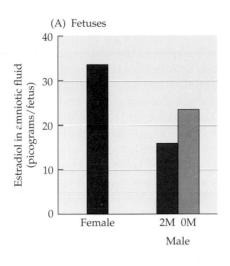

(A) Fetuses

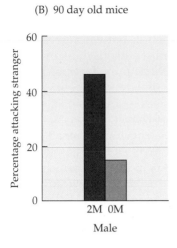

(B) 90 day old mice

1 Effects of hormones on the development of male behavior in mice. (A) Levels of estradiol, a female hormone, are higher in the amniotic fluid surrounding embryonic 0M males (sandwiched between two sisters) than in that surrounding 2M males (between two brothers). (B) At 3 months of age, 0M males are less likely than 2M males to attack another mouse when placed in an arena with the stranger for eight 10-minute trials spread over 16 days. These tests of aggression were conducted with males that had been castrated as newborns, but later received testosterone implants. After vom Saal et al. [1173].

between grown males could be attributed solely to the effects of early uterine position on brain development. 2M males, whose embryonic brains had received a little extra testosterone from their brothers, behaved more aggressively than 0M males when they were given replacement testosterone (Figure 1). Apparently the brains of 2M males were more strongly masculinized early in life than the brains of 0M males, which had experienced the feminizing influence of estradiol (a form of estrogen) derived from their embryonic sisters [1173].

Likewise, females that have been exposed embryonically to different levels of hormones behave differently as adults. In populations of wild house mice living on unmowed grassy patches formed by highway cloverleafs, 2M females occupied significantly larger areas than did 0M females (Figure 2) [1287]. Since male house mice typically have larger home ranges than females, these results suggest that females bathed in relatively high concentrations of testosterone in their mother's uterus are behaviorally masculinized as a result. Further support for this conclusion comes from observations that 2M females are more aggressive and less sexually attractive to males than female littermates that did not develop between two embryonic brothers [219].

Thus, the position of the mouse embryo in utero is an environmental variable that has developmental effects on adult behavior. The developmental effect of the environment, in the form of externally supplied sex hormones, stems from the ability of certain genes to respond to hormonal signals, once these chemicals have entered the mouse's body and become part of its internal cellular environment. The gene–hormone interactions that take place within mouse cells produce either a masculinized or a feminized brain.

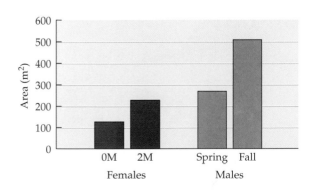

2 Home range sizes of female mice that developed embryonically between two sisters (0M females) and between two brothers (2M females), compared with the home range sizes of males in the spring and fall. After Zielinski et al. [1287].

When to Become a Forager in a Honey Bee Colony

An equally clear illustration of the behavioral effects of a hormonally mediated interaction between genes and environment comes from an insect.
Although most persons think of insects as highly inflexible automatons, many
species, including the familiar honey bee, exhibit considerable flexibility in
their behavioral development. The adaptability of the honey bee first became
evident when researchers began to study the division of labor within honey
bee colonies. These colonies usually contain tens of thousands of sterile female
worker bees laboring on behalf of a single queen, who lays most or all of the
eggs in her hive. At first, the young workers feed the larvae that hatch from
the queen's eggs (most of which will become new workers), but after about 3
weeks of nurse duty they graduate to the job of collecting pollen and nectar
outside the hive (Figure 3) [1034].

Research by Gene Robinson and his colleagues indicates that the age-related
transition between nursing and foraging is regulated by hormonal changes.
Young nurse workers have very low concentrations of juvenile hormone in their
blood, whereas older foragers have much higher concentrations of this hormone. If young bees are treated with juvenile hormone, they become precocious
foragers [973]. In contrast, the onset of foraging behavior is delayed by removal
of the corpora allata, the glands that produce juvenile hormone; furthermore,
bees without corpora allata that receive hormone treatment regain normal timing of the switch to foraging [1109]. Bees that collect pollen and nectar also have
bigger mushroom bodies, an anatomical feature of the honey bee brain (Figure 4). Neither juvenile hormone [375] nor foraging experience is necessary

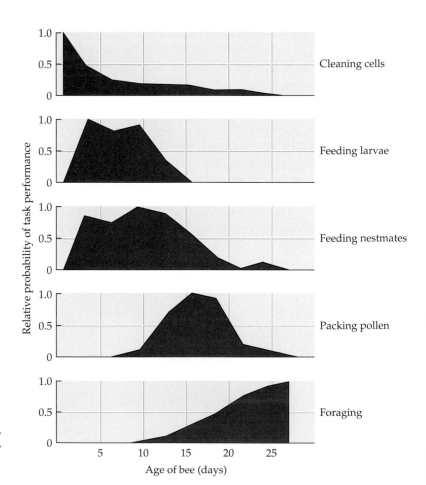

3 **Development of worker behavior in honey bees.**
The tasks adopted by worker bees are linked to their age,
as demonstrated by following marked individuals over
their lifetime. After Seeley [1034].

(A)

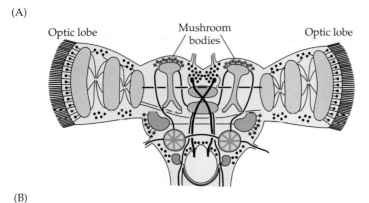

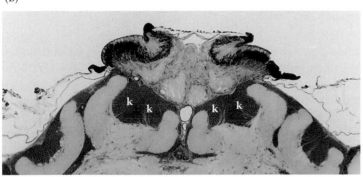

(B)

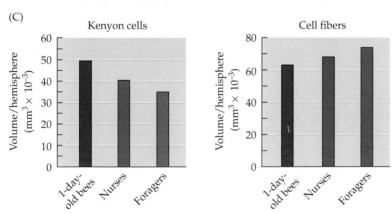

(C)

4 Developmental changes in worker bee behavior are correlated with changes in the bee's brain over time. (A) The location of the mushroom bodies in the brain of the honey bee. (B) A photomicrograph of a bee's mushroom body, showing the dark Kenyon cell region (k) and the fibers leading to other regions of the brain (the lighter stalk of the mushroom body). (C) Changes in the volume of the Kenyon cell bodies (left) and fibers (right) of the mushroom body are linked with changes In the tasks performed by worker bees. A, after Fahrbach and Robinson [375]; B, photograph courtesy of Gene Robinson; C, after Withers et al. [1260].

for this change in brain structure to occur, although both factors appear to influence the rate at which the mushroom bodies grow. The modification of the mushroom bodies takes place in anticipation of the new needs of foragers, which must learn spatial landmarks so they can travel back and forth between the hive and distant patches of flowers [1260].

But what causes the changes in juvenile hormone concentrations that take place within the bodies of honey bee workers? As it turns out, these hormonal changes are not absolutely fixed with worker age. This conclusion is based on experiments with colonies that have been manipulated so that all the workers are the same, relatively young, age. Under these conditions, a division of labor still manifests itself, with some individuals remaining nurses much longer than usual while others start foraging as much as 2 weeks sooner than average. As a result, the larvae are cared for continuously while the colony also receives food supplies.

What enables bees to make these developmental adjustments? One hypothesis is that a deficit in social encounters with older foragers may stimulate the

5   Social environment and task specialization by worker honey bees. In experimental colonies composed exclusively of young (resident) workers, the young bees do not forage if older forager bees are added to their hive. But if young (non-foraging) bees are added instead, the young residents develop into foragers very rapidly. After Huang and Robinson [556].

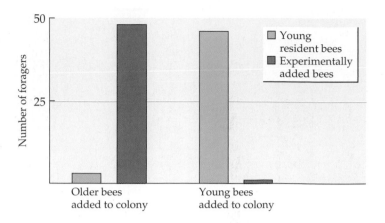

developmental transition from nurse to forager behavior. This possibility has been tested by adding groups of older foragers to experimental colonies made up of only young workers. The higher the proportion of older bees, the lower the proportion of young nurse bees that undergo an early transformation into foragers (Figure 5) [556]. The behavioral interactions between the young residents and the *older* transplants must inhibit the development of foraging behavior, because transplants of *young* workers have no such effect on young resident bees. Thus, the social environment of young honey bee workers influences their behavioral development by regulating the release of a key hormone, which then becomes part of their cellular environment and modifies the properties of neurons. After these hormonally induced changes in brain structure have occurred, workers change their behavior to match the special needs of their hive.

The interaction between genetic information and the environmental causes of task switching in the honey bee has been highlighted by a study of how genetic variation affects the rate of behavioral development in this insect. If one keeps genetically different lineages of honey bees in identical conditions, some genotypes respond to exactly the same hive environment differently by making the transition from nurse to forager more quickly (or slowly) than others. Why? If we have correctly identified the physiological causes of task switching (rising juvenile hormone concentrations caused by changes in the social environment), then the genes of fast-developing lineages can be predicted to influence things like the production schedule for juvenile hormone or the sensitivity of individuals to the age composition of their colony. As it turns out, some genotypes do tend to make more juvenile hormone sooner than usual, while other genotypes are especially sensitive to increases in the numbers of older workers in their hives [436]. Thus, several different physiological routes can lead to fast (or slow) behavioral development in the honey bee, each dependent on a particular mix of genetic and environmental factors.

When to Become a Territorial Male

Another striking example of how external social and internal hormonal environments affect behavior comes from a cichlid fish that lives in Lake Tanganyika in Africa. Males of *Haplochromis burtoni* come in two types: (1) a brilliantly colored black, yellow, blue, and red male that engages in territorial displays to repel male rivals and attract females to a nest in the lake bed, and (2) a nondescript pale brown male that is relatively inactive and definitely not territorial. Males of the second type are called "satellites" because they hang around the territories of the reproductively active, brightly colored males, but flee from these individuals when they attack (Figure 6).

6   The two male color forms of *Haplochromis burtoni.*
Males of this African cichlid fish have the ability to change their color, brain cells, and behavior in response to changes in their social status. Photograph by Russell Fernald.

The external differences in the appearance and behavior of the two forms of males are caused by internal differences in their brains, testes, and hormones. In particular, certain distinctive brain cells, the GnRH neurons, in one region of the hypothalamus are six to eight times larger in the aggressive territorial males than in the submissive satellite males (Figure 7). These cells release a hormone that stimulates development of the testes, which in turn produce male sex hormones that cycle back to the brain, promoting aggressive behavior by modulating the activity of certain cells in that organ.

Russell Fernald and colleagues have found, however, that males are not locked into one behavioral role by their current hypothalamic–testicular condition. If a territory owner is ousted after a battle with an intruder, he changes

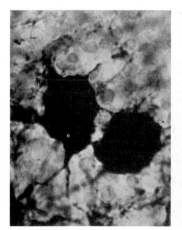 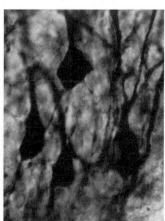 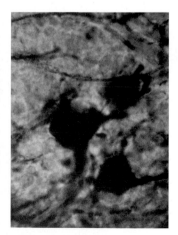 

7   Changes in brain neurons are linked to changes in social status. Representative GnRH neurons in the forebrains of male *Haplochromis burtoni:* (left to right) a territorial male; a male that has lost his territory; a nonterritorial male; and a male that was once nonterritorial but that has acquired a territory. Photographs courtesy of Russell Fernald, from Francis et al. [399].

color, almost instantly adopting the drab hues of a satellite, and moves in with the satellite crowd on the edges of the nesting area. His brain changes as well, with the GnRH neurons shrinking dramatically in size [382]. But should there be a new territorial opening, as when (for example) a territorial male is killed by a predator, a satellite can quickly become intensely aggressive, acquiring big GnRH neurons in the process. Thus, brain development in *Haplochromis burtoni* is supremely flexible, with changes in neural structure and function arising in response to cues from both the physical and social environment [399].

Experience and Behavioral Development

We have seen that social experience influences brain and behavioral development in certain fish, mice, and honey bees, and the same is true for many other species. Perhaps the classic example of this phenomenon is **imprinting,** in which a young animal's early social interactions, usually with its parents, have important developmental and behavioral consequences. Konrad Lorenz helped establish this point by rearing a clutch of greylag geese, which he encouraged to waddle after him soon after they had hatched (Figure 8). By virtue of this experience, the goslings formed an attachment to Lorenz, learning to recognize him as an individual, just as they would normally have learned to identify and follow their mother in nature. In addition to these short-term effects of imprinting, some remarkable long-term consequences occurred as well. As adults, the male greylag geese that imprinted on Lorenz courted human beings—including, but not specifically limited to, Lorenz—in preference to members of their own species. The experience of following a particular individual early in life must somehow alter those regions of the goose's nervous system responsible for sexual recognition and courtship.

The relation between imprinting and adult sexual behavior in zebra finches has been studied by Dave Vos [1176]. Wild zebra finches have a complex color pattern, including brightly colored bills (Figure 9), but Vos eliminated this complication by breeding albino finches (individuals with all-white plumage). When the nestlings of these parents were about 8 days old, just before their eyes opened, he painted the parents' bills different colors. If he applied orange nail polish to the mother's bill, he gave the father's bill a coat of red nail polish, and vice versa. He periodically freshened up the bill paints for the next 7 weeks, after which the youngsters were removed to cages where they were held in

8   Imprinting in greylag geese. These goslings have imprinted on the ethologist Konrad Lorenz and will follow him wherever he goes. Photograph by Nina Leen.

visual isolation from other zebra finches until they became sexually mature 10 weeks later. Then he ushered each subject into the central compartment of a special cage that had two small side cages attached to it, each containing an albino zebra finch, the "stimulus" finch.

In one of Vos's experiments, one of the two stimulus finches had its bill painted red, and the other had a bill coated with orange nail polish. The young adult zebra finch being tested therefore had a choice of associating with a bird with its mother's bill color or with one whose bill was the same color as its father's. Vos recorded the total time per 20-minute trial that the young adults spent near each of the two cages. In one of his experiments, he found that 12 of 14 male subjects preferred to approach and sing to the stimulus bird that had the same bill color as their mother, whether or not that bird was a female; all that counted was its bill color. In fact, in another experiment in which the young birds were given a choice between males with the same bill color as their mother versus females with same bill color as their father, many male subjects spent more time courting stimulus males than stimulus females (Figure 10). Thus, male zebra finches in the nest somehow attend to the beak color of their mother, incorporating information about this cue in ways that influence their mating behavior much later.

No such mechanism exists in young female zebra finches, which in Vos's experiments consistently approached and observed the male when the side cages contained a male and a female stimulus bird. The bill colors of the stimulus birds were not a factor in female preferences, which were apparently much more influenced by the behavior of males than by their appearance. In similar choice studies with naturally plumaged birds, females spent their time next to cages with males that sang frequently, rather than preferring red-billed or orange-billed males [240]. Thus, the effect of early visual experience on the development of adult sexual behavior can vary markedly between the sexes, presumably because male and female zebra finches differ genetically, which affects the development of the imprinting mechanisms in their bodies.

Early Experience and Recognition of Relatives

When a baby goose or male zebra finch imprints on its mother, it uses a form of associative learning as a means of recognizing her, gathering the relevant information about her identity during the period when it associates with and becomes familiar with her [644]. More generally, animals that can recognize

9   Female (left) and male (right) zebra finches. Photograph by Tony Tilford.

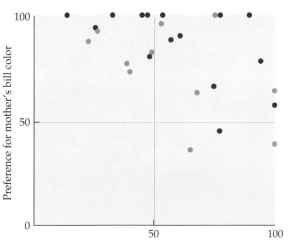

10   Imprinting in zebra finches. Male zebra finches reared in captivity with mothers whose bills had been painted red or orange were given the opportunity to court individuals (males or females) with (or without) their mother's bill color. When they had equal opportunity to court a *male* with their mother's bill color or a *female* with a different-colored bill, the imprinted males usually courted the male more often than the female. Orange circles represent birds given a choice between a female with an orange bill and a male with a red bill; red circles, between a female with a red bill and a male with an orange bill. Half the subjects in the experiment had been reared with a mother with a red bill, and half with a mother with an orange bill. After Vos [1176].

their relatives can engage in **kin discrimination,** defined as the differential treatment of members of their species in a way that depends on their genetic relatedness to the discriminating individual [1047]. The ability to respond differently to individuals of varying degrees of genetic relatedness occurs in many species that exhibit parental care (see Chapter 13) and in social species in which relatives cooperate (see Chapter 14). As we shall discover in the chapters ahead, kin discrimination often underlies behaviors that may help propagate the discriminating individual's genes. Here our focus is on the special proximate mechanisms that promote the development of this adaptive ability.

In addition to some birds, a good many mammals attend to the distinctive sensory cues provided by their companions early in life—which are usually siblings—so that they too employ associative learning as the proximate mechanism for kin discrimination. Thus, for example, as they grow up, young captive spiny mice prefer to huddle with the individuals that were in their litter, as opposed to strangers. In nature, this preference for familiar early companions would ordinarily lead siblings to remain together. But if one experimentally creates a litter composed of nonsiblings that are cared for by the same female, these unrelated littermates will treat each other as if they were siblings [932]. This discrimination error shows that a simple rule of thumb regulates the behavior of the mice: "Associate with familiar littermates; avoid unfamiliar individuals." This simple rule works in nature because normally babies from one litter never move into another nest with unrelated companions.

Almost the converse of this rule of thumb applies to the naked mole-rat, a bizarre burrowing mammal (see Figure 30, p. 451) that lives in large colonies composed primarily of relatives in an elaborate network of tunnels. In this species, reproductively active females follow the rule, "mate with males whose scents are *unfamiliar.*" As a result, females tend to breed with unrelated outsider males that have recently joined their colonies, rather than with the brothers and sons with whom they have a long association [222].

David Pfennig and his colleagues suspected that paper wasp females learn what their natal nest and its other occupants smell like shortly after the females emerge from the nest's brood cells (Figure 11) [911]. Learning of this sort would, under natural conditions, enable females to recognize and respond positively

11 **Nests of *Polistes* paper wasps** contain odors that adhere to the bodies of the wasps reared in them, providing a proximate cue for kin recognition in these insects. Note the larvae with their bodies pressed up against the cell walls. Photograph by the author.

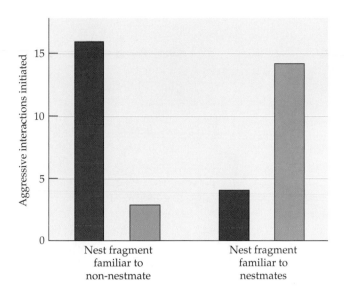

12 **Context-dependent aggression against nonkin** in experiments involving three paper wasps, one of which did not share a nest with the other two. The presence of a familiar nest fragment caused individuals to respond more aggressively to non-nestmates. Data for the average member of a nestmate pair is shown in green; the mean aggressive response of non-nestmates is shown in purple. After Starks et al. [1094].

to sisters, which share the natal nest odor, and to behave less tolerantly toward those adults that did not smell like the familiar nest. If this hypothesis is correct, it should be possible to fool paper wasps into tolerating nonkin by exposing newly emerged females for a few hours to the odor of another nest occupied by unrelated females. As predicted, transferred queens were less likely to fight with the unrelated females in the nest they had smelled as young adults than with their genetic relatives from what was an unfamiliar nest to the experimental queens. Thus, paper wasps employ a rule of thumb that reads, "Respond unaggressively to those individuals that smell like the nest that you smell soon after becoming an adult."

The actual presence of a nest fragment (with its allied odors) can be important in regulating the extent of aggression between individual paper wasps reared together or apart. In the paper wasp *Polistes dominulus*, researchers formed experimental groups of three wasps, only two of which were nestmates. If the odd wasp out had access to a nest fragment with familiar odors, she pushed the other two wasps around (Figure 12). But if the trio of wasps was housed with a piece of nest familiar to the two nestmates, they reacted aggressively toward their non-nestmate companion. The wasps used odor cues associated with a valuable resource, the nest, to calibrate their aggressiveness against individuals from other nests [1094].

Another proximate mechanism based on odor may promote kin discrimination in Belding's ground squirrels, judging from a study in which the newborn offspring of captive females were switched around at birth, creating four classes of individuals: (1) siblings reared apart, (2) siblings reared together, (3) nonsiblings reared apart, and (4) nonsiblings reared together. After having been reared and weaned, the juvenile ground squirrels were placed in an arena in pairs and given a chance to interact. In most cases, animals that were reared together, whether siblings or not, treated each other nicely, whereas animals that had been reared apart were likely to react aggressively to each other. This experiment shows that the little squirrels learned something from the experience of growing up together, and that they used this information in their social relations [544], as do other ground squirrels [543].

But perhaps the most remarkable finding of this study was that biological sisters *reared apart* engaged in fewer aggressive interactions than nonsiblings reared apart (Figure 13). In other words, the squirrels had some way of rec-

Belding's ground squirrel

13 **Kin discrimination in Belding's ground squirrels.** Sisters reared apart display significantly less aggression toward each other than other combinations of siblings reared apart, which are as aggressive to each other when they meet in an experimental arena as nonsiblings reared apart. After Holmes and Sherman [544].

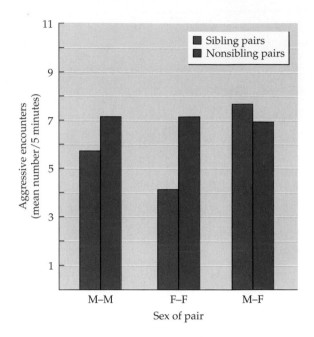

ognizing their sibs—perhaps through an odor similarity—that was not dependent on the experience of sharing a mother and a burrow [542, 544]. The mechanism underlying kin discrimination by female Belding's ground squirrels and other animals like them has been labeled **phenotype matching** by Warren Holmes and Paul Sherman. These creatures apparently know something about their own phenotype—their appearance or odor or vocal qualities—and use this information as a reference point to compare with the traits of others [545].

Phenotype matching probably underlies a striking example of kin recognition and discrimination in the red fire ant, which forms large colonies of workers who sometimes live with and work for several egg-laying queens. As new queens attempt to join the egg-laying contingent, workers with the *Bb* genotype systematically track down and kill any queen with the *BB* genotype. The killers evidently can recognize queens that lack the *b* allele, probably because of an odor difference between carriers of the *BB* and *Bb* genotypes. (Incidentally, there are few individuals homozygous for the *b* allele because they tend to die before reaching adulthood.) If the body of a *BB* queen is rubbed over a *Bb* worker, other *Bb* workers will then attack it, almost certainly because the unfortunate worker has acquired the odor of a *BB* queen [607].

The protein encoded by the *B* gene in fire ants has yet to be identified, but more is known about the gene–protein connection in some vertebrates in which odors also play a key role in learned kin recognition. In some species, differences in the scents of individuals arise because of genes belonging to the major histocompatibility complex (MHC) [170]. The MHC genes encode a special class of proteins, the MHC glycoproteins, which occupy positions in cell membranes. MHC glycoprotein chemistry is such that these proteins respond differently to cells depending on their origin. The body's own cells are ignored, but invader organisms, such as viruses and bacteria, trigger a reaction that leads to a defensive immune response.

Although the MHC genes presumably evolved in the context of fighting disease, they may now also be involved in odor-based kin recognition in some animals [895]. The MHC genes come in many alleles (as many as 50 per gene in human beings and house mice), and no one allele predominates. As a result, different individuals tend to have different MHC genotypes. Because MHC alleles can affect the odors an animal produces, different members of the same species often have their own distinctive blend of scents. These differences can be detected by others in some species, including fish [855], mice [895], and possibly even humans, despite our modest olfactory skills [1195].

Kin recognition based on MHC genes can be used to make decisions about which individuals to associate with [855] or with whom to reproduce [897]. The proximate basis for kin discrimination based on MHC-coded odors could conceivably involve either familial imprinting, as has been shown for house mice [896], or phenotype matching, also known in some circles as the "armpit effect" because it involves comparing one's own odor with those of other individuals. Clear evidence for phenotype matching comes from golden hamsters, which, like Belding's ground squirrels, can identify family members that they meet for the first time as adults. These little mammals have glands on their flanks that are specialized for producing scents, which are sometimes applied to other individuals during aggressive encounters.

To examine the role of flank gland scents in kin discrimination, hamsters that had been reared by foster mothers with unrelated littermates were given a chance to inspect flank gland scents from both unrelated strangers and their genetic siblings, neither of whom they had ever met. Sexually receptive adult female hamsters were quicker to investigate, and spent more time sniffing, the odors of unrelated males than those of their siblings. In addition, the females exposed to sibling odors were much more likely to rub their flank glands over the sibling odors (Figure 14), applying their scents as a warning signal to these other individuals not to come near them. Nonrelatives, especially unfamiliar males (which are potential mates), did not stimulate as much aggressive marking behavior. Thus, golden hamsters are able to match the odors of others against their own scents, using this ability to mate with nonrelatives, the better to avoid the negative effects of inbreeding [768].

(A)

(B)

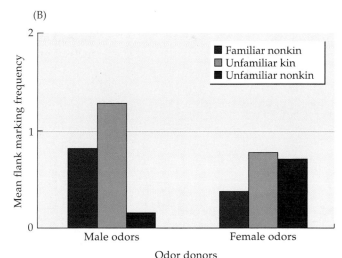

14 Functional significance of phenotype matching in the golden hamster.
(A) Flank marking by a golden hamster, which is pressing its head and shoulder against the vertical wall of its cage as it applies chemical scents from the flank gland to the wall. (B) Females more frequently applied their scent to odors of brothers that they had never met (green bar) than to those of either familiar nonkin (genetically unrelated males with whom they had been reared–purple bar) or to unfamiliar nonkin (unrelated males that they had never met–red bar). After Mateo and Johnston [768]; Photograph by Bob Johnston and Jill Mateo.

Learning as Behavioral Development

The male zebra finch that has imprinted on his mother's bill color and the Belding's ground squirrel that knows what its nestmates smell like have both learned something about their social environment. The information these animals acquire leads to long-lasting changes in their behavior, presumably through permanent modifications of certain neural mechanisms. Therefore, imprinting and learned kin discrimination can both be considered a kind of behavioral development based on cellular changes [375].

This fundamental similarity between developmental processes and social learning applies to all other forms of learning as well, including spatial learning. Many animals, including beewolf wasps (see p. 7), memorize visual landmarks in order to move efficiently from place to place. The spatial memory of a black-capped chickadee, for example, enables it to relocate hundreds of seeds or small insects that it has hidden in bark crevices or patches of moss scattered over a wide area. To study this little bird's memory, David Sherry provided captive chickadees with a chance to store food in holes drilled in small trees placed in an aviary. After the chickadees had placed a sunflower seed in 4 or 5 of 72 possible storage sites, they were shooed into a holding cage for 24 hours. Sherry removed the seeds in the interim and closed each of the 72 storage sites with a Velcro cover. When the birds were released back into the aviary, they spent much more time inspecting and pulling at the covers at their hoard sites than at sites where they had not stored food 24 hours before (Figure 15). Because the storage sites were empty and covered, there were no olfactory or visual cues provided by stored food to guide the birds in their search; they had to rely solely on their memory of where they had hidden food [1049]. In nature, these birds store only one food item per hiding spot and never use the same location twice, yet they can relocate their caches as many as 28 days after the event [522].

Clark's nutcrackers may have an even more impressive memory, for they scatter as many as 9000 caches of pine seeds (1–10 seeds per cache) over entire hillsides (Figure 16). The bird digs a little hole for each store of seeds, then completely covers the cache. A nutcracker does this work in the fall, and then relies on its stores through the winter and into the spring, so that it may be months before it comes back to retrieve the seeds from a particular cache [62].

It could be that nutcrackers do not really remember where each cache is, but instead rely on a simple rule of thumb, such as "caches will be made near little tufts of grass." Or they might only remember the general location where food

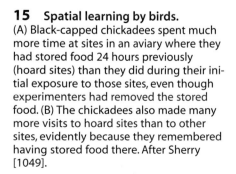

Black-capped chickadee

15 Spatial learning by birds.
(A) Black-capped chickadees spent much more time at sites in an aviary where they had stored food 24 hours previously (hoard sites) than they did during their initial exposure to those sites, even though experimenters had removed the stored food. (B) The chickadees also made many more visits to hoard sites than to other sites, evidently because they remembered having stored food there. After Sherry [1049].

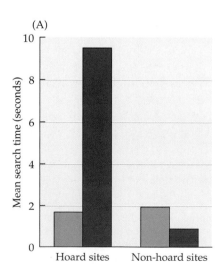

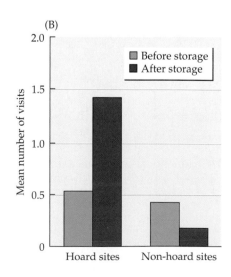

was stored, and once there, look around until they saw signs of caching. But experiments similar to those performed with chickadees show that the birds do remember exactly where they hid their food. In one such test, a nutcracker was given a chance to store seeds in a large outdoor aviary, after which it was moved to another cage. The observer, Russ Balda, mapped the location of each cache, then removed the buried seeds and swept the cage floor, removing any signs of cache-making. No visual or olfactory cues were available to the bird when it was permitted to go back to the aviary a week later and hunt for the food. Balda mapped the locations where the nutcracker probed with its bill, searching for the nonexistent caches. The bird's spatial memory served it well, for it dug into as many as 80 percent of its ex-cache sites, while only very rarely digging in other places [62]. Moreover, some of the apparent errors made by nutcrackers almost certainly occur because the birds are examining spots they know do not contain caches of their own, but might contain food for some other reason [82]. Other long-term experiments on nutcracker memory have demonstrated that nutcrackers can remember where they have hidden food for at least 6, and perhaps as long as 9, months [63].

The ability of nutcrackers and chickadees to store spatial information in their brains is surely related to the ability of certain brain mechanisms to change biochemically and structurally in response to certain kinds of sensory stimulation. One anatomically distinct region in the brain, the hippocampus, appears to be especially important in memory storage, given that food-storing bird species have larger, more neuron-rich hippocampi than those that do not scatter food caches through their environment [224]. The hippocampus may not only be capable of storing information about hoard site locations, but may actually require the sensory experiences associated with food storing behavior in order to develop fully. To test this possibility, Nicky Clayton and John Krebs hand-reared some marsh tits—close relatives of the black-capped chickadee—in the laboratory. They gave some of the birds opportunities to store whole sunflower seeds at three stages after hatching, while others were always fed powdered sunflower seeds, which the birds eat, but cannot and will not store [225]. Marsh tits, and birds of several other related species, that have opportunities to store food have more cells in the hippocampus than those birds unable to gain storing experience, a case of "use it or lose it" (Figure 17).

Marsh tit

Coal tit

Mountain chickadee

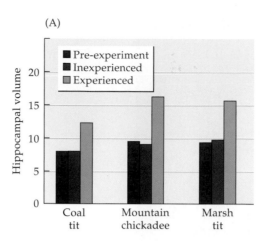

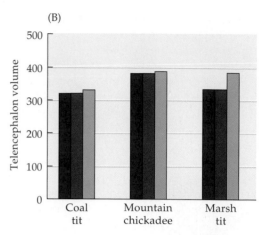

17 **Changes in hippocampal volume as a result of food storing experience** in coal tits, mountain chickadees, and marsh tits. (A) The volume of the hippocampus was greater in birds that had had the opportunity to store food than in young birds whose brains were examined before the experiment began or in birds that had little experience in storing food. (B) The volume of the telencephalon, another brain structure not involved in spatial learning, did not vary for these three categories of birds. After Clayton [224].

Evolution and Behavioral Development

Our review has covered only a few selected examples of the broad range of environmental effects on behavioral development. Change the *material environment* by altering the availability of amino acids or fats or carbohydrates in the diet of a young animal, and the gene–environment interactions that occur will change as well, potentially affecting the structural development of the brain or the activity of hormone-releasing organs, with assorted behavioral consequences for the affected individual. Alter the *experiential environment* by changing the sensory inputs from the physical environment or from social interactions with other animals, and behavioral development will shift as well. A male white-crowned sparrow has to hear certain sounds as a fledgling or young adult if he is to mimic these acoustical stimuli and eventually sing like his neighbors. By the same token, young honey bees are developmentally sensitive to the social experiences they have with their older colonymates, while the interactions a male zebra finch nestling has with his mother shape the development of his adult mate preferences.

These and other similar cases raise an evolutionary question about behavioral development: why is it reproductively advantageous to link behavioral development to social experience in the white-crowned sparrow, honey bee, zebra finch, and so many other species? In Chapter 2, we presented the hypothesis that male white-crowns able to alter their song to resemble that of a neighbor might have enjoyed a reproductive edge, thanks to their ability to communicate more effectively with the other male. A male white-crown cannot enter life knowing exactly who his territorial neighbors will be and precisely how they will sing. These elements of unpredictability in the species' social environment could favor individuals able to adjust to that unpredictability.

We can apply this same argument to other cases of developmental flexibility. Honey bee colonies that have lost a substantial portion of their foragers through bad luck or the vagaries of the weather probably would have fared better if their surviving nurse workers could quickly fill the missing foragers' places. Likewise, imagine a cichlid fish whose males possessed neural and endocrine systems that permitted only all-out territorial behavior, even in weakened individuals faced with rivals with ample energy reserves. (The fact that satellite males grow more rapidly than territorial individuals demonstrates the energetic costs of being an aggressive territory defender [524].) Now imagine that in such a species, a mutant male appeared that could adopt a camouflaged color and an energy-saving, growth-enhancing, risk-reducing, wait-and-see behavioral tactic when confronted with a formidable replacement [399]. Such a male seems far more likely to live to compete successfully another day than the one-response types. If so, the proximate mechanisms of the flexible response should have spread through the species over time, enabling today's males to adjust to changes in their social status and competitive environment.

In general, when members of a species face somewhat unpredictable environmental conditions, individuals may benefit if they have several different developmental and behavioral options. This argument seems to work for a tiger salamander species known for having two types of individuals: (1) a typical aquatic larva, which eats small pond invertebrates such as dragonfly nymphs, and (2) a cannibal type, which grows larger, has much more powerful teeth, and feeds on other tiger salamander larvae unfortunate enough to live in ponds with it (Figure 18). The development of the cannibal type, with its distinctive form and behavior, depends on certain factors in the salamanders' social environment. For example, cannibals develop only when many salamanders live together [238]. Moreover, they appear more often when the larvae in a pond (or aquarium) differ greatly in size, with the largest individual much more likely to take the developmental pathway to cannibalism than its smaller companions [751]. In addition, the cannibal form is more likely to develop when the population consists largely of unrelated individuals than when the tank holds the offspring of a single mating—in other words, a group of siblings (Table 1) [910]. If a larger-than-average salamander lives with many other young salamanders that do not smell like its close relatives [913], its development may well shift from the typical track to the one that turns it into a giant, fierce-toothed cannibal.

What benefit do tiger salamanders derive from having two potential developmental pathways and a switch mechanism that enables them to change their growth pattern and behavior? If numerous salamander larvae occupy a pond, and if most are smaller than the individual that becomes a cannibal, then the

18   Tiger salamanders occur in two forms. The typical form (being eaten by its companion) feeds on small invertebrates and grows more slowly than the cannibal form (which is doing the eating). Cannibals have broader heads and larger teeth than their insect-eating companions. Photograph by Tim Maret, courtesy of James Collins.

TABLE 1 The effect of genetic relatedness on the development of cannibal forms of the tiger salamander in aquaria with equal larval densities

Composition of aquarium population	Cannibal develops	No cannibal develops	Total experiments
Siblings only	31 (40%)	46 (60%)	77
Nonsiblings present	67 (84%)	12 (16%)	79

Source: Pfennig and Collins [894]

cannibal gains access to an abundant food source that is not being exploited by its fellows, and so grows quickly. Rapid growth enables cannibals to metamorphose into adults before their fellow pond dwellers, an important advantage in ponds that evaporate in the summer. Flexibility in making the switch is a plus because in some ponds, would-be cannibals would starve because their prey would be too scarce, or too similar to their own size to be captured. In addition, potential cannibals can lose genetic fitness if they harm their relatives (see p. 435). Because salamanders have no way of knowing in advance whether their environment will be loaded with suitable prey of the same species and right degree of nonrelatedness, selection has presumably favored individuals that happened to have the ability to become cannibals only if conditions happened to be appropriate.

The Adaptive Value of Developmental Flexibility

The idea that important but unpredictable variation in an animal's social environment favors flexibility in its behavioral development can be applied beyond the social realm. Many nonsocial but variable aspects of the environment can have major reproductive consequences for individuals. For example, the development of running capacity in the lizard *Anolis sagrei* depends on where it grows up. This species lives in trees in the Bahamas, where the lizards dash about on trunks, branches, and twigs while competing with rivals, foraging for food, and escaping from predators. In this and other species of *Anolis* (and many occur on the Caribbean islands), the maximum speed attained by sprinting lizards is a function of the length of the animal's legs. Those species that consistently live on broad tree trunks have long legs, which make it possible for them to sprint at great speeds; those species that forage primarily on small branches and twigs have short legs, which reduce the speed at which they can move, but which make it easier for them to maintain a grip on narrow branches.

The lizard *A. sagrei* can occupy the full spectrum of habitats available in trees, from trunk to twigs. This observation suggested to Jonathan Losos and his colleagues that the development of this species might be influenced adaptively by the nature of the environment that the young animals experienced [722]. Losos therefore reared hatchling *A. sagrei* in terraria that contained either broad (trunklike) surfaces or narrow (branchlike) surfaces. As they grew in size, the two groups of lizards acquired different kinds of hind legs (Figure 19). Lizards that moved about on broad surfaces developed into longer-legged types than those that grew up maneuvering about on narrow branches. In nature, animals that had grown up with ample trunk space would benefit from the capacity to run rapidly on this easily negotiated surface, whereas those confined to small branches and twigs would do better with the alternative phenotype.

Note that the kind of structural and behavioral plasticity that selection has evidently favored in *A. sagrei* is far from open-ended. Instead, developmental circumstances of a particular sort lead to the production of a particular phenotype appropriate for those circumstances. Under different conditions, a differ-

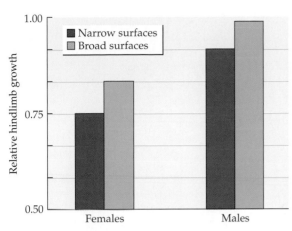

19 Developmental effect of rearing environment on the lengths of the hind legs of *Anolis sagrei*. The lizards were reared in two different environments, one with broad perching surfaces and the other with narrow twiglike perches. After Losos et al. [722]; photograph by Jonathan Losos.

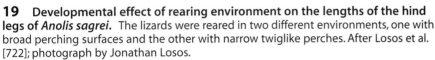

ent habitat-appropriate type develops. This same principle applies to learning, which, when it occurs, tends to help an individual deal adaptively with a biologically relevant situation, rather than producing flexibility for flexibility's sake. For example, the reason why marsh tits may make hippocampal expansion and spatial learning ability contingent upon having experience with storing food may have something to do with the frequency with which their environments differ in the amount of food available for storage. If, in a given winter or particular habitat, little food suitable for storage exists, then there is little advantage in being able to learn the location of food stores, and no gain for an investment in extra hippocampal tissue. But when food of the right sort for storage is abundant, then it can pay to have the ability to make a particular kind of response—not any old change, but one that results in a larger hippocampus and greater spatial learning ability.

To illustrate the connection between environmental variation and the evolution of biased learning, let's examine certain Australian thynnine wasps known for their apparent behavioral *inflexibility*. Male thynnines spend their adult lives searching for receptive wingless females, which announce their readiness to mate by releasing a sex pheromone (a special volatile chemical) while perched on a stem or twig (Figure 20). Males fly to scent-releasing females and carry them away to copulate elsewhere [1103].

However, male wasps will also attempt to mate with the bizarre flowers of certain small orchids that occur where they live. The flowers of these wasp-attracting orchids produce an odor that mimics the sex pheromone released by a receptive thynnine female (although the details of the chemical mimicry have not been documented in this group as well as they have for some equally devious European orchids [130]). Male wasps track this odor to an orchid, which has a special petal that vaguely resembles the body of a wingless female wasp. The male pounces upon and tries (unsuccessfully) to pick up the petal (Figure 21). In the course of the attempt, the male's body comes into contact with the pollinia, or pollen-bearing sacs, of the flower. The sticky pollinia adhere to the male, and when he finally gives up the futile task of trying to carry the "female" away, the pollinia go with him. Should he be drawn to another individual of the orchid species that deceived him, he will transfer pollen to that plant, fertilizing it rather than a female of his own species.

20   **Male thynnine wasps use olfactory and visual cues** to locate mates. (Left) A wingless female thynnine wasp that is releasing a sex pheromone to attract males. (Right) A winged male thynnine wasp has grasped a scent-releasing female and is in the process of initiating copulation. Photographs by the author.

Thus, thynnine males appear to be remarkably obtuse automatons, programmed by their internal mechanisms to rush about responding to simple cues in their environment—so simple that plants can mimic those cues and cause the insects to waste their time and energy. But one almost never sees wasps actually visiting orchids in nature, so perhaps thynnines are not quite as thickheaded as they seem. In order to observe wasp–orchid interactions, Australian botanists usually cut some orchids, put them in water-filled jars, and move them to a new location. Only then will male wasps rush in to pounce upon the decoy petal of the orchid. After a brief frenzy of pseudo-copulations, the number of males coming to a sample of cut orchid specimens falls off sharply (Figure 22).

These results suggest that male thynnines are attracted to an orchid shortly after its flower opens and the mimetic scent is released for the first time. But after being deceived and grappling with a particular orchid's decoy petal, the

21   **A male thynnine wasp** attempting to carry off the female decoy petal of the elbow orchid. Note the yellow pollinia stuck to the male's back. Photograph by the author.

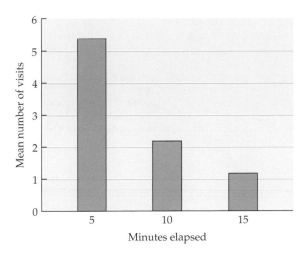

22   Learning by male thynnine wasps. When deceptive orchids are placed in a new location, they initially receive many visits from wasps, but the visits soon decrease. After Peakall [893].

male stores information about the experience somewhere in his brain and thereafter avoids responding sexually to the scent *coming from that particular orchid* [893]. The reproductive benefits of the male wasp's behavioral flexibility are obvious. Male wasps cannot be programmed in advance to know where female wasps and deceptive orchids will appear on any given day. By using experience to learn to avoid specific orchids while remaining responsive to novel sources of sex pheromone, the male wasp saves time and energy and improves his chance of encountering a receptive female that has begun to release sex pheromone somewhere in his searching zone.

All proximate mechanisms that confer behavioral flexibility on individuals surely come with a price tag. For one thing, the systems that make it possible to develop two body forms with different feeding tactics, or neurons that record experience and make learning possible, require calories for their construction. The view that learning ability requires costly neural tissue is supported by a study of male long-billed marsh wrens from the western and eastern United States. West Coast marsh wrens have more song types in their repertoire than their East Coast counterparts. When permitted to listen to a tutor tape in captivity, young West Coast wrens learned nearly 100 songs, whereas their East Coast counterparts incorporated only about 40 songs into their repertoire [658]. When the birds' brains were examined, the song control systems of the West Coast wrens weighed on average 25 percent more than the equivalent nuclei of the East Coast wrens.

If the large brain of human beings is related to our prodigous ability to learn, as many persons have argued, then the cost of human learning mechanisms may be great indeed. Although the brain makes up only 2 percent of our total body weight, it demands 15 percent of all cardiac output and 20 percent of the body's entire metabolic budget [42] (but see [784]). Furthermore, the costs of learning mechanisms include not just the expense of developing and maintaining them, but their damaging effects when they malfunction. For example, animals sometimes use experience to modify their behavior in a way that lowers, not raises, their fitness. Suppose, for example, that an individual observes others and adopts their solution to a behavioral problem (e.g., the path to take to a feeding site) when it could do better by adopting a different solution (e.g., it could travel to the site more economically by another route [673]). For another example of a fitness-reducing effect of learning, consider that if a rat or human being eats a novel food and then becomes ill for some other reason, that rat or human will often mistakenly form a learned aversion to a perfectly nutritious item [211].

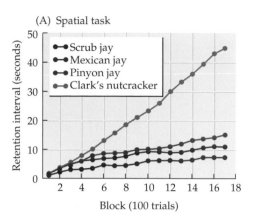

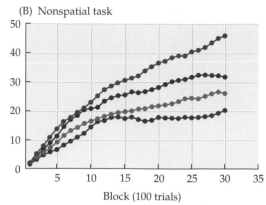

23 **Spatial learning abilities differ among species.** (A) Captive Clark's nutcrackers performed much better than three other corvid species in experiments that required the birds to retain information about the location of a circle. (B) But when the birds' ability to remember the color of a circle was examined, the nutcrackers did not excel in this nonspatial learning task. After Olson et al. [856].

Given the costs of behavioral flexibility, we would expect it to evolve only when an investment in the underlying mechanisms is repaid. Thus, for example, the capacity to develop a larger memory apparatus and better spatial learning skills should evolve in species that regularly store food, but not in species that cache food less often, or not at all. This prediction has been tested by comparing the spatial learning abilities of four bird species that are all members of the same family, the Corvidae, but vary in their predisposition to store food. As we have seen, Clark's nutcracker is a food-storing specialist, and it has a large pouch for the transport of pine seeds to storage sites. The pinyon jay also has a special anatomical feature, an expandable esophagus, for carrying large quantities of seeds to hiding places. In contrast, the scrub jay and Mexican jay lack special seed transport devices and appear to hide substantially less food than their relatives.

Individuals from the four species were tested on two different learning tasks in which they had to peck a computer screen to receive rewards. One task required the birds to remember the *color* of a circle on the screen (a nonspatial learning task), and the other required memory of the *location* of a circle on the screen (a spatial task). When it came to the nonspatial learning test, pinyon jays and Mexican jays did substantially better than scrub jays and nutcrackers. But in the spatial learning experiment, the nutcracker went to the head of the class, followed by the pinyon jay, then the Mexican jay, and, finally, the scrub jay (Figure 23). These results show that the birds have not evolved all-purpose learning abilities that apply equally to all tasks; instead, their learning skills are designed to promote success in solving the special problems that they face in their natural environments [856].

Sex Differences in Spatial Learning Ability

If behavioral flexibility evolves only when special environmental demands favor versatility, then if the sexes within a species differ in the size of their home ranges, the sex that moves over a wider area ought to exhibit superior spatial learning skills. Steve Gaulin and Randall FitzGerald tested this prediction in studies of three species of small rodents, all belonging to the genus *Microtus*. Males of the polygynous meadow vole have home ranges more than four times as large as those occupied by each of their several mates. In contrast, males and

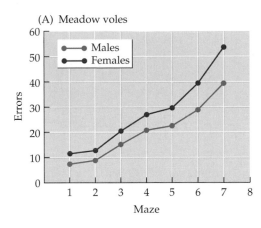

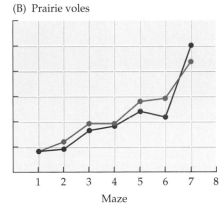

(A) Meadow voles

(B) Prairie voles

- Males
- Females

Errors

Maze

Maze

24 Sex differences in spatial learning ability are linked with mating system. (A) Polygynous male meadow voles consistently made fewer errors on average than females when learning how to get through seven different types of mazes of increasing complexity. (B) In contrast, females matched male performance in the monogamous prairie vole. After Gaulin and FitzGerald [421].

females of both the prairie vole and the pine vole are monogamous and share the same-sized living space. When tested in a variety of mazes, which the animals had to solve in order to receive food rewards, males of the wide-ranging meadow vole consistently made fewer errors than females of their species (Figure 24). In both the prairie and pine voles, however, males and females did equally well on these spatial learning tests; in these species, the two sexes have similar home ranges and so are confronted with equivalent spatial learning problems in their natural lives [420, 421].

Given the role of the hippocampus in spatial learning, it is not surprising that male meadow voles invest more heavily in this structure, as measured by the proportion of brain volume it occupies, than do females of their species. In contrast, no differences in hippocampal size exist in male and female pine voles [572]. Moreover, the differences in spatial learning ability and hippocampal size between male and female meadow voles appear to be expressed only during the summer breeding season, when males are searching widely for mates. When the animals are not breeding, males can gain no reproductive advantage from special learning skills, but would pay the metabolic price of maintaining an enlarged hippocampus. By having a brain that shrinks when the extra tissue is not useful, a meadow vole can reduce the costs of a neural mechanism required for learning [571].

In a species in which females face greater spatial challenges than males, we would expect females to make larger investments in the expensive neural foundations of spatial learning. The brown-headed cowbird is such a species, because females search widely for nests of other birds to parasitize. They must also remember where potential victims have started their nests in order to return to them one to several days later when the time is ripe for the cowbird to add her egg to those already laid. In contrast, male cowbirds do not confront such difficult spatial problems. As predicted, the hippocampus (but no other brain structure) is considerably larger in female brown-headed cowbirds than in males. No such difference occurs in some nonparasitic relatives of this species [1050] (Figure 25). Similarly, among the South American cowbirds, breeding females of two parasitic species also make larger investments in hippocampal tissue than males [957], but only during the breeding season, when the benefits to females of spatial learning ability are high [226].

Meadow vole

Prairie vole

The Evolution of Associative Learning

We have focused on the kind of flexibility that spatial learning confers, finding that the ability to learn the locations of key features in one's environment evolves in response to the special challenges associated with locating widely

25 Sex differences in the hippocampus. Female cowbirds have a larger hippocampus than males, as would be expected if this brain structure promotes spatial learning and if selection for spatial learning ability is greater on female than on male cowbirds. Red-winged blackbirds and common grackles do not exhibit this sex difference. After Sherry et al. [1050].

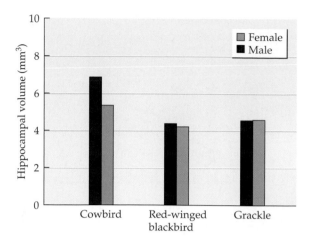

and unpredictably dispersed resources or mates. But are the many other kinds of learning similarly linked to special ecological problems?

Consider **operant conditioning,** in which an animal learns to associate a voluntary action with the consequences that follow from performing it [1067]. Operant conditioning (or trial-and-error learning) does occur outside psychology laboratories (Figure 26), but it has been studied most extensively in Skinner boxes, named after the psychologist B. F. Skinner. After a white rat has been

26 Trial-and-error learning occurs in nature when a tree frog attempts to capture a bombardier beetle. The tree frog is quickly educated by the beetle, which fires a hot chemical repellent onto its tongue. After one experience with this prey, the tree frog will avoid it on sight thereafter. Photographs by Tom Eisner.

27   Rat in a Skinner box. The operantly conditioned rat approaches the bar (top left) and then presses it (top right). The animal awaits the arrival of a pellet of rat chow (bottom left), which it consumes (bottom right), so that the bar-pressing behavior is reinforced. Photographs by Larry Stein.

introduced into a Skinner box, it may accidentally press a bar on the wall of the cage (Figure 27), perhaps as it reaches up to look for a way out. When the bar is pressed down, a rat chow pellet pops into a food hopper. Some time may pass before the rat happens upon the pellet. After eating it, the rat may continue to explore its rather limited surroundings for a while before again happening to press the bar. Out comes another pellet. The rat may find it quickly this time, and then turn back to the bar and press it repeatedly, having learned to associate this particular activity with food. It is now operantly conditioned to press the bar.

Operant conditioning techniques have been used with success to guide the learning of responses far more complex than pressing a bar, as demonstrated by the ability of Skinnerians to train pigeons to play table tennis with their beaks or to communicate symbolically with one another [370]. These same procedures were used to get nutcrackers and jays to peck at circles on computer screens in the experiments described above. Operant approaches have also been employed to condition human beings and other animals to regulate internal processes such as heart rate or brain electrical activity, which at one time were thought to lie outside of conscious control [111].

Although Skinnerian psychologists once believed that one could condition with equal ease almost any operant (defined as any action that an animal could perform), researchers soon began to discover contradictory cases. For example, one can readily condition a white rat to do some things, such as running in a running wheel, by sounding a warning noise and giving it an electric shock if it fails to get going. A rat that has had some experience with hearing the sound and receiving a shock while standing in the wheel, but not while running, will make the appropriate association and start running whenever it hears the warning sig-

28 Biased learning. Learning curves of white rats for three operants (running, turning, rearing) that were equally rewarded (the rats were not shocked when they performed the appropriate operant). Rats failed to learn to rear up on their hind legs to avoid a shock. After Bolles [120].

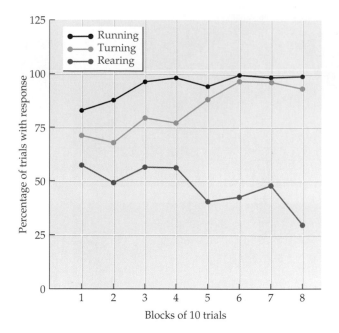

nal [120]. But Robert Bolles found that a rat cannot be conditioned to stand upright in the running wheel using the same procedures. Rats that happen to rear up just after the sound and are not shocked nevertheless fail to make the association between this behavior and avoidance of the shock. In fact, the frequency with which they stand up when they hear the warning signal actually *declines* over time (Figure 28).

Likewise, John Garcia and his colleagues have found that white rats can learn to avoid some, but not all, sensory cues that are associated with certain consequences [417, 418]. For example, if a rat feeds on or drinks a novel, distinctively flavored food or liquid and then is exposed to X-rays, the animal becomes ill, because even tiny doses of radiation cause a buildup of toxic chemicals in irradiated tissues and body fluids. Subsequently, the rat often refuses to touch whatever it ate or drank before it became sick. The degree to which the food or fluid is avoided is proportional to (1) the intensity of the resulting illness, (2) the intensity of the taste of the substance, (3) the novelty of the substance, and (4) the shortness of the interval between consumption and illness [418]. But even if there is a long delay (up to 7 hours) between eating a distinctive food and exposure to radiation and consequent illness, the rat still links the two events and uses the information to modify its behavior. Most kinds of associative learning will not take place when the interval between a cue and its consequence lasts more than a few seconds.

The specialized nature of this taste aversion learning is further shown by the rat's complete failure to learn that a distinctive sound (a click) precedes internal illness. Rats can learn to associate clicks with shock punishment and will learn to take appropriate action to avoid shocks upon hearing the click. Yet when the punishment is a nausea-inducing treatment, they fail to learn to link the sound with the treatment. In addition, rats have great difficulty in making the association between a distinctive taste and shock punishment. If, after drinking a sweet-tasting fluid, the rat receives a shock on its feet, it often remains as fond of the fluid as it was before, as measured by the amount drunk per unit of time, no matter how often it is shocked after drinking this liquid. Thus, the nature of cue and consequence determine whether a rat can learn to modify its behavior (Figure 29).

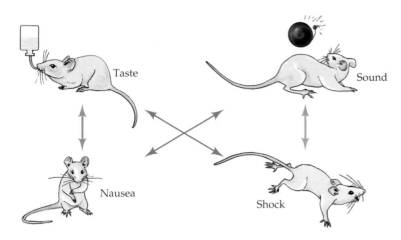

29 Biases in taste aversion learning. Although white rats can easily learn that certain taste cues will be followed by sensations of nausea and that certain sounds will be followed by skin pain caused by shock, they have great difficulty forming learned associations between taste and consequent skin pain or between sound and subsequent nausea. After Garcia et al. [418].

Why is it that white rats are so adept at learning to avoid novel foods with distinctive tastes that are associated with illness, even hours after ingesting the food? The white rat is a domesticated laboratory variant of the wild Norway rat. Under natural conditions, a Norway rat becomes completely familiar with the area around its burrow, foraging within that area for a wide variety of foods, plant and animal [719]. New plants and insects are constantly coming into season and then disappearing. Some of these organisms are edible and nutritious; others are toxic and potentially lethal. A rat cannot clear its digestive system of toxic foods by vomiting. Instead, the animal takes only a small bite of anything new. If it gets sick later, it *should* avoid this food or liquid in the future because eating large amounts might kill it [418]. This case demonstrates that even what appears to be a general, all-purpose form of learning is actually a specialized response to particular kinds of associations that occur in nature.

The Evolution of Developmental Homeostasis

Our analysis of the specialized nature of animal learning is based on the idea that learning abilities in particular—and more generally, all forms of developmental flexibility—can evolve only when the reproductive costs associated with those abilities, such as the energetic expenses of building the underlying mechanisms, are outweighed by the reproductive benefits of flexibility in particular environments. The trick is to determine what kinds of environments favor developmental adaptability and what kinds do not, if we are to explain why so many behaviors are not especially flexible. Note, for example, that white-crowned sparrows almost always develop a fully functional species-specific song, and almost all honey bee workers begin life as nurses rather than foragers, while tiger salamanders regularly develop into cannibals *or* the typical form, rather than something halfway in between.

One possible reason why the developmental process tends to generate certain outcomes far more often than others may stem from the ability of the process to compensate for many potential genetic and environmental deficits. Take, for example, the reaction to an alarm call by baby Belding's ground squirrels, which usually stop what they are doing to look around when they hear the call. The development of this behavior might be expected to be dependent upon the babies being able to interact with their mothers in this highly maternal species. Yet ground squirrels housed in an enclosure during the day without their mothers react with increased vigilance to taped alarm calls nearly as often as youngsters who have the benefit of living with their mothers in the same kind of enclosure [767]. Likewise, baby rats that were separated from their mother and siblings

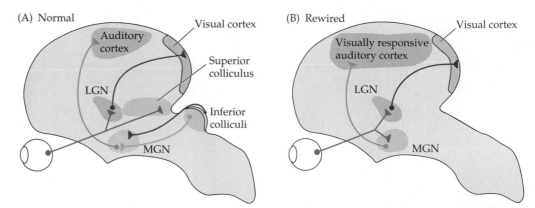

30   Rewiring the ferret brain. (A) The optic nerve normally runs from the retina of the eye to connect with the lateral geniculate nucleus (LGN) and superior colliculus; the LGN then sends messages on to the visual cortex of the brain. (B) In some experimental subjects, researchers were able to reroute the optic nerve to the medial geniculate nucleus (MGN), which normally receives auditory inputs from the inferior colliculi (these connections had been severed in the experimental animals). In the rewired animals, the now visually activated MGN sent signals to a part of the brain that serves as auditory cortex in unaltered ferrets, providing this structure with visually derived information rather than auditory information. After Sur et al. [1111].

when less than 2 days old and fed artificially for 3 weeks still approach scents from the anal excreta of female rats, just as naturally reared rat pups do when temporarily separated from their mothers [415]. Male crickets that live isolated from their fellows sing a normal species-specific song despite their severely restricted social and acoustical environment [91]. Captive hand-reared female cowbirds that have never heard a male cowbird sing nevertheless adopt the appropriate precopulatory pose when they hear cowbird song for the first time, if they have mature eggs to be fertilized [623].

The ability of animals to acquire the correct proximate foundation for normal behavior, even under suboptimal conditions, has been called **developmental homeostasis.** This phenomenon has been demonstrated experimentally in several ways. For example, a team of neurobiologists managed to manipulate the developing optic nerve of a newborn small mammal, the ferret, so that the nerve split and grew toward two destinations, the visual cortex (its customary contact region, which processes visual information) and the auditory cortex (Figure 30). The auditory cortex typically makes sense of sounds, but it could not form its usual connection to the auditory nerve from the ferret's ears in those animals manipulated by the researchers. The question was, what perceptions would the auditory cortex provide in response to sensory signals from the optic nerve? The researchers found that the rewired ferret used its auditory cortex for visual analysis, not for hearing [1040].

On the one hand, this result is stunning testimony to the plasticity of behavioral development and its potential sensitivity to experience. Cells that ordinarily would have been employed in the perception of sounds were able to change their apparent destiny, becoming part of the visual system of the ferret. The genes in the cells of the presumptive auditory portion of the brain clearly responded to visual messages relayed via the optic nerve and subsequently developed along a much different pathway than if they had been exposed to signals from the auditory nerve. Both the structure and the function of the "auditory" cortex in the experimental subjects differed greatly from the typical pattern [1040]. However, the study also paradoxically revealed developmental homeostasis at work. Brain cells that would normally never receive a flow of visual information from the optic nerve nevertheless had the capacity to respond to that information *in a functional manner,* forming circuitry that provided the ferret with visual images, not chaotic gobbledygook. The ferret's developmental mechanisms had the ability to overcome a highly abnormal experimental environment while generating a brain system capable of carrying out an important task, the visual analysis of optic nerve signals.

The same ability to cope effectively with an extremely unusual environment is evident in the development of the social behavior of young rhesus monkeys reared in extraordinary social deprivation by Margaret and Harry Harlow [493, 494]. (The Harlows' experiments were conducted in an era when animal rights were not the issue that they are today; readers can decide for themselves whether their harsh treatment of infant monkeys yielded information of sufficient importance to justify the research.) In one such study, the Harlows separated a young rhesus from its mother shortly after birth. The baby was placed in a cage with an artificial mother (Figure 31), which might be a wire cylinder or a terry cloth figure with a nursing bottle. The baby rhesus gained weight normally and developed physically in the same way that nonisolated rhesus infants do. However, it soon began to spend its days crouched in a corner, rocking back and forth, biting itself. If confronted with a strange object or another monkey, the isolated baby withdrew in apparent terror.

The isolation experiment demonstrated that a young rhesus needs social experience to develop normal social behavior. But what kind of social experience—and how much—is necessary? Interactions with a mother are insufficient for full social development of rhesus monkeys, since infants reared alone with their

31 **Surrogate mothers used in social deprivation experiments.** This isolated rhesus infant was reared with wire cylinder and terry cloth dummies as substitutes for its mother. Photograph courtesy of Harry Harlow.

32   Socially isolated rhesus infants that are permitted to interact with one another for short periods each day at first cling to each other during the contact period. Photograph courtesy of Harry Harlow.

mothers failed to develop truly normal sexual, play, and aggressive behavior. Perhaps normal social development in rhesus monkeys requires the young animals to interact with one another. To test this hypothesis, the Harlows isolated some infants from their mothers, but gave these infants a chance to interact with three other such infants for just *15 minutes* each day [493]. At first, the young rhesus monkeys simply clung to one another (Figure 32), but later they began to play. In their natural habitat, rhesus babies start to play when they are about 1 month old, and by 6 months they spend practically every waking moment in the company of their peers. Even so, the 15-minute play group developed nearly normal social behavior. As adolescents and adults, they were capable of interacting sexually and socially with other rhesus monkeys without exhibiting the intense aggression or withdrawal of animals that had no social contacts as infants.

The Adaptive Value of Developmental Homeostasis

Studies of rhesus monkeys have provided important information on the resilience of the developmental mechanisms of that primate. Naturally one wonders about the relevance of these studies for another primate, human beings, whose intellectual development is often said to be dependent upon the early experiences that children have with their parents and peers. But is this true? Consider, for example, the results of a study of Dutch teenagers who were born or conceived at a time when their mothers were being starved as a result of the Nazi transport embargo during the winter of 1944–1945, which prevented food from reaching the larger Dutch cities during this time [1096]. Deaths from starvation were common, and for most of the famine period, the average caloric intake was about 750 calories per day. As a result, women living in cities under famine conditions produced babies of very low birth weights. In contrast, rural women were less dependent on food transported to them, and their babies weighed much more than urban infants born or conceived at the same time.

(A)

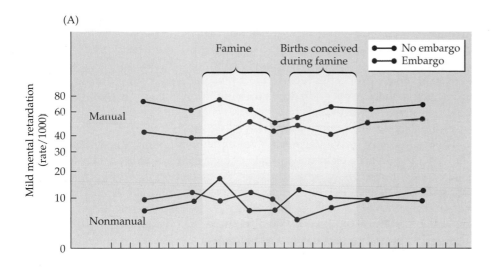

(B)

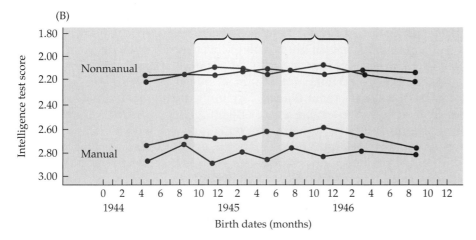

33   Developmental homeostasis in humans. Maternal starvation has surprisingly few effects on intellectual development in humans, judging from a study on (A) rates of mild mental retardation and (B) intelligence test scores among 19-year-old Dutch men whose mothers lived under Nazi occupation while pregnant. (The lower the intelligence test score, the greater the intelligence of the subject.) The subjects were grouped according to the occupations of their fathers (Manual or Nonmanual) and whether their mothers lived or gave birth to them in a city subjected to food embargo by the Nazis (Embargo) or in a rural area unaffected by the embargo (No embargo). Individuals who were conceived or born under famine conditions exhibited the same rates of mental retardation and the same level of test scores as men conceived by or born to unstarved rural women. After Stein et al. [1096].

One would think that normal brain development absolutely requires adequate nutrition during pregnancy, when much of brain growth occurs. However, Dutch boys who were born in urban famine areas did not exhibit a higher incidence of mental retardation at age 19 than rural boys whose early nutrition was far superior (Figure 33). Nor did those boys born to food-deprived mothers score more poorly than their relatively well nourished rural counterparts when they took the Dutch intelligence test administered to draft-age men. Children born at low weights apparently suffered no permanent intellectual damage [1096]. These results are buttressed by the discovery that the mortality rates of Finnish adults who experienced severe nutritional shortfalls in utero (during a 19th century famine) lived just as long on average as those who were born after the famine was over [599].

No one believes that pregnant women should be deprived of food [819], and some rare forms of mental illness were more common in Dutch children whose mothers had been starved [1113]. However, the development of the fetus and young person evidently can proceed successfully even under some highly adverse conditions, perhaps because our developmental systems evolved in past environments in which episodes of starvation were not terribly uncommon. The occurrence of normal development in stressful environments testifies to the adaptively guided, structured nature of the process, which can be thrown off course only by extremely unusual environmental shortfalls or severe genetic deficits. Given this fact, it is not surprising that many gene knockout experiments (see p. 61) have resulted in little or no effect on behavioral development, probably because many genetic systems have built-in redundancy as well as the capacity to compensate for shortfalls in the production of key products [617, 914]. These properties reduce the chance of devastating developmental disruption in the face of certain transitory environmental deficits and defective genes. As a result, individuals have a better chance of acquiring the characteristics associated with reproductive success.

The reproductive advantages of developmental homeostasis would be particularly great in species in which mate choice is based on body symmetry. In the barn swallow, for example, females have been reported to prefer males whose long outer tail feathers are the same length [807], while females of the Iberian rock lizard associate preferentially with males endowed with a symmetrical distribution of pheromone-releasing pores on their thighs [762]. As for humans, some researchers have reported that both men and women find symmetry in facial features appealing (Figure 34) [557, 967]. Perhaps prospective mates in these and other species respond positively to body or facial symmetry because these attributes announce the individual's capacity to overcome challenges to normal development [810]. Disruptions to development caused by mutations or by an inability to secure critical material resources early in life are thought to generate asymmetries in appearance. If body asymmetry reflects suboptimal development of internal mechanisms as well, then a preference for symmetrical traits (or attributes closely allied with them) could enable the selective individual to acquire a partner with "good genes" to be transferred to their offspring. Alternatively, the benefit to the choosy individual in a parental species could be the acquisition of a mate in excellent physiological condition who could provide superior care for its young. In keeping with this prediction, the faces of human females with a history of health problems are judged less attractive, and the same is true of the faces of men who grew up in households with relatively little income [557].

However, debate exists on all aspects of this scenario. Although, as noted, some researchers report that asymmetrical individuals have indeed experienced developmental difficulties and deficits [55, 808], other researchers disagree [110, 340]. Moreover, although symmetrical individuals apparently enjoy a mating advantage in some species (see above), in other species, no such advantage has been observed [698, 1147] (see also [881]). Finally, in some species in which a mate preference for symmetry has been reported, the differences between preferred symmetrical and rejected asymmetrical individuals are often so slight that it seems unlikely that the degree of symmetry per se provides the basis for making the choice. Starlings, for example, have been shown to be simply incapable of perceiving the kinds of very small differences that characterize most naturally occurring body asymmetries in their species [1115]. Furthermore, when women are asked to rate photographs of men's faces in terms of their attractiveness, their rankings do correlate with male facial symmetry, but the same rankings emerge when the women are provided with photographs of only the left or right side of the face, thereby eliminating information about facial symmetry. These results suggest that facial symmetry must correlate with some other feature that women actually use to make their judgments [1021]. Likewise, some female scorpionflies and crickets choose symmetrical males without even looking at them. Instead, these animals select mates on the basis of pheromonal or acoustical signals that happen to be correlated with body symmetry [1061, 1131].

In a few species, however, all the criteria for visual mate choice based on body symmetry have been documented. In the brush-legged wolf spider, for example, males wave their hairy tufted forelegs at females during courtship. Males vary in the degree to which the tufts on the right and left foreleg are the same size; those spiders with unevenly sized (asymmetrical) tufts tend to be smaller

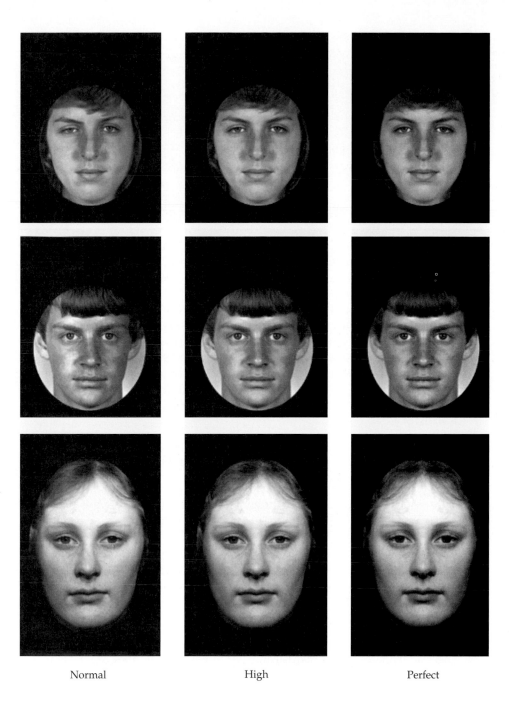

| Normal | High | Perfect |

34 Facial symmetry and attractiveness. When asked to rate these faces, most persons find the image on the far right to be most attractive. From Rhodes et al. [967].

and in poorer condition than those whose tufts are symmetrical. To determine whether female wolf spiders of this species attend to tuft symmetry, George Uetz and Elizabeth Smith took advantage of the willingness of wolf spiders to respond to video images played on a tiny Sony Watchman microtelevision (Figure 35). When a female wolf spider sees a video of a courting male and finds him acceptable, she will respond by adopting a precopulatory pose, which generally involves lowering the front of her body and elevating the abdomen. Uetz and Smith recorded the responses of females to digitized videotapes of a courting spider that were identical in every respect except for the degree to which the male's foreleg tufts were symmetrical. The control tape showed a male with a high degree of tuft symmetry; the experimental videos showed the same images in which one tuft had been digitally reduced or enlarged, creating

35   Testing mate choice in a female wolf spider. The female (on the floor of the arena, to the left), responds to a moving image of a displaying male on the screen of a tiny television (to the right of the female). Photograph by George Uetz.

asymmetrical forelegs. Female spiders found the video of the control male far more sexually stimulating than the experimentally asymmetrical images [1162]. At least in species of this sort, developmental homeostasis seems highly likely to confer a reproductive advantage by virtue of increasing the odds that individuals will be able to attract mates and leave descendants.

Summary

1. The development of any trait is the result of an interaction between the genotype of a developing organism and its environment, which consists not only of the material resources it receives and the metabolic products, especially hormones, produced by its cells (the material environment), but also its sensory experiences (the experiential environment). Any of these factors can act in ways that have long-term developmental and behavioral consequences, in part by affecting which genes are activated within the cells of the developing individual.

2. Various kinds of developmental mechanisms confer behavioral flexibility on individuals. In some cases, cues in the environment activate alternative developmental pathways leading to different structural and behavioral phenotypes. Learning mechanisms are fundamentally similar, with individuals using certain kinds of experience to modify their neural systems and, thus, their behavior. The flexibility to make behavioral adjustments provides fitness advantages for individuals in environments where they are likely to encounter somewhat unpredictable, but biologically important, conditions during their lifetimes.

3. The developmental process often ignores or overcomes certain environmental (and genetic) shortfalls that might conceivably disrupt the production of certain traits. This developmental homeostasis is seen in the development of normal physiological and behavioral traits in animals growing up in challenging, less than optimal environments. The resilience of the developmental process may help individuals develop key characteristics, such as symmetrical body features, that promote or are correlated with reproductive success.

Discussion Questions

1. "Some researchers suspect that male homosexuality in humans arises in part from exposure of the fetus to relatively high levels of male sex hormones. But others refuse to rule out the possibility that childhood interactions with a domi-

neering mother or weak father have some effect on the later expression of the trait. If there is a connection between fetal hormone exposure and homosexuality, then androgen levels in utero should increase over successive pregnancies in those women who have a series of sons, given that male homosexuality occurs more often in men who have relatively many older brothers. It is also known that women are more likely to be lesbians if they exhibit the masculine pattern of finger lengths (in which the index finger is typically somewhat shorter than the ring finger), suggesting that their homosexuality also arises from early exposure to masculinizing hormones. Finally, some recent studies have uncovered no difference in the nature of mother–son relationships for homosexual and heterosexual men." In this make-believe press report, based partly on an actual study [1242], which sentences constitute hypotheses, predictions, tests, or conclusions (if any)?

2. Recall the case of the cichlid fish whose males can adopt two different reproductive roles, thanks to developmentally flexible hypothalamic neurons. Is there any role for developmental homeostasis in this system? Likewise, review the case of the honey bee and its mushroom bodies for evidence of developmental homeostasis.

3. Design an experiment to test whether body symmetry is an important factor in mate choice in the zebra finch. Take advantage of the fact that you can place colored plastic bands of various sorts on the legs of captive finches. Formally present your causal question, hypothesis, and prediction(s). Secure the data you need for your test from [1116]. What scientific conclusion is justified based on the test evidence?

4. Juveniles of various species of salmon and other related fish generally spend their early years in the freshwater river where they hatched. At some point, the juveniles then move out into the ocean, only to return some years later to breed in the very river from which they came. However, overfishing and dam building have reduced many stocks of salmonid fishes. One possible way to boost breeding populations of these endangered fishes is to rear the young in a hatchery, increasing their chances of survival during the period when they are highly vulnerable to predators. Then the larger juveniles can be released into the river their parents came from to join the cohort of naturally reared fish in the hope that eventually some will migrate to sea and then return to increase the size of the breeding stock. What sorts of questions would be raised about this conservation approach by persons knowledgeable about behavioral development?

5. In a study in which men and women were asked to sit at a computer and navigate through a virtual maze (see the figure on the right), the men were able to complete the task more quickly and with fewer errors over five trials than the women [802]. (Let me add that in other tests involving language skills, women score higher on average than men.) What possible proximate developmental mechanisms might be responsible for the sex differences in navigational ability? Keeping in mind the evolutionary explanation for sex differences in spatial learning ability in voles, what prediction can you make about the nature of human mating systems over evolutionary time?

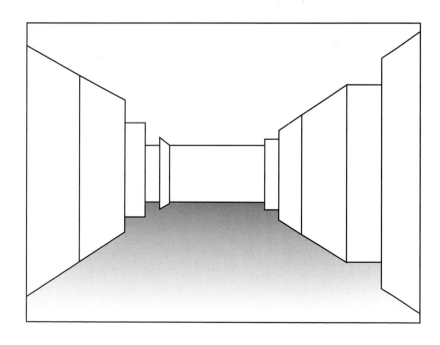

Suggested Reading

Gene Robinson discusses the integration of molecular biology, developmental biology, neurobiology, and evolutionary biology with the goal of understanding both proximate and ultimate causes of social behavior [973, 974]. David Pfennig and Paul Sherman review the fascinating subject of kin recognition [912]. Russell Fernald has written an excellent article on the environmental basis of neural changes in a cichlid fish [382]. The study of body symmetry and its detection by mate-choosing animals is full of interesting challenges [810].

particular sensory inputs processed, or, what neural circuits provide the commands sent to a set of muscles?

Both microanalyses and macroanalyses have helped build our current understanding of such things as how moths flying at night can detect and avoid hungry bats rushing in for the kill, how the star-nosed mole uses its amazing nose to locate tasty worms in its underground tunnels, and how homing pigeons can return directly to their loft after they have been dropped off 100 kilometers from home in a strange place. As we look at these and other findings, we will see that, despite the differences in neural machinery that help explain why different animal species behave differently, the neurons and nervous systems of all animals share two key properties in common. At the proximate level, the elements within nervous systems exhibit great specificity of action. Certain sensory receptors, for example, specialize in the detection of specific kinds of stimuli, even though this means that they fail to report a lot of other potential information to the brain. At the ultimate level, the special features of information-gathering systems and selective responses to sensory inputs help individuals of a given species overcome special obstacles to reproductive success. The close match between the properties of animal nervous systems and the demands placed on species by their environments demonstrates that selection has shaped the proximate mechanisms of behavior.

How Neurons Control Behavior

When animals behave, they often provide clues about the decision-making properties of their neurons and nervous systems. I study the behavior of bees other than the honey bee, with a special research interest in whether body size influences a male's mating chances. Therefore, I have had more opportunities than the average person to capture and separate mating pairs of bees, which I do in order to measure the male's head width, a close correlate of male body weight. When I was working with a bee called *Centris pallida,* I found that if, after finishing my measurements, I placed the male whose copulation I had so rudely interrupted on my upturned thumb, he would grasp it firmly and stroke it with his legs and antennae as if he were holding a female of his species. (Incidentally, male bees don't sting, so my experiment required no courage on my part.) Even though my thumb has only a slight similarity, at best, to a female *Centris pallida,* some males made repeated attempts to copulate with it before giving up and flying away (Figure 1, left).

1 Complex responses to simple stimuli. Male bee (left) attempting (unsuccessfully) to copulate with the author's thumb. An Australian beetle (right) with everted penis probing a discarded beer bottle, which has some color pattern similarities to the wing covers of the female beetle. Left, photograph by the author; right, photograph courtesy of Darryl T. Gwynne.

It was evident to me, even though I am not a neurophysiologist, that the bee's nervous system was designed with some special operating rules. Apparently, when a sexually motivated male bee grasps almost any object approximately the size of a female of its species, the sensory signals generated by its tactile receptors are sent to other parts of its nervous system, where cells activate a complex series of muscle commands. The behavioral result is the sequence of movements that constitute courtship in *Centris pallida*. That these activities can be stimulated by my thumb instead of a female bee indicates that the nervous system of a male *Centris pallida* processes and reacts to simple key cues, rather than performing a complete sensory analysis of a complex object before initiating a response.

Having read the work of the pioneering ethologists Niko Tinbergen and Konrad Lorenz, I was amused, but not surprised, by the obtuse behavior of my subjects, which by no means are the only male animals to attempt to copulate with something other than a female of their species (Figure 1, right). Tinbergen and Lorenz described many cases in which animals responded with an elaborate behavior pattern to stimuli that barely resembled the naturally occurring object that normally triggers the behavior. A classic study of this sort involved the begging behavior of newly hatched herring gull chicks. Tinbergen knew that begging gull chicks peck at the red dot toward the end of their parent's bill (Figure 2), which causes the adult to regurgitate a half-digested fish or some other mouth-watering morsel for its offspring [1141]. But he found that two-dimensional cardboard models of gull heads, and even painted sticks with bands on the end, also stimulated begging behavior in newborn herring gulls (Figure 3) [1141]. From these experiments, Tinbergen deduced that when a young gull looks at its parent's beak, it attends to a few simple cues, which activate sensory signals that are relayed to its brain. Within the brain, other neurons generate a set of motor commands that cause the chick to peck at the effective stimulus— whether it is located on its mother's beak or a piece of cardboard.

Tinbergen and Lorenz collaborated on another famous experiment that identified a simple stimulus capable of triggering a complex behavior. They found that if they removed an egg from under an incubating greylag goose and put it a half meter away, the goose would retrieve the egg by stretching its neck forward, tucking the egg under its lower bill, and rolling the egg carefully back into its nest. If they replaced the goose egg with almost any roughly egg-shaped

2 An instinct. A young gull pecks at its parent's beak, an action that may induce the parent to regurgitate food for the youngster. Photograph by Frederick Atwood.

3 Effectiveness of different visual stimuli in triggering begging behavior of young herring gull chicks. Two-dimensional cardboard cutouts of the head of a gull with a red dot on its beak (A) are no more effective in eliciting begging behavior in a gull chick than models of the beak alone (B), provided the red dot is present. Moreover, a model of a gull head without the red dot (C) is a far less effective stimulus than is an unrealistically long "beak" with contrasting bars at the end (D). After Tinbergen and Perdeck [1141].

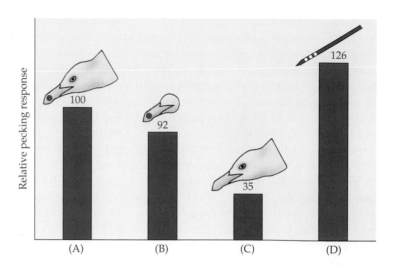

object, the goose would invariably activate its egg retrieval response. And if the researchers removed the object as it was being retrieved, the bird would continue pulling its head back just as if an egg were still balanced against the underside of its bill [1136]. From these results, Tinbergen and Lorenz concluded that the goose must have a special perceptual mechanism that reacts to certain visual cues associated with eggs. Moreover, that perceptual mechanism must send its input to an associated neural mechanism that automatically activates an invariant motor program for egg retrieval.

The gull chick's pecking response and the greylag goose's retrieval behavior are only two of dozens of instincts that Tinbergen, Lorenz, and other ethologists have studied. An **instinct** can be defined as a behavior pattern that appears in fully functional form the *first time* it is performed, even though the animal may have had no previous experience with the cues that elicit the behavior. Tinbergen and company labeled such instinctive responses **fixed action patterns,** or **FAPs.** They called the key component of the object that activates an FAP a **sign stimulus,** or a **releaser** (if the sign stimulus was a signal from one individual to another) (Figure 4). The hypothetical neural system responsible for triggering an FAP in response to the appropriate sensory input was called an **innate releasing mechanism** (Figure 5).

The simple relationship between sign stimulus or releaser, innate releasing mechanism, and FAP is highlighted by the ability of some species to exploit the FAPs of other species, a tactic known as code-breaking [1228]. We have already discussed the orchids whose flower petals provide the visual and olfactory releasers that elicit approach and mating behavior by certain male wasps (see Figure 21, p. 96). Similarly, when a sexually aroused male desert bee (not *Centris pallida* in this case) pounces upon clusters of the tiny larvae of a blister beetle, the bee is almost certainly responding to a mimetic sex pheromone produced by the beetles [473]. Some of the deceptive larvae grasp the bee and are eventually transferred to female bees that copulate with the parasite-burdened male. The transferred larvae may then go on to feed on the food that female bees store in underground tunnels for their offspring.

Yet another code breaker is a rove beetle that lays its eggs in the nests of the ant *Formica polyctena* [535]. The beetle larvae produce an attractant pheromone that causes ant workers to carry them into their brood chambers, where the grubs feast on ant eggs and ant larvae. Not content with this larder, the parasites also mimic the begging behavior of ant larvae by tapping a worker ant's mandibles with their own mouthparts. This action is a releaser that triggers

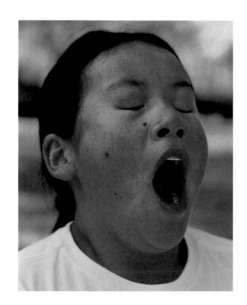

4 A releaser. A yawn triggers yawning in other humans that observe the yawner. Photograph by Christopher Small.

regurgitation of liquid food (an FAP) by the ant. The larvae eventually metamorphose into adult beetles, which can still elicit food transfers from adult worker ants by producing the right releasers for this situation (Figure 6).

Likewise, the chicks of avian brood parasites, such as the European cuckoo and the North American cowbird, take advantage of a simple rule of thumb that parent birds use to make decisions about which nestlings get fed [487, 1228]. When an adult songbird returns to its brood with food, it favors cheeping youngsters that are able to reach up high, with head moving and mouth gaping. Cuckoo and cowbird nestlings grow rapidly and become larger than their hosts' own offspring, and therefore can generate these releasers of parental feeding better than their smaller nestmates [695]. Because the parasite begs for food very effectively, it gets more than its fair share, eventually growing into a demanding youngster larger than its foster parents (Figure 7).

Sensory Receptors and Survival

Although researchers today rarely employ ethological terms such as fixed action pattern or innate releasing mechanism, many behavioral biologists have studied well-defined responses to simple stimuli in an effort to figure out how certain neurons contribute to an adaptive behavior. Consider the classic work of Kenneth Roeder on the ability of night-flying moths to evade predatory bats, something that you too can observe on warm evenings in temperate North America if you happen to possess "a minimum amount of illumination, perhaps a 100-watt bulb with a reflector, and a fair amount of patience and mosquito repellent" [976]. With all these items in place, you may see a bat burst into view and swoop down upon a moth attracted to the light. But you may also sometimes see a moth turn abruptly or dive straight down just before a bat shows up, evidence that at least some moths are able to detect and avoid these predators. Moreover, you can sometimes trigger an abrupt turn or precipitous dive by a flying moth simply by jangling a set of keys as it flies into view. This response suggests that the moth is detecting a simple acoustical cue that provides the trigger for a particular behavior, just as simple visual cues are sufficient to activate begging behavior in a baby gull.

As it turns out, the hypothesis that acoustical signals trigger the turning or diving behavior of moths is correct, but the sounds that the moth hears when the keys are jangled together are not the sounds that you and I hear. Instead, the moth detects the ultrasonic stimuli produced by the clashing keys, which makes sense when you consider that hunting bats produce ultrasonic vocalizations with sound frequencies between 20 and 80 kilohertz (kHz)—well outside the hearing range of humans, but not moths.

Bats use ultrasonic calls to navigate at night, something that was not suspected until the 1930s, when researchers with ultrasound detectors were able

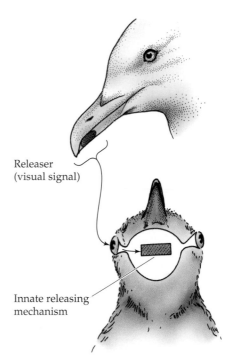

Releaser (visual signal)

Innate releasing mechanism

5 Releasers, innate releasing mechanisms, and fixed action patterns. The red dot on a parent gull's bill is a releaser of the gull chick's begging behavior. The chick's nervous system has an innate releasing mechanism that responds to the releaser with a fixed action pattern—its begging response.

6 A code breaker. This rove beetle mimics the begging signals produced by its hosts by tapping a worker ant with its antennae before touching the ant's mouthparts with its forelegs. These actions constitute a releaser of regurgitation by the ant, which provides food to the parasitic beetle. Adapted from a drawing by Turid Forsyth.

7 An exploiter of instinct. A young cuckoo begs for food from its foster parent, a reed warbler, which provides for the cuckoo at great cost to itself and its own offspring. Photograph by Ian Wyllie.

to eavesdrop on the steady stream of high-frequency sounds produced by flying bats. Donald Griffin suggested that the bats made these sounds in order to listen for weak echoes reflected back from objects in their flight path [459]. This echolocation hypothesis for bat navigation initially had many skeptics, but they were convinced when they read about Griffin's experiments with some captive little brown bats, a common North American species. When Griffin placed the bats in a dark room filled with fruit flies and wires strung from ceiling to floor, his subjects had no trouble catching the insects while negotiating the obstacle course—until Griffin turned on a machine that filled the room with high-frequency sound. As soon as the machine-produced ultrasound bombarded the bats, they began to collide with obstacles and crash to the floor, where they remained until Griffin turned off the jamming device. In contrast, loud sounds of 1–15 kHz (which humans can hear) had no effect on the bats because these stimuli did not mask the high-frequency echoes they listen to as they fly in the dark. Griffin rightly concluded that the little brown bat employs a sonar system to avoid obstacles and detect prey at night.

As Roeder watched moths evade echolocating bats, he guessed that the insects might be able to hear pulses of bat ultrasound. If he was right, he knew he should be able to find ears somewhere on some moths—and he did (Figure 8). A noctuid moth has two ears, one on each side of the thorax. Each ear consists of a thin, flexible sheet of cuticle—the tympanic membrane, or tympanum—lying over a chamber on the side of the thorax. The tympanum is attached to two neurons, the A1 and A2 sensory receptors. These receptor cells are deformed when the tympanum vibrates, which it does when intense sound pressure waves sweep over the moth's body. Roeder decided to focus his attention on these sensory receptors, using a cellular approach to get at the proximate basis of moth behavior.

The moth's A1 and A2 receptors work much the same way that most neurons do: they respond to the energy contained in selected stimuli by changing the permeability of their membranes to sodium ions. The effective stimuli for a moth's acoustical receptors are provided by the movement of the tympanum, which mechanically stimulates the receptor cell, opening stretch-sensitive chan-

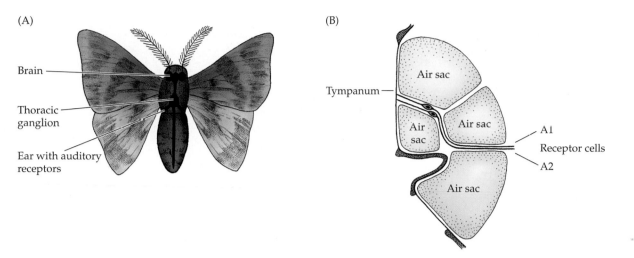

(A)

Brain

Thoracic ganglion

Ear with auditory receptors

(B)

Tympanum

Air sac

Air sac

Air sac

A1

Receptor cells

A2

Air sac

8   Noctuid moth ears. (A) The location of the ear. (B) The design of the ear, which features two sensory receptors (A1, A2) linked to a tympanum that vibrates when exposed to sounds. After Roeder [976].

nels in its membrane. As positively charged sodium ions flow in, they change the electrical charge inside the cell relative to the charge on the other side of the membrane. If the inward movement of ions is sufficiently great, a substantial, abrupt, local change in the electrical charge difference across the membrane may occur and spread to neighboring portions of the membrane, sweeping around the cell body and down the axon, the "transmission line" of the cell (Figure 9). This brief, all-or-nothing change in electrical charge, called an **action potential,** is the signal that one neuron uses to communicate with another.

When an action potential arrives at the end of an axon, it may cause the release of a neurotransmitter at this point. This chemical signal diffuses across the narrow gap, or **synapse,** separating the axon tip of one cell from the body of the next cell in the network. Neurotransmitters can affect the membrane permeability of the next link in a chain of cells in ways that increase or decrease the probability that this neuron will produce its own action potential(s). If a neuron fires in response to stimulation provided by the preceding cell in the network, the message may be relayed on to the next cell, and on and on, sometimes producing a chain reaction in which action potentials from distant receptors have excitatory (or inhibitory) effects that reach deep into the nervous system. Neurons far from the source of the initial action potentials can be stimulated by volleys of relayed signals. They may continue the chain reaction by providing a message in the form of patterned action potential outputs that reach the animal's muscles and cause them to contract.

In the case of the noctuid moth studied by Roeder, the A1 and A2 receptor cells pass on information via relay cells called **interneurons,** whose action potentials can change the activity of other cells in a large cluster of neurons called the thoracic ganglion and in the brain (Figure 10). As messages flow through these parts of the nervous system, certain patterns of action potentials fired by cells in the thoracic ganglion trigger other interneurons, whose action potentials in turn reach motor neurons that are connected with the wing muscles of the moth. When a motor neuron fires, the neurotransmitter it releases at the synapse with a muscle fiber changes the membrane permeability of the muscle cell. These changes initiate the contraction or relaxation of the muscle, which drives the wings and thereby affects the moth's movements.

9 Neurons and their operation. The diagram illustrates the structure of a generalized neuron with its dendrites, cell body, axon, and synapses. Electrical activity in a neuron depends first on the effect of certain stimuli on the dendrites. Electrical changes in a dendrite's membrane can, if sufficiently great, trigger an action potential that begins near the cell body and travels along the axon toward the next cell in the network.

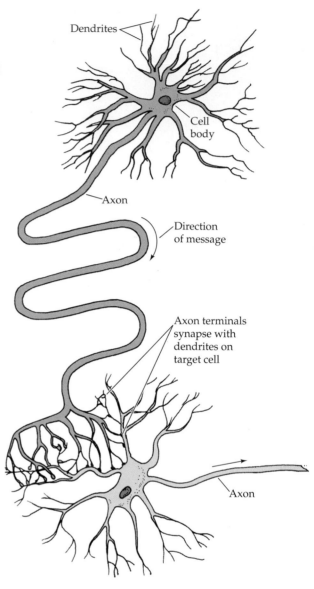

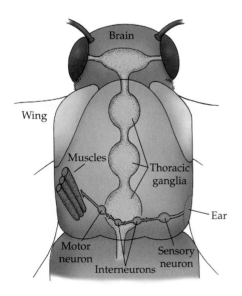

10 Neural network of a moth. Receptors in the ear relay information to interneurons in the thoracic ganglion, which communicate with motor neurons that control the wing muscles.

Thus, the moth's behavior, like that of any animal, is the product of an integrated series of chemical and biophysical changes in a network of cells. Because these changes occur with remarkable rapidity, an animal can react to stimuli in its environment in fractions of a second. An American cockroach, for example, can begin to move away from approaching danger, such as a hungry toad lunging toward it or a flyswatter wielded by a cockroach-loathing city dweller, in as little as a hundredth of a second after the air pushed in front of the toad's head or the descending flyswatter reaches the roach's body [191].

Although the neurons of many animals share similar features, the acoustical receptors of noctuid moths perform very special sensory tasks. The moth's A1 and A2 receptors are specialists, finely tuned to collect information about ultrasound, as Roeder determined by attaching recording electrodes to these receptors in living, but restrained, moths [976]. When he projected a variety of sounds at the moths, the electrical response of the receptors was relayed to an oscilloscope, which produced a visible record. These recordings revealed the following features (Figure 11):

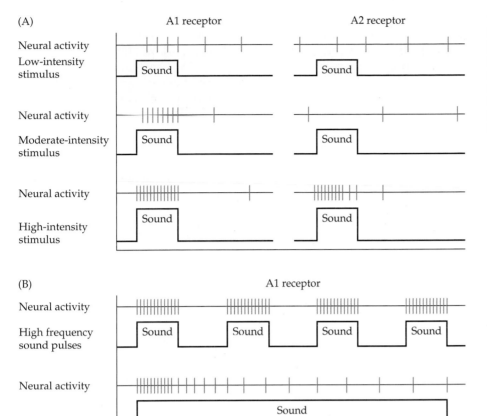

11 Properties of the auditory receptors of a noctuid moth. (A) Sounds of low or moderate intensity do not generate action potentials in the A2 receptor. The A1 receptor fires sooner and more often as sound intensity increases. (B) The A1 receptor reacts strongly to pulses of sound, but ceases to fire after a short time if the stimulus is a constant sound.

1. The A1 receptor is sensitive to ultrasounds of low to moderate intensity, whereas the A2 receptor begins to produce action potentials only when an ultrasound is loud.

2. As a sound increases in intensity, the A1 receptor fires more often, and with a shorter delay between the arrival of the stimulus at the tympanum and the onset of the first action potential.

3. The A1 receptor fires much more frequently to pulses of sound than to steady, uninterrupted sounds.

4. Neither receptor responds differently to sounds of different frequencies over a broad ultrasonic range. Thus, a burst of 30 kHz sound elicits much the same pattern of firing as an equally intense sound of 50 kHz.

5. The receptor cells do not respond at all to low-frequency sounds, which means that moths are deaf to stimuli that we can easily hear. Of course, we are deaf to sounds that moths have no trouble hearing.

Although the moth's ears have just two receptors each, the amount of information they can provide to the moth's nervous system about echolocating bats is impressive. The key property of the A1 receptor is its great sensitivity to pulses of ultrasound. It is so sensitive that the cries of a little brown bat generate action potentials when the predator is 30 meters away, long before the bat can detect the moth. Because the rate of firing in the A1 cell is proportional to the loudness of the sound, the insect has a system for determining whether a bat is flying toward it.

In addition, the moth's ears gather information that it can use to locate the bat in space. If a hunting bat is on the moth's left, for example, the A1 receptor in its left ear will be stimulated a fraction of a second sooner and somewhat more strongly than the A1 receptor in its right ear, which is shielded from the sound by the moth's body. As a result, the left receptor will fire sooner and more often than the right receptor. The moth's nervous system could also detect whether a bat is above it or below it. If the predator is higher than the moth, then with every up-and-down movement of the insect's wings, there will be a corresponding fluctuation in the rate of firing of the A1 receptors as they are exposed to, then shielded from, bat cries by the wings. If the bat is lower than the moth, there will be no such fluctuation in neural activity (Figure 12).

As neural signals initiated by the receptors race through the moth's nervous system, they may ultimately generate motor messages that cause the moth to turn and fly directly away from the source of ultrasonic stimuli [978]. When a moth is moving away from a bat, it exposes less echo-reflecting area than if it were presenting the full surface of its wings to the bat's vocalizations. If a bat receives no echoes from its calls, it cannot detect prey. Bats rarely fly in a straight line for long, and therefore the odds are good that a moth will remain undetected if it can stay

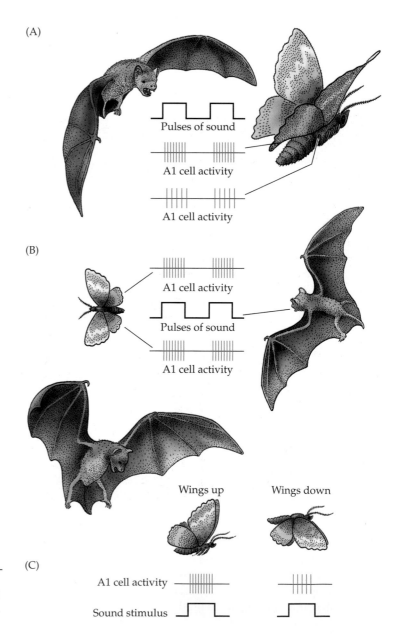

12 How moths might locate bats in space. (A) When a bat is to one side of the moth, the A1 receptor on the side closer to the predator fires sooner and more often than the shielded A1 receptor in the other ear. (B) When a bat is directly behind the moth, both A1 receptors fire at the same rate and time. (C) When a bat is above the moth, activity in the A1 receptors fluctuates in synchrony with the moth's wingbeats. (Figures not drawn to scale.)

out of range for a few seconds. By then the bat will have found something else within its 3-meter moth detection range and will have veered off to pursue it.

In order to employ its antidetection response, a moth need only orient itself so as to synchronize the activity of the two A1 receptors. Differences in the rate of firing by the A1 receptors in the two ears are probably monitored by the brain, which relays neural messages to the wing muscles via the thoracic ganglion and allied motor neurons. The resulting changes in muscular action steer the moth away from the side of its body with the ear that is more strongly stimulated. As the moth turns, it will reach a point at which both A1 cells are equally active, at which time it will be facing away from the bat and flying away from danger (see Figure 12).

Although this reaction is effective if the moth has not been detected, it is useless if a speedy bat has come within 3 meters. At this point, a moth has at most a second, and probably less, before the bat reaches it [596]. Therefore, moths in this situation do not try to outrun their enemies, but instead employ drastic evasive maneuvers, including wild loops and power dives, that make it relatively difficult for bats to intercept them. A moth that executes a successful power dive and reaches a bush or grassy spot is safe from further attack because echoes from the leaves or grass at the moth's crash landing site mask those coming from the moth itself [978]. Other nocturnal insects have independently evolved the capacity to sense ultrasound and to take similar kinds of evasive action against approaching bats (Figure 13) [793–795, 1272].

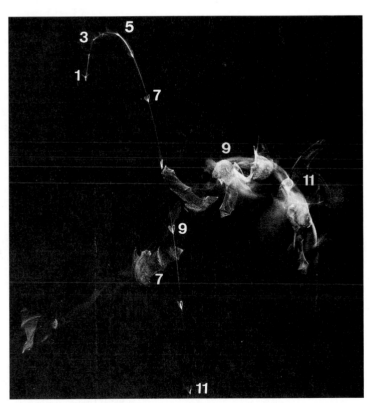

13 **Bat ultrasonic cries trigger evasive behavior** in nocturnal insects. In normal flight, a mantis holds its forelegs close to the body (top left), but when it detects ultrasound, it rapidly extends its forelegs (bottom left), which cause the insect to loop and dive erratically downward. Likewise, a lacewing may employ an anti-interception power dive (right) when approached by a hunting bat. The numbers superimposed on this multiple-exposure photograph show the relative positions of a lacewing and a bat over time. (The lacewing survived.) (Left, top and bottom) After Yager and May [1272]; photographs courtesy of D. D. Yager and M. L. May. (Right) Photograph by Lee Miller.

Roeder speculated that the physiological basis for this erratic escape flight lies in the neural circuitry leading from the A2 receptors to the brain and back to the thoracic ganglion [977]. When a bat is about to collide with a moth, the intensity of the sound waves reaching the insect's ears is high. It is under these conditions that the A2 cells fire. Their messages are relayed to the brain, which in turn may shut down the central steering mechanism that regulates the activity of the flight motor neurons. When the steering mechanism is inhibited, the moth's wings begin beating out of synchrony, irregularly, or not at all. As a result, the insect does not know where it is going—but neither does the pursuing bat, whose inability to plot the path of its prey may permit the insect to escape.

This case study shows how important knowledge of natural history and the real-world problems confronting an animal can be in helping a researcher formulate hypotheses about a species' sensory mechanisms. Kenneth Roeder knew that noctuid moths live in a world filled with echolocating moth killers; he searched for, and found, a specialized proximate mechanism that helps moths cope with these enemies. Likewise, persons interested in bat hearing have searched for, and found, neurons that are specialized for the detection of echoes returning to bats from their prey [858, 930].

Relaying and Responding to Sensory Input

We have focused thus far primarily on how the noctuid moth's sensory receptors collect information about ultrasound. Of course, if the insect is to act on that information, its receptors must be able to forward messages to those parts of its central nervous system that can process acoustical inputs and activate responses. Just how this is accomplished has not been worked out for noctuid moths, but we know something about the proximate mechanisms of information relay in another insect, the cricket *Teleogryllus oceanicus*, which also flees from ultrasound producers [803]. Just as is true for noctuid moths, the cricket's ability to avoid bats begins with the firing of certain ultrasound-sensitive acoustical receptors in its ears, which are found on the cricket's forelegs. Sensory messages from these receptors travel to other cells in the cricket's central nervous system. Among the receivers is a pair of sensory interneurons called *int-1*, one of which is located on each side of the insect's body. Ron Hoy and his co-workers established that *int-1* plays a key role in the perception of ultrasound by playing sounds of different frequencies to a cricket while recording the responses of its *int-1* cells. These cells became highly excited when the cricket's ears were bathed in ultrasound. The more intense a sound in the 40–50 kHz range, the more action potentials the cells produced, and the shorter the latency between stimulus and response—two properties that exactly match those of the A1 receptor in noctuid moths.

These results suggest that the *int-1* cell is part of a neural circuit that helps the cricket respond to ultrasound. If this is true, then it follows that if one could experimentally inactivate *int-1,* ultrasonic stimulation should not generate the typical reaction of a tethered cricket suspended in midair, which is to turn away from the source of the sound by bending its abdomen (Figure 14). As predicted, crickets with temporarily inactivated *int-1* cells do not attempt to steer away from ultrasound, even though their acoustical receptors are firing. Thus, *int-1* is necessary for the steering response.

The corollary prediction is that if one could activate *int-1* in a flying, tethered cricket (and one can, with the appropriate stimulating electrode), the cricket should change its body orientation as if it were being exposed to ultrasound, even when it was not. Experimental activation of *int-1* is sufficient to cause the cricket to bend its abdomen [844]. These experiments convincingly establish a causal relationship between *int-1* activity and the apparent bat evasion response of flying crickets.

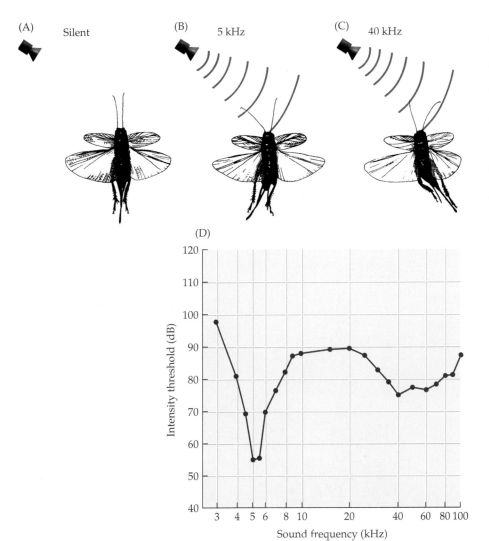

(A) Silent

(B) 5 kHz

(C) 40 kHz

(D)

14 Avoidance of and attraction to different sound frequencies by crickets. (A) In the absence of sound, a flying tethered cricket holds its abdomen straight. (B) If it hears low-frequency sound, the cricket turns toward the source of the sound. (C) If it hears high-frequency sounds, it turns away. Males of this species produce sounds in the 5 kHz range; some predatory bats produce high-frequency calls of about 40 kHz. (D) Crickets are especially sensitive to these two classes of sounds, which do not have to be particularly intense to elicit a turning response. After Moiseff et al. [803].

The *int-1* cells are sensory interneurons, part of the relay apparatus that controls which sensory messages move forward to higher processing centers; these cells do not innervate the cricket's locomotory muscles directly. What is the proximate motor mechanism that enables a flying cricket to carry out orders from its brain to steer away from a source of ultrasound? This problem attracted the attention of Mike May, who began his study not by conducting a carefully designed experiment, but instead by "toying with the ultrasound stimulus and watching the responses of a tethered cricket" [772]. As May zapped the cricket with bursts of ultrasound, he noticed that the beating of one hindwing seemed to slow down with each application of the stimulus. Crickets have four wings, but only the two hindwings power cricket flight. If the hindwing opposite the source of ultrasound really did slow down, that would reduce power or thrust on one side of the cricket's body, with a corresponding turning (or yawing) of the cricket away from the stimulus (Figure 15).

On the basis of his informal observations, May constructed a formal hypothesis: namely, that by lifting a hind leg into a hindwing, the cricket altered the beat of that wing and thereby changed its flight path. To test this hypothesis, May took a number of high-speed photographs of crickets with and without hind legs. Without the appropriate hind leg to act as a brake, both hindwings continued beating unimpeded when the cricket was exposed to ultrasound. As

15   How to turn away from a bat—quickly. (A) A flying cricket typically holds its hind legs so as not to interfere with its beating wings. (B) Ultrasound coming from the right causes the leg on the opposite side of the body to be lifted up into the wing. (C) As a result, the beating of the left wing slows, and the cricket turns away from the source of ultrasound as it dives to the ground. Drawings by Virge Kask, from May [772].

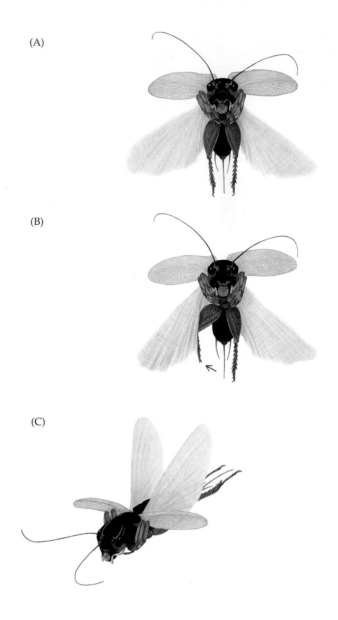

(A)

(B)

(C)

a result, crickets without hind legs required about 140 milliseconds to begin to turn, whereas intact crickets started their turns in about 100 milliseconds. May concluded that neurons in the ultrasound detection network order the appropriate motor neurons to induce muscle contractions in the opposite-side hind leg of the cricket. As these muscles contract, they lift the leg into the wing, interfering with its beating movement, thereby causing the cricket to veer rapidly away from an ultrasound-producing bat [772].

Central Pattern Generators

In crickets and moths, ultrasound stimulation leads to receptor response, and sensory signals from receptors are carried by interneurons to other neurons within the central nervous system. When these cells respond, they generate signals that turn on certain motor neurons. The end result is that certain muscles contract or relax in ways that cause the moth to stop beating its wings or the cricket to lift one leg upward, braking the activity of a wing. The escape dives of moths and the evasive swerves of flying crickets pursued by bats are effective one-step responses triggered by simple releasing stimuli in these animals' environments. Most behaviors, however, involve a coordinated series of

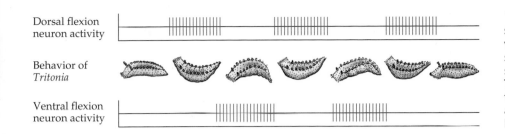

Dorsal flexion
neuron activity

Behavior of
Tritonia

Ventral flexion
neuron activity

16   Escape behavior in *Tritonia* sea slugs. Alternating bouts of dorsal and ventral muscular contractions cause the slug to undulate away from a noxious stimulus, the scent of a predatory sea star. The dorsal and ventral muscles are controlled by two different neurons whose activity is correlated with the alternating bouts of muscle contraction.

muscular responses, which cannot result from a single command from a cell or cell network. Consider, for example, the escape behavior of a sea slug of the genus *Tritonia*. When it comes into contact with a releaser—chemicals associated with the body of a predatory sea star—the slug begins to swim, in the ungainly fashion of sea slugs, by bending its body up and down [1243]. If all goes well, it will move far enough away from the sea star to live another day.

How does *Tritonia* manage its multi-step escape response, which requires from two to twenty alternating bends, often continuing long after the slug is no longer in contact with the chemical stimulus that triggered the reaction? The slug bends its body through alternating contractions of its dorsal and ventral muscles (Figure 16). When the dorsal muscles contract, the slug's body is pulled upward into a U-shaped position; when the ventral muscles contract, the slug's body is pulled downward. These two sheets of muscles are regulated by two large motor cells, the dorsal flexion neuron (DFN) and the ventral flexion neuron (VFN), which fire in alternating sequence (Figure 17A). Electrical stimulation of the DFN causes the slug's body to bend upward; stimulation of the VFN generates the opposite response [1243]. Because the slug's escape behavior requires first an upward, then a downward bend, Dennis Willows suggested that a third neuron might control the two flexion neurons to ensure that they carry out their activities in the required alternating sequence. According to this control model (Figure 17B), a general excitatory neuron (GEN) is responsible for producing an alternating cycle of activity in the DFN and in the VFN.

This hypothesis was tested by P. A. Getting, who ultimately rejected it because he found that *several* interneurons control the two flexion neurons [426, 427]. One of these interneurons, labeled the C2 cell, has the capacity to excite certain other cells, the dorsal swim interneurons, that control the DFN, while inhibiting still other interneurons linked to the VFN (Figure 17C). As a result, when the C2 cell fires, the DFN receives commands to fire, causing dorsal flexion. But the interaction between the C2 cell and the dorsal swim interneurons is such that after a period of excitation, the C2 cell begins to *inhibit* the DFN while exciting the interneurons that stimulate the VFN. As a result, the DFN ceases to fire at the same time the VFN begins its burst of activity. The situation then reverses. Alternating bouts of activity in the interneurons regulating the DFN and VFN lead to alternating bouts of DFN and VFN firing, and thus alternation of dorsal and ventral bending.

The capacity of the C2 cell to impose order on the two cells under its command means that this neuron qualifies as a key component of a **central pattern generator.** Cells of this sort play a preprogrammed set of messages—a motor tape, if you will—that helps organize the motor output underlying movements of the sort that ethologists might label fixed action patterns. Interestingly, the motor program of a central pattern generator can sometimes be modified by communications received from the motor cells it commands. Activity in the dorsal swim interneurons of *Tritonia*, for example, alters the release of a neurotransmitter by the C2 cell for a period afterward [604]. So, for example, if the C2 cell is at first weakly stimulated, it may produce just two bursts of activity cor-

(A)

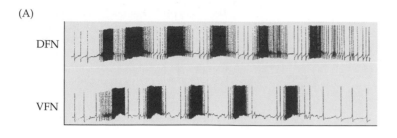

(B)

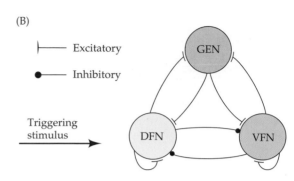

(C)

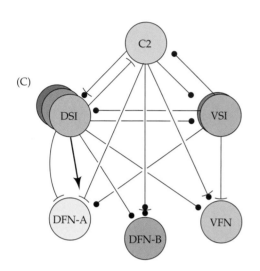

17   Neural control of escape behavior in *Tritonia*: Two hypotheses.
(A) Recordings of dorsal flexion neuron (DFN) and ventral flexion neuron (VFN) activity, showing that activity in one cell somehow inhibits activity in the other. (B) One model put forth to explain this mutual inhibition proposed that the DFN and VFN are directly linked, so that signals from one cell directly inhibit the other. Both flexion neurons are also connected to a general excitatory neuron (GEN), which becomes active when the DFN begins to fire in response to a triggering stimulus. An active GEN excites the VFN, which begins to fire, thereby shutting down the DFN temporarily, and so on. (C) After rejection of this model, a new, more complex hypothesis was developed in which cerebral neuron 2 (C2), three dorsal swim interneurons (DSI), and two ventral swim interneurons (VSI) together constitute a central pattern generator that regulates the activity of the flexion neurons (DFN-A, DFN-B, and VFN), which do not communicate directly with each other. The DSIs modulate the strength of the connection between C2 and DFN-A (arrow). Several of the synapses in this model are multicomponent synapses that are capable of sequential excitatory and inhibitory activity; for example, when active, the synapse between C2 and DFN-B is first inhibitory, then excitatory, then inhibitory again. A, B after Willows [1243]; C after Katz and Frost [604].

responding to two cycles of dorsal and ventral bending. The dorsal swim interneurons also undergo two bursts of activity, but continue to fire at a relatively low rate then for some time. If shortly thereafter the same weak stimulus is applied again to C2, this time C2 will fire more vigorously, initiating a corresponding increase in the activity of the dorsal swim interneurons, and ultimately a longer period of swimming (Figure 18) [603]. Thus, the central pattern generator of *Tritonia* can change its properties as cells modulate the excitability of other cells, thereby adjusting the strength of the escape response in relation to the recent experience of the sea slug. Something fundamentally similar happens whenever animals learn something, thanks to the ability of some neurons and neural networks to change their operating rules in response to certain patterns of stimulation.

18   Neuromodulation in the central pattern generator of *Tritonia*. After the first weak stimulation of interneurons C2 and DSI, the motor program generates two cycles of response, but the DSI remains active. When exactly the same stimulus is applied two minutes later, the result is a stronger, four-cycle response by the motor program. After Katz [603].

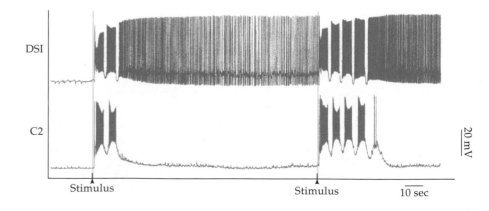

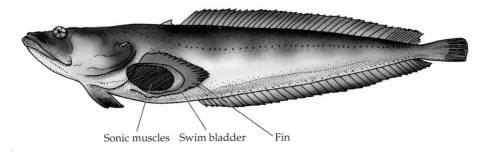

19 Song-producing apparatus of the male midshipman fish. The sonic muscles control the movement of the swim bladder, thereby controlling the fish's ability to hum. After an illustration by Margaret Nelson, in Bass [79].

Sonic muscles   Swim bladder   Fin

The Song of the Midshipman Fish

Let's look at a central pattern generator in a vertebrate, the plainfin midshipman, a fish that sings by contracting and relaxing certain muscles in a highly coordinated fashion [79]. Only the large males of this rather grotesque fish sing, and only at night during spring and summer while guarding certain rocks. The humming song they produce is so loud that houseboat owners in the Pacific Northwest can hear them. The male fish sings to attract females of his species; the fish spawn at the defended rocks, and the male guards the eggs his mates place in his territory.

How do the male fish produce their songs? When Andrew Bass and his coworkers inspected the anatomy of the fish's abdomen, they found a large, air-filled swim bladder sandwiched between layers of muscles (Figure 19). The bladder serves as a drum; rhythmic contractions of the muscles "beat" the drum, generating vibrations that other fish can hear. Muscle contractions require signals from motor neurons, which Bass found connected to the sonic muscles. He applied a cellular dye called biocytin to the cut ends of these neurons, which absorbed the material, staining themselves brown. And the stain kept moving along, crossing the synapses between the first cells to receive it and the next ones in the circuit, and so on, through the whole network of connected cells. By cutting the brain into fine sections and searching for cells stained brown by biocytin, Bass and his colleagues mapped the sonic control system. In so doing, they discovered two discrete collections of interrelated neurons that generate the signals controlling the coordinated muscle contractions required for midshipman humming. These two clusters located in the upper part of the spinal cord near the base of the brain (Figure 20), containing some 2000 neurons each, are the paired sonic motor nuclei. The long axons of their neurons travel out

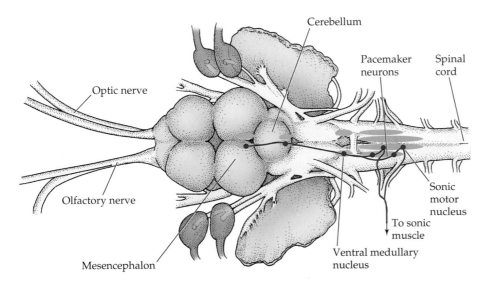

Cerebellum

Pacemaker neurons   Spinal cord

Optic nerve

Olfactory nerve

Mesencephalon

Ventral medullary nucleus

Sonic motor nucleus

To sonic muscle

20 Neural regulation of the sonic muscles in the midshipman fish. Signals from the central region of the brain (the mesencephalon) travel by way of the cerebellum and ventral medullary nuclei to the sonic motor nuclei in the upper part of the spinal cord. The firing of the pacemaker neurons regulates the frequency of firing by the neurons in the sonic motor nuclei; these signals in turn set the rate of contraction of the sonic muscle, and thus the frequency of the sounds produced by the fish. After an illustration by Margaret Nelson, in Bass [79].

from the brain, fusing together to form the occipital nerves, which reach the swim bladder muscles.

In addition to these components, two other anatomically distinct elements of the nervous system ally themselves with the sonic motor nuclei. First, next to each nucleus lies a sheet of pacemaker neurons—a multicellular central pattern generator, if you will—that adjusts the activity of the sonic motor neurons so that the sequence of muscle contractions will yield a proper song. Second, in front of the pair of sonic motor nuclei are some special neurons that appear to connect the two nuclei, probably to coordinate the firing patterns coming from the left and the right nucleus, so that the muscle contractions and relaxations will be synchronous, the better to produce a humming sound [79].

The Proximate Basis of Stimulus Filtering

We have just examined examples of research on the attributes of individual neurons and neural clusters that are involved in the detection of certain kinds of sensory information, the relaying of messages to other cells in the nervous system, and the control of motor commands as they are sent to muscles. A universal feature of neurons and nervous systems that contributes to the effective performance of these basic functions is **stimulus filtering.** This term refers to the ability of neurons to ignore—to filter out—vast amounts of potential information in order to focus on biologically relevant elements within the barrage of stimuli reaching an animal.

Stimulus Filtering by Auditory Receptors

The noctuid moth's auditory system offers an object lesson on the operation and utility of stimulus filtering. First, the A1 receptors are activated only by acoustical stimuli, not by other forms of stimulation. Moreover, these cells completely ignore sounds of relatively low frequency (in the range that you and I can hear). Finally, even when the A1 receptors do fire in response to ultrasound, they react in exactly the same way to different ultrasonic frequencies of the same intensity (whereas human auditory receptors produce varying messages in response to sounds of different frequency, which is why we can tell the difference between C and C-sharp). The noctuid moth's sensory apparatus appears to have just one task of paramount importance: the detection of cues associated with its echolocating predators. To this end, its auditory capabilities are tuned to pulsed ultrasound at the expense of all else. Upon detection of these critical inputs, the moth can take effective action.

Noctuid moths are not the only animals with biased auditory systems designed to filter out the irrelevant and focus on the truly important. The relationship between stimulus filtering and a species' special obstacles to reproductive success is evident in every animal whose sensory systems have been carefully examined. Consider the females of certain parasitoid flies, which have an acoustical mechanism that helps them locate male crickets. These *Ormia* flies deposit their larvae on the unlucky crickets they find. The little maggots burrow into the crickets and proceed to devour them from the inside out. Larvae-laden female flies can find food for their offspring because they have ears that can hear singing crickets, as researchers discovered when they found *Ormia* coming to loudspeakers that were playing tapes of cricket songs at night.

The unique ears of the female fly consist of two air-filled structures with tympanic membranes and associated acoustical receptors on the front of the thorax. Vibration of the fly's tympanic "eardrums" activates the receptors, just as in noctuid moth ears, and thus provides the fly with information about sound in its environment—but not every sound. As predicted by a trio of evolutionary biologists, Daniel Robert, John Amoroso, and Ronald Hoy, the female fly's audi-

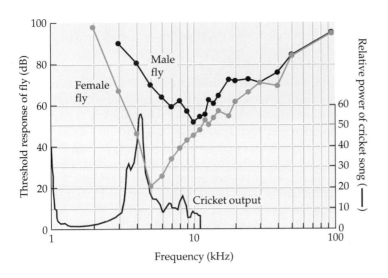

21 Tuning curves of a parasitoid fly. Females, but not males, of the fly *Ormia ochracea* find their victims by listening for the calls of male crickets, which produce sound with a frequency–power spectrum that peaks between 4 and 5 kHz. The female fly, unlike the male fly, is maximally sensitive to sounds around 5 kHz. After Robert et al. [972].

tory system is tuned to (i.e., most sensitive to) the dominant frequencies in cricket songs (Figure 21). That is, the female can hear sounds of 4–5 kHz (the sort produced by crickets) more easily than sounds of 7–10 kHz, which have to be much louder if they are to generate any response [972]. The tuning curve of the male *Ormia*, however, does not reflect a special sensitivity to sounds of 4–5 kHz. Although males can hear, they do not depend on finding singing male crickets, and as a result, their acoustical system has evolved its own distinctive properties. The differences in sensitivity to various sound frequencies within and between the sexes of this fly provide another clear example of stimulus filtering.

Another fly parasitoid related to *Ormia* tracks down singing male katydids, *Poecilimon veluchianus*, whose ultrasonic mate-attracting calls fall largely in the 25–35 kHz range [1107]. What sound frequencies should elicit maximal response in the ears of this katydid-hunting parasitoid, if stimulus filtering enables the animal to achieve goals that are biologically relevant? As you can see (Figure 22),

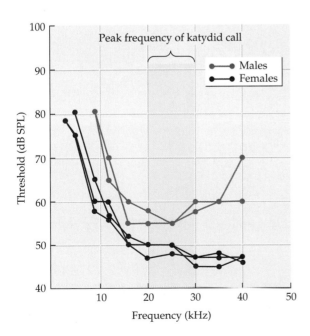

22 Tuning curves of a katydid killer. Females of the fly *Therobia leonidei* parasitize male katydids, whose stridulatory calls contains most of their energy in the range of 20–30 kHz. The female fly is much more sensitive than the male fly to sounds in this range. After Stumpner and Lakes-Harlan [1107].

the ears of *Therobia leonidei* are indeed tuned to the song frequencies of its host, which are pitched much higher than the sounds that female *Ormia* hear best [972].

Stimulus Filtering in the Tactile Mode

Stimulus filtering is not limited to acoustical systems, but is characteristic of every neural mechanism, all of which respond unequally to different kinds of stimulation. We shall illustrate this point by examining the operating rules of the neural system that provides the star-nosed mole with the sensation of touch. The star-nosed mole lives in wet, marshy soil, where it burrows about in search of earthworms and other prey. In its dark tunnels, earthworms cannot be seen, and indeed, the mole's eyes are greatly reduced in size, so that it largely ignores visual information even when light is available. Instead, the mole relies heavily on touch to find its food, using its strange and wonderful nose to sweep the tunnel walls as it moves forward. Its two nostrils are ringed by 22 fleshy appendages, 11 on each side of the nose (Figure 23). These

23   **The star-nosed mole's nose** (top left) differs greatly from that of the eastern mole (top right), let alone those of its more distant relatives, which include the African hedgehog (bottom left) and the masked shrew (bottom right). All four species, however, rely on tactile information to a considerable degree in locating prey, which range from insects to earthworms. Photographs by Ken Catania.

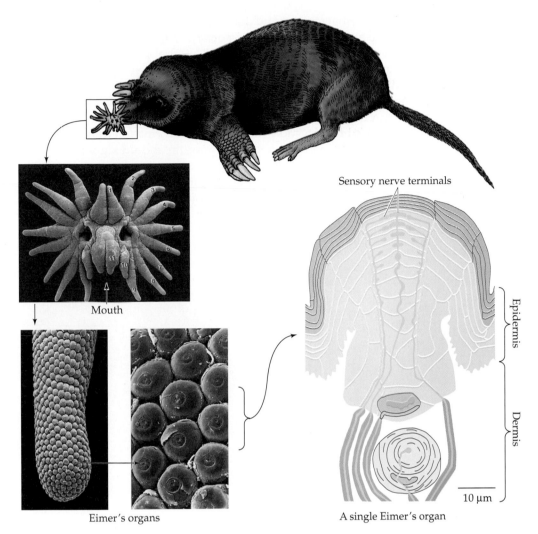

Sensory nerve terminals

Epidermis

Dermis

10 µm

Mouth

Eimer's organs

A single Eimer's organ

24 A special tactile apparatus. The 22 appendages of the nose of the star-nosed mole are covered with thousands of Eimer's organs. Each organ contains a variety of specialized sensory cells that respond to mechanical deformation of the skin above them. After Catania and Kaas [206]; photographs by Ken Catania.

appendages cannot grasp or hold anything, but instead are covered with a thousand or so tiny sensory devices called Eimer's organs. Each of these organs contains several receptor cells that react to changes in pressure (Figure 24). With its vast array of receptors, the animal can collect extremely complex patterns of information about the surfaces touched by the nose [206].

Whenever the mole brushes an earthworm with, say, appendage 5, it instantly sweeps its nose over the prey so that the two projections closest to the mouth (appendage 11) come in contact with the object of interest. The tactile receptors on appendage 11 generate a volley of signals, which are carried by nerves to the brain of the animal. Although the two appendage 11s contain only about 7 percent of the Eimer's organs on the star nose, more than 10 percent of all the nerve fibers relaying information from the nose's touch receptors to the brain come from these two appendages. In other words, the mole uses relatively more neurons to relay information from appendage 11 than from any other appendage. Not only is the relay system biased toward inputs from appendage 11, but the animal's brain also gives extra weight to signals from this appendage. Information from the nose travels through nerves to the somatosensory cortex, the part of the brain that receives and decodes sensory signals from touch receptors all over the animal's body. Of the portion of the somatosensory cortex that is dedicated to decoding inputs from the 22 nose appendages, about 25 percent deals exclusively with messages from the two appendage 11s (Figure 25) [207]. This discovery was made by Kenneth Catania and Jon Kaas when they exposed the cerebral cortex of an anesthetized mole and recorded the responses of cortical neurons as they touched different parts of the animal's nose [206]. Perhaps the mole's brain is "more interested in" information from appendage 11 because of its location right above the mouth; should signals from this appendage activate a cortical order to capture a worm and consume it, the animal is in the right position to carry out the action immediately [206].

The disproportionate investment in brain tissue to decode tactile signals from one part of the nose is mirrored on a larger scale by the biases evident in the

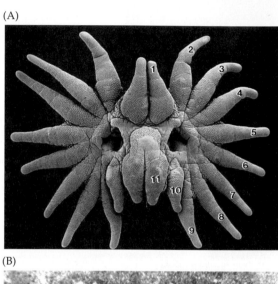

(A)

(B)

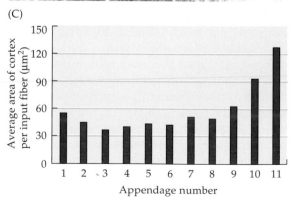

(C)

25   **The cortical sensory map of the star-nosed mole's tactile appendages** is disproportionately weighted toward appendage 11. (A) The nose of the mole with each appendage numbered on one side. (B) A section through the area of the somatosensory cortex that is responsible for analyzing sensory inputs from the nose. The cortical areas that receive information from each appendage are numbered. (C) The amount of cortex devoted to the analysis of information from each nerve fiber carrying sensory signals from the different nose appendages. After Catania and Kaas [207]; photographs by Ken Catania.

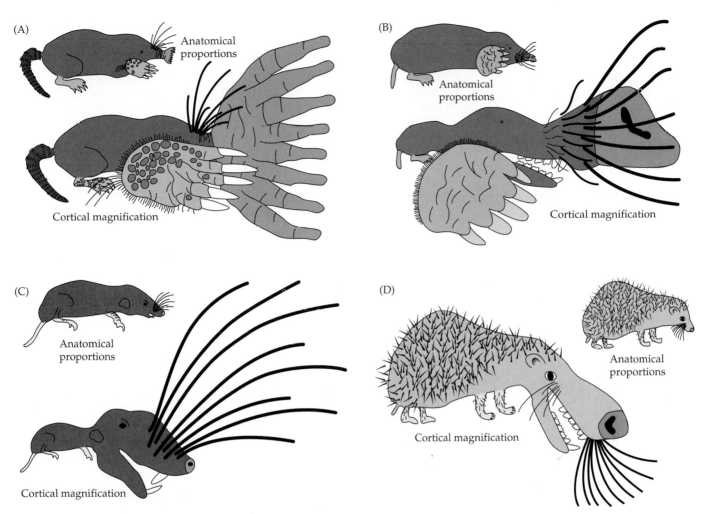

26 Sensory analysis in four insectivores. In each case, the smaller drawing shows the actual anatomical proportions of the animal; the larger drawing shows how the body is proportionally represented in the somatosensory cortex of the animal's brain. (A) The star-nosed mole devotes much more somatosensory cortex to processing inputs from its nose and forelimbs than it does to signals coming from receptors on other parts of its body. (B) Cortical magnification in the eastern mole focuses on sensory inputs from the feet, nose, and sensory hairs, or vibrissae, around the nose. (C) Cortical magnification in the masked shrew also reveals the importance of the vibrissae. (D) Sensory signals from the vibrissae are magnified cortically to a lesser degree in the African hedgehog. After Catania and Kaas [206] and Catania [205].

somatosensory cortex as a whole. In the star-nosed mole, the somatosensory cortex is given over largely to analyzing signals from the nose and hands at the expense of other parts of the body. This bias makes adaptive sense for this species because of the biological importance of the mole's hands for burrowing and the mole's nose for locating prey (Figure 26A).

Adaptive Proximate Mechanisms of Behavior

Our brief survey has revealed that both individual cells and neural networks have special operating rules that filter the information an animal receives, relays, and processes. It has also reminded us that the biases in the ways individuals perceive their environments differ from species to species in ways that make adaptive sense. Even though the acoustical systems of bat-evading moths and cricket-attacking flies are both characterized by specializations based on stimulus filtering, the moths focus on bat-produced ultrasound, whereas the flies are especially sensitive to sounds well within the range of human hearing—namely, those generated by calling crickets. The same match between environmental demands and perceptual specialization occurs in the tactile mode. Note that other insectivores that are closely related to the star-nosed mole exhibit their own distinctive patterns of cortical magnification (Figure 26B–D), which represent different evolutionary responses to the different ecological problems each species confronts and solves by tactile means [205].

Research on the major components of visual systems confirms the point that nervous systems focus on biologically significant items and events, especially stimuli associated with food, predators, and mates. In some species, for example, ultraviolet light provides information relevant to an individual's chances of surviving and reproducing, and these animals possess a correlated sensitivity to wavelengths of light in the ultraviolet range (which we humans cannot perceive). Ultraviolet light is reflected from many flowers visited by bees, which can find the nectar source within a flower more quickly if they can detect the flower's ultraviolet-reflecting patterns (Figure 27, top). Likewise, the ability of

27 **Ultraviolet-reflecting patterns have great biological significance for some species.** In both sets of photographs, the image on the left shows the organism as it appears to humans while the image on the right shows the organism's UV-reflecting (pale) surfaces. (Top) The ultraviolet pattern on this daisy advertises the central location of food for insect pollinators. (Bottom) Only males (top specimens) of this sulphur butterfly have UV-reflecting patches on their wings, which helps signal their sex to other individuals of their species. Top, photographs by Tom Eisner; bottom, photographs by Randi Papke and Ron Rutowski.

28 A bird that can sense ultraviolet light. The bluethroat's throat feathers appear purely blue to us, but not to bluethroats themselves, which also see the ultraviolet light reflecting from the feathers. Photograph by Bjørn-Aksel Bjerke, courtesy of Jan Lifjeld.

some butterfly species to see UV light helps them to respond to the UV-reflecting patterns on the wings of other butterflies, which announce species membership and sexual identity (Figure 27, bottom) [997].

Nor is ultraviolet light perception limited to the insects [90]. Many, if not all, songbirds can see UV light, including the bluethroat of northern Europe and Asia. The bird's common name refers to a patch of blue feathers located, as you may have suspected, on the male's throat (Figure 28). These feathers reflect light that we perceive as blue, but they also reflect ultraviolet light that is invisible to us. If bluethroats can see UV light, then perhaps UV reflectance from the throat area, which is displayed during courtship, affects male mating success. This hypothesis generates the prediction that males whose blue throats have been altered to absorb, rather than reflect, ultraviolet radiation will become less attractive to females. To check this prediction, a Scandinavian research team placed captured male bluethroats of the same age in pairs in an aviary. One member of each pair had sunscreen plus a fatty substance from an avian preening gland rubbed onto its display patch, while the other received an application of preening gland oil only. The chemicals in the sunscreen absorbed the UV wavelengths, whereas the glandular secretions did not. In 13 of 16 paired trials, a female bluethroat approached the UV-reflecting male more often than the male whose ornament had been altered to become UV-absorbing [36]. Moreover, when similar manipulations were performed on free-living bluethroats early in the breeding season, the UV-reflecting males acquired a mate and began nesting sooner than their UV-absorbing rivals, and they were also more likely to secure copulations with females other than their primary partners, showing that UV signals influence a male's ability to attract mates (Figure 29) [586].

How might female bluethroats benefit from having the sensory equipment to see UV light? Perhaps a preference for bluethroat males whose throat patches reflect more UV light leads females of this songbird to choose older partners, since older males typically have more UV-reflecting feathers than younger males [36]. Older male bluethroats have demonstrated their capacity to survive, and so may be able to pass on "good genes" to their offspring (see p. 350). Likewise, the ability to use UV light when foraging may give the hunting bird one more

29 The ultraviolet reflectance of the throat patch is related to mate choice in the bluethroat. (A) Males with experimentally reduced ultraviolet reflectance were slower to attract mates than control males. (B) Males with reduced ultraviolet reflectance also obtained fewer copulations with females that were not their primary mates. After Johnsen et al. [586].

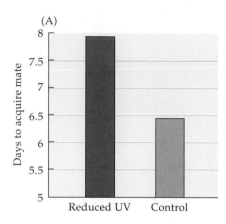

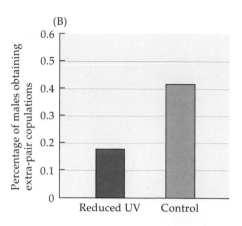

cue to work with, all the better to overcome the camouflaged color patterns of so many of its edible prey (see p. 193) [216].

Even though the studies I have just described have not involved direct inspection of the nervous systems of bluethroats or butterflies, they tell us something about the properties of the sensory receptors and information decoders possessed by these animals. Bluethroats and butterflies have nervous systems with perceptual capabilities that help individuals manage their special environments. If we were to apply this principle to our own species, we could predict that selection should have endowed us with proximate mechanisms suited to the ecological problems we face. Indeed, much evidence suggests that our highly specialized acoustical skills, for example, have evolved to match the human social environment, which is dominated by language [920]. Moreover, our visual perception complements our acoustical analysis of language in a most interesting way. As it turns out, a listener's perception of spoken language is heavily influenced by visual cues provided by the moving lips of a speaker. When someone sees a video of a person mouthing the nonsense sentence, "My gag kok me koo grive," while an audiotape synchronously plays the nonsense sentence, "My bab pop me poo brive," the viewer/listener will hear quite clearly, "My dad taught me to drive." This result demonstrates that our brains have circuits that integrate the visual and auditory stimuli associated with speech, using the visual component to alter our perception of the auditory channel [766]. Even though most people are totally unaware of their lip-reading skills, the ability increases the odds that they will understand what others say to them.

Lip reading depends on specific neural clusters within the visual cortex. In particular, a region of the brain called the superior temporal sulcus becomes especially active when we view moving mouths, hands, and eyes [24] (Figure 30). People are remarkably good at detecting even subtle movements of these body parts because this skill enables the viewer not only to read lips, but also to deduce the intentions of other individuals, an adaptive ability for a highly social species. Cells in the superior temporal sulcus are also activated by certain static visual stimuli, especially those associated with faces. They perceive, for example, the direction in which a companion's eyes are gazing, which says something about his current focus of interest. Because neurons within the superior temporal sulcus are "tuned" to particular facial stimuli, such as eye position, we can better predict the actions of those around us [24].

Nor is the superior temporal sulcus the only brain area devoted to the analysis of faces. An area called the fusiform gyrus, which is on the underside of the cerebral cortex, plays a critical role in our capacity to identify familiar faces quickly and accurately. And we are exceptionally good at this task. If an exper-

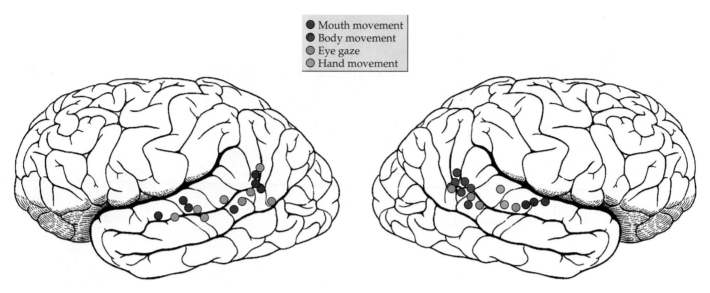

30 **Socially relevant movements of the lips, mouth, hands, and body** activate neurons in different parts of the superior temporal sulcus in the human brain. The left and right hemispheres are shown on the left and right, respectively. Each circle represents findings by a particular research team. After Allison et al. [24].

imenter gives someone 5 seconds to inspect each of 50 photographs of unfamiliar faces, odds are that when the subject is tested later, he will be able to pick out about 90 percent of the previously glimpsed faces from a large collection of photographs of faces [196]. We know that face recognition is dependent on an intact fusiform gyrus because people who have suffered an injury to this part of the brain are no longer able to recognize people by their faces [357], a phenomenon that gave Oliver Sacks the title for his book, *The Man Who Mistook His Wife for a Hat* [1006]. Some persons with a damaged fusiform gyrus can see things perfectly well, they can name many objects, they can identify particular individuals by the sound of their voice or by familiar clothing, but when shown pictures of the faces of their friends, their spouses, even themselves, they are at a complete loss [309]. Interestingly, other slightly brain-damaged persons have just the opposite problem. They cannot identify ordinary objects when they see them, but have no difficulty recognizing particular faces, suggesting that the special neural mechanisms dedicated to face recognition are still intact in their brains [86].

Various techniques, including functional magnetic resonance imaging, can be used to identify exactly which parts of the cerebral cortex light up when the subject sees a picture of a face (Figure 31). Brain imaging reveals that cells in a small part of the posterior fusiform gyrus fire only when a person looks at a face. This neural module, called the facial fusiform area, does not respond to pictures of inanimate objects, although another nearby region of the brain does, providing strong evidence of task specialization by modules that provide biologically relevant perceptions for their owners [482, 601].

Yet another approach to identifying which part of the brain does what has been to place electrodes directly on the surface of the cerebral cortex and record the electrical activity of neurons. This technique is used to map the brains of epileptic patients prior to operations designed to remove the tissue responsible for their epileptic seizures. Needless to say, the idea is to destroy as little of the brain as possible, which means finding the dysfunctional tissue in order to leave the rest alone when the surgery is performed. In the course of recording the electrical activity in different parts of the brains of these subjects, who remain conscious, researchers have been able to locate those parts of the brain that react to

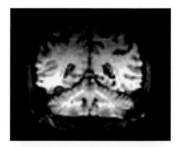

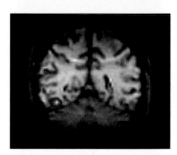

31 **A special purpose module in the human brain: the face recognition center.** Magnetic resonance images of the brains of two persons looking at a photograph of a human face. The active part of the brain, the facial fusiform area, is shown in red. After Kanwisher et al. [600].

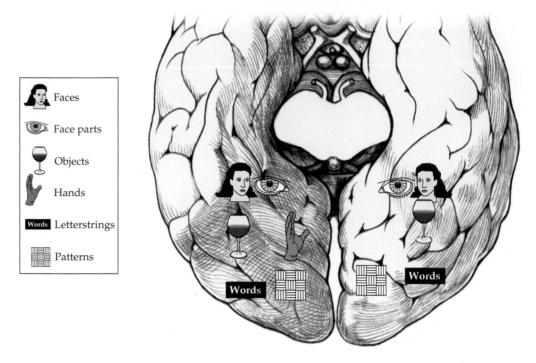

32   Specialization of function in different parts of the visual cortex of humans. Different circuits perform different analyses of images in our environment. The ability to remember faces, for example, is dependent upon a specialized site in the fusiform gyrus. After Puce et al. [940].

different kinds of visual stimuli. This approach has confirmed that different sections on the underside of the brain are devoted to different kinds of visual analyses. The sites that respond strongly to images of entire faces are different from those that react primarily to parts of faces, such as eyes, which in turn are different from those that are activated when the subject is shown a picture of an object (Figure 32) [940]. Our brains are computers with stimulus-filtering circuits that enable us to perceive some things much more readily than others.

Adaptive Mechanisms of Navigation

Our visual system is biased toward gathering biologically significant information from our social environment. Persons in the past who were better than average at remembering who was friend and who was foe, and at deducing the intentions of others by examining their faces, have in effect left us their brains, with all their adaptive filters, oddities, and quirks, as their legacy. The point that nervous systems have been shaped by selection can be reinforced by considering the relationship between the brains of animals and their ability to travel to destinations of their choice. In Chapter 4 we discussed the capacity of some birds to remember where they have hidden food (p. 90). As you may recall, a specific component of the brain, the hippocampus, appears to play a major role in spatial memory, enabling some chickadees and nutcrackers to memorize the landmarks associated with their food caches and thus to find them again when needed.

Humans also can form mental maps based on visual landmarks, maps that must have been vital to the survival of foraging hunters and gatherers who traveled out from and back to a home camp. For us, as for many other animals, the hippocampus plays a critical role in navigating from point A to point B [740]. Activity in the hippocampus can be monitored in fully conscious persons who are engaged in various tasks, such as playing computer games. In one such study, each player was permitted to explore a virtual town with a complex maze of streets before he was given the task of reaching a goal point as quickly as possible. Individuals able to navigate accurately and quickly evidently used the right hippocampus and right caudate nucleus, judging from the high levels of activity observed in these parts of the brain (Figure 33).

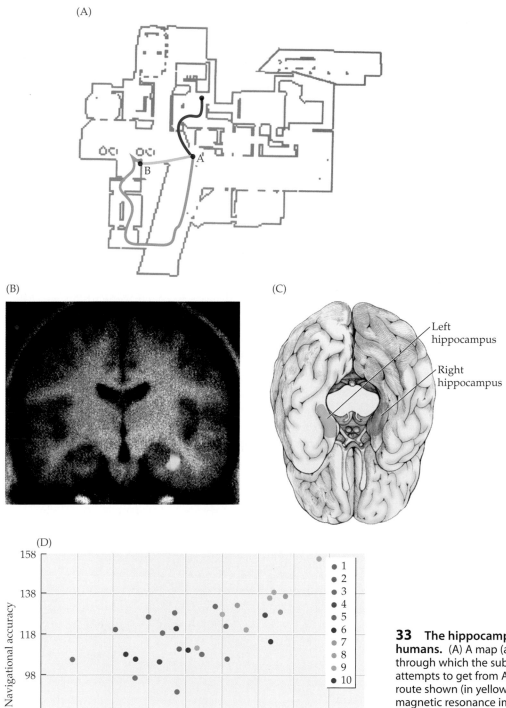

33 The hippocampus is essential for navigation by humans. (A) A map (aerial view) of the virtual town through which the subjects navigated. Three examples of attempts to get from A to B are illustrated. The shortest route shown (in yellow) was the most accurate. (B) Average magnetic resonance image for 10 subjects while performing the navigation task, showing the location of peak activity (bright yellow), which lies in the right hippocampus. (C) The location of the right hippocampus in a brain viewed from the underside. (D) The accuracy of the 10 navigators was a function of the intensity of the activity in their right hippocampus. After Maguire [740] and Carter [202].

Wandering albatross

Cataglyphis ant

If it is true that our learned mental maps are stored within the hippocampus, then persons with complex maps formed while navigating through varied environments, such as cities, should have more hippocampal tissue than persons who have less experience in acquiring spatial representations of their surroundings. One group of experienced navigators is the licensed taxi drivers of London, who must demonstrate that they can remember the locations of thousands of addresses before receiving a license. As predicted, the average posterior hippocampus size in a sample of London taxi drivers was indeed larger than that in a comparable group of men who did not drive taxis for a living [741]. Navigational experience itself appears to be responsible for the development of a large hippocampus capable of storing a great deal of spatial information. The more years of taxi driving, the larger the posterior hippocampus, as revealed by magnetic resonance imaging.

Although the performance of these experienced taxi drivers in negotiating the streets of London is impressive, if you were to take a London cabbie and drop him off in Detroit with the requirement that he get to 16th and Ash without assistance from others, he would almost certainly be totally lost. Nor would he set off confidently if he were given a compass by a bystander. A compass would help only if he knew where he was and where he was headed, which requires a map of some sort. Although humans are handicapped when it comes to navigating across unfamiliar terrain without a map and compass, many other animals show no such disability (Figure 34). In some of these species, the hippocampus may be involved in one way or another, as suggested by the observation that (for example) the size of the hippocampus increases in garden warblers that migrate from Europe to tropical Africa [501]. However, just exactly what sensory information contributes to the amazing navigational capacity of garden warblers and other navigating species remains uncertain, even after years of intense study.

(A)

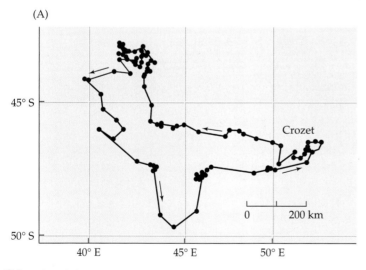

(B)

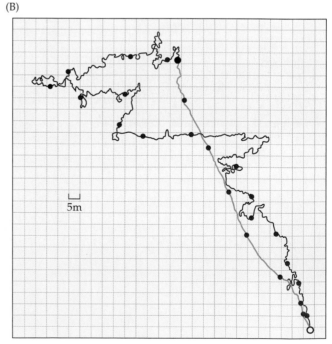

34 The ability to navigate over unfamiliar terrain requires both a compass sense (knowing in what direction to move) and a map sense (knowing the location of home or some other goal). (A) The flight path taken by a wandering albatross on a foraging journey of over 4000 kilometers from its nest in the Crozet Islands in the southern Indian Ocean, north of Antarctica, and back again. (B) A trip of 592 meters by a foraging ant out from its nest (at the large open circle) and then directly back home 140 meters away after capturing a prey at the spot marked with a large black circle. A after Weimerskirch et al. [1203]; B after Wehner and Wehner [1200].

We do know something about the proximate mechanisms underlying the navigational sense of a few species, such as the honey bee and the homing pigeon. Both species are skilled navigators, as demonstrated by a honey bee's ability to make a beeline back to its hive after a meandering outward journey in search of food, and by a homing pigeon's ability to make a pigeonline back to its loft after having been released in a distant and strange location. Because honey bees and homing pigeons are active during the daytime, they both, as one might suspect, use the sun's position in the sky as a directional guide [606, 1175]. Even we can do this to some extent, knowing that the sun rises in the east and sets in the west— provided that we also know approximately what time of day it is. Every hour the sun moves 15 degrees on its circular arc through the sky. Therefore, one has to adjust for the sun's movement if one is to use its position as a compass.

A honey bee leaving its hive notes the position of the sun in the sky relative to the hive as it flies off on a foraging trip. It might spend 15 to 30 minutes on its trip and might move into unfamiliar terrain in search of food. If on the homeward flight it oriented as if the sun had not shifted, the bee would get lost. In reality, honey bees rarely get lost, in part because they can compensate for the sun's movement, thanks to an internal clock mechanism (see p. 159) [706]. This ability can be demonstrated by training some marked bees to fly to a sugar-water feeder some distance away from their hive (say, 300 meters due east of the hive). One can then trap the bees inside the hive and move it to a new location. After 3 hours have passed, the hive is unplugged, and the bees are free to go in search of food. They do not have familiar visual landmarks to guide them, and yet some marked individuals remember that food is found 300 meters due east. They fly 300 meters due east of the new hive location, to the spot where the food source "should be." They have compensated for the 45-degree shift in the position of the sun that has taken place during their 3-hour confinement. Here again we have a special ability that depends on a special proximate mechanism designed to do something of biological importance for its owner.

Pigeons can also be tricked into demonstrating how important a clock sense is if they are to orient accurately by the sun [1181]. The birds can be induced to reset their biological clocks by placing them in a closed room with artificial lighting and then shifting the light and dark periods so that they are out of phase with sunrise and sunset in the real world. If sunrise is at 6:00 A.M. and sunset at 6:00 P.M., for example, one might set the lights in the room to go on 6 hours earlier (midnight) and off 6 hours earlier (noon). A pigeon exposed to this routine for several days would become *clock-shifted* 6 hours out of phase with the natural day. If taken from the room and released at 6:00 A.M. at a spot some distance from its loft, the bird will behave as if the sun has been up for 6 hours (as if it is noon), which will cause it to orient improperly. For example, let's say that the pigeon is released at a place 30 kilometers due west of its loft. Its map sense somehow tells it this, and it attempts to orient itself to fly east (Figure 35). As Charles Walcott points out: "To fly east at 6:00 A.M., you fly roughly toward the sun, but because your clock tells you it is really noon, you know that the sun is in the south and that to fly east, you must fly 90 degrees to the left of the sun. And this is exactly what the birds do, although they presumably do not go through the reasoning process I have described."

An Olfactory Map

Through sophisticated experiments, researchers have been able to deduce the properties of the control mechanisms underlying the astonishing navigational capabilities of some animals, many of which employ a sun compass and a clock mechanism. The sun compass, however, is by no means the only compass mechanism used by pigeons and other long-distance travelers in the animal kingdom [1182]. Pigeons and many other animals can, for example, detect

35 Clock-shifting and altered navigation in homing pigeons. The results of an experiment in which pigeons were released at Marathon, New York (about 30 kilometers east of Ithaca, where their home loft was located). On sunny days, birds in the control group generally headed west, back toward the home loft. But pigeons whose biological clocks had been shifted by 6 hours usually misoriented so that, on average, they headed north. After Keeton [605].

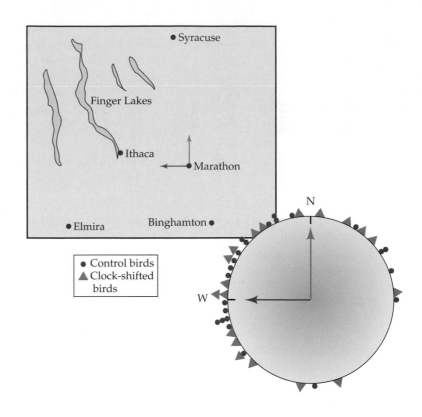

the weak lines of magnetic force created by the earth's magnetic field, as is apparent when experimental alteration of the field disorients navigating subjects [445]. In addition, several migratory songbirds have evolved the capacity to use the plane of polarized light at sunset as an orientation guide when beginning their journeys [1] and to use patterns of star positions as a compass after sunset [1253], two cues that are helpful to birds that set off in the evening for a long night of travel.

Let's turn from animals' directional compasses to the mechanisms that underlie their map sense, the other requirement for navigation. Among the various candidate map mechanisms that have been explored (e.g., [475]), one possibility is that some animals construct olfactory maps, which they then use to smell their way home [2, 985]. This hypothesis was first advanced for the homing pigeon by the Italian ornithologist Floriano Papi. Birds are generally thought to have a poor sense of smell, although exceptions exist among the petrels and albatrosses, seabirds with an unusually large number of olfactory receptors and a correspondingly large amount of brain tissue that deals with olfactory inputs. Some of these seabirds are attracted by volatile chemicals, such as dimethyl sulfide, when these compounds are added to vegetable oil and dumped overboard to create a scented oil slick. Dimethyl sulfide is naturally released when small marine invertebrates consume floating phytoplankton. Therefore, the special ability of some seabirds to smell these chemicals enables them to locate patches of edible marine organisms, which are scattered widely over vast stretches of ocean [837]. Moreover, Gabrielle Nevitt suggests that because concentrations of dimethyl sulfide, and perhaps other chemicals, differ from place to place over the South Atlantic Ocean, some seabirds could make use of the olfactory landscape to locate and return repeatedly to rich foraging areas as they travel to and from their nesting grounds in trips that routinely cover thousands of kilometers (Figure 34A).

Does the humble homing pigeon also have an olfactory sense that it can use to find its way home when it is released in an area that it has never visited

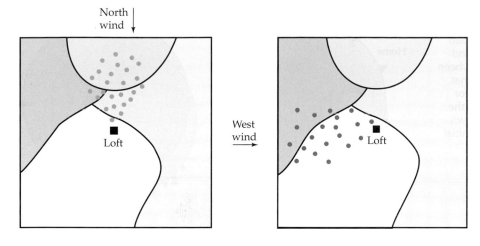

36 An olfactory map? If different locations have different odors, a north wind and a west wind would each have a distinctive odor. Pigeons at their home loft might be able to store this information to construct an olfactory map for later use when navigating back to the loft. After Papi [886].

before? Papi hypothesized that homing pigeons do form a mental map of odors (Figure 36), learning which scents always come on winds from the north, which from the west, and so on [887]. If the birds use this information to navigate home, then interference with the olfactory system, either by cutting key olfactory nerves or by applying a temporary local anesthetic to olfactory receptors, should interfere with their homing success. Indeed, experimental manipulations of the pigeon's olfactory network often do reduce the accuracy of the birds' initial orientation upon release (Figure 37).

An alternative explanation for these results, however, is that the experimental manipulations decrease the birds' motivation and attention, so that they have less interest in homing. But smell-blind pigeons that are released in familiar places, and so know the local landmarks, orient normally when released and zip home with the same efficiency as birds with intact olfactory nerves (Figure 38) [106]. Thus, manipulations of a pigeon's olfactory system apparently do not reduce the bird's readiness to fly home. Still, it would be helpful to test the olfactory map hypothesis without assaulting the birds' sensory equipment.

If the olfactory map hypothesis is correct, then if one were to deflect the route taken by winds arriving at a loft, the homing pigeons living there should construct a deflected olfactory map, which would throw them off when they were released away from home. For example, if winds from the west were forced

(A)   (B)

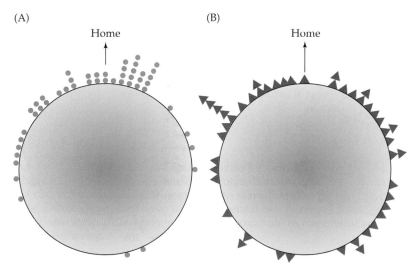

37 Orientation by homing pigeons depends on olfaction. Each symbol represents the mean initial orientation of the pigeons tested in a given study. (A) Orientation of control pigeons with unaltered olfaction. Note the clustering of symbols around the home direction. (B) Pigeons with blocked olfaction were much less likely to orient accurately toward the home loft on release. After Papi [887].

2. The classic ethological approach to the study of nervous systems showed that some animals possess neural elements shaped by natural selection to detect key stimuli and to order appropriate species-specific responses to these biologically relevant cues. Modern neurophysiological studies confirm that nervous systems contain highly specialized sensory receptors whose design facilitates the acquisition of critical information from the environment, as well as interneurons and motor cells with unique properties that contribute to particular behavioral abilities. Moreover, central pattern generators within the central nervous system can produce a programmed series of messages to selected muscles, facilitating complex patterned responses to certain stimuli.

3. Stimulus filtering, the filtering out of irrelevant information, occurs at all levels in all nervous systems. Sensory receptors ignore some stimuli in favor of others, while interneurons relay some, but not all, of the messages they receive. Within the central nervous system, many cells and circuits are devoted to the analysis of certain categories of information, although this means that other inputs are not as thoroughly processed.

4. The proximate mechanisms of stimulus filtering are the basis for adaptive perception and adaptive responses. By discarding some potential information, animals are able to focus on the biologically relevant stimuli in their environment, increasing the odds of a prompt and effective reaction to these key stimuli. Because the obstacles to reproductive success differ among species, selection has resulted in the evolution of proximate mechanisms with different design features. That is why some animals have perceptual abilities not possessed by humans, such as the capacity to hear ultrasound, or to see ultraviolet light, or to form olfactory maps useful for navigating home across unfamiliar terrain.

Discussion Questions

1. After Kenneth Roeder discovered that some noctuid moths avoid bats by detecting ultrasound, he established that the A1 receptor in certain noctuids does not discriminate between ultrasounds of different frequencies. He made this discovery by formulating a hypothesis, developing predictions, performing tests of the predictions, and coming to a conclusion. Formally reconstruct this process.

2. When cockroaches are attacked by a toad, they turn and run away. A roach has wind sensors that react to puffs of air pushed ahead of approaching predators. These sensors are concentrated on its cerci, two thin projecting appendages at the end of its abdomen. One cercus points slightly to the right, the other to the left. Use what you know about moth orientation to bat cries to suggest how this simple system might provide the information the roach needs to turns away from the toad, rather than toward it. How might you test your hypothesis experimentally? See [191] after answering.

3. Males of the Australian katydid *Sciarasaga quadrata* sing at an unusually low frequency (around 5 kHz) and at an unusually low intensity (about 20 decibels lower than other similar katydids). This katydid can control the size of the openings to its ears, which are located in its forelegs. When the ears are fully open, the katydid's auditory receptors fire most often in response to (i.e., they are tuned to) sounds of 15–20 kHz. Why is this surprising? What prediction would you make (and why) about the tuning curve of the auditory receptors when the ear opening is partly closed? How might selection pressure from parasitic flies have contributed to the evolution of this katydid's signal-producing and signal-receiving apparatus? What predictions follow from your antiparasite hypothesis? See [983] after answering.

4. Fiddler crabs live on tidal flats, where male crabs court females and where predators, such as gulls, hunt the crabs. Crabs often approach or ignore one another, but a gull flying toward them sends the crabs dashing for their tunnels in the mud. Crabs keep their eye stalks perpendicular to the flat ground on which they roam (see the photograph on the next page). How would you test the hy-

pothesis that fiddler crabs hold their eye stalks in this position in order to use the following simple rule of thumb: Flee from any moving stimulus that appears above the level of the eye (at the top of the stalk) and approach or ignore any moving stimulus that appears below eye level? After answering, check [687]. Apply ethological terminology to this system by identifying the releaser, the fixed action pattern, and the innate releasing mechanism. Photograph by John H. Christy, Smithsonian Tropical Research Institute.

5. Cortical magnification occurs in human beings as well as in all other mammals. The drawing below illustrates a representation of the amount of cortical tissue devoted to tactile inputs from different parts of the human body. Provide some hypotheses on why differences exist between our cortical map and those of the insectivores discussed earlier in this chapter.

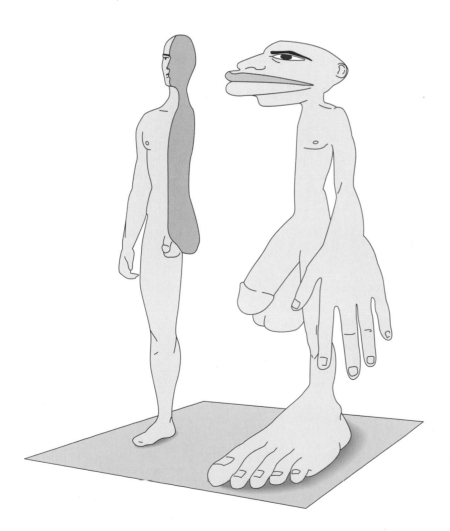

Suggested Reading

Kenneth Roeder's *Nerve Cells and Insect Behavior* [976] is a classic on how to conduct research on the physiology of behavior. Modern textbooks on the neurophysiology of behavior have been written by Jeffrey Camhi [191], Peter Simmons and David Young [1063], and Thomas Carew [194]. For analysis of a highly specialized sensory ability not covered in the text, read about the electric sense in certain fishes and the uses to which electroreception is put [548, 597, 598]. Good general reviews of animal navigation can be found in [1199]. *Mapping the Brain* by Rita Carter [202] is a beautiful book about the operation of the human brain. *Evolving Brains* by John Allman focuses on the evolution of the human brain, but contains a wealth of fascinating information on other species as well [25].

6

The Organization of Behavior: Neurons and Hormones

*I*n the previous chapter, we talked about animals as if they were endowed with neural computers designed to detect key stimuli, discriminate among patterns of inputs, and order adaptive reactions. So, for example, when a flying moth is exposed to ultrasound, its receptors react, setting in motion a chain of neural events, which lead to an adaptive response to this acoustical stimulus. The capacity of neural mechanisms to filter out irrelevant information, to perceive some things very reliably, and to order effective reactions makes biological sense.

But nervous systems do more than just activate response X in the presence of stimulus Y. Imagine a flying male moth hot on the scent trail of a distant pheromone-releasing female. If the male's nervous system operated simply by switching on flight behavior whenever female sex pheromone was present in the air, the scent-tracking moth would be unable to avoid bats. But the moth's nervous system does not operate in such a simple-minded way. Instead, a male moth usually aborts his flight toward a source of female sex pheromone when he hears loud ultrasonic pulses [3]. Thus, the male gives precedence to avoiding bats, thanks to his neural networks.

◀ *These male red-sided garter snakes emerging from hibernation are ready to mate, despite the fact that they have almost no testosterone in their blood. Photograph by François Gohier*

mechanism with a built-in cycling schedule that acts independently of any cues from the animal's surroundings. That such environment-independent timing mechanisms might exist should be plausible to anyone who has flown across several time zones and then tried to adjust immediately to local conditions. The second theory is that animals alter the relationships between command centers in their nervous systems strictly on the basis of information gathered by mechanisms that monitor the surrounding environment. Such devices would enable individuals to modulate their behavior in response to certain changes in the world around them.

Let's consider these two possibilities in the context of the calling cycle of male *Teleogryllus* crickets. Each day's calling bout could begin at much the same time because the crickets possess an internal timer that measures how long it has been since the last bout began; they could use this environment-independent system to activate the onset of a new round of chirping each evening at dusk. Alternatively, the insect's neural mechanisms might be designed to initiate calling when light intensity or some other environmental variable reached a particular threshold level. If this second hypothesis is correct, then crickets held under constant environmental conditions that provide no time cues should show no cyclical pattern of calling. But in fact, laboratory crickets held in rooms in which the temperature stays the same and the lights are on (or off) all day long still continue to call regularly for several hours each day. Under conditions of constant light, calling starts about 25 to 26 hours later than it did the previous day (Figure 5). A cycle of activity that is not matched to environmental cues is called a **free-running cycle**. Because the free-running cycle of cricket calling deviates from the many 24-hour environmental cycles caused by the earth's daily rotation about its axis, we can conclude that the cyclical pattern of cricket calling is caused in part by an environment-independent internal **circadian rhythm** ("circadian" means "about a day").

Now let's place our crickets in a regime of 12 hours of light and 12 hours of darkness. The switch from light to dark offers an external environmental cue that the crickets can use to adjust their timing mechanism. In a few days, the males will all start to call about 2 hours before the lights go off, accurately anticipating nightfall, and they will continue until about 2.5 hours before the lights

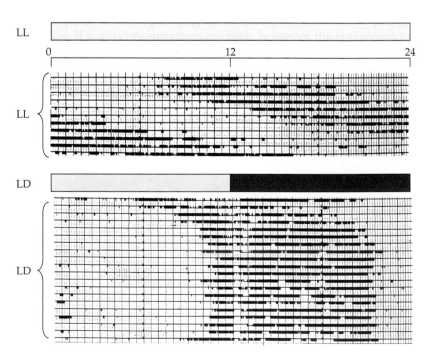

5 Circadian rhythms in cricket calling behavior. Each horizontal line on the grid represents one day; each vertical line represents a half hour on a 24-hour time scale. Dark marks indicate periods of activity—in this case, calling. The bars at the top and middle of the figure represent the lighting conditions; thus, for the first 12 days of this experiment, male crickets were kept in constant light (LL), and for the remainder, they were subjected to 12-hour cycles of light and dark (LD). Male crickets held under constant light exhibit a daily cycle of calling and noncalling, but the calling starts later each day. The onset of "nightfall" on day 13 acts as a cue that reschedules the calling rhythm, which soon stops shifting and eventually begins an hour or two before the lights are turned off each day. After Loher [716].

go on again in the "morning" (see Figure 5). This cycle of calling matches the natural one, which is synchronized with dusk; unlike the free-running cycle, it does not drift out of phase with the 24-hour day, but is reset each day so that it begins at the same time in relation to lights-out [716]. From these results we can conclude that the complete control system for cricket calling has both environment-independent and environment-dependent components: an environment-independent timer, or **biological clock,** set on a cycle that is not exactly 24 hours long, and an environment-activated device that synchronizes the clock with local conditions.

How Do Circadian Mechanisms Work?

In their studies of the circadian mechanisms of crickets and many other animals, investigators have been able to infer something about the properties of these systems by examining how they react to various environmental manipulations, usually involving changes in light and dark regimes, as just illustrated with crickets. But behavioral biologists have also employed a technique similar to that used by Roeder in his studies of mantis nervous systems—namely, they have surgically disconnected various parts of the nervous system in order to test predictions about the possible roles played by the different components. Thus, if one cuts the nerve carrying sensory information from the eyes of a male cricket to the optic lobes of his brain (Figure 6), depriving him of his vision, he enters a free-running cycle. Visual signals are evidently needed to reset the daily rhythm to local conditions, but a rhythm persists in the absence of this information. If, however, one separates both optic lobes from the rest of the brain, the calling cycle breaks down completely; the cricket will now call with equal probability at any time of the day. These results are consistent with the hypothesis that a master clock mechanism (Figure 7) resides within the optic lobes, sending messages to other regions of the nervous system [587, 880].

The search for the location and operating rules of a putative master clock in mammals and other vertebrates led researchers to look within the hypothalamus,

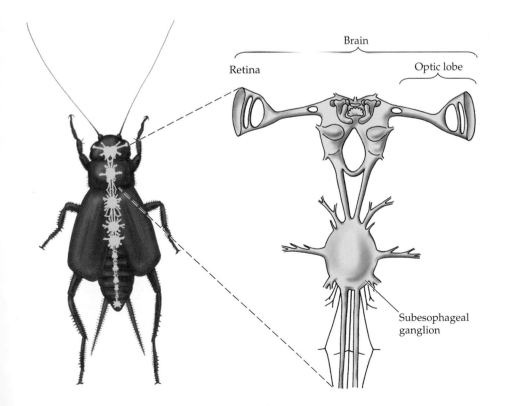

Retina

Brain

Optic lobe

Subesophageal ganglion

6   The cricket nervous system. Visual information from the eyes is relayed to the optic lobes of the cricket's brain. If the optic lobes are surgically disconnected from the rest of the brain, the cricket loses its capacity to maintain a circadian rhythm. Based on diagrams by F. Huber and W. F. Shurmann.

7 A master clock may, in some species, act as a pacemaker that regulates the many other mechanisms controlling circadian rhythms within individuals. After Johnson and Hasting [587].

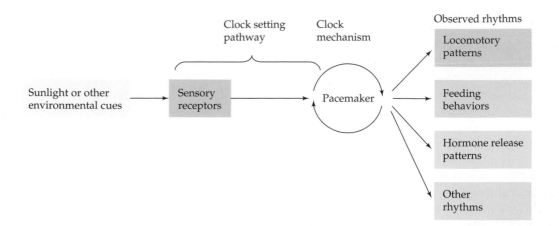

which was known to be a key regulatory area of the brain. The hypothalamus contains a structure called the suprachiasmatic nucleus (SCN), which consists of a pair of neuron clusters that receive inputs from nerves originating in the retina. The SCN is therefore the kind of mechanism that could conceivably secure information about day and night length, information that could be used to regulate a master biological clock.

If the SCN contains a pacemaker that is critical for maintaining circadian rhythms, then damage to the SCN should cause individuals to lose those rhythms. Such an experiment has been done by selectively destroying SCN neurons in the brains of hamsters and white rats, which subsequently exhibit arrhythmic patterns of hormone secretion, locomotion, and feeding [1288]. If arrhythmic hamsters receive transplants of SCN tissue from fetal hamsters, they sometimes regain their circadian rhythms, but not if the tissue transplants come from other parts of the fetal hamster brain [310].

Perhaps the SCN clock operates via rhythmic changes in gene activity, and perhaps that rhythm can be influenced by external stimuli. A key gene in this regard appears to be the *per* gene, which codes for a protein (PER) whose production varies over a 24-hour schedule in concert with the product of another mammalian gene called *tau* (Figure 8). The product of *tau* is an enzyme whose production is turned on when PER is at peak abundance in the cell. The enzyme degrades PER, contributing to a 24-hour cycle in which PER first increases in abundance and then falls [1275].

A striking feature of this system, whose complexity I have barely hinted at, is that the key genes regulating cellular circadian rhythms in mammals are also present in insects. *Drosophila* fruit flies and honey bees also have the *per* gene, a chain of DNA composed of somewhat over 3500 base pairs, which provides the information needed to produce the PER protein chain of about 1200 amino acids. Alterations in the base sequence involving as little as a single base substitution can result in dramatically different circadian rhythms in fruit flies (Figure 9), as well as in humans (carriers of one *per* mutation typically fall asleep at 7:30 in the evening and arise at 4:30 in the morning [1144]). These results strongly suggest that the gene's information plays a critical role in enabling circadian rhythms.

If the *per* gene does play this role, then animals in which the *per* gene is relatively inactive should behave in an arrhythmic fashion. The *per* gene is relatively inactive in young honey bees, which generally remain within the hive to care for eggs and larvae. And young honey bees are in fact just as likely to perform these tasks at time X as at time Y over a 24-hour period. In contrast, older honey bees, which forage for food during the daytime only, exhibit highly defined circadian rhythms, leaving the hive to collect pollen and nec-

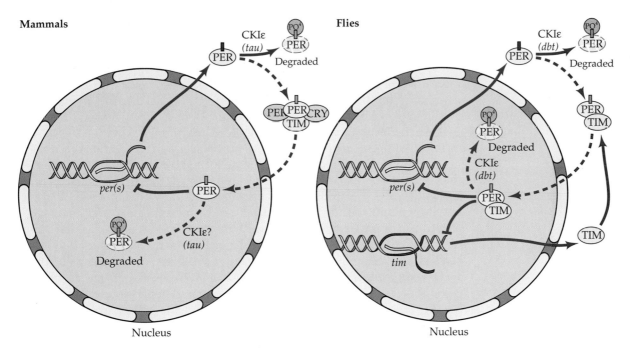

Mammals

Flies

8 **The genetics of biological clocks in mammals and fruit flies.** In both groups, a set of three key genes produces proteins that interact to regulate the activity of certain other genes on a cycle lasting approximately 24 hours. One of the genes (*per*) codes for a protein (PER) that gradually builds up inside and outside the cell nucleus over time. Another key gene is *tau* in mammals and *dbt* in flies, which code for an enzyme that helps break down PER, slowing its rate of accumulation in the cell. But during peak periods of production of PER, more PER is available to bond with another protein (TIM) coded for by a third gene (*tim*). The PER-TIM complex does not degrade, so more PER re-enters the nucleus, where it blocks the activity of the very gene that produces it, though only temporarily. Then a new cycle of *per* gene activity and protein production begins. After Young [1275].

tar only during that part of the day when the flowers they seek are most likely to be resource-rich. Foragers have almost three times as much of a particular chemical in their brain cells that is produced as a result of an active *per* gene than do young nurse bees [1146].

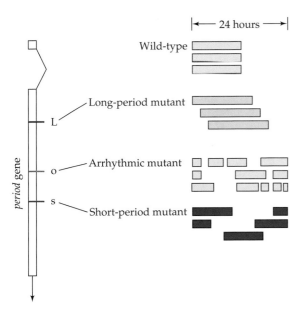

9 **Mutations of the *per* gene affect the circadian rhythms of fruit flies.** On the left is a diagram of the DNA sequence that constitutes the *per* gene. The locations of the base substitutions present in three mutant alleles found in fruit flies are indicated on the diagram. The activity patterns of wild-type flies, and those associated with each mutation, are shown on the right. After Baylies et al. [81].

The fact that the fruit fly, the hamster, the honey bee, and you and I all have the same gene serving much the same function indicates that we inherited this gene, as well as some others involved in the regulation of activity patterns, from a very ancient animal that lived perhaps 550 million years ago [1275]. In mammals and other vertebrates, but not insects, the *per* gene plays its role within the neurons of the SCN, which communicates with the pineal gland, a small organ tucked into the brain. Here too the communication has genetic effects [398, 1095]. In brief, chemical signals from SCN neurons activate a specific gene (CREM) in cells of the pineal gland, a gene whose protein product (ICER) is presumably involved in some way in the manufacture of a substance called melatonin. As the ICER protein builds up in the pineal cells, however, it begins to inhibit activity in the very gene needed to make more ICER. (Note that this is another case of self-regulating negative feedback of the sort that appears to occur in the SCN clock mechanism.) However, the sensitivity of the CREM gene to feedback inhibition from its protein product varies depending on the photoperiod (hours of daylight in a 24-hour period) that the individual has experienced in the previous few days. Animals that have been exposed to a series of "nights" that are 8 hours long show peak activity in the CREM gene about 6 hours after the lights go off; in contrast, animals that have been exposed to 12-hour nights for a couple of weeks reach peak CREM activity about 10 hours after the lights go off (Figure 10). In other words, the pineal gland alters its biochemical output in relation to the recent history of photoperiods that the animal has experienced. This ability enables the pineal to adapt to seasonal changes in day length, thereby helping the animal adjust its daily schedule to these changes.

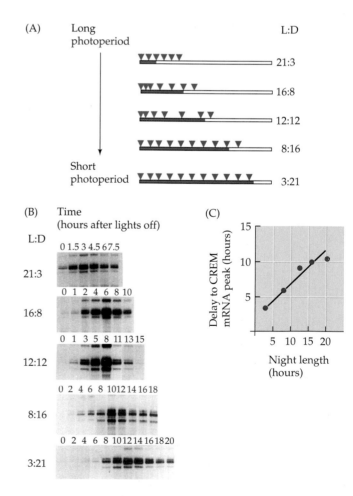

10 Gene activity varies with photoperiod. The number of hours after dark before the CREM gene reaches peak activity depends on the number of hours of light an animal experienced during the preceding days. (A) White rats were placed under several experimental light:dark (L:D) regimes. Samples of rats were killed at the times indicated by the arrows. (B) The rats' pineal glands were analyzed to determine how much of the messenger RNA coded for by the CREM gene was present. (C) The peak production of this mRNA occurred from 3 to 12 hours after lights out, depending on the photoperiod that the animals had recently experienced. After Foulkes et al. [398].

Thus, in some mammals, the SCN apparently contains a central pacemaker that sends signals to the pineal gland. The pineal cyclically changes its production of melatonin with adjustments for shifts in photoperiod length, integrating the environment-independent and environment-dependent elements that regulate daily changes in behavior. (Melatonin's likely role in controlling daily activity patterns has led to its widespread use in capsule form by travelers wishing to avoid jet lag.) The adaptive value of the environment-independent components of such mixed systems may be that they enable individuals to alter the timing of their behavioral and physiological cycles without having to constantly check the environment. At the same time, the presence of an environment-dependent element permits individuals to fine-tune their cycles in keeping with subtle variations in their particular environments. As a result, a nocturnal mammal will become active at about the right time each night, but will gradually shift its activity cycle to accommodate the changes in day length as spring becomes summer, or summer becomes fall.

It would be intriguing to study the genetics and physiology of the SCN and pineal gland in a mammal that is largely divorced from the day–night cycle. The naked mole-rat is such an animal. Naked mole-rats live in colonies in an extensive network of underground tunnels (see p. 451). The animals almost never come to the surface, but instead feed on roots and tubers collected from their tunnels. They have no special dependence on what is going on above them during the day or night. As you might predict, naked mole-rats exhibit no circadian rhythms of any sort (Figure 11). Instead, individuals scatter generally brief episodes of wakefulness among longer periods of sleep, with the pattern changing irregularly from day to day [300].

Long-Term Cycles of Behavior

Because of their unusual lifestyle, naked mole-rats do not have to deal with cyclically changing environments, and they have apparently lost their circadian clocks as a result. But almost all other creatures confront not just daily changes in access to food or risk of predation, but also changes that cover periods longer than 24 hours, such as the annual cycles of seasonal changes that occur in many parts of the world. If circadian clocks enable animals to prepare physiologically and behaviorally for certain predictable daily changes in the environment, might not some animals possess a circannual clock that runs on an approximately 365-day cycle [467]? Such a mechanism might be similar to the circadian master clock, with an environment-independent timer capable of generating a circannual rhythm in conjunction with a mechanism that keeps the clock set to local conditions.

Testing the hypothesis that an animal has a circannual rhythm is technically difficult because individuals must be maintained under constant conditions for at least 2 years after their removal from the natural environment. One successful study of this sort involved the golden-mantled ground squirrel [894] of temperate North America, which in nature spends the late fall and winter hibernating in an underground chamber. Five members of this species were born in captivity, then blinded and held thereafter in constant darkness and at a constant temperature while supplied with an abundance of food. Year after year, these ground squirrels entered hibernation at about the same time as their fellows living in the wild (Figure 12).

In another study, several nestling stonechats were taken from Kenya to Germany to be reared in laboratory chambers in which the temperature and photoperiod were always the same. Needless to say, these birds, and their offspring, never had a chance to encounter the spring rainy season in Kenya, which heralds a period of insect abundance and is the time when Kenyan stonechats produce their offspring, which must be fed many insects. The transplanted stonechats, despite their unnatural constant environment, also exhibited an

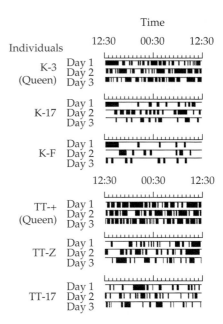

11 Naked mole-rats lack a circadian rhythm. Patterns of activity are shown for six individuals from two captive colonies held under constant low light. Dark bars indicate periods when the individual was awake and active. After Davis-Walton and Sherman [300].

12   Circannual cycle of the golden-mantled ground squirrel. Animals held in constant darkness and at a constant temperature nevertheless entered hibernation (green bars) at certain times year after year. After Pengelley and Asmundson [894].

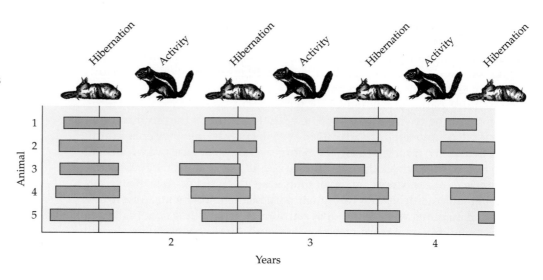

Stonechat

13   Circannual rhythm in a stonechat. When transferred from Kenya to Germany and held under constant conditions, this male stonechat still underwent a regular long-term cycle of testicular growth and decline (purple lines) as well as feather molts (bars). The cycle was not 12 months long, however, so that the timing of molting and testicular growth shifted over the years (see the dashed lines that angle upward from left to right). After Gwinner and Dittami [468].

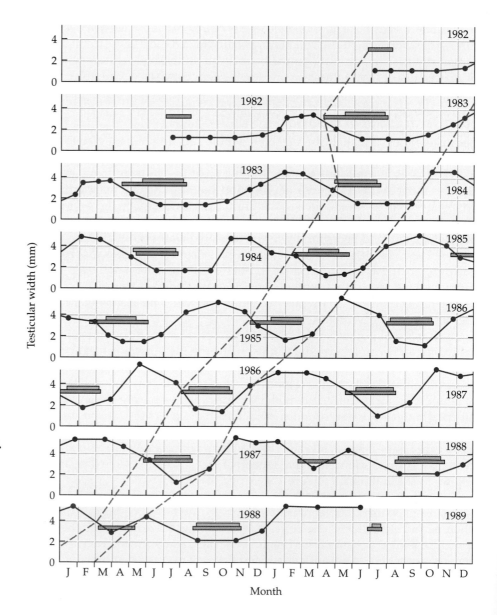

annual cycle of reproductive physiology and behavior, but one that shifted out of phase with that of their Kenyan compatriots over time (Figure 13). One male, for example, went through nine cycles of testicular growth and decline during the 7.5 years of the experiment. Evidently, the stonechat's circannual cycle is generated in part by an internal, environment-independent mechanism, just as is true for the golden-mantled ground squirrel [469].

The Physical Environment Influences Long-Term Cycles

In nature, environmental cues reset circadian and circannual clocks so as to produce behavioral rhythms that match the particular features of the animal's environment, such as the times of sunrise and sunset, or the onset of the rainy season in a given year, or the increasing day lengths associated with spring. This fine-tuning of behavioral cycles involves mechanisms of great diversity that respond to a full spectrum of environmental influences. These environmental cues vary from species to species according to their ecological circumstances.

The banner-tailed kangaroo rat, for example, is an animal whose readiness to forage for food changes over the course of the lunar cycle [713, 714]. Robert Lockard and Donald Owings reached this conclusion by monitoring the activity of free-living kangaroo rats in a valley in southeastern Arizona. To measure kangaroo rat activity, Lockard invented an ingenious food dispenser/timer that released very small quantities of millet seed at hourly intervals. To retrieve the seeds, an animal had to walk through the dispenser, depressing a treadle in the process. The moving treadle caused a pen to make a mark on a paper disk that turned slowly throughout the night, driven by a clock mechanism. When the paper disk was collected in the morning, it carried a temporal record of all nocturnal visits to the dispenser.

Data collection was sometimes frustrated by ants that perversely drank all the ink or by Arizonan steers that trampled the recorders. Nevertheless, Lockard's records showed that when the kangaroo rats had accumulated a large cache of seeds in the fall, they were selective about foraging, usually coming out of their underground burrows only at night when the moon was not shining (Figure 14). Because the predators of kangaroo rats (coyotes and owls) can see their prey more easily in moonlight, banner-tails probably are safer when foraging in complete darkness. For this reason, the kangaroo rats apparently possess a mechanism that enables them to shift their foraging schedule in keeping with nightly moonlight conditions. Perhaps kangaroo rats simply avoid foraging under bright moonlight, or perhaps the rodents have an internally driven circalunar rhythm that approximates the lunar cycle.

White-crowned sparrows also undergo dramatic changes in behavior in keeping with seasonal changes in the physical environment. In the spring, males return from distant wintering grounds to their summer headquarters in the northern United States, Canada, and Alaska. There they establish breeding territories, fight with rivals, and court sexually receptive females. In concert with these striking behavioral changes, the gonads of the birds grow with dramatic rapidity, regaining all the weight lost during the winter, when they shrink to 1 percent of their breeding-season weight. In order to properly time the regrowth of their gonads and the onset of their reproductive activities, the birds must somehow anticipate the spring breeding season. How do they manage this feat?

The sparrows' ability to change their physiology and behavior depends on their capacity to detect changes in the photoperiod, which grows longer as spring advances in temperate North America [379]. One hypothesis on how such a system might work proposes that the clock mechanism of white-crowns exhibits a daily cyclical change in sensitivity to light, with a cycle that is reset each morning at dawn. During the initial 13 hours or so after the clock is reset, this mechanism is highly insensitive to light; this insensitivity then steadily gives way to increasing sensitivity, which reaches a peak 16 to 20 hours after the start-

14   Lunar cycle of banner-tailed kangaroo rats. Each thin black mark represents a visit made by a banner-tail to a feeding device with a timer. From November to March, the rats were active at night only when the moon was not shining. A shortage of seeds later in the year caused the animals to feed throughout the night, even when the moon was up, and later still to forage during all hours of the day. After Lockard [713].

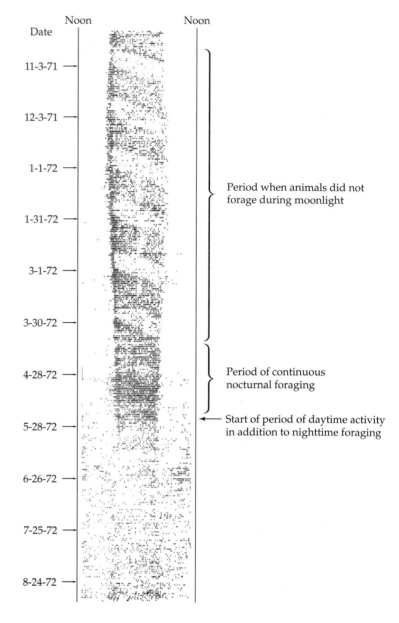

Banner-tailed kangaroo rat

ing point in the cycle. Photosensitivity then fades very rapidly to a low point 24 hours later, at the start of a new day and a new cycle. Therefore, if the days are 12 or 13 hours long and the nights 12 or 11 hours long, the system will never become activated because no light is present during the light-sensitive phase of the cycle. However, if the days are 14 or 15 hours long, light will reach the bird's brain during the photosensitive phase, initiating a series of hormonal changes that lead to the development of its reproductive equipment and the drive to reproduce.

If this model of the photoperiod-measuring system is correct, it should be possible to deceive the system. William Hamner, working with house finches [488], and Donald Farner, in similar studies with white-crowned sparrows [378], stimulated testicular growth by exposing captive birds to light during the hypothesized photosensitive phase of their circadian rhythms. In Farner's experiment, birds that had been on a regular schedule of 8 hours of light and 16 hours of darkness (8L:16D) were shifted to an 8L:28D schedule. Because the light periods were now out of phase with a 24-hour cycle, these birds sometimes received light during the time when their brains were predicted to be highly photosensitive. The male birds' testes grew under these conditions, even though there

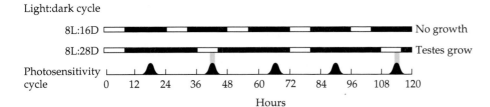

Light:dark cycle

15 A cycle of photosensitivity. An experiment with white-crowned sparrows tested the hypothesis that these birds possess a clock mechanism that is especially sensitive to light between hours 17 and 19 of each day. The lower line represents these hypothetical periods of photosensitivity. The open and solid sections of the two upper horizontal bars show the light and dark periods of two different light–darkregimes. Only sparrows under the 8L:28D experimental regime were exposed to light during the supposed photosensitive phase of the cycle, and only they responded with testicular growth. After Farner [378].

was a lower ratio of light to dark hours than under the 8L:16D cycle, which did not stimulate testicular growth (Figure 15) [378].

Crossbills exhibit a much less constrained pattern of seasonal breeding than white-crowned sparrows do. Craig Benkman has shown that food intake, not photoperiod length, appears to be the primary determinant of breeding in both the white-winged and the red crossbill [89]. He found that these birds will breed in most months of the year, as long as they can secure enough pine kernels to sustain themselves and a brood of offspring.

Thomas Hahn, however, noticed a break in crossbill breeding in December and January (Figure 16). Therefore, even though these birds are more flexible and opportunistic than the average songbird, perhaps they too have an underlying reproductive cycle dependent upon photoperiod. When Hahn held red crossbills at a constant temperature with unlimited access to their favorite food, while letting the birds experience natural photoperiod changes, he found that male testis length fluctuated in a cyclical fashion (Figure 17), becoming much reduced during October through December, even when the birds had all the pine seeds a crossbill could hope for [476]. In addition, free-living red crossbills undergo declines in sex hormone concentrations in the fall and a marked decrease in gonad size, even in areas where their food is abundant [477]. Therefore, the reproductive opportunism of the bird is not absolute, but rather is superimposed on the proximate photoperiod-driven mechanism characteristic of temperate-zone songbirds.

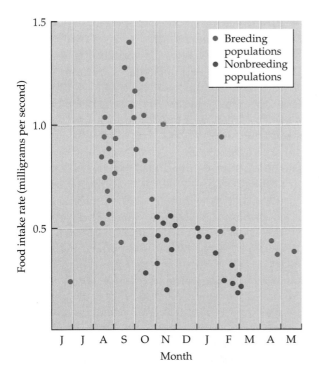

White-winged crossbill

16 Food intake and reproductive timing in the white-winged crossbill. Breeding populations usually occur in areas with relatively high food availability. Nonbreeding populations generally occur in areas where the birds have low food intake. Note, however, the absence of breeding populations in December and January. After Benkman [89].

17   Photoperiod affects testis size in the red crossbill. Six captive birds were held under natural photoperiods, which changed over the seasons, but the temperature and food supply were held constant. The curve represents the average testis length among these birds at different times during the year. After Hahn [476].

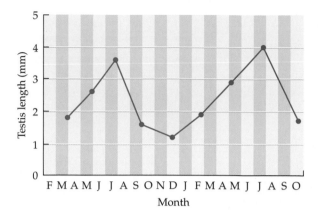

Red crossbill

The persistence of the standard timing system in the flexibly breeding crossbills could be explained at the evolutionary level in at least two ways: (1) the photoperiod-driven mechanism might be a nonadaptive holdover from the past, or (2) crossbills might derive reproductive benefits from the retention of a physiological system that reduces the likelihood that they will attempt to reproduce at times when factors other than food supply (such as cold temperatures) make successful reproduction unlikely.

Changing Priorities in Changing Social Environments

As we have seen, different features of the physical environment, such as moonlight, day length, or food supply, are used by different species to regulate their behavioral priorities. In addition, some animals possess physiological mechanisms that permit them to adjust behaviorally to changes in their social environment. Thus, for example, when Hahn and several co-workers performed another experiment on crossbills in which some captive males were caged with their mates, while others were forced into bachelorhood but were kept within sight and sound of the paired crossbills in a neighboring aviary, the bachelor males experienced a slower return to reproductive condition after the winter break than did the paired males [478].

The social environment also affects the behavioral priorities of house mice. After a male house mouse mounts a female and ejaculates, he immediately becomes highly aggressive toward mouse pups, which he will kill if he finds them. He remains prone to commit infanticide for almost 3 weeks, but then becomes more and more paternal. At this point he will protect and care for any young pups he encounters. When about 7 weeks have passed since ejaculation, he becomes infanticidal once again [900].

This remarkable cycle has clear adaptive value. After a male transfers sperm to a partner, 3 weeks pass before she gives birth. Attacks on pups during these 3 weeks will invariably be directed against a rival male's offspring, with all the benefits attendant upon their elimination (see p. 14). After 3 weeks, a male that switches to the paternal mode will almost always care for his own neonatal offspring. After 7 weeks, his weaned pups will have dispersed, so that once again he can practice infanticide advantageously.

At the proximate level, what kind of mechanism could enable a male to switch from infanticidal Mr. Hyde to paternal Dr. Jekyll 3 weeks after a mating? One possible explanation involves an internal timing device that records the number of days since the male last copulated. If such a sexually activated timing mechanism exists, then an experimental manipulation that either increases or decreases the length of a "day," as perceived by the mouse, ought to have an

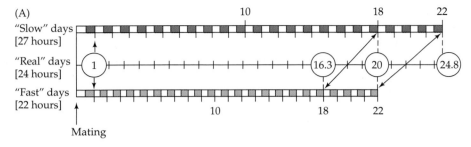

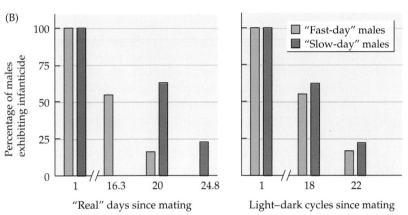

(B)

Percentage of males exhibiting infanticide

"Real" days since mating

Light–dark cycles since mating

- "Fast-day" males
- "Slow-day" males

18 Regulation of infanticide by male house mice. (A) Male mice were held under artificial "slow-day" and "fast-day" experimental conditions. (B) Most of the males held under fast-day conditions had stopped being infanticidal by 20 real days (= 22 fast days) after mating; males experiencing slow days did not show the same decline until nearly 25 real days had passed. After Perrigo et al. [900].

effect on the absolute amount of time that passes before the male makes the transition from killer to caregiver.

Glenn Perrigo and his co-workers manipulated day length by placing groups of mice under two laboratory conditions, one with "fast days," in which 11 hours of light were followed by 11 hours of darkness (11L:11D) to make a 22-hour "day," and another with "slow days" (13.5L:13.5D) that lasted for 27 hours. As predicted, the total number of light–dark cycles, not the number of 24-hour periods, controlled the infanticidal tendencies of males (Figure 18). Thus, when male mice in the fast-day group were exposed to mouse pups 20 real days after mating, only a small minority committed infanticide because these males had experienced 22 light–dark cycles during this period. In contrast, more than 50 percent of the males in the slow-day group attacked neonates at 20 real days after mating because these males had experienced only 18 light–dark cycles during this time. These results demonstrate that the males' timing devices register the number of light–dark cycles that have occurred since mating, and that this information provides the proximate basis for the control of the infanticidal response [900].

The Effects of Copulation in the Green Anole

Copulation alters the behavioral priorities of many animals, not just house mice. In the green anole, for example, females become sexually receptive in the spring, after the winter hibernation. The process is gradual and can be affected by the degree to which the female lizards are courted by dewlap-displaying males (see Figure 19, p. 95). Those females that are often courted secrete more pituitary gonadotropic hormones, which speed ovarian development and the consequent production of estrogen by the mature ovaries [259]. Estrogen travels from the ovaries via the bloodstream to the brain, affecting various endocrine and neural target cells there. Through the action of specific hormones, the female's brain eventually becomes primed to activate precopulatory neck-arching behavior in response to a territorial male's courtship signals. Should such a female encounter an adult territorial male that flags her with his

Green anoles

extended dewlap as he bobs his head up and down, she is likely to copulate with him.

The female anole provides a fine example of the multiplicity of effects exerted by hormones on the physiological foundations of behavior. As we saw in Chapter 4, hormones play major roles in the *development* of structures, such as the brain and ovaries, that play direct and indirect roles in controlling an animal's behavior. These chemicals also have a host of *activational* effects, such as altering the neural networks that respond to the communication signals that female anoles receive from males, priming the females to do some things rather than others.

If the amount of estrogen in the female's bloodstream is indeed a key proximate factor that lowers her threshold for sexual behavior, then removal of an adult female anole's estrogen-producing ovaries should abolish sexual receptivity (which it does), and implantation of an estrogen pellet in an ovariectomized, unreceptive female should restore her sexual receptivity (which it does). But once a female has mated, her receptivity ceases within 5 to 7 minutes. Thereafter, she will ignore, run from, or even attack any male that dares court her over the next 10 to 14 days, until she has a mature egg once more (Figure 19) [259, 785].

This dramatic postcopulatory change in the female's receptivity (at a time when she still has high concentrations of estrogen in her blood) might be caused by stimuli associated with mating. For example, the experience of being mounted and bitten on the neck by a copulating male might provide sensory stimulation that serves to reset the sexual receptivity command center, assuming that one exists in female anoles. If this hypothesis is correct, then if a female is courted and then mounted by a male anole whose double penis has been surgically removed, she should become unreceptive even though a functional mating has not occurred. Contrary to this prediction, females retain their willingness to copulate after the penis-deprived male dismounts.

Instead, female anoles apparently require mechanical stimulation of the genital tract during copulation, to which they respond by producing prostaglandin (PG), a hormone that becomes more abundant in a female's bloodstream immediately following a mating. If PG somehow acts as an internal cue that activates a change in sexual receptivity, then receptive, unmated females that receive an injection of PG ought to become unwilling to copulate. In fact, shortly after receiving such an injection, the experimentally treated females react to courting males by dashing away or attacking (Figure 20) [1145]. Note that the behavioral effects of PG are far more rapid than the slow priming action of estrogen on female responsiveness to male courtship displays.

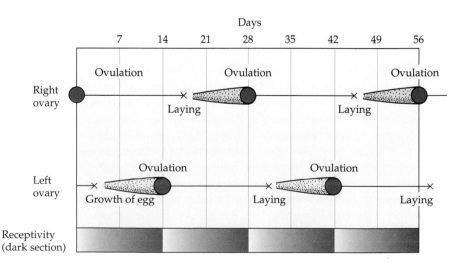

19 **Cycle of sexual receptivity in the female green anole during the breeding season.** One egg matures at a time, in either the left or the right ovary. As an egg matures, the female gradually becomes receptive and will mate, after which she abruptly becomes unreceptive. After Crews [259].

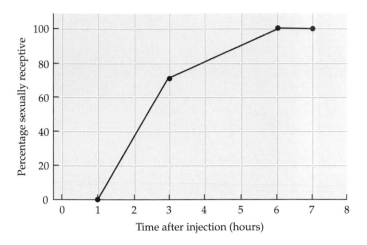

20 Prostaglandin controls sexual receptivity in female green anoles. When injected into seven females, prostaglandin temporarily abolished their readiness to copulate. After Tokarz and Crews [1145].

Hormones Help Organize Social Behavior

We have reviewed just a few examples of the physical and social cues that can influence an animal's behavioral priorities, particularly with respect to the annual cycle of reproduction. These environmental signals, whatever they may be, often exert their effects by changing the hormonal state of the individual. In such cases, hormones act as intermediaries between the environment and behavioral command centers, shifting priorities among those centers so that individuals react appropriately to changing conditions. As we have just seen, PG is a hormone that links the completion of copulation with a swift change in sexual receptivity at the time when the female green anole no longer requires sperm to fertilize her egg. In a host of other species, hormones also underlie the integrated shifts in physiology and behavior that promote reproduction at times when environmental, social, and internal physiological conditions are most favorable.

Hormonal regulation of reproduction is common, particularly in vertebrates, leading to the view that hormones are the key arbiters of sexual behavior, serving as a means of communication between an animal's external environment and its various internal organs of reproduction. According to this theory, certain hormones provide the proximate basis of reproductive behavior, leading to an **associated reproductive pattern** in which gamete production and sexual activity are linked by or associated with increases in particular hormones (Figure 21A).

The male green anole offers a classic example of an associated reproductive pattern, with testosterone concentrations in the blood rising when males in the southern United States are making mature sperm, defending territories, and mating with females. The same pattern applies to red deer stags living in Great Britain, where they mate during September and October. Stags that have been living peacefully with one another all summer become aggressive as they begin to court females. At this time, their testes generate sperm and testosterone, whose behavioral importance has been demonstrated by castrating adult males prior to the rut. The castrated individuals show little aggression and do not try to mate with sexually receptive females. If the behavioral differences between castrated and intact males stem from an absence of circulating testosterone in the castrated stags, then testosterone implants should restore aggressive and sexual behavior during the mating season. Testosterone implants do have these effects, showing that changes in testosterone concentrations promote seasonally adaptive shifts in physiology and behavior [704].

Although many species studied to date appear to possess associated reproductive patterns, the theory that hormonal control underlies reproductive

21 Associated and dissociated reproductive patterns. (A) The traditional view, according to which environmental cues trigger internal hormonal changes, which are necessary to activate behavioral responses. (B) An alternative view, in which mating can either be associated with a surge of gonadal hormones or be dissociated in time in relation to gonadal (and hormonal) activity. After Crews [260].

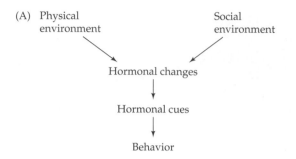

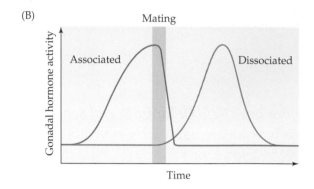

behavior continues to be tested, as it should be [266]. If testosterone is required for sexual behavior in birds, then castration should eliminate male sexual behavior because removal of the testes eliminates a major source of testosterone. This prediction has been confirmed for some species, including the Japanese quail, which if castrated stops reproducing but if given a testosterone implant is back in business [873], just like the red deer. But the prediction fails for white-crowned sparrows. Even without his testes, a male white-crown will mount females that solicit copulations, provided that he has been exposed to long photoperiods [816]. Moreover, some populations of white-crowned sparrows raise more than one brood per breeding season. Those male white-crowns that mate with females to produce a second or third clutch of fertilized eggs in the summer have relatively low testosterone concentrations in their blood at that time (Figure 22), further evidence that testosterone cannot be essential for male sexual behavior in this species.

Nor is the white-crowned sparrow unique among birds. Males of the spotted antbird, a tropical species, are quite capable of courting females when their testosterone concentrations are exceedingly low [1235]. So what is the function of testosterone if it does not increase sexual drive? Perhaps it motivates males of these species to be aggressive. If so, then we can predict that in seasonally territorial birds, testosterone concentrations should be especially high early in the breeding season, when males are defending a territory and engaging in aggressive interactions with rivals. This phenomenon has been observed in white-crowned sparrows and some other birds [1256]. Even the spotted antbird, which does not exhibit elevated testosterone concentrations in any particular seasonal pattern, does boost its testosterone concentrations in response to apparent challenges from outsiders. Broadcasting recorded songs of spotted antbirds in an antbird territory results in a rapid buildup of testosterone in the birds' blood [1235].

The hypothesis that testosterone is a promoter of aggression also leads to another prediction, namely, that testosterone concentrations should remain high longer in males of polygynous species because these individuals attempt to

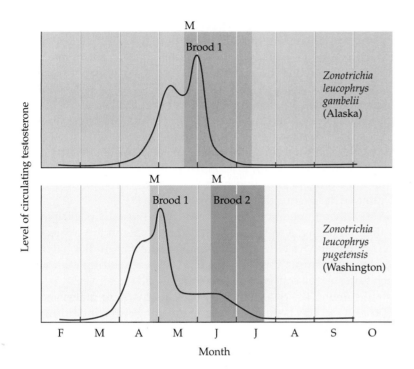

22   **Hormonal and behavioral cycles in single- and multiple-brooded populations of white-crowned sparrows.** Testosterone concentrations in the blood of male white-crowns peak shortly before the time when the males mate with females (M) in their first breeding cycle of the season. In populations that breed twice in one season, however, copulation also occurs during a second breeding cycle, at a time when testosterone concentrations are declining. After Wingfield and Moore [1256].

acquire a series of mates over time. For this reason, polygamous males may have to fight with rivals for longer periods than males of monogamous species, who devote themselves primarily or exclusively to one mate and her brood of offspring. This prediction too has received some support [954].

Even in species in which testosterone facilitates adaptive sexual or aggressive responses, it is the rule that testosterone concentrations fall to almost nothing outside the breeding season or after the likelihood of territorial challenges subsides. Why should this be? Because testosterone has a multiplicity of effects (Figure 23), the hormone may carry a price, actually reducing fitness at certain times or in certain situations. Testosterone appears to interfere with the immune

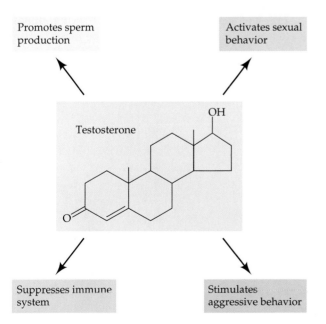

23   **The chemical structure of testosterone** and its diverse effects on physiology and behavior. After Wingfield et al. [1255].

system and other cellular mechanisms that combat disease, which may explain why males of so many animals are more likely than females to be infected by viruses, bacteria, and parasites [629]. High testosterone concentrations may also contribute to high corticosterone concentrations, which may contribute to or be reflective of greater physiological stress [1257], and thus greater vulnerability to disease-causing organisms. In addition, the direct behavioral effects of testosterone can be costly. Males under the influence of testosterone, for example, often become so involved in trying to mate with females or in fighting with rivals that they become sitting ducks for predators or parasites. When male impalas (an African antelope) attempt to acquire herds of females, they become so distracted that they often neglect to groom themselves, which they need to do in order to remove dangerous ticks from their bodies [817].

Moreover, in socially monogamous birds whose males usually participate in the rearing of their offspring, a testosterone-saturated male might neglect his young in favor of defending his territory or singing to attract an additional mate. If so, then if one were to implant testosterone pellets in breeding males of a species in which testosterone concentrations normally fall after an early period of territory establishment, the paternal care provided by these males should decline. This experiment has been done on several songbirds, including the dark-eyed junco [619]. As predicted, testosterone-implanted males behaved differently from control birds in several ways (Figure 24). Testosterone-implanted males fed their young less often than control males, and the reproductive success of these males' partners declined as a result, suggesting that the reduction in parental care had negative consequences for their broods. On the other hand, male juncos with extra testosterone were more likely than control males to mate with females other than their social mates and sire offspring in that way. In experiments done to date, the overall annual reproductive success of testosterone-implanted and control males has been the same. Nor do the testosterone-implanted males experience higher mortality (if their implants are removed after the breeding season is over). Thus, it is possible that the negative effects of testosterone on juncos are matched by certain benefits of the hormone.

We can say with assurance that the pattern of hormonal control of reproduction and territorial aggression varies greatly from species to species, showing that selection has resulted in many modifications of hormonal mechanisms of behavior (Figure 25). In the song sparrow, for example, males defend territories long after the breeding season is over and gonadal sex hormones have essentially disappeared. But these birds produce nongonadal steroids, especially estrogen, which plays a critical role in the post-breeding territorial phase. Males given a substance called fadrozole (FAD), which blocks the manufacture

Dark-eyed junco

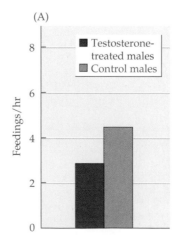

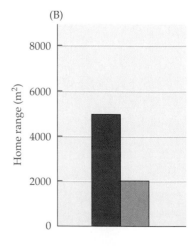

24 **Experimental application of testosterone to male dark-eyed juncos has multiple effects.** (A) Testosterone-treated males feed their broods less often than control males. (B) Treated males also range over larger areas than control males. After Ketterson [619].

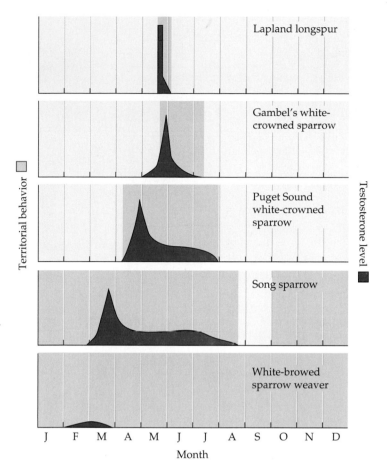

Lapland longspur

Gambel's white-crowned sparrow

Puget Sound white-crowned sparrow

Song sparrow

White-browed sparrow weaver

Territorial behavior □

Testosterone level ■

J F M A M J J A S O N D

Month

25 Testosterone and territorial behavior. No one pattern exists for the relationship between testosterone concentrations and the duration of male territoriality. In some bird species (top three panels), a surge of testosterone occurs at the onset of territoriality and breeding, but in other birds (bottom two panels), males are territorial at times when they have little or no circulating testosterone. After Wingfield et al. [1255].

of estrogen, sing less often and stay farther away from a simulated intruding rival than untreated controls or males that have received both FAD and replacement estrogen (Figure 26) [1084].

The red-sided garter snake offers yet another demonstration that hormonal systems of behavior control are not the same for every vertebrate. This cold-blooded reptile lives as far north as southern Canada; consequently, it spends much of the year dormant in a sheltered underground hibernaculum, which may house thousands of snakes. On warm days in the late spring, the snakes begin

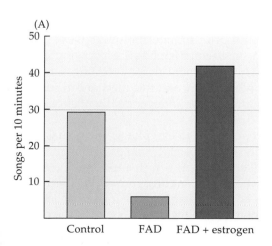

(A)

Songs per 10 minutes

50
40
30
20
10

Control   FAD   FAD + estrogen

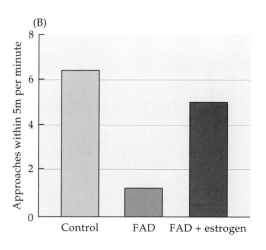

(B)

Approaches within 5m per minute

8
6
4
2
0

Control   FAD   FAD + estrogen

26 Estrogen and territorial behavior. Male song sparrows treated with fadrozole (FAD), a chemical that blocks production of estrogen, were significantly less likely to sing (A) or to come within 5 meters of a simulated intruder (B) than were control males or males given both FAD and estrogen. After Soma et al. [1084].

27 Spring mating aggregation of red-sided garter snakes. The males search for and avidly court emerging females. Photograph by Nic Bishop.

to stir, and soon emerge from their hibernaculum en masse (Figure 27). Before going their separate ways, they engage in an orgy of sexual activity, with males slithering after females and attempting to copulate with them. Although males compete for females by trying to contact receptive partners before their fellow males do, they do not fight with one another for the privilege of mating.

Examination of the sex hormone concentrations in their blood reveals that these nonaggressive snakes have almost no circulating testosterone or any other equivalent substance. Yet they have no trouble mating, so red-sided garter snakes, like white-crowned sparrows, are animals with a **dissociated reproductive pattern** (see Figure 21B) [265]. Various hormonal manipulations have been performed on adult male garter snakes without effect on their sexual behavior. Removal of the pineal gland prior to hibernation, however, produces male snakes that almost always fail to court females the following spring [266]. The pineal gland provides a critical mechanism for detecting temperature increases following a period of hibernation, and this mechanism suffices to activate sexual behavior in male garter snakes independently of circulating testosterone.

This does not mean that testosterone has no role to play in the sexual cycle of the garter snake. High concentrations of testosterone are present in males in the fall and contribute to the production of sperm, which are stored internally over the winter in anticipation of the spring mating frenzy. Furthermore, although temperature increases may be the activational cue for sexual activity, testosterone may play an organizational role in the development of the mech-

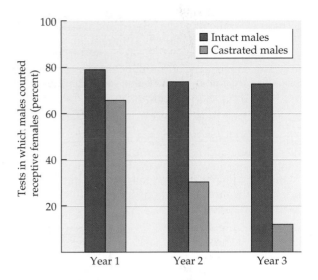

28 **Testosterone and the long-term maintenance of mating behavior.** Male garter snakes whose testes were removed shortly before the breeding season in year 1 remained sexually active during that breeding season, despite the absence of testosterone. But in years 2 and 3, these males became less and less likely to court receptive females compared with males that still possessed their testes. After Crews [262].

anisms underlying reproductive behavior in the red-sided garter snake, as it does in so many other vertebrates. Evidence on this point comes from experiments on adult male snakes that have been castrated. Without their testes, these individuals cannot produce testosterone, but they still exhibit courtship behavior after a period of hibernation under laboratory conditions. If, however, the castrated snakes are tested again after a second bout of hibernation, their sexual activity falls sharply. These results suggest that the surge of testosterone that occurs prior to hibernation primes the neural systems that are needed for sexual behavior the following spring (Figure 28).

This hypothesis receives more support from the finding that testosterone implants given in the summer before the male snake's first hibernation turn 1-year-old males into sexually active animals, although snakes of this age are normally sexually immature. Therefore, testosterone production, which normally begins in the snake's second or third year, appears to be necessary for the full development and maintenance of those mechanisms that control sexual behavior [262]. However, high concentrations of testosterone are not necessary for copulation to occur in the spring, another example of the rule that there is no one hormonal mechanism that regulates the sexual behavior of all animals, or even all vertebrates, in exactly the same way [4, 263].

Summary

1. Because an animal's environment provides various stimuli that could trigger contradictory responses, and because its physical and social environment often change over time, animals gain by having mechanisms that set priorities for their different behavioral options. One such proximate system includes behavioral command centers that mutually inhibit one another, so that animals do not try to do several things simultaneously.

2. As the environment changes, the nature of the inhibitory relationships between neural command centers may also change adaptively. Devices to achieve this end include the various pacemaker or clock mechanisms that regulate nervous system functioning and hormonal output in cycles that typically last either 24 hours or 365 days. Circadian and circannual clocks have environment-independent components, but they can also adjust their performance by acquiring information from the environment about local conditions, such as the time of sunrise or sunset.

3. Hormones constitute a third class of mechanisms for establishing behavioral priorities. In many animals, changes in the physical environment (such as seasonal changes in photoperiod) and in the social environment (such as the presence of potential mates) are detected by neural mechanisms and translated into hormonal messages. The consequent hormonal changes can help set in motion a cascading series of physiological and behavioral changes that make reproductive activity the top priority in the appropriate season of the year.

4. The precise roles played by hormones in effecting behavioral change vary from species to species. Male sexual behavior, for example, may or may not be dependent on high testosterone concentrations in the blood. Species often differ in the mechanisms organizing behavior, which tells us that natural selection has produced a variety of proximate solutions to the problem of regulating command centers in an adaptive manner.

Discussion Questions

1. Produce at least one ultimate hypothesis for why kangaroo rats might use a circadian rhythm to time their daily activity, rather than simply checking from time to time on whether it is dark outside the burrow.

2. In studying the hormonal control of behavior, it is common to remove an animal's ovaries or testes and then inject the creature with assorted hormones to see what behavioral effects they have. What advantage does this technique have over another approach, which is simply to measure the concentrations of specific hormones in the blood of animal subjects from time to time? The far less invasive direct measurement approach would show, for example, whether mating usually occurred when circulating testosterone or estrogen concentrations were elevated.

3. In the guinea pig, individual males vary in their sex drive, as measured (for example) by the number of times males will ejaculate when given access to receptive females for a standard period of time. One hypothesis for this variation is that male sex drive correlates with circulating testosterone concentrations. What prediction follows from this hypothesis? The graph shown below presents data on an experiment with three guinea pigs in which male sex drive was measured. All three males were castrated, after which their sex drive continued to be monitored, until finally, after some weeks, the three males were given supplemental testosterone and their sex drive measured again at intervals. What is the relevance of these data for the hypothesis in question? What scientific conclusion can you reach based on these results?

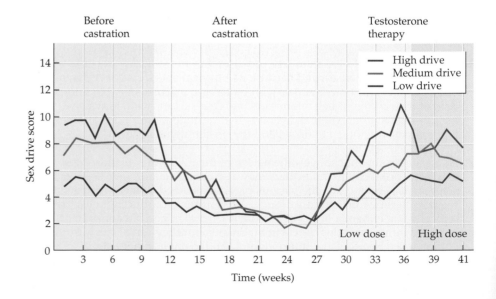

4. In the Mongolian gerbil and the Florida scrub jay, adults live in groups in which there are adult breeders and adult nonbreeding helpers who assist the breeding pair in rearing offspring (see p. 438). Develop a hormonal hypothesis (based on your reading of this chapter) for the difference in reproductive behavior between the two categories of individuals in these social units. Then generate some predictions from your hypothesis before getting data from [220, 1025].

5. Two species of silkmoths differ in the timing of their emergence from their pupal cocoons, with species A emerging in the middle of the night and species B entering the world as adults at dawn. Emergence behavior in the two species is clearly under the control of a biological clock of some sort, but where might this clock be? As it turns out, removal of the brain of a pupal moth does not kill the moth or prevent its metamorphosis to adulthood, but it does destroy the species-typical emergence pattern for both species. How could you further test the hypothesis that the clock mechanism is contained within the silkmoths' brains? (Take advantage of the fact that you can successfully transplant the brain to the abdomen of a silkmoth pupa.) Check [1157] for relevant data after developing your tests.

6. In many mammals, individuals infected with disease-causing parasites or microorganisms become generally inactive and unresponsive, even to the point of not eating or drinking. How could you determine whether this change was an adaptive change in behavioral priorities designed to help the animal overcome its illness or a maladaptive by-product of the toxic effects of infectious agents that benefit, or at least do no harm to, the bacteria or parasite? Of what relevance to this issue are the following findings: (1) sexual activity is suppressed in female rats, but not males, that are ill because their immune systems have been experimentally challenged with infectious agents [51], and (2) rats mildly infected with a single-celled parasite (*Toxoplasma gondii*) are more likely to visit compartments in their cages that contain cat urine than those where rabbit urine has been applied (see the figure below from [95]). (Note that cats are an important host of *T. gondii*.)

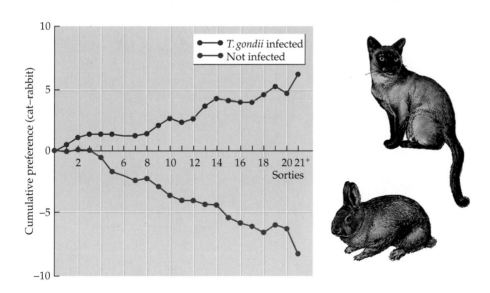

Suggested Reading

Kenneth Roeder's classic *Nerve Cells and Insect Behavior* [976] and Vincent Dethier's *The Hungry Fly* [314] discuss how some animals avoid conflicting responses and structure their behavior over the short haul. Randy Nelson has written a fine textbook that covers all the topics in this chapter in much more detail [836]. In just a few pages, Michael Young explains the exceedingly complex molecular basis of circadian rhythms about as clearly as possible [1275]. David Crews and his colleagues are responsible for a beautifully detailed picture of the organization of anole behavior [259] as well as for a review of diversity in the mechanisms controlling behavior [264].

7 | Adaptation and Antipredator Behavior

*I*n this chapter, our focus shifts from the proximate causation of behavior to an examination of behavior's ultimate causes. The behaviors we will analyze here from an evolutionary perspective are those that may help individuals cope with predators so that they or their offspring have a better chance of survival and, thus, a better chance of passing on their genes.

Predators are, needless to say, good at detecting, attacking, capturing, and finally consuming prey. Interactions that end in a meal for the predator are, however, rarely advantageous for the eaten animal. With their immediate survival at stake, we would expect to find prey species interfering with the attempts of their enemies to detect, attack, capture, and consume them. Recall, for example, that noctuid moths have such great sensitivity to ultrasound that they can sometimes avoid detection by hungry bats by turning away from their enemies before they can be identified as prey. Moreover, even if a moth is under attack, it may escape, thanks to anti-interception power dives and loops that help it live to reproduce another day.

Noctuid moths are not alone in possessing adaptive defensive behavior. Here we shall consider many examples of possible anti-detection, anti-attack, anti-capture, and anti-consumption devices. Before we begin this survey, however, let's decide just exactly what we mean when we say that a behavioral trait is an adaptation.

The Meaning of Adaptation

When an animal is overlooked by its predators, it obviously benefits. As we have seen, parent black-headed gulls help their chicks secure this advantage by removing conspicuous broken eggshells from the nest, thereby eliminating a visual cue associated with their otherwise well-camouflaged offspring. Nesting gulls may also help their offspring in another way by collectively mobbing predators that enter their summer breeding colonies, where dozens or hundreds of pairs of gulls nest in loose association on the ground (Figure 1). Should a fox, crow, badger, hawk, or human appear near a colony of black-headed gulls, the breeding adults respond with a volley of loud cries. If the intruder continues to move into the colony, groups of gulls fly toward it, calling raucously and defecating profusely (Figure 2).

As noted in Chapter 1, two different but interrelated evolutionary questions can be asked about something like mobbing behavior: (1) What was the original mobbing trait like, and how has it changed over time? (2) Was or is the trait an adaptation, and if so, what was or is its adaptive value? For the moment, let's focus on the second issue, which requires that we define "adaptation." Although different definitions exist [496, 686], all the widely used ones emphasize the role of natural selection in the evolution of adaptations. Thus, we shall define an **adaptation** as a heritable trait that (1) either spread because of natural selection and has been maintained by selection to the present or (2) is currently spreading relative to alternative traits because of natural selection. In all such cases, the trait in question has conferred and continues to confer (or is just beginning to confer) higher genetic success (on average) to individuals that have it compared with other individuals with different alternative traits. This definition permits us to test hypotheses about possible adaptations by focusing on the *current* benefits of a given trait, which has some major practical advan-

1 A breeding colony of black-headed gulls. Pairs nest on the ground quite close to other pairs. Note the chick's camouflaged appearance. Photograph by Roger Tidman.

2 Black-headed gulls mobbing a trespasser in their colony.

tages over trying to test whether certain traits provided genetic benefits to individuals in the past [960].

Persons who accept this definition of adaptation, which owes much to George C. Williams [1239], have a powerful tool with which to identify behavioral puzzles worth solving: namely, traits that carry substantial reproductive costs for the individuals that have them. Consider mobbing behavior by black-headed gulls. These birds expend valuable time and energy diving at their enemies, and they are sometimes injured or even killed as a result, just as are other birds that mob dangerous predators [890]. Could the reproductive costs of mobbing be outweighed by certain benefits? This question lies at the heart of the cost-benefit approach employed by adaptationists who seek to understand how costly traits might nevertheless be favored by natural selection. The approach requires the production of hypotheses on why certain reproductive or fitness benefits of a trait might more than compensate individuals for the fitness costs of the attribute. (In evolutionary biology, "fitness" is a measure of an individual's reproductive or genetic success, so that "fitness benefit" refers to the positive effect of a trait on the number of surviving offspring produced by an individual or the number of genes it contributes to the next generation. "Fitness cost" refers to the damaging effects of the trait on these measures of indivdual genetic success).

TABLE 1 Three reasons why not all current traits are adaptations

Reason	More examples
1. The trait evolved under conditions that no longer exist, but persists because insufficient time or the absence of appropriate mutations prevented the replacement of the now non-adaptive trait. Some arctic moths fly in regions where bats are absent, but still cease locomotion upon exposure to an experimental ultrasonic stimulus [1005].	Arctic ground squirrels live where snakes do not, but when they are experimentally exposed to snakes, they show some of the same responses as other ground squirrel species whose ranges overlap with dangerous predatory snakes [245].
2. The trait develops as a maladaptive side effect of an otherwise adaptive proximate mechanism—that is, one that generally causes an adaptive outcome. Female rodents living in a communal nest may sometimes give milk to offspring other than their own as a by-product of their strong parental drive, which usually results in adaptive care of their own genetic offspring [500].	A strong, generally adaptive drive to care preferentially for genetic offspring may lead some human stepparents to engage in criminal child abuse that seems certain to reduce the fitness of the child abuser [284]. (See p. 482 for more on this case.)
3. The trait is expressed as a maladaptive consequence of a very recent change in the environment (a hypothesis that combines elements of hypotheses 1 and 2). Certain beetles attempt to copulate with discarded beer bottles (see p. 116).	Sea turtles sometimes die as a result of eating plastic bags, which resemble jellyfish sufficiently to trigger a feeding response. In the past, that response would have led to the adaptive ingestion of jellyfish, not the potentially lethal consumption of plastic debris [672].

If we are to apply the cost-benefit approach to mobbing by gulls, we must figure out how the behavior could yield a *net* fitness gain (benefits > costs) to individuals that risk their lives when mobbing an enemy. Many possibilities exist [274], but the great ethologist Niko Tinbergen favored the idea that mobbing gulls confuse and distract predators intent on finding gull eggs or chicks, thereby raising their reproductive success sufficiently to compensate for the fitness-lowering risks taken by gulls trying to keep predators from finding their offspring[1138].

I too find the distraction hypothesis highly plausible, based on personal experience in a gull rookery where I spent most of my time trying to get out of the way of dive-bombing gulls and their excreta, not always successfully. The roar of wind through a gull's wings as it passed just overhead was unnerving, and the thwack of a gull's foot on the top of my head was downright painful. Had I been hunting for gull chicks or eggs, I would have been most ineffectual because of my eagerness to avoid the defecating dive-bombers overhead.

Although the predator distraction hypothesis makes sense, we must test it before accepting it, because we could be wrong. Many traits are not adaptations; nonadaptive or maladaptive characteristics can arise for a variety of reasons (Table 1) [257]. A test of the hypothesis that mobbing adaptively distracts predators begins with some predictions. For example, if the hypothesis is correct, predators searching for gull eggs should be less successful at finding these food items when they are being mobbed by nesting gulls. This prediction has been tested simply by observing what happens when mobbing black-headed gulls interact with nest-robbing predators, such as carrion crows [662]. Egg-eating crows can avoid swooping gulls, but only if they continually face their attackers. Thus, while being mobbed, they cannot look around comfortably for nests and eggs, a fact that offers modest support for the distraction hypothesis. Because distracted crows are probably less likely to find their prey, a mobbing gull gains a reproductive benefit that probably outweighs the costs of the activity, especially since crows do not attack adult gulls.

Another, more challenging prediction from the predator distraction hypothesis is that the degree of success experienced by mobbing gulls in protecting their eggs should be proportional to the degree to which predators are actually mobbed. Hans Kruuk used an experimental approach to test this prediction [662]. He placed ten hen eggs, one every 10 meters, along a line running

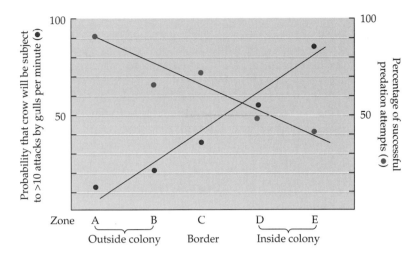

3 Does mobbing protect eggs? When hen eggs were placed outside and inside a gull nesting colony, crows searching for the eggs within the colony were subject to more attacks by mobbing gulls (red circles), and as a result, they discovered fewer hen eggs (blue circles). After Kruuk [662].

Black-headed gull

from the outside to the inside of a gull nesting colony. The eggs placed outside the colony, where mobbing pressure was low, were much more likely to be found and gobbled up by carrion crows and herring gulls than eggs inside the colony, where the predators were intensely harassed (Figure 3).

Therefore, both observational and experimental tests support the hypothesis that mobbing helps protect the eggs and young of black-headed gulls. Note that these tests did *not* involve measuring gull reproductive success by counting the number of surviving offspring produced by individuals in their lifetimes. Kruuk looked instead at the number of hen eggs that were uneaten, on the reasonable assumption that had they been gull eggs in a gull nest, they would have had a chance to become surviving offspring for the gull parent, in which case they would constitute part of the parent's lifetime genetic contribution to the next generation.

Behavioral ecologists often have to settle for an indicator or correlate of reproductive success or genetic success when they attempt to measure fitness. In the chapters that follow, "fitness" or "reproductive success" is used more or less interchangeably with such measures as egg survival (Kruuk's measure), young that survive to fledging, or number of mates inseminated, or even more indirectly, the quantity of food ingested per unit of time, the ability to acquire a breeding territory, and so on. The reader should keep in mind, however, that personal reproductive success in the last analysis is measured by the number of offspring (or sometimes, grandoffspring) that reach the age of reproduction produced by an individual over its lifetime, and that correlates of this measure will be accurate to varying degrees.

The Comparative Method for Testing Adaptationist Hypotheses

Experiments are highly valued in science, so much so that most persons believe that scientific research can be performed only in high-tech laboratories by officious white-coated researchers. As Kruuk's work shows, however, good experimental science can be done in the field. Moreover, the manipulative experiment is only one of several ways in which predictions from hypotheses can be tested. We just saw how it was possible to test the predator distraction hypothesis on mobbing by using nonmanipulative observations to find out whether certain predictions are correct (e.g., do hunting crows stop searching for eggs while they are being mobbed?).

Biologists have another technique for testing adaptationist hypotheses: the **comparative method.** This approach involves testing predictions about the evolution of an interesting trait by looking at animals other than the species whose characteristics are under investigation [232]. We used the comparative method informally earlier when considering the evolution of infanticide (p. 17), and again when dealing with the response to ultrasound by moths and other insect prey of bats (p. 125). Here we shall use it to test the adaptationist explanation for mobbing by black-headed gulls. The comparative method as applied to this case yields the following prediction: if ground-nesting black-headed gulls mob certain predators to protect their eggs and young, then other gull species that lack egg predators should *not* exhibit mobbing behavior.

The rationale behind this prediction is as follows. Gulls, of which there are about 50 species, possess many similarities in structure and behavior, which is why they are all placed within the same family of birds, the Laridae (which also includes the terns). The similarities among gulls exist in part because they have a relatively recent common ancestor from whom they inherited many genes. This ancestral battery of genes gradually changed—a new allele here, a new allele there—in various gull populations. The changes came about through chance events, including mutation and genetic drift (the loss of rare alleles in small populations through accidents of various sorts), and through natural selection, which acted upon the mutations available within the different populations. As the populations derived from the ancestral gull became increasingly genetically differentiated, some geographically isolated groups evolved into new species, incapable of breeding with members of other descendant populations. Each of these populations evolved its own distinctive features overlain on the ancestral attributes that it inherited and retained from the ancestral gull.

How can we tell which features of gulls are ancestral and which are more recent innovations? One way to get started is to examine how widely certain characteristics appear among the 50 modern gull species. Most gulls nest on the ground in grassy or brushy areas, often on islands, suggesting that this trait was present in the ancestral species and was passed on to most modern species. However, a few of today's species nest on cliff ledges rather than on the ground, perhaps because they are the descendants of a fairly recent cliff-nesting gull that evolved from a ground-nesting ancestor. The alternative possibility that the ancestral gull was a cliff nester requires that the trait be lost and then regained, which produces an evolutionary scenario that requires more changes than the competing one (Figure 4). Most evolutionary biologists, although not all, believe that simpler scenarios involving fewer transitions are more proba-

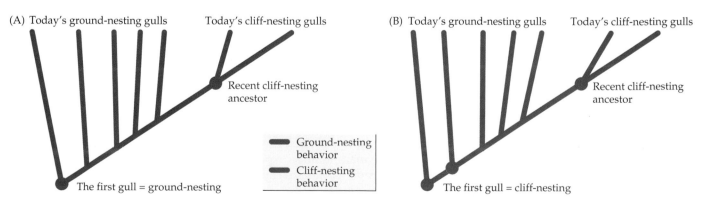

4   Gull phylogeny and two scenarios for the origin of mobbing behavior. Hypothesis A requires one switch (from ground-nesting to cliff-nesting); hypothesis B requires two switches (from cliff-nesting to ground-nesting and back again).

ble than more complicated alternatives. In this, the majority accepts a commonly held philosophical principle, Occam's Razor, which holds that simpler explanations are more likely to be correct than complex ones.

The Importance of Divergent Evolution

Note that our comparisons have cast some light on the possible origins of certain traits—an evolutionary matter somewhat different from the question of adaptation, but important for a variety of reasons. If our evolutionary scenario about the ancestral nature of ground nesting in gulls is correct, then those gull species that have diverged from the ancestral pattern because selection directly favored the divergent trait (and not for any other reason, such as random genetic drift) must have done so because of a change in their environment. Such a change would have altered the selection pressures acting upon these gulls, causing a behavioral shift away from the ancestral pattern. By finding such cases of **divergent evolution** and identifying the change in selection, one could, in principle, establish why the ancestral trait has been retained in most modern gull species but has been modified in others.

As noted already, not all gulls nest in groups on the ground, where their eggs are vulnerable to predators. The kittiwake, for example, is a gull that nests on nearly vertical coastal cliffs (Figure 5), where its eggs cannot be reached by mammalian predators and where even other gulls and hawks rarely venture because of dangerous swirling sea winds. The small, delicate kittiwakes have clawed

5 Not all gulls nest on the ground. Cliffs are utilized for nesting by kittiwake gulls, which appear in the lower half of this photograph. Photograph by Bruce Lyon.

6   Adaptations for cliff nesting. (Left) Kittiwakes build their nests on small cliff ledges and their young face away from the cliff edge. (Right) Adults have clawed feet that help them cling to even the narrowest of perches. Photographs by the author.

feet, and they can safely land and nest on tiny ledges (Figure 6). As a result, predator pressure on kittiwake eggs and young has been greatly reduced compared with that affecting their ground-nesting relatives. As predicted, nesting adult kittiwakes do not mob their predators, despite sharing a fairly recent common ancestor and many other ancestral features with black-headed gulls and other ground-nesting species. The kittiwake's behavior has become less like that of its close relatives, providing a case of divergent evolution that supports the hypothesis that mobbing by black-headed gulls is an adaptation that evolved in response to predator pressure on the eggs and babies of nesting adults [270].

The Importance of Convergent Evolution

The other side of the comparative coin is that species from different evolutionary lineages that live in similar environments, and therefore experience similar selection pressures, can be predicted to evolve similar traits yielding cases of **convergent evolution.** Such species converge upon the same adaptive solution to a particular environmental obstacle to reproductive success, despite the fact that their different ancestral species had different genes and different attributes (Figure 7). We can predict, for example, that if mobbing by colonial, ground-nesting gulls is an adaptation to distract predators from vulnerable offspring, similar behavior should have evolved in other, totally unrelated animals that nest or breed in groups and are often visited by predators intent upon eating their offspring.

Mobbing behavior has evolved convergently in other bird species, including the bank swallow and the barn swallow. Like gulls, these species nest in colonies, which are visited by predators, including snakes and blue jays, that enjoy eating swallow eggs and babies [547]. The common ancestor of swallows and gulls occurred long, long ago, with the result that the lineages of the two groups have evolved separately for eons—a fact recognized by the placement of swallows in a different family of birds (the Hirundidae). However, despite their evolutionary and genetic differences from gulls, both barn and bank swallows behave like gulls when it comes to harassing their enemies, swirling around them in groups and thereby sometimes distracting the predators that would destroy their offspring.

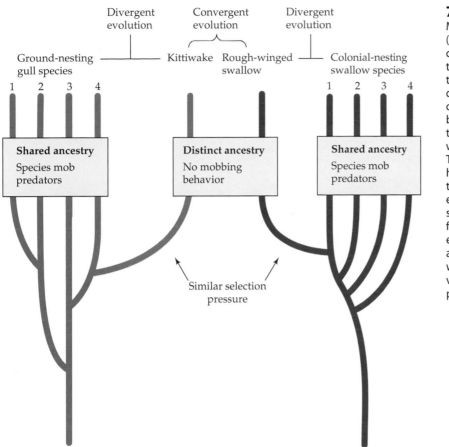

Divergent evolution   Convergent evolution   Divergent evolution

Ground-nesting gull species   Kittiwake   Rough-winged swallow   Colonial-nesting swallow species

1   2   3   4

Shared ancestry
Species mob predators

Distinct ancestry
No mobbing behavior

Shared ancestry
Species mob predators

1   2   3   4

Similar selection pressure

Evolutionary lineage: Gulls   Evolutionary lineage: Swallows

7 The logic of the comparative method. Members of the same evolutionary lineage (e.g., gull species of the family Laridae) share a common ancestry, and therefore often share the same traits, such as mobbing behavior. But the effects of shared ancestry can be overridden by a novel selection pressure. The absence of nest predators has led to divergent evolution by the kittiwake, which no longer mobs potential enemies. The same is true of the rough-winged swallow, which does not nest colonially. The black-headed gull and the bank swallow, however, both mob nest predators, even though they are unrelated birds from separate evolutionary lineages. In response to shared selection pressure from predators that have fairly easy access to their colonies, black-headed gulls and bank swallows have converged on a similar antipredator response just as kittiwakes and rough-winged swallows have convergently evolved into species that do not mob potential predators.

Even some colonially nesting mammals have evolved the mobbing trait [508, 874]. Adult California ground squirrels, which live in groups and dig burrows in the ground, react to a hunting rattlesnake by gathering around it and kicking sand in its face, thereby preventing the snake from exploring nest burrows for prey (Figure 8). Incidentally, these squirrels have evolved substantial biochemical resistance to the venom of the northern Pacific rattlesnake, the species with which they coexist uneasily, more so than to the different venom of the western diamondback, a rattlesnake that lives elsewhere and so has not exerted selection on the squirrels [104]. The squirrel's antivenin is not perfect (the chemicals that inhibit certain destructive consequences of snake venom have side effects that exacerbate other damaging properties of the venom), as one might expect in a coevolving system in which the two parties, predator and prey, have divergent interests (see Chapter 8). However, some resistance is probably better than none, and if so, the trait qualifies as an adaptation, which has the effect of reducing the costs of mobbing by adult California ground squirrels.

Because mobbing behavior has evolved independently in several unrelated species whose adults can sometimes protect their vulnerable offspring by distracting predators, we can conclude tentatively that mobbing is an antipredator adaptation. But what if I presented only supportive examples while ignoring many other colonial species in which parents do not mob the consumers of their youngsters (and do not have alternative means of protecting their offspring)? If for every colonial species in which mobbing occurred under the expected conditions there were two in which it did not, you would be skeptical of the predator distraction hypothesis, and rightly so. For this and other reasons, researchers increasingly require that the comparative method be used in

8   Colonial California ground squirrels have evolved mobbing behavior. One squirrel kicks sand at a rattlesnake, while others give a variety of alarm calls. Courtesy of R. G. Coss and D. F. Hennessy.

a statistically rigorous fashion [496]. For our purposes, however, the point is that one can, in principle, test adaptationist hypotheses by predicting that particular cases of divergent or convergent evolution will have occurred, a prediction about the past that can be checked by making disciplined comparisons among species living today.

The Diversity of Antipredator Adaptations

When baby gulls hear their parents' alarm cries and mobbing screams, they duck for cover or crouch low in the nest. Their camouflaged gray and brown plumage almost certainly makes them hard for predators to find. Many other cryptically colored animals also hide by choosing the "right" background when seeking cover. This phenomenon is especially common among the many edible insects that attract the attention of visually hunting birds (Figure 9).

A classic demonstration of the importance of matching one's camouflage to an appropriate resting place involves the peppered moth, *Biston betularia*. In some parts of Great Britain and the United States, the melanic (black) form of this moth, once extremely rare, almost completely replaced the once abundant black-dotted whitish form in the period from about 1850 to 1950 [451] (Figure 10). Most biology undergraduates have heard the standard explanation for the initial spread of the melanic form (and the special allele associated with its mutant color pattern). The idea was that industrial soot had darkened the color of forest tree trunks in urban regions. This change made whitish moths living in these places more conspicuous to insectivorous birds, which ate the white form and thereby removed the genetic basis for this color pattern. Despite some recent claims to the contrary [250], this story remains largely valid [450], especially with respect to the significance of H. B. D. Kettlewell's famous experiments [620]. Kettlewell

9   Camouflage is effective only if the appropriate background is selected. These four cryptically colored grasshoppers choose resting places suitable for their color patterns. In addition, the species at bottom right uses its legs to draw pebbles over its back, further enhancing its camouflage. Photographs by the author.

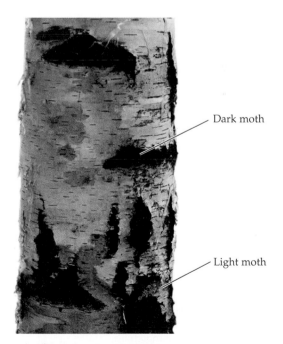

Dark moth

Light moth

10   The salt and pepper moth *Biston betularia*. One typical (light coloration) individual and one melanic (black) individual are shown in each photograph. Photographs (left) by Michael Tweedie; (right) photograph by Bruce Grant.

11 Predation risk and background selection by moths. Specimens of typical and melanic forms of the peppered moth were pinned to tree trunks or limb joints. Although the moths on limb joints were less likely to be found and removed by birds, overall, melanic forms were discovered by birds less often in polluted (darkened) woods, while typical forms "survived" better in nonpolluted woods. After Howlett and Majerus [552].

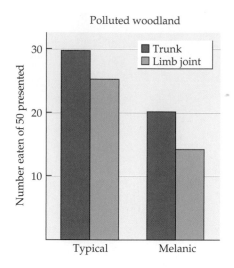

Polluted woodland   Nonpolluted woodland

Number eaten of 50 presented

■ Trunk
■ Limb joint

Typical   Melanic   Typical   Melanic

placed the two forms of the peppered moth on dark tree trunks and on pale tree trunks, finding that whichever form was more conspicuous to humans was taken by birds much more quickly than the other form. Thus, paler individuals were at special risk of attack when they perched on dark backgrounds.

It is true that in nature the moths may not often perch on tree trunks and instead may tend to select the shaded patches just below the junction of a branch with the trunk. When R. J. Howlett and M. E. N. Majerus glued samples of frozen moths to open trunk areas and to the undersides of limb joints, they showed that birds were particularly likely to overlook moths on shaded limb joints [552]. Even so, the two researchers essentially replicated Kettlewell's findings (Figure 11).

Other work done long after Kettlewell's experiments has also shown that where moths sit during the day affects their vulnerability to bird predators. The whitish moth *Catocala relicta* usually perches head-up, with its whitish forewings over its body, on white birch and other light-barked trees (Figure 12). When given a choice of resting sites, the moth selects birch bark over darker backgrounds [1015]. Alexandra Pietrewicz and Alan Kamil used captive blue jays, photographs of moths on different backgrounds, and operant conditioning techniques (Figure 13) to test whether the moth's behavior is adaptive [917]. They trained the blue jays to respond to slides of cryptically colored moths positioned on an appropriate background. When a slide flashed on a screen, the jay had only a short time to react. If the jay detected a moth, it pecked at a key, received a food reward, and was quickly shown a new slide. But if the bird pecked incorrectly when shown a slide of a scene without a moth, it not only failed to secure a food reward, but had to wait a minute for the next chance to evaluate a slide and get some food. The caged jays' responses showed that they saw the moth 10 to 20 percent less often when *C. relicta* was pinned to pale birch bark than when it was placed on darker bark. Moreover, the birds were especially likely to overlook the moth when it was oriented head-up on birch bark. Thus, the moth's preference for white birch resting places and its typical perching orientation appear to be anti-detection adaptations that thwart visually hunting predators such as blue jays.

12 Cryptic coloration and body orientation. The orientation of a resting *Catocala* moth determines whether the dark lines in its wing pattern match up with the dark lines in birch bark. Photograph by H. J. Vermes, courtesy of Ted Sargent, from [1015].

The Value of Body "Decorations"

Not all prey have beautifully camouflaged color patterns, but some can make themselves less conspicuous by adding certain materials to their bodies (Figure 14). In the case of the rock ptarmigan, the extra something is dirt, which the pure white males apply, probably by bathing in mud puddles (Figure 15).

13 Does cryptic behavior work? Images of moths on different backgrounds and in different resting positions are shown to a captive blue jay, which is rewarded for detecting a moth. Photograph by Alan Kamil.

During the winter breeding season, the males' brilliant white plumage is inconspicuous against the snow, but it becomes a problem during the interval between snowmelt and the annual molt, in which they replace their white feathers with cryptic mottled brown plumage. During this in-between time, males can make themselves substantially less visible by soiling their feathers. Human observers can often spot the all-white males on dark ground when they are on the order of 1000 meters distant, whereas they detect dirtier males only about 20 meters away on average [812]. If falcons also find dirty ptarmi-

14 Certain additions may improve the camouflage of a prey species, such as this geometrid moth caterpillar, which hooks flower petals onto its back. Photograph by Ron Rutowski.

15   Mud bathing as camouflage. (Left) The male rock ptarmigan has brilliant white plumage during the spring which it dirties (right) with mud after the breeding season is over. Photographs by Bruce Lyon.

Decorator crab

gan harder to spot, as seems likely, mud bathing has clear adaptive value for the ptarmigan.

The camouflage hypothesis for altering one's appearance has also been tested for the decorator crab, which adorns its back with living algae. In areas in which the alga *Dictyota menstrualis* occurs, juvenile crabs use this plant in preference to all others when making an algal cloak (Figure 16) [1089]. If this preference is adaptive, we can predict that crabs decorated with the favored alga will be less likely to be killed by fish than crabs that are unable to use this alga. In exper-

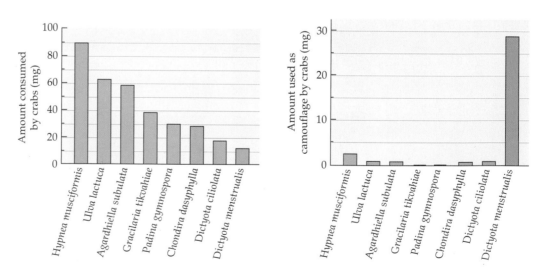

16   Selection of algae as food versus camouflage by the decorator crab. In the laboratory, the crabs rarely eat *Dictyota menstrualis,* but they use it almost exclusively for their camouflage decorations. After Stachowicz and Hay [1089].

iments with tethered crabs, individuals that were forced to make use of an algal species not normally used by the crabs disappeared at five times the rate of those crabs that were able to use the preferred alga for a defensive coat [1089].

Further evidence that the crab's choice is adaptive comes from studies of the same species of prey in areas in which the alga *Dictyota* does not occur. Decorator crabs from outside the range of *Dictyota* show no special preference for it when experimentally offered the opportunity to construct their cloaks from this alga. Selection in the area of overlap between crab and alga must be actively favoring individuals that pick *Dictyota,* whereas no such selection occurs in regions where the plant is absent. Here we have evidence for the capacity of natural selection to fine-tune an animal's adaptations to match local conditions.

However, is the added protection gained by crabs decorated with *Dictyota* due to enhanced camouflage or something else? The alga happens to be endowed with a potent chemical defense that repels omnivorous fish, which refuse to eat it [1089]. In effect, the crab is exploiting the defense mechanism of the alga for its own purposes. Even if it is detected by a crab-eating fish, the predator may leave the crab alone because of its toxic covering [1090].

The Value of Warning Behavior

Other animals that are chemically protected forego camouflage altogether in favor of color patterns that may be positively flamboyant (Figure 17). The monarch butterfly, for example, features a conspicuous orange and black pattern on its wings. The larvae of this species feed upon poisonous milkweeds, and they sequester and save the extremely potent plant poison in their tissues [163]. In dealing with the monarch butterfly, you would be wise to ignore the recommendation of the lepidopterist E. B. Ford, who wrote, "I personally have made a habit, which I recommend to other naturalists, of eating specimens of every species which I study" [390]. Although few humans have been

17 Warning coloration and toxins. Animals that are chemically defended typically behave in a conspicuous manner. (Top left) The monarch butterfly's body and wings may contain potentially lethal cardiac glycosides, which it sequesters from its milkweed diet. (Top right) Sawfly larvae regurgitate repellent eucalyptus oils when threatened. (Bottom) Blister beetles possess blood laced with cantharidin, a highly noxious chemical. Photographs by the author.

18   **Effect of monarch butterfly toxins.** A blue jay that eats a toxic monarch vomits a short while later. Photographs by Lincoln P. Brower.

tempted to take up the Ford challenge when it comes to monarch butterflies, the same is not true for butterfly-eating birds. Those birds that make the mistake of consuming a toxic monarch usually survive, but they find the experience most unpleasant and highly educational (Figure 18). After once vomiting up a noxious monarch, a blue jay will usually avoid this species religiously thereafter [161].

You may legitimately ask how it can be advantageous for a monarch that has been eaten to induce vomiting, given that regurgitated monarchs rarely fly off into the sunset. However, vomiting and the learning it promotes may be protective adaptations of birds, which rid themselves of ingested toxins and avoid a second mistake. From the monarch's perspective, the adaptive value of its ability to store poisons from its food plants is not to make it emetic, but to make itself so bad-tasting that most birds release it immediately [163]. "Better things for better living through chemistry," as the old DuPont Chemical Company slogan had it. Monarchs whose chemistry helps them survive gain a fitness benefit from their investment in toxin sequestration.

Because nauseated predators learn to avoid the color pattern of the prey that made them sick, the door to deception is opened. Many edible prey have taken advantage of this opportunity to deceive educated predators into leaving them alone [1228]. These deceptive species, which look like truly bad-tasting animals, are called **Batesian mimics** (Figure 19), after Henry Bates, an English naturalist who long ago discovered palatable species of butterflies masquerading as poisonous ones in Brazil.

Many Batesian mimics employ behavioral traits that amplify the effect of their similarity to a protected species. For example, some harmless insects and spiders have evolved a physical resemblance to stinging ants, and they improve their mimicry by walking about in a lively and active manner with antennae waving and abdomen bobbing up and down, just like ants [783]. Thus, the jumping spiders that look like ants do not jump in the manner of related spiders, but run from place to place. Similarly, the droneflies that look like honey

19 **Batesian mimics resemble other species that are protected against predation.** Although these insects look like (left) a bee, (middle) a yellow-jacket wasp, and (right) a paper wasp, they are actually all harmless flies. Photographs by the author.

bees (see Figure 19, left) fly like them too, even visiting particular kinds of flowers for the same amount of time as the bee (Figure 20) [439].

A special case of behavioral mimicry designed to deter predators has evolved in a tephritid fly that superficially resembles one of its potential consumers, a jumping spider, which waves its conspicuous legs at fellow spiders during territorial disputes (Figure 21). The fly possesses wing markings that resemble the legs of jumping spiders. When the fly waves its wings, something it does habitually, the visual effect is similar to the aggressive leg-waving displays of a jumping spider. If the behavior of the fly is indeed a mimetic adaptation that causes it to be mistaken by predatory spiders for a territorial conspecific that should be left alone, then tephritids with intact wings should induce jumping spiders to back off, a prediction that has been tested by two different teams of biologists [457, 769].

In order to carry out a test of the mimicry hypothesis, one group of researchers became expert at fly surgery. Armed with scissors, Elmer's glue, and steady hands, the fly surgeons exchanged wings between clear-winged houseflies and pattern-winged tephritid flies. After the operation, the tephritid flies behaved normally, waving their now plain wings and even flying about their enclosures. But these modified tephritids with their housefly wings were soon eaten by the jumping spiders in their cages. In contrast, tephritids whose own wings had been removed and then glued back on repelled their enemies in 16 of 20 cases.

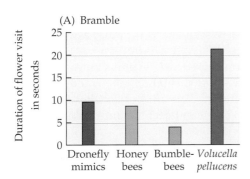
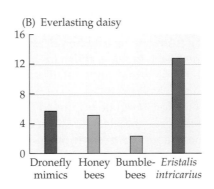
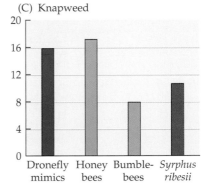

20 **Behavioral Batesian mimicry.** Droneflies that mimic honey bees spend about the same amount of time as honey bees feeding on the flowers of various plants. The mimics do not match bumblebee visit times. Other flies that do not resemble honey bees (see the fourth column in each graph) do not match honey bee visit times. After Golding and Edmunds [439].

21 Does behavioral Batesian mimicry work? The tephritid fly (right) has been waving its banded wings as a predatory jumping spider (left) approaches. The spider eventually retreated. Photograph by Erick Greene.

Houseflies with tephritid wings gained no protection from the spiders, showing that the combination of leglike color pattern *and* wing movement enables the tephritid fly to deceive its predators into treating it as a dangerous opponent rather than a meal with wings [457].

How To Stop a Pursuer

Warning coloration and behavior advertise the potential noxiousness or dangerousness of a prey, thereby making it advantageous for a predator to accept the information and not attack—except when the signal is a false one provided by a Batesian mimic. But even perfectly palatable, nonmimetic prey may be able to deter a pursuer by signaling that they cannot be easily caught. When a prey quickly and accurately communicates this point to a predator, both signaler and receiver benefit: the prey by not having to outrun the predator, the predator by saving its time and energy to invest in pursuit of more vulnerable prey.

One possible advertisement of prey unprofitability occurs when Thomson's gazelles spot a predator, such as a cheetah. The gazelles start to run away, but they may slow up a bit in order to stot—that is, to jump about a half meter off the ground with all four legs held stiff and straight and with their white rump patch made fully visible (Figure 22). Stotting might be a means of communicating to the predator that it has been seen by a gazelle that is ready and able to flee. Predators may choose to let alert gazelles go because they are so hard to catch.

The unprofitability advertisement hypothesis is not the only possible explanation for stotting, however. Perhaps a stotting gazelle sacrifices speed in escaping from one detected predator in order to scan ahead for other as yet unseen enemies lying in ambush (as lions often do) [923]. The anti-ambush hypothesis predicts that stotting will *not* occur on short-grass savanna, but will instead be reserved for tall-grass or mixed grass and shrub habitats, where predator detection could be improved by jumping into the air. But gazelles feeding in short-grass habitats do stot regularly, so we can reject the anti-ambush hypothesis and turn to some others [198, 199].

- *Alarm signal hypothesis:* Stotting might warn conspecifics, particularly offspring, that a predator is dangerously near. This could increase the survival of the signaler's offspring and relatives, thereby improving the stotter's fitness (see p. 435).
- *Social cohesion hypothesis:* Stotting might enable gazelles to form groups and flee in a coordinated manner, making it harder for a predator to cut any one of them out of the herd.

22   An advertisement of unprofitability? Stotting behavior by a springbok, a small antelope that leaps into the air when threatened by a predator, just as do Thomson's gazelle. Photograph by Jen and Des Bartlett.

- *Confusion effect hypothesis:* By stotting, individuals in a fleeing herd might confuse and distract a following predator, keeping it from focusing on one animal.

Table 2 lists the predictions that are consistent with the hypotheses under review. Because the same prediction sometimes follows from two different hypotheses, we must consider multiple predictions from each one if we are to discriminate among them.

Tim Caro learned that a solitary gazelle will sometimes stot when a cheetah approaches, an observation that helps eliminate the alarm signal hypothesis (if the idea is to communicate with other gazelles, then single gazelles should not stot) and the confusion effect hypothesis (because the confusion effect can occur only when a group of animals can flee together). We cannot rule out the social cohesion hypothesis on the grounds that solitary gazelles stot, because there is the possibility that solitary individuals stot in order to attract distant gazelles to join them. But if the goal of stotting is to communicate with fellow gazelles, then stotting individuals, solitary or grouped, should direct their conspicuous white rump patch toward other gazelles. Stotting gazelles, however, orient their

TABLE 2 Predictions derived from four alternative hypotheses on the adaptive value of stotting by Thomson's gazelle

Prediction	Alarm signal	Social cohesion	Confusion effect	Signal of unprofitability
Solitary gazelle stots	No	Yes	No	Yes
Grouped gazelles stot	Yes	No	Yes	No
Stotters show rump to predator	No	No	Yes	Yes
Stotters show rump to gazelles	Yes	Yes	No	No

23 Cheetahs abandon hunts more often when gazelles stot than when they do not, supporting the hypothesis that these predators treat stotting as a signal that the gazelle will be hard to capture. After Caro [199].

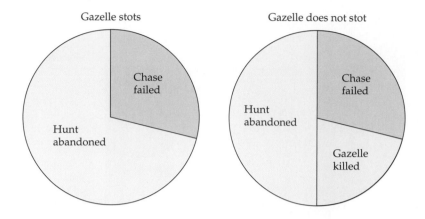

Gazelle stots

Gazelle does not stot

rumps toward the predator. Only one hypothesis is still standing: gazelles stot to announce to a predator that they will be hard to capture. Cheetahs get the message, since they are more likely to abandon hunts when the gazelle stots than when the potential victim does not perform the display (Figure 23) [199].

The Value of Vigilance

We have reviewed a few examples of traits that increase the probability that a predator will either overlook a prey or choose not to carry out a full attack. But what can an animal do if an attack is already under way? One weapon in the anti-capture arsenal of some species is vigilance—staying alert so as to detect an attacking enemy before it is too late to take effective action. Vertebrate prey generally keep an eye scanning, or nose sniffing, or ear listening at all times, no matter what else they are doing. In fact, some ducks can even keep one eye scanning while they are asleep. These animals have the option of sleeping while one of their two cerebral hemispheres remains "awake." The eye that feeds its input into the awake half of the brain remains open, and if this eye views a threatening visual stimulus, the bird takes action promptly, even though it appeared to be asleep—and in some sense, was indeed asleep—less than a quarter-second previously. The hypothesis that this unusual kind of sleeping is an adaptation for predator detection yields the prediction that birds more exposed to predators will spend more of their sleep time in this special half-awake mode. When four mallard ducks sleep in a row, the two birds on the end are in fact both more likely to use this sleep pattern than are the birds in the middle, and the eye that stays open is the one that looks out away from the group [956].

The fact that mallards often do sleep in groups may be because many eyes are better than one for perceiving danger. Groups may enable an individual that has not personally spotted a predator to escape by reacting to the flight of its companions from an enemy they have detected [702]. One way to test the many-eyes hypothesis would be to examine whether scanning rates fall as group size increases, as one would predict if group members can afford to rely on the reactions of others to detect an approaching predator. In many animals, ranging from shorebirds [258] to zebras [1019], the larger the group, the less often any one individual looks up to scan for predators. And the more scanners, the quicker the response to danger, as G. V. N. Powell showed. He released an artificial hawk that traveled along a line over an aviary containing either a single starling or a group of ten of these despised but adaptable birds [954]. The average reaction time of the solo bird was about a second longer than that of a bird in a flock, despite the fact that single birds spent nearly half their foraging time looking out for danger, whereas birds in groups spent only about 10 percent of their time scanning for predators. R. E. Kenward took this test one step closer to reality by

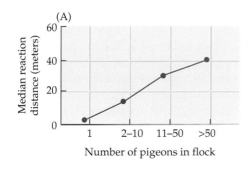

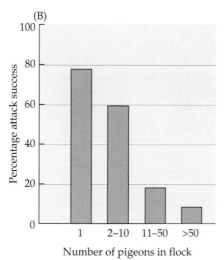

24   Vigilance and group size. (A) The larger the group of wood pigeons, the sooner the flock spotted and flew away from an approaching goshawk, and (B) the poorer the predator's chances of making a kill. After Kenward [616].

releasing a hungry, trained goshawk at a standard distance from flocks of wild wood pigeons [616]. As predicted, the larger the flock, the sooner the prey took flight, lowering the probability that the goshawk would make a kill (Figure 24).

Although this study provides support for the hypothesis that animals in groups gain vigilance benefits, clumped prey may also pay a price for living together via increased competition for food. By taking this cost into account, Mark Elgar predicted that house sparrows would not form groups when the risk of predation was low. As expected, house sparrows produced few of the "chirrup" calls that attract others to them when foraging at feeders close to safe cover and far from a potential predator, the observer (Figure 25) [355].

Goshawk with wood pigeon

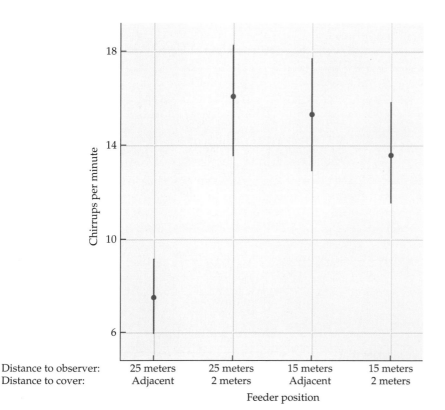

Distance to observer: 25 meters   25 meters   15 meters   15 meters
Distance to cover: Adjacent   2 meters   Adjacent   2 meters

Feeder position

English sparrow

25   Social costs and defensive benefits. House sparrows were less likely to call for companions to join them at a food source when the risk of predation was low. The number of chirrup calls given per minute was lowest when the feeder was distant from an observer and close to cover. After Elgar [355].

The Dilution Effect and the Selfish Herd

Perhaps the simplest antipredator advantage of living in a group comes from overwhelming the killing capacity of the local predators. If five predators are hunting in an area, and each kills two prey per day, the risk of death of a prey individual in a group of 1000 is 1 percent per day, whereas it is ten times higher for members of a group of 100. Is this why some mayflies synchronize their transition to adulthood [1118]? Mayflies are apparently tasty insects, much enjoyed by various predators at all stages of life, but especially when they change from aquatic nymphs into winged airborne adults. This metamorphosis is one that entire populations may undertake over a few hours on a few days each year.

To determine whether synchrony in emergence was an anti-capture adaptation, Bernard Sweeney and Robin Vannote first had to get a count of how many individuals emerged on a given evening from a segment of stream. To this end, they placed nets in streams to catch the cast-off skins of mayflies, which molt on the water's surface as they change into adults, leaving the discarded cuticle to drift downstream on the current. Counts of the molted cuticles revealed how many adults emerged on a particular evening. The nets also caught the bodies of females that had laid their eggs and then died a natural death; a female's life ends immediately after she drops her clutch of eggs into the water, provided a nighthawk or a whirligig beetle does not consume her first. Sweeney and Vannote measured the difference between the number of cast skins of emerging females and the number of intact corpses of spent adult females that washed into their nets on different days. The greater the number of females emerging together on a given day, the stronger the dilution effect, and the better the chance each mayfly had to live long enough to lay her eggs before expiring (Figure 26) [1118].

Thus, the dilution effect may contribute to the formation of groups whose members are really using one another to improve their odds of survival. This view of sociality underlies W. D. Hamilton's characterization of some groups of animals as **selfish herds** [485]. Imagine a population of antelope grazing on

Mayfly

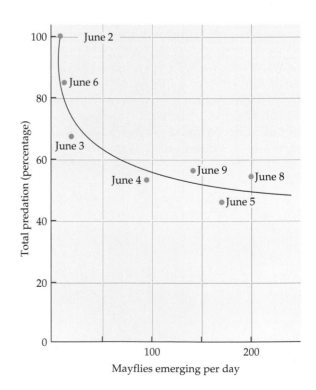

26   The dilution effect in mayflies. The more female mayflies that emerged together on a June evening, the less likely any individual mayfly was to be eaten by a predator. After Sweeney and Vannote [1118].

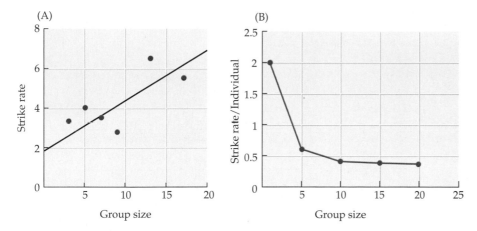

Whirligig beetles

27 Safety in numbers carries a cost for individuals. (A) Large groups of whirligig beetles are more likely to be attacked by fish than are small groups, at least under laboratory conditions. (B) On the plus side, however, the chance of attack per individual is less for members of larger groups. After Watt [1188]. Drawing by Barbara Terkanian.

an African plain in which all individuals stay well apart, reducing their conspicuousness to their predators. Now imagine that a mutant individual arises in this species, one that approaches another animal and positions itself so as to use its companion as a living shield for protection against attacking predators. The mutant that employs this tactic would incur some costs; for example, two animals may be more conspicuous to predators than one, and so attract more attacks than scattered individuals, as has been demonstrated in some cases [1189]. But if these costs were consistently outweighed by the survival benefit gained by having a reduced chance of attack and capture (Figure 27), the mutation could spread through the population. If so, eventually all the members of the species would be aggregated, with individuals jockeying for the safest position in their groups, not merely content to dilute their chances of being unlucky in a predation lottery, but actively attempting to improve their odds at someone else's expense. The result would be a clumping of individuals, a selfish herd whose members would actually be safer if they all could agree to spread out and not try to take advantage of one another. But since populations of such isolated individuals would be vulnerable to invasion by an exploitative mutant that takes fitness from its companions, the exploitative tactic could spread through the species—a clear illustration of why we define adaptations in terms of their contribution to the fitness of individuals compared with that of other individuals with alternative traits.

The selfish herd hypothesis is based on **game theory,** the theory that the fitness payoffs for an individual as it attempts (usually unconsciously) to maximize its reproductive success depends on the actions of the other animals around it, particularly those belonging to its own species. Thus, instead of focusing solely on the interaction between a prey and its predators, Hamilton asked how outcomes of prey–predator interactions might be influenced by what the other prey do.

We can ask this question about the behavior of Adélie penguins as they leave their breeding grounds to go out to sea to feed. By the water's edge, the penguins often gather in groups before jumping into the water together to swim out to the foraging grounds. The potential value of this social behavior becomes clearer when one realizes that a leopard seal may be lurking in the water near the jumping-off point (Figure 28) [246]. The seal can capture and kill only a certain small number of penguins in a short time. By swimming in a group through

28   Selfish herds may evolve in prey. Adélie penguins have formidable predators, like this leopard seal. While the seal disposes of one penguin, others can more safely enter the water and escape to the open ocean. Photograph by Gerald Kooyman/Hedgehog House.

the danger zone, many penguins will escape while the seal is engaged in dispatching one or two unfortunate ones. If you had to run a leopard seal gauntlet, you would probably do your best to be neither the first nor the last into the water. If penguins behave as we would, then the flocks that form on the ice near water qualify as selfish herds.

The selfish herd hypothesis generates testable predictions that can be applied to Adélie penguins or to any other prey species that forms groups. A key prediction is that individuals in these groups should compete for the safer positions. Bluegill sunfish breed in colonies (see p. 426), with each male defending a nesting depression in the lake bottom against rivals [463]. As predicted from a selfish herd model, males compete intensely for central territories, and larger, generally older, males are more likely to win. Thus, it is a younger, smaller, peripheral male who first confronts a foraging bullhead or cannibalistic bluegill, and as a result, his eggs are twice as likely to be eaten by a predator as those of a male with a central nest.

If peripheral bluegills are at such a disadvantage, why do they nest on the edge of a colony instead of going elsewhere? If a younger or weaker individual has no reasonable probability of forcing a more powerful rival to yield his superior position, then the subordinate animal has two options: to nest on the outskirts of the group, or to nest alone. In the bluegill, solitary nesters (which do sometimes occur in this species) experience higher snail infestation rates, and must chase predators more often, than peripheral colonial males (Table 3). Note

TABLE 3 Predation pressure on bluegill nests in relation to nest position

Position of nest	Mean number of snails per nest	Mean number of chases of egg predators per hour	Percentage of predator chases by two bluegills
Center of colony	6.9	1.5	50
Edge of colony	13.7	8.7	8
Away from colony	29.7	10.4	0

Source: Gross and MacMillan [463]

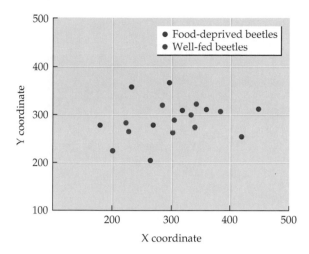

29 **Whirligig beetles face a trade-off between getting food and being safe from predators.** Food-deprived beetles tend to move to the outside of the group, where food is more accessible, even though the risk of predatory attack is greater there. The X- and Y-coordinates provide a measure of the position of a beetle in the experimental tank. After Romey [984].

also that although mobbing of a predator is more likely to occur in the center of the colony, peripheral males also sometimes practice communal defense of their nests, an option unavailable to the solitary nester. Therefore, a subordinate fish may benefit slightly by nesting in a group, even though more dominant males use him as protection against their enemies [463].

Likewise, in schools of fish (see Figure 6, p. 425), one imagines that individuals swimming on the front line would be at special risk of attack, as has been empirically demonstrated in experiments with lake chub (the schooling prey species) and rock bass (the predator) [178]. If individuals leading the way are prone to attack, one can predict that schools of fish should be elongate, with relatively few fish on the point and most streaming along behind. Three-dimensional analysis shows that schools of roach have this predicted shape. Once again, one can ask, why do any fish get ahead of the others? Presumably there are certain trade-offs that can make it advantageous for some individuals under some circumstances to run a higher risk of predation in return for, perhaps, an improved chance of finding food. In whirligig beetles, which form aggregations on the surface of ponds, streams, and lakes, those that have been deprived of food are more likely to take up a position on the edge of the group, where they are far more likely to reach floating food particles than beetles near the center of the pack (Figure 29) [984].

Fighting Back

Although animals have evolved a variety of tactics that decrease their chances of being taken by a predator, even captured animals are not necessarily helpless. Consider the notodontid moth caterpillar that I found one day. When I went over to take a closer look, the big larva suddenly dropped its head and abdomen from the leaf on which it had been resting and formed an inverted U. Undeterred, I reached for the animal, at which point it everted a complex, red, antlerlike sac from the underside of its body and sprayed me with a strongly scented aerosol of formic acid (Figure 30, left). I withdrew my hand quickly.

I do not know what natural predators the moth larva douses with formic acid—perhaps birds or ants—but it is a reasonable hypothesis that sometimes when it is grasped, its chemical defenses cause the predator to release it. The success of such defenses has been carefully documented in other similar cases (Figure 31). Consider, for example, the black widow spiders that live in my toolshed. I, like most Arizonans, treat these spiders with a mixture of fear and

30 Chemical defenses against predators. (Left) This moth larva sprays formic acid from the red osmeterium on the underside of its body. (Right) This moth, an adult of another species, releases a froth from its thoracic glands; the fluid contains toxic alkaloids that repel spiders. Photograph (left) by the author and (right) by Tom Eisner.

respect because of their toxic venom. However, some animals, such as deer mice, are willing to take a chance that they can kill the plump and edible spider with a quick bite before it can react with a defensive bite of its own. Despite the speed and agility of an attacking deer mouse, about half the interactions between this predator and black widow spiders in a laboratory arena ended with the mouse repelled and the spider alive [1168]. In a large majority of these cases, the spider survived because of its rapid secretion and deployment of a strand of silk that it produced as soon as it was jostled (Figure 31). The spider applied this silk

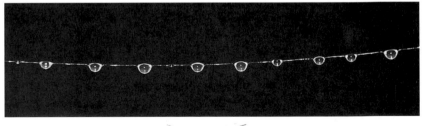

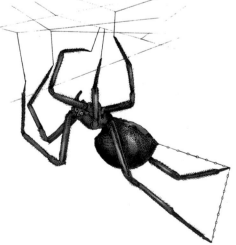

31 Silken defenses. Black widow spiders apply a sticky strand of silk (above) to predators, causing small mammalian enemies to retreat. Courtesy of Rick Vetter.

strand, adorned with droplets of a profoundly adhesive substance, to its attacker's face. The mouse typically recoiled and attempted to rid itself of the material, sometimes by rolling frantically on the ground. In nature, the spider could have used this reprieve to scramble from its web to a safe crevice.

Rick Vetter tested the adaptive value of the sticky silk strand for black widows by waxing over the spinnerets of some spiders. When attacked by mice, these individuals could not defend themselves with adhesives, and were three times as likely to be killed as unplugged spiders. Although the spider's glue is not toxic (consumption of the viscous substance did not affect the mice), it is sticky, forcing the mouse to remove it at once or suffer damage to its fur.

A Tactic of Last Resort?

Once a black widow has been bitten by a deer mouse, it's all over for the spider. But sometimes prey can survive for a time in the clutches of a predator, during which time they may produce extraordinarily loud and piercing screams. Rabbits, for example, respond to being captured in this fashion, despite the fact that they are normally the quietest of animals. Surely the scream is simply a nonadaptive reaction to stimulation of the prey's pain receptors—or is it? In a paper entitled "Adaptation unto Death," Göran Högstedt tested multiple hypotheses on the possible function of vocalizations of this sort, including the possibility that a loud scream might so startle a predator that it would release its captured prey [529]. Högstedt noted that although fear screams are noisy, as required by the startle hypothesis, nevertheless the calls persist as long as the victim remains alive, which should reduce their potential to startle (but see [241]).

An alternative to the startle hypothesis is that the screams evolved to warn others of danger, which they might do because fear screams typically contain both high and low frequencies of sound, features that enable other animals to locate the source of the sound easily. Högstedt pointed out that if this second hypothesis were correct, one would expect conspecifics to respond to the cries by taking some action. Yet prey largely ignore the calls of captured neighbors, a reaction that makes a certain amount of sense, since a predator with a captured prey will be occupied with that victim for some time rather than hunting for uncaptured specimens nearby.

But perhaps fear screams are given by young animals to enlist the aid of their parents. This hypothesis produces the prediction that the calls should be given only by dependent young, and this is not the case. Adult songbirds captured in mist nets are just as likely to scream as juveniles when handled by a human being (Figure 32).

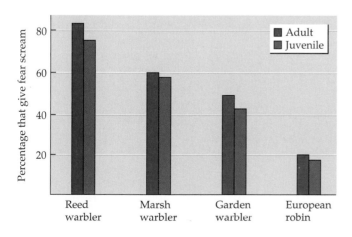

32 Fear screams and age in four European songbirds. In many species, juveniles are no more likely to give fear screams when handled by a human than are adults. This observation suggests that the function of the call cannot be to enlist parental assistance. After Högstedt [529].

The evidence points to a fourth hypothesis: namely, that the captured animal's screams attract other predators to the scene, predators that may turn the tables on the prey's captor, or at least interfere with it, sometimes enabling the prey to escape in the confusion. This hypothesis requires that predators be attracted to fear screams, and they are, as Högstedt showed by broadcasting taped screams of a captured starling, which brought hawks, foxes, and cats to the recorder. (On the other hand, in another study, when coyotes attracted to starling distress calls came toward a fellow coyote with captured starling, the attacker intensified its effort to dispatch the bird, which was not good news for the starling [1259].)

Finally, the attract-competing-predators hypothesis produces the prediction that birds living in dense cover should be more prone to give fear screams than birds of open habitats. In areas where vision is blocked, a captured bird cannot rely on other nearby predators to see its mortal struggle. Therefore, as a final effort to avoid death, it may attempt to call competing predators to the scene. Mist-netted birds of species that live in dense cover are in fact more likely to give fear screams when handled than are species that occupy open habitats [529].

Predator-attracting signals may also occur in some fish, such as the fathead minnow. As is true of many fish, fathead minnows that have been attacked release chemicals from specialized cells in the skin. Traditionally, these chemicals have been considered alarm signals designed to alert other members of the species to the presence of a predator, and in some cases, fish exposed to chemicals extracted from the skin of conspecifics do indeed appear to hide [214]. The question raised by observations of this sort, however, is how an injured fish can benefit from helping others of its species to escape from a predator. Perhaps injured fish do not release these special chemicals to benefit others, but rather to help themselves by attracting additional predators that may interfere with attacker number one, occasionally resulting in the release of the captured prey [213].

To test this hypothesis, researchers used divided aquaria with a juvenile pike in each compartment. After a fathead minnow had been placed in one compartment and was captured by the resident pike, the panel separating the two predators was removed. The second pike often approached its neighbor. (Other studies have shown that the "alarm odor" in and of itself can attract predators from a distance.) Once near its neighbor, the intruding pike tried to steal the minnow by various means. As a result of its efforts, the time required for the first pike to swallow its victim increased threefold, compared with control trials in which no second pike occupied the other compartment (Figure 33). More-

Pike with fathead minnow

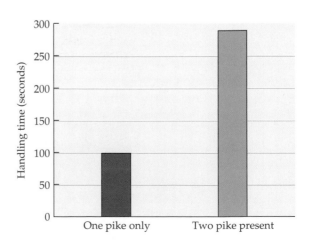

33 A predator attraction signal. By releasing chemicals from the skin when bitten by a pike, fathead minnows can attract other pike to the scene. The presence of a second pike increases the time it takes a pike to position a minnow properly for swallowing. After Chivers et al. [213].

over, in 5 of 13 trials, the minnow escaped, at least temporarily, something that never occurred during the control trials. These findings confirm the utility of the "predator attraction signal" as a last-ditch survival aid for the signaler, rather than an alarm message for its fellow prey.

Summary

1. An adaptation can be defined as a trait that raises the fitness of individuals more than other forms of the characteristic in question. The adaptationist approach provides a way to identify traits of interest (those with substantial fitness costs) and structures the hypotheses proposed to resolve these puzzles by identifying possible fitness benefits associated with the traits.

2. To test adaptationist hypotheses, a scientist must check the validity of predictions derived from these potential explanations. Evidence can be gathered through field observations, controlled manipulative experiments, or natural experiments involving comparisons among living species.

3. The comparative method yields two key predictions: (1) divergent evolution will occur among related species that face different selection pressures, despite their having a common ancestor and thus a similar genetic heritage, and (2) convergent evolution will occur among unrelated species that share similar selection pressures, despite their having a different genetic heritage.

4. Prey species have evolved many different kinds of adaptations designed to make their detection, attack, capture, or consumption by predators less likely. Thwarting detection often involves cryptic behavior, in which an animal enhances the effectiveness of its camouflage by remaining motionless on an appropriate background. Blocking an attack often involves honest advertisement of either unprofitability or unpalatability, systems that can be exploited by deceitful Batesian mimics. Prey may prevent their capture through vigilance, the dilution effect, selfish herd formation, or fighting back.

5. The selfish herd hypothesis is derived from the application of game theory to the analysis of behavior. In game theory, the adaptive value of a behavioral option to an individual depends upon what the other members of its population are doing.

Discussion Questions

1. Many people think that an adaptation is a trait that improves the survival chances of an organism. Under what circumstances would such a trait be an adaptation? Under what other circumstances would a survival-enhancing attribute be selected against?

2. For many evolutionary biologists, the term "adaptation" is reserved for characteristics that provide "current utility to the organism and [have] been generated historically through the action of natural selection for its current biological role" [80]. What *could* "current utility" mean, and what do you think it *should* mean? Make use of the terms "fitness benefits" and "fitness costs" in your answer. What practical problems would arise in attempting to apply adaptation in this sense to traits observable only in modern organisms? If a trait originated for function X and later took on a different but still adaptive biological role Y, what would you call the trait if you couldn't call it an "adaptation?" If you go back far enough in time, will the ancestral form of any current trait have the same function that it does now?

3. Consider the following situation (with thanks to Jack Bradbury). In a population of prey animals, most individuals are solitary and stay well apart from others. But some mutant types arise that search out others and use them as living shields against predators. The mutants take fitness from the would-be

solitary types by making them more conspicuous to their predators. Set the fitness payoff for solitary living at *P* in a population composed only of solitary individuals. But when a solitary individual is found and used by a social type, the solitary animal loses some fitness (*B*) to the social type. There is a cost (*C*) to being social in terms of the time required to find a solitary individual to hide behind, and a cost arising from the increased conspicuousness to predators of groups composed of two individuals rather than one. When two social types interact, we will say that they each have one chance in two of being the one that happens to hide behind the other when a predator attacks. A game theory diagram (see below) summarizes these interactions. If *B* is greater than *C*, what behavioral type will come to predominate in the population over time? Now compare the average payoff for individuals in populations composed entirely of solitary versus social types. If the average fitness of individuals in a population composed only of social types is less than that of individuals in a population composed only of solitary types, can hiding behind others be an adaptation?

	Opponent	
	Solitary	Social
Focal animal Solitary	P	$P-B$
Social	$P+B-C$	$P+\frac{B}{2}-\frac{B}{2}-C=P-C$

4. The ability to hear ultrasound in one species of noctuid moth is considered an adaptation because it apparently enables individuals to hear and avoid nocturnal, ultrasound-using bats. Imagine that you wished to test this hypothesis with the comparative method. Identify the utility of each of the following lines of evidence on the hearing abilities of other insect species. Specify whether these cases involve convergent evolution, divergent evolution, or neither.

   a. Almost all other species of noctuid moths also have ears that respond to ultrasound.
   b. Almost all the species in one evolutionary lineage that includes the noctuid moths and many other moths belonging to several superfamilies also have ears that respond to ultrasound [1271].
   c. Some diurnal noctuid moths have ears, but are largely or totally incapable of hearing ultrasound [409].
   d. Almost all butterflies, which belong to the same large evolutionary grouping as the noctuids but are usually active during the day, lack ears and so cannot hear ultrasound [410].
   e. Members of one small group of nocturnal butterflies have ears on their wings and can hear ultrasound; they respond to ultrasonic stimulation by engaging in unpredictable dives, loops, and spirals [1271].
   f. Lacewings and praying mantises fly at night and have ears that respond to ultrasound (see p. 125).

5. In my front yard, I sometimes find several hundred male native bees clustering in the evening on a few bare plant stems (see photo at left, photograph by the author). An assassin bug sometimes approaches the cluster and kills some bees as they are settling down for the night. Devise at least three alternative hypotheses on the possible anti-assassin bug value of these sleeping clusters, and list the predictions that follow from each hypothesis.

6. Stephen J. Gould and Richard Lewontin claim that adaptationists make the mistake of believing that every characteristic of living things is a perfected adaptive product of natural selection [448], when in reality many attributes of living things are not adaptations (see Table 1). Moreover, in their eagerness to explain everything as an adaptation, adaptationists have, according to Gould and Lewontin, invented fables as absurd as the fictional "just-so" stories of Rudyard

Kipling, who made up amusing explanations for the leopard's spots and the camel's hump. How might adaptationists defend themselves against these charges? Do adaptationists have the means to discover whether they have been wrong to hypothesize that trait X is an adaptation?

Suggested Reading

Two books, one by Wolfgang Wickler [1228] and the other by Rod and Ken Preston-Mafham [936], contain many amazing examples of animal coloration and behavioral defenses. John Endler has provided a modern review of the interrelationships between predator adaptations and prey counteradaptations [368]. The cost–benefit approach to antipredator behavior is described by Steven Lima and Lawrence Dill [703]. Tim Caro's papers on gazelle stotting [198, 199] illustrate the adaptationist approach very well, as does his review of how the pursuit deterrence hypothesis has been tested [200]. S. J. Gould and R. C. Lewontin's attack on adaptationism [448] is worth reading—critically. Berni Crespi succinctly reviews the various definitions for adaptation and the several reasons for the occurrence of maladaptations in nature [257].

8 The Evolution of Feeding Behavior

A bullfrog has many dietary choices to make. Does its selection of nightcrawlers maximize the calories it gains? Photograph by Kenneth H. Thomas

*L*iving prey have evolved a wide spectrum of responses to the risk of being detected, attacked, captured, and consumed by a predator. Having now read about how skilled prey species are at postponing their demise, you may have developed a certain sympathy for foraging predators, which have to overcome a whole series of obstacles if they are to get something to eat. On the other hand, if you have ever observed a cactus wren or a blue jay plucking one insect after another from their hiding places in dense vegetation—a katydid here, a caterpillar there, then another caterpillar—you might rightly conclude that predators can take care of themselves. In a way, the hunting skills these animals exhibit are the product of their prey, whose amazing camouflage and other protective devices have favored the evolution of predatory counteradaptations in a kind of arms race that continues to this day. As a result, animals hunting for food are generally good at detecting prey, good at deciding when to attack, good at capturing prey they have chosen to pursue, and good at consuming the animals that they have captured. This chapter examines each of these aspects of the arms race between prey and predator, with the additional goal of illustrating

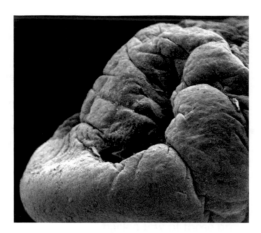

3 **The evolutionary history of the nose of the star-nosed mole** may be revealed by comparing the noses of modern mole species. (Top left) The appendage-free nose of the eastern mole is touch-sensitive, as is that of other moles and many other related species in the order Insectivora. (Top right) The Townsend's mole has a more complex, flanged nose covered with touch-sensitive Eimer's organs. This nose may resemble the nose of an ancestral species that gave rise eventually to the star-nosed mole. (Bottom) The star-nosed mole has a nose with many fingerlike appendages, all of which are covered with Eimer's organs. Photographs from Catania [205].

4 **A cladogram of lizard prey detection mechanisms.** Orange branches indicate that the members of this lizard family are olfactory foragers, which flick their tongues to detect prey odors. Blue branches represent ambush predators, which employ vision rather than chemoreception to find their meals. After Cooper [243].

occurred as species C and D each gave rise to two descendant lineages, one with the ancestral hunting mode and one with the derived, or novel, mode of foraging. Thus, for example, one ancestral species that split off from extinct species C went on to give rise to the many descendant species now placed in the family Iguanidae, all of which have retained the active mode of foraging associated with that distant common ancestor. Likewise, the origin of the new kind of foraging behavior in six different families on the right side of the cladogram can be traced back to a single ancestral species that split off from extinct species B. (You should be able to use the cladogram to reconstruct the history behind the foraging mode of members of the Cordylidae and the Eublephainae.)

Getting Help from Companions

Foragers do not necessarily have to rely only on themselves to track down food, but may also secure information about food location from their companions. Perhaps the most famous case of this phenomenon involves honey bees and their dances, which foraging worker bees perform when they return to the hive after having found a good source of pollen or nectar [1175]. These bees move in circuits on the vertical surface of the honey comb in the complete darkness of the hive. The dancers attract other bees, which follow them around as they move through their dance.

Researchers watching dancing bees in special observation hives have learned that their dances contain a surprising amount of information about the location of a food source (such as a patch of flowers.) If the bee executes a round dance (Figure 5), she has found food fairly close to the hive—say, within 50 meters of it. If, however, the worker performs a waggle dance (Figure 6), she has found a nectar or pollen source more than 50 meters distant. By measuring either the number of times the bee runs through a complete waggle-dance circuit in 15 seconds or the duration of the waggle-run portion of the circuit, a human observer can tell approximately how far away the food source is. The longer-lasting the waggle-run portion, the more distant the food.

Moreover, by measuring the angle of the waggle run with respect to the vertical, someone observing a waggle-dancing bee can also tell the *direction* to the food source. Apparently, a foraging bee on the way home from a distant but rewarding flower patch notes the angle between the flowers, hive, and sun. The bee transposes this angle onto the vertical surface of the comb when she performs the waggle-run portion of the waggle dance. If the bee walks directly

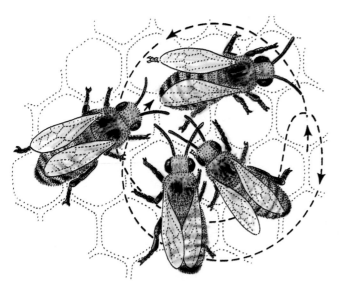

5 Round dance of honey bees. The dancer (the uppermost bee) is followed by three other workers, who may acquire information that a profitable food source is located within 50 meters of the hive. After von Frisch [1174].

6   Waggle dance of honey bees. As the worker performs the waggle-run portion of the dance, she shakes her abdomen from side to side. The duration and the orientation of the waggle runs contain information about the distance and direction to a food source. In this illustration, workers attending to the dancer learn that food may be found by flying 20 degrees to the right of the sun when they leave the hive. (A) The directional component of the dance is most obvious when it is performed outside the hive on a horizontal surface in the sunlight, in which case the bee points the waggle runs directly toward the food source. (B) On the comb, inside the dark hive, the dance is oriented with respect to gravity so that the deviation of the waggle run from the vertical equals the deviation of the direction to the food source from a line between the hive and the sun.

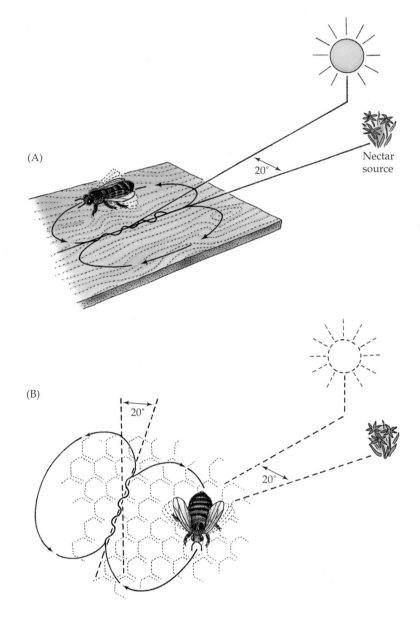

up the comb while waggling, the flowers will be found by flying directly toward the sun. If the bee waggles straight down the comb, the flower patch is located directly away from the sun. A patch of flowers positioned 90° to the right of a line between the hive and the sun is advertised with waggle runs pointing 90° to the right of the vertical on the comb. In other words, when outside the hive, the bees' directional reference is the sun, whereas when the bees are inside, the reference is gravity.

The conclusion that the dance displays of bees contain information about the distance and direction to good foraging sites was reached by Karl von Frisch after several years of careful experimental work [1175]. His basic research protocol involved training bees (which he daubed with dots of paint for identification) to visit feeding stations, which he stocked with concentrated sugar solutions. By watching the dances of these trained bees, he saw that their behavior changed in highly predictable ways depending on the distance and direction to a food source. More importantly, his dancing bees were able to direct other bees to a food source they had found (Figure 7), leading him to believe that bees use the information in the dances of their hivemates to find good foraging sites.

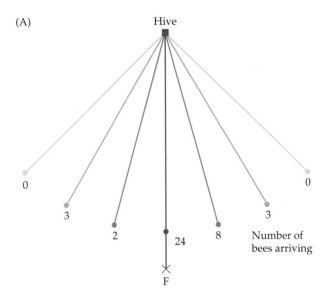

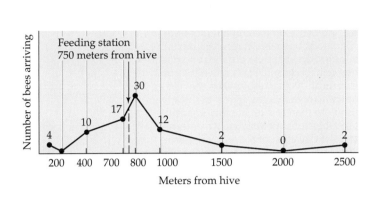

7 Testing directional and distance communication by honey bees. (A) A "fan" test to determine whether foragers can convey information about the direction to a food source they have found. After training scout bees to come to a feeding station at F, von Frisch collected all newcomers that arrived at seven feeding stations with equally attractive sugar water. Most new bees arrived at the station in line with F. (B) A test for distance communication. Scouts were trained to come to a feeding station 750 meters from the hive. Thereafter, all newcomers arriving at feeding stations placed at various distances from the hive were captured. In this experiment, 47 newcomers were captured at the two stations closest to 750 meters, far more than were caught at any two other stations. After von Frisch [1174].

Over the years, von Frisch's conclusions have been debated, with some scientists arguing that the workers recruited to the food source rely exclusively on the flower odors present on the dancing bee's body as a guide for their search, rather than deriving information from the dance movements per se [1206]. However, when foragers from a hive were trained to two feeders that were given the same scent but were located in opposite directions from the hive, and only one feeder was advertised by waggle dances (because only that feeder contained a concentrated sugar solution), the bees recruited from this hive arrived primarily at the feeder advertised by the dancers. This experiment offers strong evidence in favor of the proposition that workers gain useful information from the movements of dancing bees [1035].

Wolfgang Kirchner and Andreas Grasser have provided more support for this conclusion by comparing the performance of recruits from a hive that could be turned on its side or held upright in the standard position. When the bees were forced to dance in the darkness on the horizontal comb of a hive turned on its side, the dance followers gained no directional information from the dances, whereas when the hive was returned to an upright position and the comb was vertical, most recruits appeared at the feeder that the scouts had visited (Figure 8). The effect held, however, only for feeders located 100 or more meters from the hive, when the waggle dance was used. When the round dance was used by foragers returning from a feeder placed 10 meters from the hive, the recruits from the upright hive did no better than those from the hive turned on its side, as one would expect, since the round dance does not contain information about the direction to a specific food source. Thus, foraging bees can convey the locations of distant food sources to their hivemates via dance movements, provided the dancers have access to a vertical dance floor (which they always do in nature) [624].

8 Differences in successful recruitment to a food source when a honey bee hive is upright versus when it is placed on its side. Although scout bees perform a waggle dance on the comb under both conditions, bee recruits receive usable information about the distance and direction to a food source only when the comb is vertical. (A) As a result, when the hive is upright, more recruits appear at the advertised site than at a set of other unadvertised locations. (B) Moreover, the mean angular deviation of the recruits' search directions is much less when the hive is upright, meaning that the recruits are much more tightly bunched around the advertised site than when the hive is on its side. After Kirchner and Grasser [624].

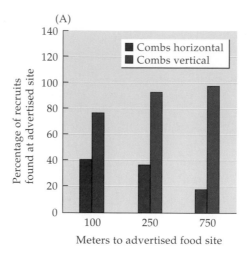
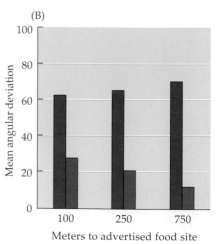

The History of Honey Bee Dances

The evolutionary basis of honey bee dances can be analyzed using both the historical approach and the adaptationist approach. A historical analysis was provided by Martin Lindauer, who used comparisons among living bee species to reconstruct the possible antecedents of this highly specialized behavior [706]. Lindauer found that three other members of the genus *Apis* perform dance displays identical to those of the familiar honey bee (*Apis mellifera*), except that in one species, *A. florea,* the bees dance on the horizontal surface of a comb built in the open over a tree branch (Figure 9). To indicate the direction to a food source, a worker simply orients her waggle run directly at it. Because this is a less sophisticated maneuver than the transposed pointing done in the dark on vertical surfaces by *A. mellifera,* it is probably similar to a form of dance communication that preceded the dances of *A. mellifera.*

9 The nest of an Asian honey bee, *Apis florea,* is built out in the open around a branch. Dancing workers on the flat upper surface of the nest (two nests are shown here) can run directly toward the food source when performing the waggle dance. Photograph by Steve Buchmann.

Lindauer then looked to bees not belonging to the genus *Apis* for recruitment behaviors that might give hints about the steps that might have preceded the first waggle dancing. Stingless bees of the tropics employ diverse communication systems. Lindauer organized those systems into the following evolutionary scenario, which is based on the premise that evolution proceeds gradually through the accumulation of many small changes (see p. 283):

- *The possible first stage:* Workers of some species of stingless bees in the genus *Trigona* run about excitedly, producing a high-pitched buzzing sound with their wings, when they return to the nest from flowers rich in nectar or pollen. This behavior arouses their hivemates, which detect the odor of the flowers on the dancer's body. With this information, they leave the nest and search for similar odors. The actions of the dancing bee do not provide specific signals indicative of the direction or distance to the desirable food.

- *A possible intermediate stage:* Workers of other species of *Trigona* do convey information about the location of a food source. In these species, a worker that makes a substantial find marks the area with a pheromone produced by her mandibular glands. As the bee returns to the hive, she deposits pheromone on tufts of grass and rocks every few yards. Inside the hive entrance, bees wait to be recruited. The successful forager crawls into the hive and produces buzzing sounds that stimulate these individuals to leave the hive and follow the scent trail that she has made (Figure 10).

- *A still more complex pattern:* A number of stingless bees in the genus *Melipona* convey separate distance and directional information. A dancing forager communicates information about the distance to a food source by producing pulses of sound; the longer the pulses, the farther away the food is. In order to transmit directional information, she leaves the nest with a number of followers and performs a short zigzag flight that is oriented toward the source of nectar. The scout returns and repeats the flight a number of times before flying straight off to the nectar site, with the recruited bees in close pursuit.

10 Communication by scent marking in a stingless bee. In this species, workers that found food on the opposite side of a pond could not recruit new foragers to the site until Martin Lindauer strung a rope across the pond. Then the scouts placed scent marks on the vegetation hanging from the rope and quickly led others to their find. Photograph by Martin Lindauer.

Since Lindauer's pioneering work on bee communication, James Nieh and his colleagues have provided much more information on the complexity of the recruitment signals in one species of *Melipona* [841]. The bee *M. panamica* provides one kind of buzzy acoustical signal when unloading food in a funnel-like entryway to the nest and then generates another kind of signal when dancing in the entryway. The sounds made while unloading food inform recruits whether the resource is high in the canopy of the tropical forest or lower down near the forest floor (Figure 11). The second bout of signaling conveys information about the *distance* to the food source advertised by the dancer. However, in order to convey information about the *direction* to the food source, the scout must leave the nest. If she is experimentally prevented from doing so, then equal numbers of untrained workers will turn up at the site previously visited by the scout and at a site located in exactly the opposite direction. Recruits that have access to all three sources of information about the location of a flower patch will be able to narrow their search still further when within 10 meters of the site because of scent marks deposited at the spot by the scout before she returned to the hive to buzz, dance, and lead [840].

Although, judging from *M. panamica*, some stingless bees have communication systems every bit as sophisticated as that of the honey bee, not all do. The diversity of communication systems among stingless bees suggests that communication about *distance* to a food source by an ancestor of the honey bee probably at first involved only agitated movements by a food-laden worker. Other workers that were stimulated by the activity of the returning forager would then have left the hive to find food. In some species, selection may have subsequently favored standardization of the sounds and movements made by successful foragers, as in *Melipona*. This trend leads to the round and waggle dances of *Apis* bees that contain symbolic information about how far food is from the hive [706, 1246].

In contrast, communication about the *direction* to a food source appears to have originated with personal leading, with a worker guiding a group of recruits directly to a nectar-rich area. Here the evolutionary sequence has involved less and less complete performance of the guiding movements as generations of queens produced workers with a greater and greater tendency to perform incomplete leading. At first this may have taken the form of partial leading (as in some *Melipona*) and later involved simply pointing in the proper direction with a waggle run on a horizontal surface (as in *A. florea*). The final step is the transposed pointing of *A. mellifera*, in which the direction of flight relative to the sun is converted into a signal (the waggle run) oriented relative to gravity.

11 Acoustical communication of the height of a food source by the bee *Melipona panamica*. When a scout that has been trained to collect food from a feeder high in a tree interacts with her nestmates (left panel), she produces a specific kind of acoustical signal (shown as a black sonogram in the panel beneath the bees). After listening to these sounds, recruits are much more likely to locate the training feeder than a control feeder placed at the base of the tree. After Nieh [841].

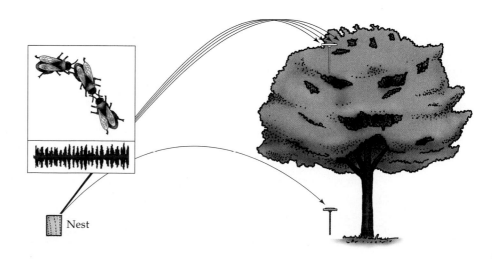

Nest

The Adaptive Value of Honey Bee Dances

Having outlined a possible historical sequence leading to the wonderfully complex honey bee dances, we now ask, what is the adaptive value of this behavior? Tom Seeley and Kirk Visscher proposed three hypotheses on how the time and energy costs of dancing might be more than repaid by fitness benefits to the queen whose workers perform this behavior [1037]. Since worker honey bees are sterile females, their activities cannot promote their own reproductive success, but their behavior can raise the fitness of family members that are able to reproduce—most notably their mother (see p. 445). The most obvious possibility is that a dancing scout helps her nestmates find food supplies more quickly than they would otherwise, reducing the costs of food location for workers and thus increasing the caloric cost–benefit ratio for the colony. This hypothesis predicts that, on average, a scout bee, who hunts for new food sources without information from other bees, will need more time to find food than will a recruit, who searches for a food source with the aid of dance information from her hivemates.

In fact, however, although many worker bees from Seeley and Visscher's hives followed dancers, a recruit typically did not locate the flower patch advertised by the dance that she had followed on her first attempt. Indeed, a recruit needed about 2 hours of searching on average before she found the advertised food source, whereas a scout required a little less than 90 minutes on average to find a new food source worth dancing about back at the hive. We must reject the time-saving hypothesis for recruitment by dancing bees.

A second possibility is that dancers direct recruits to better food sources than they could find on their own, increasing the calories gained per unit of time spent foraging by workers. Based on this hypothesis, we can predict that a recruit, returning with food from an advertised patch, will bring back more pollen or nectar on average than will a scout that is returning from a patch she discovered herself. Consistent with this prediction, Seeley and Visscher found that after having been directed to a patch, recruited foragers tended to make several highly productive trips in a row to that site, whereas scouts made less productive trips and tended to cease visiting their food sources after just one trip (Figure 12).

Finally, recruitment of a large workforce to a food patch might mean that more of its resources can be harvested before other bee colonies and competitors can deplete it. If this is true, then the buildup of bees at a flower patch after a scout discovers and begins advertising it should be faster than if each bee had to discover the patch without guidance from colonymates.

(A) Recruit

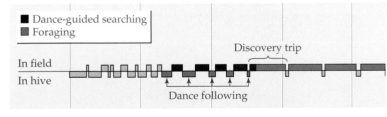

(B) Scout

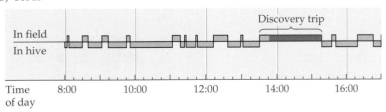

12 Foraging behavior of a representative recruit and scout honey bee. Both bees made many short flights in search of food. (A) The recruit followed a number of dancers and eventually located a new food source advertised by a nestmate. Thereafter, she foraged steadily at this source. (B) In contrast, the scout, after making a series of fruitless searches, discovered a food source and collected food there before returning to the hive. However, she then resumed searching for food elsewhere, without success. After Seeley and Visscher [1037].

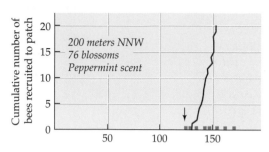
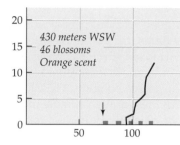
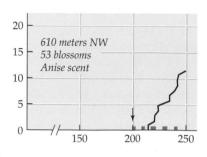

Time since flowers were set out (minutes)

13 **Rapid buildup of recruited foragers at experimental flower patches** after their discovery by scout bees. Researchers placed pots of borage flowers, treated with different scents, at three locations on an island. At each of the three locations, considerable time passed before a scout found the patch (arrow), but shortly thereafter, many additional bees arrived at the flowers. The green bars indicate the presence of the scout bee at the food patch. After Seeley and Visscher [1037].

To test this prediction, Seeley and Visscher moved a colony of bees to a barren island off the coast of Maine, along with a number of pots of plants in bloom (a portable flower patch), then moved the flowers from place to place on the island and measured the time it took scout bees to locate each patch. Scouts came by at the rate of about one bee every 1 to 3 hours. But once a scout returned to her hive and danced there, recruited bees quickly flooded the area. Even though any one recruit took an average of 2 hours to find a newly advertised food source, some of the many bees following the scout's dances arrived at the advertised site quickly, resulting in a fairly rapid buildup of recruits at a given site (Figure 13). As a result, the pollen and nectar at the food patch went to the colony of the scouts and recruits, rather than to other nectar and pollen foragers, a clearly adaptive result from the perspective of the dancers and their hivemates.

One apparently inefficient feature of waggle dances, however, is that they are far from precise, with considerable variation in the angles of the waggle runs. As a result, recruits are spread out to some degree around the site that stimulated the dancing bee to begin recruiting. Now, it could be that this imprecision in dance information is not an adaptation, but merely reflects constraints on the ability of dancers to keep their waggle runs on target. On the other hand, an adaptationist alternative is that dance imprecision is actually adaptive, since valuable flower patches typically cover a fairly large area. Therefore, by sending recruits out to sites located all around a given target within a larger food patch, a scout may actually promote more effective foraging by the colony workforce.

14 **The amount of error in the directional component of waggle dances** given by scout bees is greater for a diffuse patch of flowers than for a point resource, a new nest box. The divergence angle measures the difference between the actual direction to a feeder or hive site and the direction indicated by the dancing bee during the waggle-run component of her dance. After Weidenmuller and Seeley [1201].

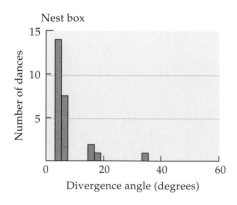
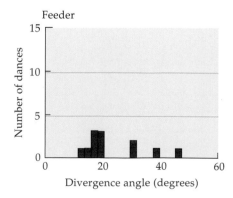

If the adaptationist explanation for dance imprecision is correct, then bees using waggle dances to advertise a resource located at a specific point, rather than a diffuse resource patch, should dance more precisely, enabling recruits to arrive closer to the point resource in question. When scouts are dancing to direct others to a potential nest site, a point resource, prior to the swarm's decision about which site to occupy (see p. 252), their waggle runs do indeed show less scatter from the "correct" angle on average than when the bees are announcing the location of a more diffuse food source (Figure 14) [1201].

The Information Center Hypothesis

Active recruitment of others to a food source is much less well developed in vertebrates than in the social insects. However, animals need not communicate intentionally with others about where a source of food is located in order to give away information about where to get a good meal. An observant member of a dense colony of nesting birds, for example, might monitor the foraging success of returning neighbors and simply follow the successful hunters back to where they had been hunting. In this way, an animal that had been unsuccessful in its own attempts to detect food could find good foraging areas.

The idea that nesting colonies serve as information centers [1183] generates several predictions. One is that foraging birds should leave their nesting colonies in groups and fly off together in the same direction. If clumping of departures does not occur, or if the birds head out in all directions, social transmission of information seems unlikely to have occurred. Some studies have documented a tendency of birds to leave the colony together, heading in the same direction. Thus, for example, when one barn swallow flies away from its nest to gather flying insects, other birds are likely to follow [503]. Moreover, the departing swallows tend to adopt similar bearings, suggesting that some individuals are following others.

If the information center hypothesis is correct, then when there are followers and leaders, the followers should have previously foraged without success, and the leaders should be birds that have done well recently. This prediction failed its test for one population of barn swallows [503]. Birds that came back empty-beaked from a foraging trip were no more likely than successful birds to follow another bird away from the colony on the next foraging trip (Figure 15). Nor were successful foragers more likely to be followed than unsuccessful ones.

The same indifference to successful foragers occurred during an experimental study of colonial black-headed gulls. Even when gulls landed by their nests

Barn swallow

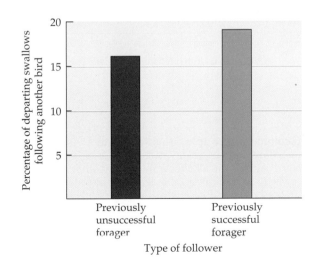

Type of follower

15 Do barn swallow colonies serve as information centers? If barn swallows gain information about the location of food from successful colonymates, then foragers that have failed to find food on the previous trip should tend to fly after a departing neighbor—but they do not. After Hebblethwaite and Shields [503].

Osprey

with fish in their beaks, neighboring birds nesting nearby showed no tendency to fly after the fish finder when it left the nest on another foraging trip, nor did they tend to fly in the same direction [35].

On the other hand, Erick Greene discovered that ospreys—large fish-eating hawks that form loose nesting colonies in some coastal areas—do watch their colonymates to learn where to find certain kinds of fish; namely, those that appear here and there in large schools [455]. Not only are ospreys more likely to begin foraging shortly after a fellow osprey returns to the colony with a schooling alewife or smelt, but they also tend to fly out in the direction taken by the successful forager. In addition, birds that see a successful hunter with prey are able to capture the same prey much more quickly than are ospreys that hunt without benefit of observing a successful forager (Figure 16).

Still more support for the information center hypothesis comes from John Marzluff and his colleagues, who watched young, nonterritorial ravens leaving communal roosts in groves of trees. On winter morning after morning, dozens of ravens left the roost in the space of a few minutes, with most individuals headed in the same direction. To find out whether naive followers could gain information about the location of food, the raven research team captured some ravens and held them for a number of days in an aviary. They then released some of the birds near roosts of ravens that had just found a dead moose (generously supplied by the researchers while the captured ravens were cooling their heels in the aviary). All 14 birds released at roosts in the evening followed their roostmates out the next morning to the bait. In contrast, only 4 of 15 captured ravens that were released away from roost sites found the food on their own [763].

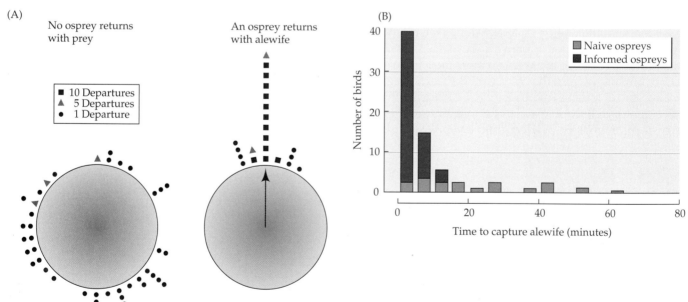

16 Do osprey nesting colonies serve as information centers? (A) When ospreys leave their nests at a time when no colonymate has returned with prey in the preceding 10 minutes, they fly off in all directions. But if another osprey has come back with an alewife (a schooling fish), the departing birds tend to fly off in the same direction (represented by an arrow) as the successful hunter. (B) Naive ospreys (those that have not seen a prey-carrying colonymate prior to departure) take much longer to find alewife schools than informed ospreys (those that have just observed a returning neighbor carrying an alewife in its talons). After Greene [455].

Locating Prey by Deceit

Many predators do not have the chance to locate prey by watching others. Some of these solitary hunters, including many spiders, use deceptive lures of various sorts to get their prey to come to them (Figure 17) [345, 570]. One spidery deception involves the construction of a yellow orb web by the golden orb-weaving spider. This web is so attractive to certain tropical stingless bees that even if they are lucky enough to break free, they will dive back into the same web they have just escaped from. Because the bees depend on yellow flowers for pollen and nectar, they are strongly attracted to yellow objects, something that the spider exploits for its own dining pleasure [252].

The webs of orb weavers are generally constructed with silk that absorbs ultraviolet light, unlike the UV-reflecting silks used by most other groups of spiders, which do not build the exposed circular webs of the orb weavers. Why this difference? Catherine Craig has suggested that since many flying insects can perceive ultraviolet light, orb webs built entirely of UV-reflecting silk would be more conspicuous to prey than those composed of UV-absorbing silk, especially since orb weavers place their traps out in open, sunny areas. In contrast, other spiders build smaller traps in dimly lit environments, where the risk that the silk will be detected by approaching prey is less.

But if this explanation is true, why do orb weavers sometimes decorate their webs with highly conspicuous zigzags of white ultraviolet-reflecting silk (Figure 18)? Perhaps these web additions attract insects that visit ultraviolet-reflecting flowers, which are commonly pollinated by UV-detecting bees. The fact that bees constitute the large majority of prey of some spiders with ornamented webs supports this hypothesis, as does the finding that in those webs in which only a single line of decorative silk has been laid down by the owner, insects are more likely to become trapped in the half of the web where the decorative silk resides than in the other half (Figure 19) [251]. Likewise, webs with more decorations capture more prey per hour than webs with fewer decorations [514]. Finally, isolated decorations cut from webs attract prey [1184].

17 Deception as an adaptive tactic for capturing prey. (Left) A juvenile female jumping spider (on the left) is plucking the web of an orb-weaving spider (on the right) in such a way as to mimic the signals of a prey item trapped in the web. When the orb-weaver comes closer, the deceptive predator will attack and kill the deceived prey. (Right) The bolas spider swings its lure, a ball impregnated with a scent identical to the sex pheromone of certain female moths. Male moths that approach the odor source are often captured by the sticky ball and then reeled in to be eaten. Photographs by (left) Robert Jackson, from Jackson and Wilcox [570] and (right) William G. Eberhard.

18   Web ornament of an orb-weaving spider. The large female spider has added three thick, conspicuous zigzagging lines of ultraviolet-reflecting silk, which radiate out from her central resting point. Photograph by Marie Herberstein.

On the other hand, it could be that spiders that have been catching many prey, thanks to a good site for their traps, are the spiders that tend to add decorations to their webs. If this is true, then the correlation between web decorations and higher rates of prey capture need not mean that the decorations *caused* the increase in prey capture. Instead, perhaps high rates of prey capture enable spiders to afford web decorations. This hypothesis can be evaluated by testing the prediction that spiders given an abundance of food should invest more in web decorations than spiders deprived of food. Indeed, in two species of orb-

19   Do web ornaments lure prey? Garden spider webs without ultraviolet-reflecting ornaments capture fewer prey per hour than those containing ornaments. Furthermore, in webs with only one ornament, more flying insects are trapped in the half of the web containing the ornament than in the half lacking these structures. After Craig and Bernard [251].

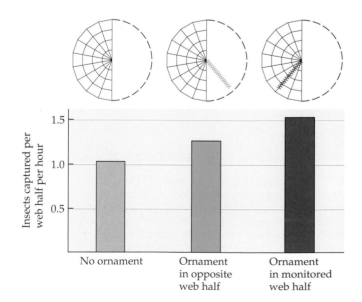

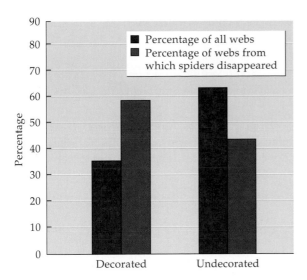

20 The price of web decorations? In an Australian orb-weaving spider, decorated webs make up about a third of the total, but these webs contributed more than half of all cases of spider disappearances. After Herberstein [514].

weaving spiders, females that were given more food made significantly larger decorations, and in one species, well-fed females were more likely to add decorations of any size to their webs than were food-deprived individuals [113].

Moreover, if one places an orb-weaving spider in a wooden frame and lets it build a web there, so that the web can be moved to sites chosen by the experimenter, then one can put two such webs side by side in a field. This done, one can compare the rate of prey capture in a web with decorations with that in another web whose decorations have been removed—something that can be easily accomplished by excising the two web lines that support the decoration (the other web also has two web lines cut, but not those that hold the decoration). Such an experiment controls for the effects of site productivity, since the webs being compared are essentially in the same location. Under these conditions, the decorated web catches about a third less prey than the undecorated one. Thus, a foraging cost, not a benefit, applies to web decorations [114].

So why add the special silk to a web, if prey are less likely to fly into webs with decorations? One possibility is that when birds see the bright decorations, they swerve to avoid colliding with the webs, the better to keep from getting covered in sticky silk. If this is true, then in the experiment above, webs with decorations should have been less often damaged by birds than webs without any "keep away" signals. And in fact, the actual results from the experiment matched the predicted ones, with decorated webs damaged by birds 45 percent less often than undecorated ones [112]. However, a study with another species revealed that spiders were more likely to disappear from decorated than from undecorated webs, suggesting that predators may actually target the occupants of more conspicuous webs (Figure 20) [514]. It may be too early to choose between the prey attraction and predator deterrence hypotheses for web-decorating behavior, especially since different species may decorate their webs for different purposes [515].

Selecting What to Eat

Once potential foods have been attracted or otherwise located, foraging animals may still be highly selective about what they actually consume. For example, in his study of howler monkeys in Costa Rica, Ken Glander discovered the following rules for leaf choices made by these leaf-eating primates (Figure 21) [437]:

1. The more common a tree species, the less likely the monkeys were to feed on its leaves. Instead, they spent considerable time searching out the scarcer species.

2. Even among the less common tree species, the howlers refused to eat leaves from most individuals available to them. For example, the monkeys took plant material from only 12 of 149 specimens of one acceptable tree species growing in the forest.

3. The monkeys preferred the scarcer, smaller new leaves to the more abundant, larger mature leaves.

4. The monkeys often ate only the petiole and dropped the larger leaf blade.

All four of these choices elevate the time and energy costs of foraging for howlers. Glander found, however, that there was a reason for each of them:

1. The most common tree species had leaves loaded with alkaloid poisons or indigestible tannins.

2. Among the scarcer, preferred tree species, howlers sought out just those individuals with especially low levels of alkaloids and tannins.

21   Food selection by howler monkeys. Although surrounded by leaves, these monkeys forage very carefully, avoiding toxic leaves and leaves low in nutritional value. Photograph by Ken Glander.

3. New young leaves contain more water and less nonnutritive fiber than mature leaves do. When the monkeys did eat mature leaves, they selected specimens whose leaves had a higher (12.4 percent) protein content than the mature leaves of the trees they rejected (which averaged only 9.4 percent protein).

4. "Wasteful" feeding occurred because the monkeys were eating the leaf part—the petiole—that is lowest in toxins while discarding the more poisonous leaf blade.

Howlers may be surrounded by leaves, but by feeding selectively, they ingest fewer toxins and gain more usable protein per unit of time.

How to Choose an Optimal Clam

But what about foods that differ only in size, not in toxicity? Howard Richardson and Nicholas Verbeek noticed that crows in the Pacific Northwest often leave littleneck clams uneaten after locating them [968]. The crows dig the clams from their burrows, but they usually discard the smaller ones on the beach and go on to open only the larger ones, which they drop on rocks. Their acceptance rate increases with prey size: the crows open and eat only about half of the 29-millimeter-long clams they find, while consuming all clams in the 32–33-millimeter range.

The two Canadian researchers determined that the most profitable clams were the largest, not because they broke open more easily, but because they contained many more calories than smaller clams. So why bother with 29- or 30-millimeter clams at all? Perhaps because there is a time cost to abandoning one clam in order to find a larger replacement. By considering the caloric benefits available from clams of different sizes and the costs of searching for, digging up, opening, and feeding on clams, Richardson and Verbeek were able to construct a mathematical model—a hypothesis—based on the assumption that the crows would select a diet that maximized their caloric intake. The model predicted that clams about 28.5 millimeters in length would be opened and eaten about half the time, given the search costs required to find clams of different sizes. The crows' behavior shows that they agree with the researchers' math (Figure 22) [968].

Richardson and Verbeek's work is based on **optimal foraging theory,** which is derived from the adaptationist assumption that foraging decisions should be optimal in the sense of maximizing the fitness of the decision maker. Thus, if diet selection by crows is indeed optimal, crows should choose those clams that contribute the most to their reproductive success. Note that Richardson and

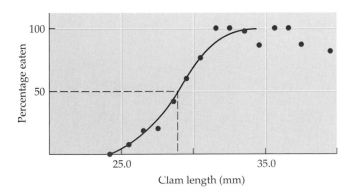

22 Optimality model of prey selection in relation to prey size. The red curve represents the predicted percentages of small to large clams that crows should eat after finding them, based on the assumption that the birds attempt to maximize their rate of energy gain per unit of time spent foraging for clams. The blue circles represent the actual observations, showing that the model's predictions were supported. After Richardson and Verbeek [968].

Verbeek did not measure fitness directly. Instead, they measured the net energy gained while crows were foraging for clams, on the assumption that by maximizing its rate of caloric intake, a crow would gain more energy for the production of offspring or more time to spend on other fitness-promoting activities.

This assumption has been tested experimentally with captive zebra finches, which were all given the same foods, but under different experimental feeding regimes that resulted in different daily energy gains for individuals. The more net energy the finches acquired per day, the better their survival and reproductive success [690]. Likewise, the biomass of prey provided by a water pipit (another songbird) to its young in the field correlates with the number of fledglings produced [407], and well-fed orb-weaving spiders produce more egg sacs than poorly fed individuals [1043].

How to Choose an Optimal Mussel

The oystercatcher is a shorebird whose foraging decisions can also be matched against predictions taken from optimality models. Two Belgian researchers, P. M. Meire and A. Ervynck, developed a calorie maximization hypothesis to apply to oystercatchers feeding on mussels [789]. They too calculated the profitability of different-sized prey, based on the calories contained in the mussels (a fitness benefit) and the time required to open them (a fitness cost). Even though mussels over 50 millimeters long require more time to hammer or stab open, they provide more calories per minute of opening time than smaller mussels. Therefore, the model predicts that oystercatchers should focus primarily on the largest mussels. But in real life, the birds do not prefer the really large ones (Figure 23). Why not?

- *Hypothesis 1:* The profitability of very large mussels is reduced because some cannot be opened, reducing the average return from handling these prey.

In their initial calculations of prey profitability, the researchers had considered only those prey that the oystercatchers actually opened (Figure 24, model A). As it turns out, oystercatchers select some large mussels that they find impossible to open, despite their best efforts. The handling time wasted on large impregnable mussels reduces the average payoff for dealing with this size class of mussels. When this factor is taken into account, a new optimality model results, yielding the new prediction that the oystercatchers should concentrate on mussels 50 millimeters in length, rather than the very largest size classes (Figure 24, model B). The oystercatchers, however, actually prefer mussels in the 30- to 45-millimeter range. Therefore, time wasted in handling large, invulnerable mussels fails to explain the oystercatchers' food selection behavior.

- *Hypothesis 2:* Many large mussels are not even worth attacking because they are covered with barnacles, which makes them impossible to open.

Oystercatcher

23 Available prey versus prey selected. Foraging oystercatchers choose mussels to attack that are larger than the average available mussel, but they do not concentrate on the very largest mussels. After Meire and Ervynck [789].

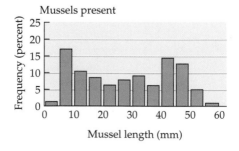
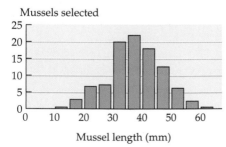

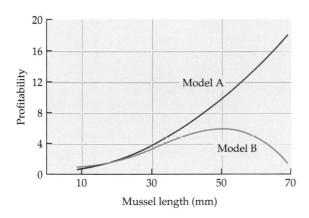

24 Two optimal foraging models yield different predictions because they calculate prey profitability differently. Model A calculates the profitability of a mussel based solely on the energy available in opened mussels of different sizes divided by the time required to open these prey. Model B calculates profitability with one added consideration; namely, that some very large mussels must be abandoned after being attacked because they are too difficult to open. After Meire and Ervynck [789].

This additional explanation for the apparent reluctance of oystercatchers to feast on large, calorie-rich mussels is supported by the observation that oystercatchers never touch barnacle-encrusted mussels. The larger the mussel, the more likely it is to have acquired a tough coat of barnacles, which makes the mussel so impenetrable that these prey should not even be considered in the calculations. According to a mathematical model that factors in (1) the effect of prey opening time, (2) time wasted in trying but failing to open a mussel, and (3) the actual range of realistically available prey, the birds should focus on 30- to 45-millimeter mussels—and they do.

Not only has an optimality approach helped explain why oystercatchers select mussels other than the very largest ones in their environment, but it has also been used to examine prey stealing as a foraging tactic. Oystercatchers can try to steal opened mussels from others, but this tactic is far more likely to be practiced by young, inexperienced birds than by older ones, even older juveniles. Individuals that devote their time and energy to attempting to grab food from other birds cannot spend this time finding suitable mussels on their own. Thus, a cost should accrue to the would-be thief. But young oystercatchers pay a lesser cost than older ones because the youngsters are so inept at opening mussels on their own that they cannot lose as many calories in the trade-off between food finding and food stealing. As they grow older, they can open mussels on their own more efficiently, reducing the relative payoff for thievery. As one would predict, older oystercatchers spend less time interfering with other foragers and more time working on their own (Figure 25) [443].

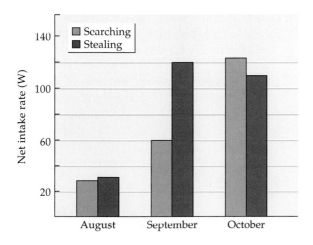

25 Change in the foraging tactics of oystercatchers with age. The net energetic gain estimated for young birds that search for prey on their own and for those that steal prey by attacking a fellow oystercatcher. The predicted returns are even in August, favor stealing in September, and shift to searching without thievery in October; the birds behave accordingly. After Goss-Custard [443].

Criticisms of Optimality Theory

By developing and testing optimality models, researchers have concluded that northwestern crows and European oystercatchers choose prey that provide the maximum caloric benefit in relation to time spent foraging. But some persons have criticized the use of optimality theory on the grounds that animals do not always behave as efficiently as possible. However, optimality models are not constructed to make statements about perfection in evolution, but to make it possible to test whether one has correctly identified the conditions that shaped the evolution of an animal's behavior. As we have just seen, the factors included in an optimality model have a large effect on the predictions that follow. If an oystercatcher is assumed to consider every mussel in a tidal flat a potential prey item, then it is predicted to make different foraging decisions than if the modeler assumes that oystercatchers flat-out ignore all barnacle-covered mussels. If the predictions of a model fail to match observations, researchers can make progress by rejecting that model and going on to develop and test alternative hypotheses that take other factors into consideration.

If ecological elements other than caloric intake affect oystercatcher prey selection, for example, then a caloric maximization model will fail its test, as it should. And for many foragers under some conditions, prey selection does have consequences above and beyond the acquisition of calories. If you suspected, for example, that predators had shaped the evolution of an animal's foraging behavior, then the kind of optimality model you might choose to construct and test would not focus solely on calories gained versus calories lost. If foraging exposes an animal to sudden death, then we might predict that when the risk of attack is high, the animal will sacrifice short-term caloric gains for long-term survival [541, 984]).

Consider the following case. Leaf-cutter ants collect plant material that they use to make fungus gardens in their underground nests. The ants feed on the fruiting bodies (mushrooms) of the fungus, which depends on the ants for its propagation. Indeed, the ants combat another undesired species of fungus that invades their gardens by adding solutions of bacteria to the gardens. The added microorganisms release antibiotic by-products that block the growth of the unwanted fungus, thereby increasing the production of the favored fungus in the ants' gardens [275]. So far, so good for the assumption that the ants are working to maximize their food intake.

However, although the ants could also increase the food available to them by collecting leaves night and day, in reality, workers of *Atta cephalotes* are primarily nocturnal. During the day, foraging occurs only at a low level, and the foragers at work then are small and relatively inefficient in cutting up and transporting sections of leaves back to the nest [866]. Why does the colony forage this way? Because *large* foragers that work *during the day* are hard-hit by a parasitic fly, which lays its eggs on them; when the fly larvae hatch, they burrow into the head of their victim, eventually killing it. In contrast, nocturnal foragers of all sizes are safe from the fly. As a result, the colony trades off short-term diurnal energy gain for a longer total foraging life for its valuable large workers (Figure 26). A simple optimality model that predicted that worker leaf-cutters of all sizes should forage around the clock would be wrong, but helpful, because testing that model would lead eventually to a better understanding of leaf-cutter foraging behavior.

The Evolution of Alternative Diets

Colonies of *Atta cephalotes* contain individuals with alternative behavioral tactics: the smaller, diurnally active leaf collectors and the larger, nocturnally active foragers. Earlier we discussed another example of this phenomenon, the tiger salamander, with its specialist cannibal types that feed upon the insect-eating forms of their own species in ponds where the two types coexist—uneasily

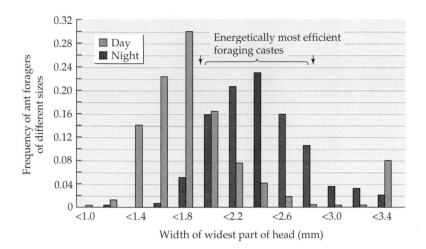

26 **Foraging efficiency is compromised when the risk of predation is high.** During the day, but not at night, foraging leaf-cutter ants tend to be smaller than the optimal size for efficient food collecting. But during the day, large foraging worker ants are at risk from parasitoid phorid flies, which ignore ants with head widths of 1.8 mm or less. After Orr [866] and Wilson [1249, 1250].

(p. 93). A somewhat similar case involves the New Mexico spadefoot toad, whose tadpoles also come in two forms: a rapidly developing type that specializes in eating small invertebrates called fairy shrimp, and a slowly developing type that feeds on odds and ends, but especially on bits of dead algae and other detritus on the bottom of the temporary desert ponds where the tadpoles live [909]. A high-protein fairy shrimp diet evidently enables the carnivore morphs to reach the toadlet stage faster than their detritus-eating companions. So why don't all tadpoles eat shrimp?

Any time alternative behavioral types coexist, we have a Darwinian puzzle to solve. One would think that if one type reproduced more than the other, then the type that left more surviving descendants would replace alternative types with lower fitness. If carnivore tadpoles grow faster than other types, then they are more likely to reach adulthood before their ephemeral ponds dry out, and so should leave more of their genes than the deceased omnivorous tadpoles they left behind in the encrusted mud. So, again, why don't all spadefoot tadpoles eat shrimp?

The solution to this puzzle begins with the discovery that the difference between the two types is not caused by a genetic difference between them, but rather by dietary differences. Therefore, the differences between the two tadpole morphs are not subject to natural selection (see Chapter 1). A young tadpole that finds many fairy shrimp can become a carnivore that specializes on this prey, but if the shrimp become scarce, the tadpole can revert to the omnivorous detritus-eating type. Thus, spadefoot toads apparently have evolved the capacity to assess the available food supply and adjust their feeding behavior accordingly. If this hypothesis is correct, then an experimental shift in the ratio of carnivore to omnivore tadpoles in a pond should cause some of the now unnaturally common form to switch to the other foraging mode, bringing the ratio of the two forms back to its original level. We have, then, a game theoretical hypothesis (see p. 205) in which the adaptive feeding mode depends on what other members of the population are eating.

David Pfennig tested this hypothesis by dividing five natural desert ponds into compartments of equal size and then changing the frequency of carnivores in some of the compartments, while leaving others untouched as controls. After the initial manipulation, he sampled the frequency of carnivores in the experimental and control compartments over a number of days. In compartments in which carnivore numbers had been boosted experimentally, carnivores gradually declined relative to omnivores until the original proportion of fairy shrimp consumers had been reached. In sections in which carnivore numbers were

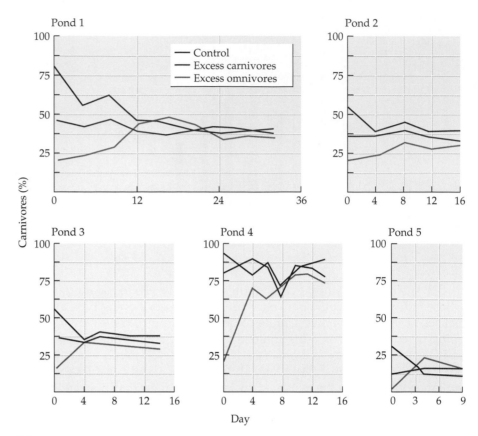

27 Frequency-dependent adjustment of food selection in spadefoot toad tadpoles. Five ponds were divided into compartments, and the ratio of carnivorous (shrimp-eating) to omnivorous tadpoles was experimentally manipulated in some compartments. Each graph represents one pond. The proportions of the two types quickly moved back toward the equilibrium level maintained in the control (nonmanipulated) compartments. After Pfennig [909].

depleted initially, carnivores became more common until reaching the same frequency as in the unmanipulated control compartments (Figure 27). These results strongly suggest that when competition for fairy shrimp is high relative to their availability in a pond, the benefit from being a shrimp-eating specialist declines, favoring individuals with the capacity to switch their selection of food accordingly [909]. What has been selected for is not a shrimp-eating phenotype versus a detritus-eating alternative, but the ability to be flexible in a very carefully defined manner.

Consuming What You Select to Eat

It is one thing to attempt to feed on shrimp or any other living prey, but it is something else again to consume that food, especially if the prey does not wish to be eaten. Certainly the nonaquatic prey of lions, wolves, African wild dogs, and spotted hyenas give every indication of wanting to remain alive, yet through the cooperative efforts of lion prides, wolf packs, and hyena clans, prey as large as Cape buffalo and moose can be taken down and killed. In fact, social carnivores can capture prey that weigh from six to twelve times as much as any one adult hunter (Figure 28) [255, 786, 1018]. In contrast, solitary species of cats, dogs, and hyenas usually forage for much smaller victims. These patterns of

28   The benefits of group hunting. A clan of spotted hyenas has captured, killed, and eaten a wildebeest, a large antelope. Photograph by Jonathan Scott.

convergence among nonrelatives belonging to three families (Felidae, Canidae, and Hyaenidae) and divergence between closely related species within these families provide comparative evidence that group hunting is an adaptation for the capture of prey too large, or too dangerous, to be subdued by predators working on their own.

This hypothesis has been subjected to optimality tests by researchers working with lions and African wild dogs. Let's assume that these carnivores hunt in groups of sizes that return the most fitness for each individual. Let's also assume that net fitness gain is positively correlated with the amount of food consumed by an individual. With these assumptions in place, we can measure the amount of food eaten by individuals in groups of different sizes to test the prediction that members of groups of the preferred size will maximize their food intake.

When Thomas Caraco and Larry Wolf analyzed George Schaller's data on lions, they found that food intake per lion per day was higher for lions hunting in pairs than for lions hunting alone. However, groups of three or more lions actually did *less* well on a per capita basis than groups of two or singletons [193]. In a subsequent study, Craig Packer, David Scheel, and Anne Pusey found that lone hunters regularly ate as much as or more meat per day than group hunters (Figure 29) [879]. However, small prides of two to four females do not disband to hunt alone during times of food scarcity, despite a very low foraging success at these times. If group hunting is not advantageous in terms of increasing net food intake per individual, then why stick together?

Scott and Martha Creel [255] point out that most persons who have applied optimality theory to lion foraging have focused only on the *benefit* of cooperative hunting as measured in terms of meat eaten per day. The models change if one considers some of the *costs* of hunting incurred by individuals in groups of different sizes. If the distance traveled to make a kill changes with group size, then the energy expended to reap the same benefit also changes, altering the cost–benefit equation for hunters in different group sizes. To determine whether cooperative hunters are really trying to maximize the difference between energy gained and energy expended in the hunt, the Creels collected data on the benefits of social foraging in terms of kilograms of meat gained by wild dogs hunt-

29 **Group size and feeding success in lions.** In times of prey abundance and scarcity, solitary lions generally ate more kilograms of meat per day than females hunting in groups. After Packer [879].

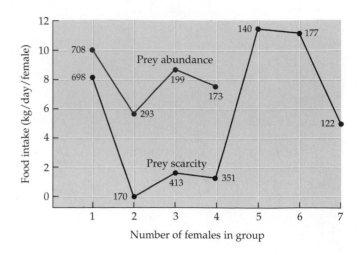

ing in packs of 3 to 20. But they also measured the costs of hunting in terms of such things as the number of kilometers that the dogs ran while rapidly pursuing their prey. With these data, they calculated two different measures of foraging success: (1) the kilograms of meat gained per wild dog each day and (2) the kilograms of meat gained per dog minus the cost of the energetically expensive chases required to secure food (Figure 30). Plotting these values against the size of the hunting pack, they showed that (1) dogs in small and large packs did better than those in packs of intermediate size when only foraging benefits were considered but that (2) dogs in larger groups tended to do better than those in smaller groups when *net* benefits (energy gained minus energy expended) were taken into account (Figure 30).

In reality the mean size of wild dog hunting packs is about 10, which tells us two things. First, the more realistic model for wild dog group size is the one that factors in the costs of chases as well as the caloric benefits of successful kills. Second, the fact that most wild dogs live in groups of 10 instead of fifteen or twenty indicates that optimal foraging considerations, at least as calculated in terms of net energetic benefits, cannot be the whole story in setting group size, a conclusion that applies to lions as well.

How to Open a Whelk

Once prey have been detected, attacked, and captured, the job of a forager is not necessarily over. Even after lions bring down a wildebeest or zebra, these

30 **Group size and feeding success in African wild dogs.** When one considers the costs as well as the benefits of foraging in groups of different sizes, the net caloric benefit tends to be greater for individuals that hunt in larger packs. The circles are mean values, with the area of the circle proportional to the number of observations for packs of a given size. After Creel [254].

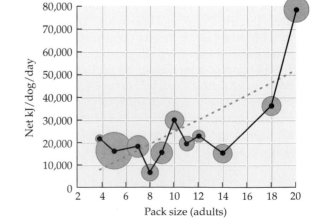

predators often have to contend with would-be thieves from outside the pride before they can enjoy the fruits of their labor. Likewise, a crow that has picked up a shellfish from a tidal flat still has to break into the prey if it is to feast upon the flesh within. When a beachcombing crow spots a clam, snail, or whelk, it often picks it up, flies into the air, and then drops its victim. If the mollusk's shell shatters on the rocks, the bird plucks out and eats the exposed body. The adaptive significance of the bird's behavior seems straightforward. It cannot use its beak to crack open the extremely hard shells of certain mollusks. Therefore, it breaks its prey by dropping them on rocks. This seems adaptive. Case closed.

But we can be much more ambitious in our analysis of this component of the crow's foraging decisions. When a hungry crow decides to consume a whelk, it has many choices to make: which whelk to select, how high to fly before dropping the prey, and how many times to keep trying if the whelk does not break on the first try.

Reto Zach observed that (1) crows picked up only large whelks about 3.5 – 4.4 centimeters long, (2) they flew up about 5 meters to drop the chosen whelks, and (3) they kept trying until the whelk broke, even if many flights were required. Zach sought to explain the crows' behavior by applying the now familiar hypothesis that the birds' decisions were optimal in terms of maximizing whelk flesh available for consumption per unit of time spent foraging [1277]. This hypothesis yields the following predictions: (1) large whelks should be more likely than small ones to shatter after a drop of 5 meters; (2) drops of less than 5 meters should yield a reduced breakage rate, whereas drops of much more than 5 meters should not greatly improve the chances of opening a mollusk; and (3) the probability that a whelk will break should be independent of the number of times it has already been dropped.

Zach tested each of these predictions in the following manner. He erected a 15-meter pole on a rocky beach and outfitted it with a platform whose height could be adjusted and from which whelks of various sizes could be pushed off to drop to the ground. He collected samples of small, medium, and large whelks and dropped them from different heights (Figure 31). He found, first, that large whelks required significantly fewer 5-meter drops before they broke than either medium-sized or small whelks. Second, the probability that a large whelk would break improved sharply as the height of the drop increased—up to about 5 meters. On still higher drops, the probability that the mollusk would shatter improved very little. Third, the chance that a large whelk would break was not affected by the number of previous drops and was instead about one in four on any drop. Therefore, a crow that abandoned an unbroken whelk after

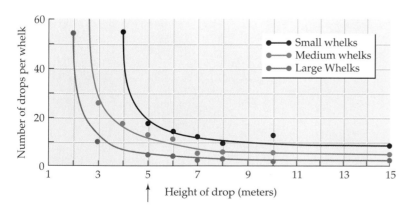

31 Optimal consumption decisions by northwestern crows when feeding on whelks. The curves show the number of drops at different heights needed to break whelks of different sizes. Northwestern crows drop only large whelks, and drop them from a height of about 5 meters, thereby minimizing the energy they expend in opening whelks. After Zach [1277].

a series of unsuccessful drops would not have a better chance of breaking a replacement whelk of the same size on its next attempt. Moreover, finding a new prey would take time and energy.

Zach went one step further by calculating the average number of calories required to open a large whelk (0.5 kilocalories), a figure he subtracted from the food energy present in a large whelk (2.0 kilocalories) for a net gain of 1.5 kilocalories. In contrast, medium-sized whelks, which require many more drops, would yield a net loss of 0.3 kilocalories; trying to open small whelks would have been even less profitable. Thus, the crows' rejection of all but large whelks was clearly adaptive, and their selection of dropping height and their persistence in the face of failure enabled them to reap the greatest possible energy return for their foraging activities [1277].

Why Do Humans Consume Alcohol, Spices, and Dirt?

Humans sometimes behave in irrational ways when it comes to diet, or so it seems, and these cases, as always, are a stimulating challenge to the evolutionary biologist. For example, a great many people in modern societies are addicted to alcohol, often at great fitness cost to themselves. Here is clearly maladaptive behavior. But is the trait an inexplicable artifact of human culture? Not according to Robert Dudley, who presents a historical hypothesis on the origins of alcoholism [339]. Dudley points out that since our closest relatives, the chimpanzees, derive most of their calories and nutrients from ripe fruits, it is likely that the ancestral species that gave rise eventually to modern chimpanzees and humans was also a frugivore. Fruit-eating chimpanzees, and primates in general, prefer ripe fruits because these have the highest concentrations of sugars (Figure 32). Ripe fruits also happen to contain a certain amount of ethanol, which is volatile and so provides a potential olfactory cue to the location of high-payoff foods, all the more so because ethanol itself is a calorie-

32 **A primate eating ripe fruit.** Photograph by Mary Beth Angelo.

rich food. Thus, it seems likely that an ancestral fondness for ethanol would have been adaptive. However, the proximate mechanisms that facilitate alcohol consumption evolved in a world without Budweiser, chardonnay, and gin. Today these same mechanisms can be employed maladaptively in ways that would have been impossible even a few thousand years ago.

Or take the addition of spices to our food, a subject explored by Jennifer Billing and Paul Sherman [105]. These researchers were attracted to the topic by the obvious disadvantages of spice use, which include the high price of many of the spices that humans add to their food. The Countess of Leicester, who lived in England in the 13th century, was willing to pay as much for a pound of cloves as she would for a cow [941]. Most of the major voyages of the early European explorers, Christopher Columbus included, were motivated by the desire for spices, as all North American schoolchildren learn. And yet the caloric and nutritive value of most spices is small, given that they are often used sparingly in any one dish. It would therefore be easy to conclude that spice use is simply a random product of cultural invention, especially since culinary traditions vary so greatly from culture to culture.

Billing and Sherman, however, proposed and tested the adaptationist hypothesis that spices served (and may continue to serve) a fitness-enhancing function by virtue of their antimicrobial properties. This hypothesis, needless to say, requires that spices kill the dangerous bacteria that sometimes contaminate our food, especially meats held under imperfect refrigeration. The microbiological literature provides evidence pertinent to this prediction (Figure 33). Moreover, the antimicrobial hypothesis also predicts that the extent to which spices are used should be a function, not of their local cultivation, but of the risk of dangerous microbial contamination, which is related to local climate. As expected, traditional recipes for meat dishes in hot, tropical countries call for more bac-

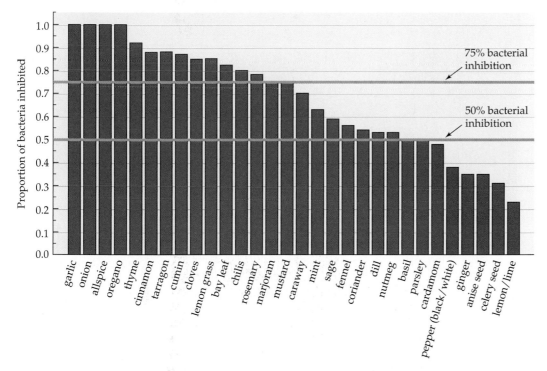

33   The anti-microbial properties of the major spices, most of which inhibit the growth of one-half or more of the kinds of bacteria against which they have been tested. After Billing and Sherman [105].

tericidal spices than do recipes for meat-containing dishes in, say, Norway or Sweden. Thus, we can conclude that spices have been used adaptively by humans even during the relatively short period of our evolution when novel spices have been made widely available.

Finally, consider another oddity: the occurrence of dirt eating, a trait that is widespread among primates [656], including humans. No one has ever claimed that the clayey soil consumed by chimps or humans has any caloric value whatsoever. Among the hypotheses advanced to explain this habit in our species is the hypothesis that clay eating is a pathological, aberrant behavior of no functional significance. In contrast, another hypothesis proposes that clays are consumed to detoxify certain kinds of food, thereby improving their nutritional value [585].

The pathology hypothesis predicts that only a relatively few, possibly somewhat deranged, individuals will eat clay. This prediction does not withstand scrutiny, since clay eating is, or was, common practice in a diverse set of cultures, including the Aymara of Andean Bolivia, the Hopi of the American Southwest, and the natives of Sardinia, an island in the Mediterranean Sea. What the people in these cultures have in common is a dietary reliance on bitter tannin-rich acorns or varieties of bitter alkaloid-laden potatoes. When potatoes are dipped in a clay slurry and then baked, or when acorns are baked into a clay-containing bread, the tannins and alkaloids in these foods are either bound to the clays or altered chemically by their components, rendering the foods more palatable and less toxic [585].

A comparative test of the detoxification hypothesis involves the prediction that animals distantly related to humans will also seek out certain clays when their diets feature foods high in tannins or alkaloids. Rhesus macaques and some other primates not especially closely related to humans do eat clay when feeding on tannin-rich vegetation [656]. Moreover, red-and-green macaws also eat clay; these huge, spectacular parrots feed on certain seeds, unripe fruits, and leaves, which are apparently high in toxins (Figure 34) [830]. The clays

34 **Clay eating has evolved in several species of macaws that feed on foods rich in tannins or toxins,** including these birds, which regularly gather at this Amazonian river bank to eat clay. Photograph by Frans Lanting.

selected by the parrots line the birds' digestive tract for hours, and bind with at least one plant alkaloid, thereby protecting the cells of the gastrointestinal lining from the toxic effects of this chemical [431]. This evidence suggests that the detoxification explanation is correct, showing again that progress can be made by testing adaptationist hypotheses about feeding behavior.

Summary

1. Animal species possess a remarkable array of tactics to overcome the defenses of prey. Predators may locate prey by means of special perceptual systems, by relying on information provided by successful foraging companions, or by deceiving prey into approaching them.

2. Foraging animals that have found potential foods may nevertheless be highly selective, taking (for example) only those items that provide large caloric payoffs. Overcoming the defenses of prey in order to capture them may be advanced by social cooperation in some cases. Even foods that have been located, attacked, and captured may still pose problems for the consumer, which may employ special devices to secure the nutrients in well-protected or toxic food items.

3. Optimality theory can be used to make predictions about what a fitness-maximizing solution to a foraging problem should be. The prediction that animals should maximize energy gained per unit of foraging time in order to maximize their reproductive success has been used extensively in the study of feeding behavior. Although some foraging animals do behave in ways that match this prediction, many other species compromise energy maximization to reduce the risk of predatory attack or to deal with some other problems that constrain the acquisition of calories. Tests of optimality hypotheses are designed to help researchers identify the factors involved in the evolution of animal behavior, not to support claims that all behavior is perfectly adaptive.

Discussion Questions

1. When sharks attack humans, they often bite the victim once and then swim away. Researcher X argues that sharks rarely eat these unfortunate persons because "sharks prefer the odor and taste of their customary prey, seals and sea lions." Researcher Y disagrees: "Sharks release humans because they are not fat enough, and therefore provide too few calories to be part of the sharks' optimal diet." Clear up the confusion here.

2. William Cooper found one species within the lizard family Scincidae that is an exception to the rule that skinks forage using olfactory cues, rather than visual ones [244]. This ambush predator shows no interest in attacking cotton swabs dipped in an extract of insect cuticle, unlike a close relative that regularly bites cotton swabs of this sort while ignoring those dipped in water. What kind of evolutionary process is illustrated by this example? What can you say about (1) the origin of ambush behavior in the ambush-foraging skink and (2) the adaptive value of olfactory chemical discrimination exhibited by the searching skink species?

3. In a Canadian lake containing two species of stickleback fish, one species typically forages on plankton in the open water, while the other specializes on insects taken from the lake bottom. One hypothesis argues that these differences evolved as a result of competition for limited resources, which favored individuals that diverged in their diets from members of other similar species. Provide testable predictions from this hypothesis with respect to (1) the diets of other pairs of closely related species that co-occur in other Canadian lakes, (2) the diet of a stickleback species when it is the sole occupant of a given lake, and (3) the

survival and growth rates of individual sticklebacks that come from single-species lakes when they are experimentally placed in enclosures with members of a population of plankton-feeding specialists. (Note: Individuals of all stickleback species vary to some extent in their dietary specialization.) Check your predictions against the data in [1022, 1023].

4. In some places, American crows open walnuts by dropping them on hard surfaces. Unlike northwestern crows working with whelks, American crows reduce the height from which they drop walnuts from about 3 meters on the first drop to about 1.5 meters on the fifth drop. If this tendency is adaptive, what prediction follows about a difference between whelks and walnuts in the likelihood of breaking on successive drops? In addition, American crows tend to drop walnuts from lower heights when other crows are present. If this trait really is an adaptation, what prediction must be true? Check your answers against data in [268].

5. The possibility that cannibalism was a regular feature of life (and death) in some human societies is highly controversial. Develop a cost–benefit analysis of the behavior and make predictions about the circumstances under which an adaptationist would expect the behavior to occur. In addition, read Jared Diamond's article on the subject [323] and reconstruct his argument: identify the question he seeks to answer, produce the alternative hypotheses, the predictions, the tests, and the conclusion.

6. Various persons have proposed that when a horse switches from slower trotting to faster galloping, it changes gaits to minimize the energetic expense of locomotion, under the assumption that animals able to minimize the energetic costs of getting from A to B will enjoy greater reproductive success than individuals that squander their energy supplies in inefficient locomotion. The energy minimization hypothesis was tested by Claire Farley and Richard Taylor with the help of three cooperative horses willing to run on a treadmill while outfitted to provide data on their oxygen consumption, a factor directly related to energy use [377]. What scientific conclusion is justified on the basis of the figure at left? Does this study support those who claim that optimality theory is not useful because the assumptions underlying particular hypotheses are often oversimplified and incorrect?

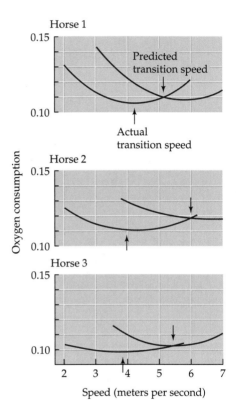

Suggested Reading

For a general review of predator decisions, see John Krebs and Alejandro Kacelnik's book chapter [655]. Their review also covers some mathematical models based on optimality theory, an approach that has often been applied to foraging behavior. The mathematics of modeling are presented in detail by Dennis Lendrem [691] and by Marc Mangel and Colin Clark [743]. Bernd Heinrich's *Bumblebee Economics* [504] is a good companion to this chapter, as it deals clearly and simply with optimality theory as applied to bumblebees. For more on the amazing lives of honey bees, read Tom Seeley's *The Wisdom of the Hive* [1035]. Reto Zach's article [1277] on optimal foraging in the northwestern crow is a model of clarity. For a critique of optimality models, see a paper by C. J. Pierce and J. C. Ollason [915], which can be contrasted with Krebs and Kacelnik's chapter [655].

9 *Choosing Where to Live*

When I was a teenager, my father and I went bird-watching all day long one weekend in May near our home in southeastern Pennsylvania. We chalked up about 100 species, including yellow warblers singing high in the sycamore trees by the White Clay Creek, blue-winged warblers flitting among the little trees in shrubby, overgrown farm fields, and common yellowthroats (another species of warbler) in the marshy spots. The male warblers calling loudly from their perches not only had made the effort to find a very specific habitat, but were also prepared to defend their turf against others of their species, as they announced over and over again (see Chapter 2). In addition to the costly investments it made in selecting a particular habitat and defending a territory, each warbler that my father and I saw had traveled hundreds or even thousands of miles from wintering grounds as far away as South America in order to set up a breeding territory in our neighborhood. In a few months, if they survived, they would head south again on a great migration that their offspring would undertake as well.

Each of the decisions that these little songbirds make about where to live comes with obvious and substantial

◀ *Monarch butterflies* *cluster in winter in only a handful of locations in Mexico and California, an example of highly refined habitat selection. Photograph by Frans Lanting*

costs, which is why these decisions are interesting to evolutionary biologists. Why should a yellow warbler refuse to nest in a cattail marsh? Why should a common yellowthroat spend hours each day singing to keep rival males informed of its willingness to fight for space in the same marsh that yellow warblers avoid? And why should a blue-winged warbler born in Landenberg, Pennsylvania, fly all the way to Honduras, only to turn around in a few months and fly all the way back again? This chapter will show how these puzzles can be investigated.

Habitat Selection

The rule that certain species live in particular places applies to all groups of animals, not just warblers, perhaps because the opportunities for successful reproduction are so much better in habitat A than habitat B for members of species X. The importance of access to appropriate habitat has been dramatically illustrated by the link between habitat destruction and the declining populations of certain animals. Loss of isolated people- and dog-free beaches has, for example, endangered the western snowy plover and the least tern, both of which nest on open beaches and small sandy islands. If the populations of these birds are falling because of habitat loss, then creation of artificial beaches in areas inaccessible to people and their dogs ought to attract breeding plovers and terns, helping local populations rebound. This experiment has been done in Batiquitos Lagoon in southern California and elsewhere. When material dredged off the sea bottom was used to produce artificial beaches and sandbars, both species of birds occupied the sites (Figure 1) and reproduced successfully [742, 933].

1 **Habitat selection and conservation.** Knowledge of the habitat preferences of nesting least terns has enabled conservation biologists to create suitable environments for this endangered species. (Left) A least tern on its nest in the open, sandy beach habitat the bird requires if it is to breed successfully. (Right) This large sandy island, built with dredged materials, has attracted many pairs of nesting seabirds, including the least tern. Photographs by (left) D. Robert Franz and (right) Troy Mallach.

Habitat restoration may help rebuild snowy plover and least tern populations. Habitat *management* is another conservation tool based on the premise that particular species are matched to particular environments. The Florida scrub jay, another endangered species, lives only in sandy, scrubby areas of Florida. Once this habitat burned irregularly but naturally as a result of lightning strikes, creating a mosaic of scrub oak woodland with open sandy patches surrounded by other plant associations. The scrub jays prefer to nest in substantial areas of open oak habitat, and their nesting success is highest there [146]. When the natural fires are quickly put out, however, the scrub oak thickets eventually become tall and dense, egg-eating blue jays move in, and scrub jays lose out [1265]. Any management scheme devised to increase the Florida scrub jay population will surely have to employ fire as a tool to shape the habitat to match the preferences and needs of the scrub jays.

Given the importance of being able to breed in specific habitat types, we would expect animals to have evolved strong preferences for some places over others, even if they are capable of reproducing in a variety of environments. The European great tit, for example, can nest either in mixed woodland or in hedgerows. But the birds prefer mixed woodland to hedgerow, as demonstrated by the shifts made by hedgerow birds into woodland sites upon the removal of breeding pairs from the favored habitat [652].

Likewise, the black-throated blue warbler of North America appears to prefer to nest in hardwood forest with a dense shrub understory, even though the birds will breed in forests with a lower shrub density. Evidence for the preference comes from the fact that older, experienced individuals occupy high-shrub-density sites, relegating yearlings to woodland with a lower shrub density. Birds able to acquire territories in the high-shrub-density forests fledge significantly more young than those unable to breed in this habitat [540], just as great tits able to nest in preferred woodlands have higher reproductive success than those nesting in hedgerows [652].

The prediction that individuals able to exercise their habitat preferences will leave more descendants than those unable to acquire prime real estate receives support from the widespread finding that members of an animal species often occupy both source habitats (where the population grows) and sink habitats (where the population declines). Poor-quality sink habitats are utilized by individuals that are unable to insert themselves into superior source habitats, often because they are excluded by older, more accomplished competitors [324] and so must make the best of a bad situation elsewhere.

The predicted relationship between habitat preference and reproductive success does not, however, always hold. In the Czech Republic, the blackcap warbler has a choice between stream-edge deciduous forests and mixed coniferous woodlots away from water. The bird prefers the stream-edge habitat, which attracts the initial colonists in spring. Yet despite the preference for stream-edge habitat, the reproductive success of breeding pairs in the two environments is essentially the same [1202]. Why? Perhaps because four times as many pairs pack themselves into the preferred habitat as into the mixed conifer alternative. This pattern suggests that when the density of competitors for space and resources reaches a certain level in prime habitats, individuals can achieve the same reproductive success by settling in second-ranked habitat types where they face less competition. In other words, the birds apparently make habitat selection decisions based not just on the nature of the vegetation and other markers of insect productivity, but also on the intensity of the competition from others of their species.

Blackcap behavior is therefore consistent with the conclusions reached by Steve Fretwell and his colleagues, who used game theory (see p. 205) to predict what animals should do when faced with a choice between alternative habitats of different quality and different levels of competition. Fretwell's algebra indi-

2 Finding a new home. A swarm of honey bees waits while scouts go off to search for a new hive site. Photograph by Kirk Visscher.

cated that as the density of resource consumers in the superior habitat increased, there would come a point at which an individual could gain higher fitness by settling for a lower-ranked habitat, which had fewer settlers of the same species and thus less competition for key resources [406].

We can apply this approach in a qualitative way to the habitat choices made by honey bee colonies in selecting a new nest site, something that happens when a colony has grown sufficiently large to split in two. The old queen and half her worker force fly off in a swarm, leaving the old hive and the remaining workers to a daughter queen. The departing swarm settles temporarily in a tree, where the workers hang from a limb in a mass around their queen (Figure 2). Over the next few days, scout workers search for chambers in the ground, in cliffs, or in hollow trees. Of the many such sites within range of the waiting swarm, only those with a volume of 30 to 60 liters cause returning scouts to perform a dance back at the swarm [1033], which communicates information about the distance, direction, and quality of the potential new home (see p. 220). Other workers around a dancing scout may be sufficiently stimulated to fly out to the spot themselves. If it is attractive to them, they too will dance and send still more workers to the area. Over a couple of days, some spots will continue to attract scouts, and some will come back to recruit for that site. Bees that have been dancing for less favored sites eventually drop out, so that at some point almost all the active recruiters will be advertising one location, at which time the swarm leaves its temporary perch and flies to the nest site that has received the most "votes" (Figure 3) [1036].

What is especially interesting about this process from the perspective of game theory is that when some bees were experimentally offered two identical high-quality nest sites at different distances from the home hive—say, 50 and 200 meters away—the bees chose the more distant of the two [706]. One might think that they would prefer the closer spot, since covering even a few hundred meters

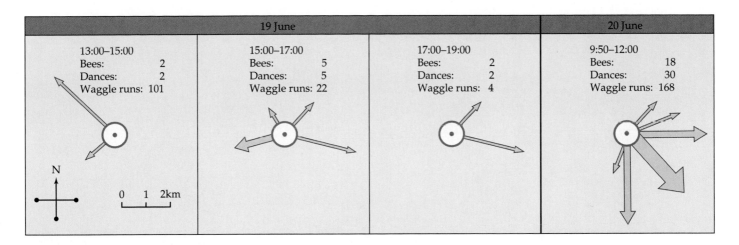

3   Changes in the numbers of honey bee scouts advertising the location of different future hive sites for their swarm (whose location is represented with a circle). The width of the arrows radiating from the circles is proportional to the number of different bees dancing for that potential hive site during the period shown at the top of each panel. After Seeley and Buhrman [1036].

can exhaust their queen, a prodigious egg layer but a poor flier. However, the value of a site is affected by its proximity to other colonies, and by moving a substantial distance from the old hive, a dispersing swarm presumably reduces competition for food, increasing the calories available for both the old queen's new colony and her daughter's hive.

We can test this hypothesis by predicting that a swarm's readiness to fly the extra distance should be correlated with the intensity of the potential competition between mother and daughter hives. As it turns out, northern European bee colonies are much larger, and thus more likely to compete for food, than those in southern Europe, a difference caused by the importance of having a large number of bees as a living insulation blanket for the queen and her core workers during cold winters [579, 580]. As expected, large honey bee colonies in cold Germany move farther when finding new hive sites than do the much smaller swarms of bees that live in warmer Italy [444].

Habitat Preferences in a Territorial Species

Honey bee colonies do not attempt to defend a foraging area as their exclusive feeding preserve, and therefore dispersing swarms can settle wherever they

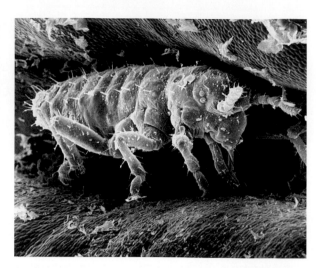

4   Poplar aphid galls. (Left) A poplar aphid gall at the base of a leaf. (Right) A scanning electron micrograph of a female inside her gall. Courtesy of Tom Whitham.

choose without direct interference from other colonies. For other animals, such as warblers, a truly free choice among habitats is not possible because some members of the species actively exclude others from superior sites. Tom Whitham examined the effects of this kind of aggressive competition on habitat selection in a tiny insect, the poplar aphid [1221–1223].

Each spring in Utah, vast numbers of aphid eggs hatch in crevices in the bark of cottonwood poplar trees. The emerging aphids walk to leaf buds on cottonwood branches. Each female—and there may be tens of thousands per tree—actively selects a leaf, settles by its midrib, almost always near the base, and in some way induces the formation of a hollow ball of tissue, a gall, in which she will live with the offspring she bears parthenogenetically (Figure 4). When her daughters are mature, the gall splits, and the aphids disperse to new plants.

Whitham found that newly emerged females quickly occupied all the very large leaves on poplars, but that since there were about 20 aphids for every large leaf on the trees he studied, these females regularly encountered competitors, with whom they fought for as long as 2 days (Figure 5), sometimes dying in the attempt to secure a territory [1224]. Given the costly nature of these fights, we would expect that major benefits come from possessing large leaves—and they do (Figure 6).

Defeated females and small ones incapable of effective fighting are forced to accept inferior habitats. Our expectation, based on optimality theory, for the defeated individuals is that they will make the best of a bad situation, accepting the best of the lower-value habitats available to them. The options for these individuals include finding a smaller, unoccupied leaf or settling with an established territorial foundress on a large leaf. If a leaf already has a resident female, the latecomer will have to form her gall farther out on the midrib, where it

5   Territorial dispute between two poplar aphids. Females may spend hours kicking each other to determine who gets to occupy a preferred leaf or the superior location on a leaf. Courtesy of Tom Whitham.

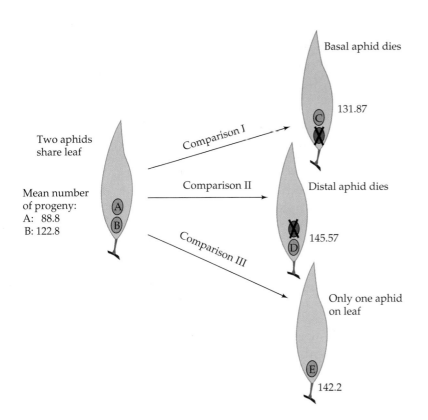

6 Territories and reproductive success. The average number of progeny produced by aphids that succeed in monopolizing a poplar leaf versus those that are forced to share a leaf of the same size with a rival. After Whitham [1224].

Two aphids share leaf

Mean number of progeny:
A: 88.8
B: 122.8

Comparison I

Comparison II

Comparison III

Basal aphid dies
131.87

Distal aphid dies
145.57

Only one aphid on leaf
142.2

will provide fewer nutrients than the gall at the prime location near the leaf petiole. If the other aphid is already enclosed in a gall, the newcomer will not have to fight, but she will have to settle for a reduction in fitness. However, the second colonist on a large leaf can do just as well as a single aphid on a medium-sized leaf, and substantially better than a lone colonist on an even smaller leaf (Table 1). As predicted, when aphid females do double up, they choose large leaves, despite their scarcity [1223].

Dispersing from One Place to Another

Once a poplar aphid finds a leaf to live on, she will spend the rest of her life there. But for other species, such as the honey bee, dispersal from one home base to another happens regularly. In order to go from A to B, animals must expend calories not only while moving, but even before dispersal, when they must invest in the development of locomotory muscles. Consider, for example, that if a cricket is to leave a deteriorating environment and move to a new and better place, it will need large flight muscles to fly away. The calories and materials that go into muscle development and maintenance pre-

TABLE 1 Effect of leaf size and position on gall on the reproductive success of female poplar aphids

Number of galls per leaf	Mean leaf size (cm)	Mean number of progeny produced		
		Basal female	Second female	Third female
1	10.2	80		
2	12.3	95	74	
3	14.6	138	75	29

Source: Whitham [1223]

sumably have to come out of the general energy budget of the animal. This means that other organ systems, such as a female's ovaries, cannot develop as rapidly as they could otherwise, which imposes a fitness cost on the flight-capable individual.

This trade-off hypothesis has been tested ingeniously in the following way. In some cricket species, two forms occur, one that has the requisite machinery for flying and another flightless form with small wing muscles and low lipid (fuel) reserves. Females that can fly do not start producing eggs as quickly as females unable to get off the ground, as one would predict from the trade-off argument [1286]. Researchers can also create flightless females in a cricket species that normally has only flight-capable individuals by dabbing young adults with a chemical similar to juvenile hormone. If it is true that the ability to fly comes at a reproductive price, then the experimentally produced flightless females should be able to invest more in their ovaries than control females not exposed to the hormone. Indeed, ovarian development occurs much more rapidly in the experimental females, which feed no more or less than the controls, but whose flight muscles deteriorate while resources are directed to their ovaries (Figure 7) [1286].

Dispersing individuals not only have to pay energetic developmental and travel costs, but are also more often exposed to predators (e.g., [567])—all of which raises the question, why are animals so often willing to leave home? This question is particularly pertinent for species in which some individuals disperse while others do not, or at least do not travel as far. In the fire ant, for example, when a new generation of queens is produced, some fly away from their colony, mate with males at distant sites, and go on to seek out a new location where they will burrow into the ground and try to found a nest on their own. Few females will succeed, but many try, especially those with a particular genotype that we will label *BB*. In contrast, queens with the genotype *Bb* leave the colony to mate, but apparently often return home, attempting to rejoin the natal colony to become an egg-laying queen there [311].

One proximate basis for the behavioral difference between the two kinds of queens is their genetic difference. One ultimate reason for the behavioral difference has to do with what happens to *BB* queens if they try to join a colony where several queens are already present. As we saw in Chapter 4, many of the workers in such a colony will have the *Bb* genotype, and they will systematically dismember any new queen that lacks the *b* allele [607]. Under these circumstances, it is not surprising that *BB* queens generally avoid established colonies and do the best they can to found colonies of their own.

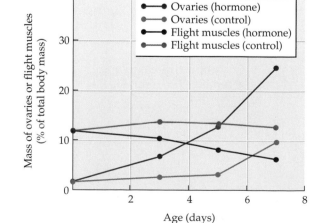

7   The trade-off between development of locomotory muscles and reproductive equipment. Female crickets treated with juvenile hormone, which creates flightless individuals with reduced wing muscle mass, followed different developmental curves than control (flight-capable) females. Note especially the difference in the growth rates of the ovaries for the two categories of females. After Zera et al. [1286].

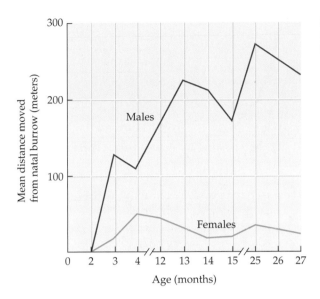

8 **Distances dispersed by male and female Belding's ground squirrels.** Males go much farther on average from their natal burrows than females. After Holekamp [531].

Belding's ground squirrel

Another species in which some individuals disperse farther than others is Belding's ground squirrel. Young male squirrels travel about 150 meters from the safety of their mother's burrow, whereas young females usually settle down only 50 meters or so from the burrow in which they were born (Figure 8) [532]. Why should young male Belding's ground squirrels go farther than their sisters? According to one argument, dispersal by juvenile animals of many species may be an adaptation against inbreeding depression [944]. When two closely related individuals mate, the offspring they produce are more likely to carry damaging recessive alleles in double doses than are offspring produced by unrelated pairs. The risk of associated genetic problems should in theory reduce the average fitness of inbred offspring, and high juvenile mortality indeed does occur in inbred populations of many animals [953]. Thus, when inbred and non-inbred white-footed mice were experimentally released into a field from which their ancestors had been captured, the inbred mice survived only about half as well as their outbred companions [583]. And even if inbred mice manage to reach adulthood, they may be less likely to reproduce than outbred individuals (Figure 9) [752].

If avoidance of inbreeding is the point of dispersing, however, then one might expect as many female ground squirrels as males to travel 150 meters from their natal burrow. But they do not, perhaps because the costs and benefits of dispersal differ for the two sexes. As Paul Greenwood has suggested for mammals generally [458], female Belding's ground squirrels may remain at or near their natal territories because their reproductive success depends on possession of a territory in which to rear their young. Female ground squirrels that remain near their birthplace enjoy assistance from their mothers in territorial defense of burrows against rival females. Thus, the benefits of remaining on familiar ground are greater for females than for males, and this difference has probably contributed to the evolution of sex differences in dispersal in this species [531].

There may, however, be another reason why male mammals disperse greater distances than females. Males (but not females) typically fight with one another for access to mates (see p. 328), and therefore losers may find it advantageous to move away from rivals they cannot subdue [815]. Although this hypothesis probably does not apply to Belding's ground squirrels, since young males have not been seen fighting with older ones around the time of dispersal, the idea deserves to be tested in other cases. Lions, for example, live in large groups, or prides, from which the young males disperse. In contrast, the daughters of the

9 **Inbreeding depression in oldfield mice** may result in the failure of inbred females to reproduce as soon as outbred females. After Margulis and Altmann [752].

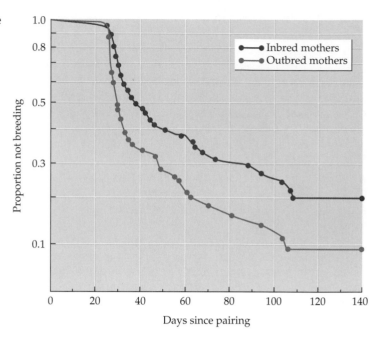

Oldfield mouse

resident lionesses usually spend their entire lives where they were born (Figure 10) [942]. The sedentary females benefit from their familiarity with good hunting grounds and safe breeding dens in their natal territory, among other things.

The departure of many young male lions coincides with the arrival of new mature males that violently displace the previous pride owners and chase off the subadult males in the pride as well. These observations support the mate

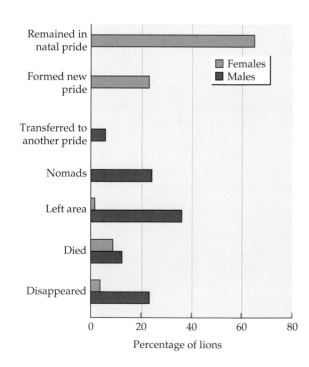

10 **Differences between the sexes in dispersal.** Female lions on the Serengeti Plain of Africa tend to remain with their natal pride, whereas males move to other prides or become nomads. After Pusey and Packer [942].

competition hypothesis for male dispersal. However, if young males are not evicted after a pride takeover, they often leave anyway, without any coercion from adult males and without ever having attempted to mate with their female relatives. Moreover, mature males that have claimed a pride sometimes disperse again, expanding their range to annex a second pride of females, at the time when their daughters in pride number one are becoming sexually mature. Proximate inhibitions against inbreeding apparently exist in lions and cause males to leave home. Ultimately, dispersing males may gain by mating with nonrelatives, even though the timing of their departure from their birthplace is not always under their control [489].

Migration

A familiar, but nonetheless amazing, form of dispersal is migration, which involves the movement away from and subsequent return to the same location on an annual basis. Nearly half of all the breeding birds of North America are migrants that take off in the fall for a trip to Mexico or Central America or South America, only to return in the spring [249]. Tiny ruby-throated hummingbirds, the weight of a penny, fly nonstop 850 kilometers across the Gulf of Mexico twice a year. Arctic terns breeding in Canada may complete a roughly 40,000-kilometer migration each year (the equivalent of seven trips across the continental United States) (Figure 11), much of it over the ocean, where the birds must stay airborne for many days and nights in a row. This is a journey that the average human would find daunting even with a frequent-flier account in hand.

Among the mammals, wildebeest, caribou, bison, seals, and whales also make prodigious round-trip journeys each year. And then there are the green sea turtles that nest on Ascension Island, a tiny speck of land a mere 8 kilome-

Arctic tern

11 The migratory route of arctic terns. These birds fly from high in the Northern Hemisphere to Antarctica and back each year. Some young birds may spend two years circling Antarctica before returning to the northern breeding grounds.

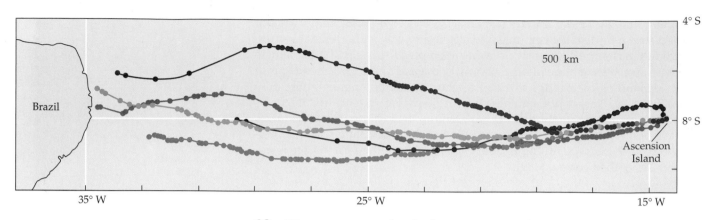

12 **Migratory routes taken by five green sea turtles** that nested on Ascension Island and then returned over 2000 kilometers to feeding areas in the ocean near Brazil. After Luschi et al. [729].

ters wide in the center of the Atlantic Ocean between Africa and Brazil. Female turtles visit the island only to deposit their eggs in beach sands. They then swim up to 2300 kilometers to warm, shallow water off Brazil, heading directly toward their goal (as ascertained by satellite telemetry of turtles outfitted with compact radio transmitters; Figure 12) [729]. The turtles feed on marine vegetation for several years before returning again to Acenscion, usually to the same beach, to lay another clutch of eggs there [201].

Migration poses a major historical problem: If sedentary species were ancestral to migratory ones, as they probably were, how could the ability to make a monumental round-trip journey evolve in a species whose members once stayed put in one place all their lives? This problem may be less severe than it seems at first glance, given that observers have been able to document rapid changes in migratory ability in modern British blackcap warblers (p. 52). Even a small genetic alteration can evidently have a large effect on the degree to which a songbird migrates [101].

More importantly, we need not propose that the ability to fly long, complex migratory routes evolved in a single step [100]. One of the lessons of history is that gradual evolutionary changes are far more probable than an abrupt jump from one trait to another very different one [304]. That migratory ability could evolve gradually is plausible given that even today, many bird species, especially in the tropics, exhibit short-distance movements of dozens to hundreds of miles, with individuals moving up and down mountainsides or from one region to another immediately adjacent one. Douglas Levey and Gary Stiles point out that short-range migrants occur in nine families of songbirds believed to have originated in the tropics. Of these nine families, seven also include long-distance migrants that move thousands of kilometers from tropical to temperate regions. The co-occurrence of short-range and long-distance migrants in these seven families suggests that short-range migration preceded long-distance travel, setting the stage for the further refinements needed for the longer and more impressive migratory trips of some species [693].

The Costs of Migration

If the long-distance migratory abilities of today's birds evolved by natural selection in stages from a sedentary ancestor, then each increase in migratory capacity must have been adaptive, with the benefits of the modified trait outweighing the costs associated with the new ability. The problem with this conclusion is that the costs of migration are not trivial. For birds, they include the extra weight that the migrant has to gain in order to store fuel for the trip. Some migrant songbirds nearly double their body weight prior to their long flights. Even moderate fuel loads dramatically alter the takeoff angles of individuals startled by a predator, almost certainly increasing the chance that the predator will catch the fleeing songbird [705]. Moreover, birds that start off with large fat reserves may nevertheless run out of gas before reaching their destination. (Lawrence Swan once saw a migrating hoopoe, a bird with the exquisite scientific name *Upupa epops*, forced by exhaustion to hop up a Himalayan pass at 20,000 feet [1117]).

An optimality approach to migration generates the prediction that migrants will evolve tactics that reduce the costs of the trip. Many European songbirds travel to central Africa by way of Spain and Gibraltar in order to cross the Mediterranean at its narrowest point [101]. This route lengthens the total journey but reduces the overwater component of the trip, perhaps decreasing the risk of drowning at sea. Likewise, red-eyed vireos migrating in the fall from the eastern United States to the Amazon basin of South America must either cross a large body of water, the Gulf of Mexico, or stay close to land, moving in a southwesterly direction along the coast of Texas to Mexico and then south. The trans-Gulf flight is shorter, but vireos that cannot fly all the way to Venezuela are dead ducks, so to speak.

In light of this danger, Ronald Sandberg and Frank Moore predicted that red-eyed vireos with low fat reserves would be less likely than those with considerable body fat to risk the long journey due south across the Gulf of Mexico. They captured migrating vireos in the fall on the coast of Alabama, classified each individual as lean or fat, and placed the birds in orientation cages similar to those described on page 53. Vireos with less than about 5 grams of body fat showed a mean orientation at sunset toward the west-northwest, whereas vireos that had been classified as having more fat tended to head due south, just as Sandberg and Moore had predicted (Figure 13) [1013].

Some songbirds even smaller than the red-eyed vireo attempt a still more impressive flight across water in the fall, one that requires a nonstop flight from Canada to South America over 3000 kilometers of ocean (Figure 14) [1241]. At first glance, a blackpoll warbler that selects this migratory route would seem to have a death wish. Surely the birds should take the safer passage along the coast of the United States and down through Mexico and Central America. But migratory blackpolls commonly appear on islands in the Atlantic and Caribbean, and they are in good condition when they arrive, demonstrating their capacity to make long nonstop ocean crossings [685].

The blindly courageous blackpoll warbler that manages this over-the-ocean trip has substantially reduced some of the costs of getting to South America.

Red-eyed vireo

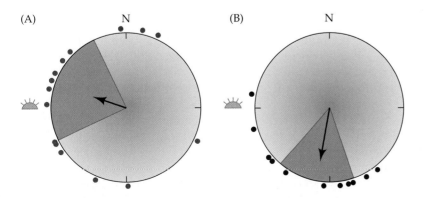

13 Body condition affects the migratory route chosen by red-eyed vireos. (A) Birds with low fat reserves do not head south toward the Gulf of Mexico, but instead head west (sunset symbol) as if to begin an overland flight toward Mexico. (B) Birds with ample energy reserves orient due south. The central arrow shows the mean orientation for the birds tested in each group. After Sandberg and Moore [1013].

14 **Transatlantic migratory path of blackpoll warblers** (arrows) from southeastern Canada and New England to their South American wintering grounds. Courtesy of Janet Williams.

Blackpoll warbler

First, the sea route from Nova Scotia to Venezuela is about half as long as a land-based trek, although admittedly it requires an estimated 50 to 90 hours of continuous flight. Second, there are very few predators lying in wait in mid-ocean or on the islands of the Greater Antilles that the blackpolls may reach. Third, the birds leave the Canadian coast only when a west-to-east-traveling cold front can push them out over the Atlantic Ocean for the first leg of the journey, after which the birds use the westerly breezes typical of the southern Atlantic to help them reach an island landfall.

The Benefits of Migration

For all their navigational and meteorological skills, migrant blackpolls and other birds cannot eliminate the costs of their travels altogether. What ecological conditions might elevate the benefits of migration enough to outweigh these costs, leading to the spread of migratory abilities by natural selection? The answer for many migratory songbirds in the Americas probably lies in the immense summer populations of protein-rich insects in the northern United States and Canada, where long days fuel the growth of plants on which herbivorous insects feed [167]. Moreover, the many hours of summer daylight mean that migrant songbirds can search for food longer each day than tropical bird species, which have only about 12 hours each day to harvest prey for their offspring.

Changes in resource availability also appear to be linked to the migrations of other vertebrates. In any number of unrelated fishes, individuals migrate from fresh water to the sea and back during their lifetimes, or alternatively, from the sea to fresh water and back. In northern latitudes, most migratory species, including the familiar salmon, spend the bulk of their lives in the ocean. In the tropics, migrants are far more likely to move into fresh water from the sea. This pattern reflects the fact that food production in the northern oceans exceeds that in bodies of fresh water, while just the reverse is true in tropical latitudes [461].

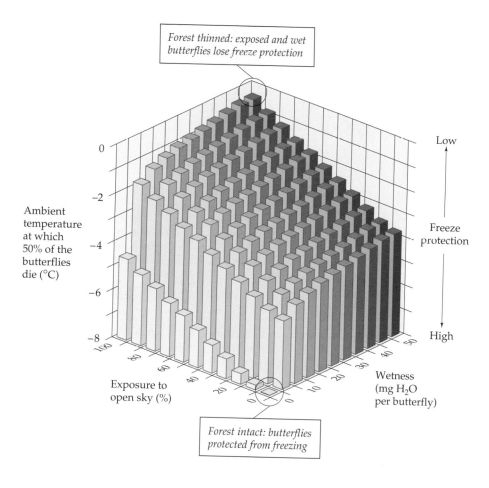

17 Habitat quality and survival of monarchs overwintering in central Mexico. Protection from freezing depends on a dense tree canopy that reduces wetting of the butterflies and their exposure to open sky, factors that greatly increase monarch mortality. After Anderson and Brower [29].

Forest thinned: exposed and wet butterflies lose freeze protection

Forest intact: butterflies protected from freezing

that they have *exactly* the same average fitness over the long haul, something very difficult to achieve.

If migratory and nonmigratory blackbirds differed because of genetic differences between them, they would be said to employ two different **strategies** [462]. In the lingo of game theory, evolved strategies are not consciously adopted game plans of the sort humans often employ, but rather behavioral traits that differ among individuals because of their hereditary makeup. The two-strategies hypothesis for blackbird behavior generates the predictions that (1) the differences between migratory and nonmigratory individuals are caused by differences in their genetic constitution, (2) individuals should not change their behavior from year to year, and (3) the lifetime fitness of the two types should be the same on average.

However, European blackbirds that migrate in one year often switch their behavior to become residents in a subsequent year (Figure 18). This finding makes it unlikely that the two-strategies hypothesis can apply in this instance. Instead, let's consider another game theoretical hypothesis, the **conditional strategy** alternative. According to this hypothesis, the genetic makeup of individuals enables them either to migrate or to remain as a resident, depending on the conditions they experience in their environment. In a conditional strategy, the behavioral differences between individuals are *not* caused by genetic differences; instead, possession of one conditional strategy permits an individual to use any of several options, or **tactics,** under different conditions. Thus, for example, the ability of individual New Mexico spadefoot toad tadpoles to adopt either a shrimp-eating or a detritus-eating diet, depending on the availability of the two foods in their desert ponds, represents a conditional strategy that controls the use of two tactics (see p. 237).

Resources other than food can also vary in availability seasonally, making migration adaptive. In the Serengeti National Park of Tanzania, over a million wildebeest, zebra, and gazelles move from south to north and back again each year. The move north appears to be triggered by the dry season, while the onset of the rains sends the herds moving south again. It might be that the herds are tracking grass production, which is dependent on rainfall. Eric Wolanski and his colleagues have, however, established that the most important factor sending the animals north is actually the decline in water supplies and an increase in the saltiness of the water in drying rivers and shrinking waterholes. If one knows the salinity of the water available to the great herds, one can predict when they will leave on their march north [1262].

Likewise, food is not the primary factor driving the movements of the best-known of all migratory insects, the monarch butterfly, which can fly from as far north as southern Canada to central Mexico [1163]. Monarch butterflies head for a safe place to spend the winter before turning around and moving north the next spring [162] (Figure 15). Many southward-bound monarchs wind up in a very few high mountain patches of Oyamel fir forest in central Mexico, where they form spectacularly dense aggregations (Figure 16). Why should some adult monarchs fly as many as 2500 kilometers to spend the winter in the cold high mountains of Mexico? Surely there are more suitable places much closer to the milkweed-producing areas that monarchs use as their spring and summer breeding grounds.

But perhaps not, since killing freezes occur regularly at night throughout the eastern United States during winter. In contrast, freezes are very rare in the Mexican mountain refugia used by the monarchs. In these forests, at about 3000 meters elevation, temperatures rarely drop below 4°C, even during the coldest winter months. Occasionally, however, snowstorms do strike the mountains,

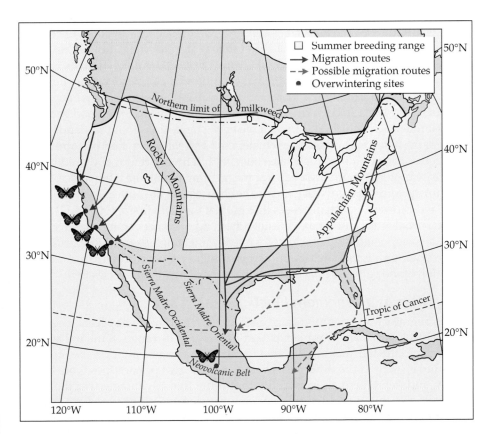

15 Monarch butterfly migration routes. Eastern and midwestern populations fly to a few small patches in the mountains of central Mexico. Western populations go to the Pacific coast. After Brower [162].

16 **Monarch butterflies spend the winter in huge clusters** that form on fir trees in a few mountains in Mexico. Photograph by Lincoln P. Brower.

and when this happens, as many as 2 million monarchs can die in a single night of subfreezing temperatures. The risk of freezing to death could be completely avoided in many lower-elevation locations in Mexico. But William Calvert and Lincoln Brower note that in warmer and drier areas, the monarchs would quickly use up their water and energy reserves. By remaining moist and cool—without freezing to death—the butterflies conserve vital resources, which will come in handy when they start back north after their 3 months in the mountains [190].

The hypothesis that the stands of Oyamel fir used by the monarchs provide a uniquely favorable microclimate that promotes winter survival may be tested shortly in an unfortunate manner. Trees from areas bordering some winter aggregations have already been cut, legally and illegally, by local firewood gatherers and lumber companies, and there have been suggestions that the stands of firs where the butterflies themselves cluster could be safely thinned by commercial loggers. Brower and his associates believe that timber removal would greatly increase butterfly mortality, even if some roosting trees were left in place. Opening up the forest canopy will increase the chances that the butterflies will become wet and exposed, which increases the risk that they will freeze (Figure 17). Thus, even partial forest cutting may destroy the conditions needed for the survival of monarch aggregations. If the loss of some Oyamel firs causes the local extinction of overwintering monarch populations, it will be a powerful but sad demonstration of the value of a particular habitat for a migrating species [29].

Migration as a Conditional Tactic

Both migratory and nonmigratory individuals occur in some species. The European blackbird is an example, since in portions of its range, some individuals migrate in the fall, leaving the area to others that overwinter on the breeding grounds [1026]. If the differences between these two phenotypes were hereditary, and one type had even a slight reproductive advantage over the other, natural selection would surely have eliminated the other type by now. The long-term coexistence of two genetically distinct kinds of individuals demands

European blackbird

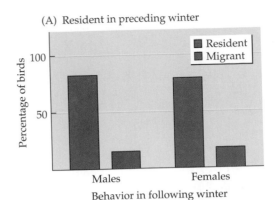
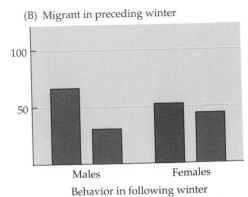

18 **Is migratory behavior the less preferred tactic for European blackbirds?**
(A) Birds that were residents in the preceding winter tended to be nonmigratory the next winter as well. (B) In contrast, birds that were migrants in the preceding winter were quite likely to switch to the resident option the following winter. After Schwabl [1026].

Applying a conditional strategy hypothesis to the blackbird case, we would expect individuals to adopt whatever tactic yields higher fitness for them. Socially dominant individuals should be in a position to select the better of the two tactics, forcing subordinates to make the best of a bad situation by adopting the option with the lower reproductive payoff (but more than they could get from futile attempts to be dominant) [302]. For example, perhaps an area can support only a few resident blackbirds during winter. Under such circumstances, subordinates faced with a coterie of more powerful residents might do better to migrate away from the competition, returning in the spring to occupy territories made vacant by the deaths of some rivals over the winter.

Given the logic of this hypothesis, we can make several predictions: (1) blackbirds should have the ability to switch between tactics, rather than being locked into a single behavioral response, (2) socially dominant birds should adopt the superior tactic, and (3) when choosing *freely* between tactics, individuals should choose the option with the higher reproductive payoff (as already illustrated by the behavior of poplar aphids when choosing leaves). In light of these predictions, it is significant that migratory European blackbirds head off in fall at times when dominance contests are increasing in frequency [727]. Moreover, blackbirds do switch fairly often from the migratory option to staying put, typically when they are older and presumably more socially dominant [728]. We do not yet know, however, whether resident blackbirds have higher reproductive success when they use the resident tactic as opposed to the migratory alternative.

Territoriality

Thus far, we have examined why animals prefer some living areas over others, and why animals might move from one location to another. We now turn our attention to yet another decision that many animals face: Having settled temporarily or permanently in one place, should the individual set up a territory that it defends against intruders, or should it coexist peacefully with others on an undefended home range? Note that yellow warblers, female poplar aphids, and Belding's ground squirrels can be extraordinarily aggressive in competing for a living area, while honey bees and monarch butterflies ignore or tolerate their fellows.

A cost–benefit approach to territoriality requires that we consider the disadvantages of aggression, one of the most obvious of which is the time cost of the

in
pr
en   10
If   9
1 r   8
ho
in   7
be   6
ho
we   5
be   4
lite
wi   3
20   2
sp
rat   1
pe

rat
wi   If
ter   of h
a l   thos
pa   able
de   male
rep   ritor
ers   terri
rise   prop
the   duct
un   territ
            etly
Te   all [1
Th   ries t
ing   Th
def   ritori
ins   cal w
sw   occu
the   foun
joi   their
me   attac
at i   secor
            inves
            advar
            birds
TA   starts
            becau
N   grour
Fo   migra
            by cla
            arrive
Te   decisi

            Territ
            In wir
            more
            earlie
To
cal
Sot

behavior. A territorial surgeonfish, for example, chases rivals away from its algae-rich turf on a Samoan reef an average of 1900 times each day [253]. The wear and tear from territorial pursuits, to say nothing of the out-and-out fighting that occurs in some territorial species, could lead to a shortened life, as has been documented for a damselfly species with both territorial and nonterritorial individuals [1159].

In addition to the risks of injury or exhaustion, other damaging effects can potentially arise indirectly from the underlying mechanisms of aggression. For example, in species in which testosterone promotes territorial defense, the effects of the hormone may exact a toll via a reduction in parental care or loss of immune function (p. 175) [1255]. Moreover, the hormone may so increase the activity level of males, even if they are not actually fighting, that they suffer as a result. Catherine Marler and Michael Moore demonstrated this point experimentally with Yarrow's spiny lizards [754, 755]. They inserted small capsules containing testosterone beneath the skin of some male lizards they captured in June and July, a time of year when the lizards are only weakly territorial. These experimental animals were then released back into a rockpile high on Mt. Graham in southern Arizona. The testosterone-implanted males patrolled more, performed more push-up threat displays, and expended almost a third more energy than controls (lizards that had been captured at the same time and given a chemically inert implant). As a result, these hyper-territorial males had less time to feed, captured fewer insects, stored less energy in fat, and died sooner than those with normal concentrations of testosterone (Figure 19).

Because territoriality is costly, we can predict that peaceful coexistence on a home range should evolve when the benefits of owning a valuable resource do

Yarrow's spiny lizard

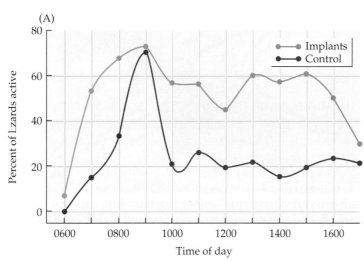

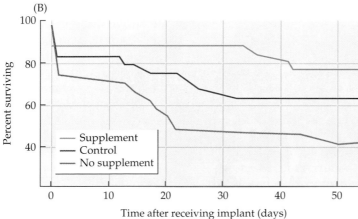

19 Energetic costs of territoriality. (A) Males of Yarrow's spiny lizard that received an experimental implant of testosterone spent much more time moving about than did control males. (B) Testosterone-implanted males that did not receive a food supplement disappeared at a faster rate than did control males. Testosterone-implanted males that received a food supplement (mealworms) survived as well as or better than controls; thus the high mortality experienced by the unfed group probably stemmed from the high energetic costs of their induced territorial behavior. After Marler and Moore [754, 755].

27   A test of the payoff asymmetry hypothesis. In tarantula hawk wasps, the more time a new resident has been on a territory, the longer the fights between that individual and the original resident (who had been temporarily removed from his territory). After Alcock and Bailey [15].

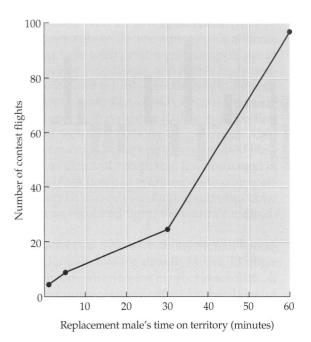

Tarantula hawk wasp

ple, one removes a male tarantula hawk wasp from the peaktop shrub or small tree that he is defending and pops him into a cooler, his vacant territory will often be claimed within a few minutes. If the ex-territory holder is quickly released, he will usually return almost instantly to his old site and oust the newcomer simply by chasing him around the tree for a few moments. But if the ex-territory holder is left in the cooler for an hour, when he is released and hurries back to his territory, a battle royale ensues. The newcomer resists eviction, and the two males engage in a long series of ascending flights, in which they climb rapidly up into the sky side by side for many meters before diving back down to the territory, only to repeat the activity again and again until finally one male—usually the replacement wasp—gives up and flies away (Figure 27).

Although cases of this sort support the payoff asymmetry hypothesis for why residents usually win, it is possible that lengthier contests occur after replacements have been on territory for some time because the removed resident has lost some resource-holding power while in captivity. This possibility was checked in a study of the European robin by taking the first replacement away and permitting a second replacement to become established before releasing the resident. In this way, Joe Tobias was able to match one-day replacements against residents that had been held captive for ten days, and under these circumstances, the ex-territory holders always won despite their prolonged captivity [1142]. In contrast, when ex-residents that had been caged for ten days went up against replacements that had been on the territory for ten days, the original territory holders always lost. Therefore, contests between replacements and ex-residents were decided by how long the replacement had been on territory, not by how long the ex-resident had been in captivity. The payoff asymmetry hypothesis appears to explain why the resident almost always wins in European robins, and it may apply to many other species as well.

Summary

1. In choosing where to live, many animals actively select certain places over others. If living in a certain kind of habitat enhances fitness, then individuals able to occupy preferred habitats should have higher fitness, a prediction that has been tested with affirmative results in a number of species.

2. The selection of living space often occurs in the context of leaving one spot for another, as when juvenile animals abandon the place where they were born to go in search of new homes. Cost–benefit analyses have helped to explain why young male mammals typically disperse farther than females of their species, probably because the costs of dispersal are greater for females than for males.

3. Migration is a special form of dispersal in which the migrant eventually returns to the place it left. The costs of migratory journeys are high; counterbalancing benefits may arise if, for example, migrants can exploit seasonal bursts of food productivity during their reproductive season.

4. An evolved strategy is an inherited behavioral mechanism that enables an individual to respond adaptively to problems it faces in its environment. A conditional strategy is a decision-making mechanism that provides its owner with the ability to select among several tactics depending on the conditions it encounters. An evolutionarily stable strategy is a behavioral mechanism that persists because it is superior to alternative strategies that have appeared by mutation in the species over evolutionary time.

5. When choosing living space, some animals invest additional time and energy in defense of the site. Territorial behavior is strongly correlated with the occurrence of valuable resources in small, economically defensible patches. Owners of defended breeding sites typically reproduce more successfully than individuals that fail to secure a territory; when some breeding territories are preferred to others, owners of the preferred sites typically enjoy greater reproductive success. When feeding territories are established, owners usually gain caloric benefits in excess of the caloric expenses incurred in defense of the site, or else they are able to secure their minimum daily caloric requirements more quickly, saving time that can be spent resting rather than foraging.

6. There are several competing hypotheses on why territory owners usually defeat challengers for control of their territories. In some cases, residents are physically stronger, more capable defenders than their would-be replacements; in other cases, residents exhibit strong defense because they have more to lose by yielding their territory than the newcomer has to gain by taking it over.

Discussion Questions

1. Some animals defend territories against members of a species other than their own. How can this behavior be adaptive? Develop at least one cost–benefit hypothesis and a set of predictions on the evolution of interspecific territoriality.

2. In American robins, as in many other migratory species, birds whose nesting attempts have failed in one year often do not return to the same spot to try again in the subsequent year. Several different hypotheses can account for this result. Deduce what some of these hypotheses might be based on the following data, which were collected to check certain predictions taken from a set of alternative

explanations. The data resulted in part from a manipulation in which the experimenter destroyed the nests of a randomly selected subsample of nesting robins and then compared their return rate in the next year with that of other robins whose nesting attempts had not been interrupted experimentally or naturally, but had instead resulted in the production of fledglings. Of the subsample of experimentally induced nest failures, only 18% of the robins came back, whereas 44% of the successful nesters returned.

3. Develop a game theoretical hypothesis to analyze the decision of a few songbirds to employ the nonterritorial route to breeding in which the pair nests surreptitiously within the territory of another pair. Come up with a two-strategies hypothesis and a conditional strategy hypothesis. What predictions follow from your two hypotheses?

4. In Belding's ground squirrels and other mammals, young males disperse while young females remain on or near the natal territory, but this pattern is exactly reversed in most birds. Why might this be so? In producing your hypotheses, consider the fitness value of a territory to a typical male mammal and a typical male bird [458].

5. As noted in the text, two kinds of individuals co-occur in some species of crickets: a flight-capable form and a flightless form. The difference between the two forms has a genetic component [979]. Why is this finding surprising? How might natural selection maintain the two forms in a cricket species? Devise tests for your alternative hypotheses.

Suggested Reading

The concept of an evolutionary strategy has been helpfully reviewed by Richard Dawkins [301, 302] and Mart Gross [462]. Tom Whitham [1221, 1223] shows how to use evolutionary theory when studying habitat selection. Dispersal in Belding's ground squirrels is analyzed at the proximate and ultimate levels by Kay Holekamp and Paul Sherman [532]. The cost–benefit approach to territoriality was first used by Jerry Brown and Gordon Orians [171], and the approach has been applied subsequently with special skill by Nick Davies and his colleagues [288, 297]. Migration has been reviewed by Hugh Dingle [328].

10 The Evolution of Communication

Watch a clan of spotted hyenas for a while and you are likely to see one hyena present its large, engorged penis to a companion, who inspects the organ while perhaps offering the other hyena a chance to nuzzle and sniff its own erect penis. Most people are surprised to learn that *females* are full participants in these greeting rituals. Indeed, female spotted hyenas possess such a realistic penis look-alike that humans cannot easily tell the sexes apart. Hyenas have no difficulty in this department, but their communication system is unquestionably odd and well worth evolutionary analysis.

Although we discussed evolutionary aspects of communication earlier, especially in the context of bird song (Chapter 2), we have not yet systematically examined how communication signals originate and evolve. How might it happen that the first giver of a signal finds a receiver that can interpret the signal appropriately? How can it possibly be advantageous for signalers to provide information when that information is so often exploited by unintended receivers? And why do receivers pay attention to the signals they are offered when signalers may take advantage of them? This chapter addresses these questions about

◀ **The message that this wolf is conveying** is easy for its fellow wolves to read. What might another wolf gain by quickly backing down from a snarling, tooth-exposing rival? Photograph by Jeff Lepore

sequence, it could be SWAJS MEIURNZMMVASJDNA YPQZK. If we kept at it, or had a computer keep churning out random sequences of letters and spaces, it would be a long, long time before METHINKS IT IS LIKE A WEASEL came up. This is just one combination of the vast number of sequences that can be created when you have 27 possibilities for the first position, times 27 for the next position, times 27 for the third…times 27 for the twenty-eighth position. The odds of getting such a phrase in a single random selection of 28 letters and spaces is, according to Dawkins, one in 10,000 million million million million million million. Those are not good odds.

However, instead of trying to get the "right" sentence in one go, let's change the rules so that we start with the random letter set shown above and use the computer to copy it over and over with an instruction to occasionally insert a new letter into one of the positions. Then we can ask the computer to look over its huge list and pick the sequence that is closest to METHINKS IT IS LIKE A WEASEL. Whatever "sentence" is closest is used for the next generation of copying, again with a few errors thrown in. The sentence in this group that is most similar to METHINKS…is selected to be copied, and so on. When Dawkins did this with a number of different starting sequences, he found that it took only 40–70 generations to reach the target sentence, not millions upon millions upon millions of attempts—a few seconds of computer time, not years [304].

Cumulative natural selection has the same effect on living systems. Some mutations induce small random changes in the genetic "sentence" possessed by individuals. Any changed genetic message that happens by accident to confer higher reproductive success on individuals propagates itself throughout the population. Once the new "sentence," with its new developmental outcome, is widespread, additional small changes that improve the reproductive output of individuals will be incorporated in the same way, one after another. The cumulative effect of this process is to rework the genetic information, and thus the phenotype, of the species. Given enough time, a species may barely resemble its ancestor, but the differences will not have arisen instantaneously.

This scenario suggests that at least one mutant allele that altered the developmental mechanisms of female embryos spread through populations of spotted hyenas in the past. All of today's spotted hyena females possess this allele, or more recently modified versions of it, and probably many other later-appearing mutant genes as well. Together, these altered genes contribute to the development of the secondary sexual characteristics of female hyenas. As a result, all of today's spotted hyena females possess an elaborate pseudopenis and the ability to use it to communicate with other hyenas.

The Adaptive Value of Past Changes

But why would the original mutant female hyena whose daughters had masculinized external genitalia have had more descendants than the other females of her era? Surely her daughters would have been harmed by the disruption of long-tested patterns of sexual development. Human females accidentally exposed to androgens as embryos not only develop an enlarged clitoris, but also become sterile in later life, illustrating just how damaging the developmental effects of hormonal changes can be. And even in modern populations of spotted hyenas, 10–20 percent of all females die as a consequence of having to give birth through the clitoris (Figure 2). Therefore, we can be confident that the spread of masculinized genitalia had to overcome some major reproductive disadvantages [401, 404].

Obviously, not all masculinized female hyenas in the past were sterile or died while giving birth to their offspring. The fitness benefits these pioneers gained from any mutant allele contributing to the development of the pseudopenis must have outweighed the negatives associated with the gene. Perhaps the

2   Cost of the pseudopenis for female spotted hyenas. The birth canal of this species extends through the pseudopenis, which greatly constricts the canal and often leads to fatal complications for the mother during birth. Drawing courtesy of Christine Drea, from Frank et al. [404].

hormone-induced pseudopenis was initially just a costly side effect of a mutant allele that had some other consequences, some so beneficial that they swamped any disadvantageous aspects of the gene.

Several persons have ideas on how females could benefit from a gene that exposed their fetuses to high levels of male sex hormones or some other developmentally potent chemical. Marion East and his co-workers suggest that the new allele originally spread not because it caused enlargement of the clitoris, but because the gene's product helped make a hyena's offspring more aggressive [343]. Hyenas bear twins, which are born with their eyes open and their canine teeth fully erupted; they sometimes begin fighting with each other soon after being born, and one pup may eventually kill the other. Surprisingly, sibling fighting, even siblicide, can sometimes raise the fitness of the parents of murderous siblings, as we will see in Chapter 13.

An alternative hypothesis is that the mutant gene spread because it helped to make adult females, rather than their infants, larger and more aggressive. The alpha female of a hyena clan wins the fierce competition for food that occurs at prey carcasses (see Figure 28, p. 239), so her offspring get to eat while the young of subordinate females are starving. Alpha females achieve their top hyena status because they are big and aggressive [401, 403]. If the origins of heightened female aggressiveness lie in a mutation that resulted in a hormonal change, then the masculinization of the external genitalia of the daughters of females with the mutation could well have been a side effect of that novel hormonal state. For reasons already discussed, the enlarged clitoris probably reduced, but did not eliminate, net fitness gains due to the heightened aggressiveness of masculinized females or their neonates. Once it had originated and spread, however, the pseudopenis may then have undergone additional changes that enabled it to contribute to, rather than harm, female reproductive success. If we could demonstrate that the modern pseudopenis now has reproductive value in and of itself, then we would have evidence that the trait and its genetic foundation are currently being maintained by natural selection for these effects.

20   Yelling is a recruitment signal. Carcasses were exploited either by nonyelling territorial singletons and pairs or by large groups of ravens, many of them yelling, most of them nonterritorial subordinates. The graph shows the percentage of days on which carcass baits were visited by various numbers of birds. After Heinrich [506].

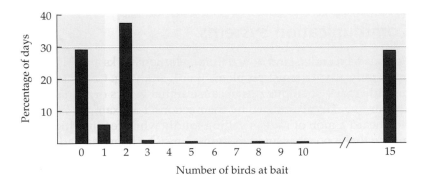

Instead, he switched his attention to an alternative explanation, an idea stimulated by seeing how cautiously some ravens approached carcasses when they first found them. Perhaps, he argued, a carcass discoverer yells to draw in other ravens so that if a predator lurks nearby, the other birds would provide possible targets, reducing the yeller's risk of being taken by a hiding coyote or fox. The incoming birds would be attracted because they would gain a wealth of food in return for taking a small risk of being the unlucky victim of a predator. However, this hypothesis generates the prediction that once a group has assembled and feeding has begun, the birds should shut up to avoid attracting more ravens, which would be unwanted and unnecessary for safety purposes. The observation that yelling continued at baits that had already acquired a retinue of actively feeding birds convinced Heinrich to discard the "dilution-of-risk-of-predation" hypothesis.

As Heinrich continued stoically lugging dead goats into the Maine woods, he came to realize that whenever he saw a single bird or a pair at a feeding site, those ravens were quiet. Yelling occurred only when three or more ravens were present, and it was then, and only then, that large numbers of other ravens came to the area. Heinrich knew that older adult ravens form pairs that defend a territory year-round. Unmated young birds usually travel solo over great distances in search of food. If a singleton attempts to feed in a resident pair's territory, the pair attacks it. It occurred to Heinrich that yelling might be a signal given by nonterritorial intruders, a signal that attracts other unmated wanderers to a food bonanza that they can exploit if they can overwhelm the defenses of the resident pair.

This "gang-up-on-the-territorial-residents" hypothesis leads to a number of predictions: (1) resident territory owners should never yell, (2) *non*resident ravens should yell, (3) yelling should facilitate a mass assault on a carcass by nonresident ravens, (4) resident pairs should be unable to repel a communal assault on their resources, and (5) a food bonanza should be eaten either by a resident pair alone or by a mob of ravens. Heinrich collected data that supported all of these predictions (Figure 20) [506]. He concluded that when a young raven yells, the consequences of providing information ("Food bonanza here") to other nonterritorial birds can include benefits (personal access to food for the yeller) that outweigh the energetic costs of yelling as well as the risk of attack by the resident ravens guarding the carcass [507]. Yelling is therefore an adaptation, currently maintained by its net fitness benefits to individuals who give the call under the appropriate circumstances.

Why Do Baby Birds Beg So Noisily for Food?

Heinrich knew how to use the adaptationist approach to clear up a mystery, and other biologists have tried to do the same with other signals that at first glance seem to impose damaging costs on signalers. A classic example is noisy begging by nestling songbirds when a parent returns to the nest with food. These loud cheeps and peeps might give the location of the nest away to a listening hawk or raccoon, resulting in the death of all the defenseless nestlings. In fact, when tapes of begging tree swallows were played at an artificial swallow nest containing a quail's egg, the egg in that "noisy" nest was taken or destroyed by predators before the egg in a nearby quiet control nest in 29 of 37 trials [688].

Further evidence for the fitness costs of begging comes from a study of differences in the begging calls of warbler species that nest on the ground versus those that nest in the relative safety of trees [497]. The young of ground-nesting warblers produce begging cheeps of higher frequencies than their tree-nesting relatives (Figure 21A). These higher-frequency sounds do not travel as far, and so may better conceal the individuals that produce them, which are especially vulnerable to predators in their ground nests. David Haskell created artificial nests with clay eggs and placed them on the ground by a tape recorder that played either the begging calls of tree-nesting or of ground-nesting warblers. The eggs advertised by the tree-nesters' begging calls were found and bitten significantly more often than the eggs associated with the ground-nesters' calls (Figure 21B).

The hypothesis that begging calls have evolved properties that reduce their potential for attracting predators yields another similar prediction: baby birds of species that experience high rates of nest predation should produce softer begging signals of higher frequency than nestlings of other species less often victimized by nest predators. This prediction was supported by data collected in one survey of 24 species from an Arizona forest [155], more evidence that predator pressure favors the evolution of begging calls that are hard to detect and pinpoint.

Given that predators can make it costly to beg for food, what benefit do begging nestlings derive from their communications? One possibility is that a noisy

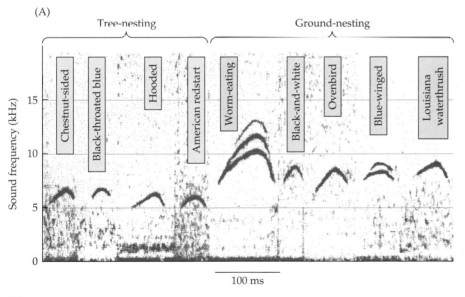
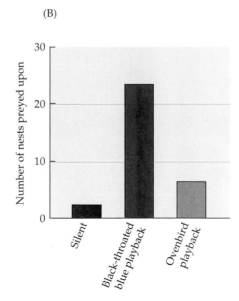

21   **Predation risk affects the evolution of begging calls in warblers.** (A) The sound frequencies of the begging calls of young ground-nesting warblers are higher than those of calls used by the young of tree-nesting warblers. (B) Experimental playback of tree-nester begging calls at artificial nests placed on the ground resulted in higher rates of discovery by predators compared with artificial ground nests "advertised" by playback of ground-nester begging calls. After Haskell [497].

baby bird provides accurate signals of its real hunger and good health, making it worthwhile for the listening parent to give it food in a nest where several other offspring are usually available to be fed. If this hypothesis is true, then it follows that nestlings should adjust the intensity of their signals in relation to the signals produced by their nestmates, which are competing for parental attention. When experimentally food-deprived baby robins are placed in a nest with normally fed siblings, the hungry nestlings beg more loudly than usual—but so do their better-fed siblings [1071].

If parent birds use begging intensity to direct food to healthy offspring capable of vigorous begging, then parents should make food delivery decisions on the basis of their offspring's calls [621]. Indeed, if you take baby tree swallows out of a nest for an hour, feeding half the set and starving the other half, then when the birds are replaced in the nest, the starved youngsters beg more loudly than those that have been fed, and the parent birds feed the active little beggars more than those who beg less vigorously [692].

In another study, one of two parent pied flycatchers was outfitted with a microchip that activated a tape recording of begging calls whenever the adult visited its nest to feed its brood. The parent that heard the extra begging signals worked harder, bringing more food to its brood; the other parent, finding the offspring relatively full and content, and not hearing the taped playback, reduced its feeding effort (Figure 22) [871]).

In all these cases, begging appears to provide a signal of need that parents use to make judgments about which offspring can benefit most from a feeding. But the question arises, why doesn't a nestling tree swallow or pied flycatcher beg loudly when it isn't all that hungry? By doing so, it might be able to secure food at a higher rate, which should result in more rapid growth or larger size, either one of which could be advantageous to it. The young of brood parasites manage this trick and benefit greatly. A young European cuckoo, for example, is able to sound like an entire nest full of baby reed warblers (Figure 23), with the result that its host parents work as hard for it as they do for a complete brood of four warbler chicks [622]. If you place another nestling in a reed warbler's nest, choosing a species comparable in size to a cuckoo—say, a baby blackbird—the young blackbird begs less intensely than the cuckoo and accordingly is fed less rapidly. But play a recording of a cuckoo's begging calls when the warblers come to the nest, and the lucky blackbird receives a roughly 50 percent increase in the number of loads of food carried to it per hour by its parentally stimulated hosts (Figure 24)[298].

Although extra-vigorous begging must come with an energetic cost, measurements of the metabolic expenses incurred by begging nestlings demonstrate that this expense is very small relative to the potential caloric gain [53]. There-

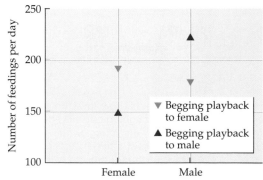

22   Begging calls as honest signals of need. When only one parent of a pair of pied flycatchers hears a playback of begging calls at its nest, that parent responds by bringing more food to its offspring. The other parent encounters less hungry, less noisy offspring and so reduces its feeding trips to the nest. After Ottosson et al. [871].

(A)

23 **The European cuckoo chick's begging call** matches that of *four* baby reed warblers. (A) A cuckoo chick begging for food from a reed warbler. The calls shown below are those of (B) a single reed warbler chick, (C) a brood of four reed warblers, and (D) a single cuckoo chick. After Davies et al. [298]; photograph by Roger Wilmshurst.

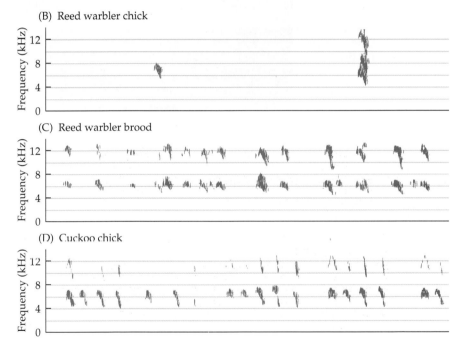

(B) Reed warbler chick

(C) Reed warbler brood

(D) Cuckoo chick

fore, the disadvantage of exaggerated begging may lie not in its energetic costs, but in the damage that any successful cheater would do to its siblings, which share genes with one another. An individual's success in propagating his or her genes can be affected by more than just his or her own personal reproductive success (p. 435). Those animals that harm their close relatives may in effect be destroying some of their own genes. Therefore, a begging nestling that happened to do very well at securing food from its parents, but at the expense of its siblings, might actually leave fewer copies of its genes overall than others that held back when begging for food.

If the sibling-protection hypothesis is correct, then the degree to which individuals moderate their begging behavior should be a function of the degree of relatedness among nest occupants. As noted above, the young of brood para-

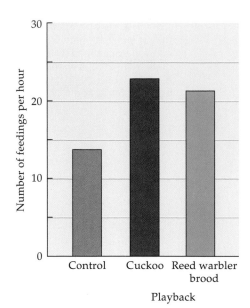

24 **The cuckoo's begging calls stimulate more frequent feeding by its hosts.** If a blackbird chick is placed in a reed warbler nest and the begging calls of a cuckoo chick are played back to the foster parents when they visit the nest, the parents deliver more food to it, matching the amount they bring to a recording of an entire brood of reed warblers. After Davies et al. [298]. Photograph courtesy of Nick Davies.

sites such as cuckoos and cowbirds, which are unrelated to their hosts and nest-mates, are much more vigorous in their begging behavior and are rewarded accordingly. In addition, in those species whose broods are often of mixed paternity, the female having mated with more than one male (see p. 370), we would expect begging to be somewhat more "selfish"—that is, louder—than in broods of highly monogamous birds whose offspring are full siblings. This prediction has been tested with positive results (Figure 25) [156], suggesting that at least part of the reason why nestling songbirds do not try to monopolize the food brought to the nest by their parents has something to do with the costs of this tactic in terms of damage to related individuals.

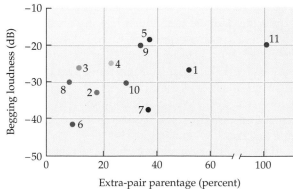

25 **Nestlings that are more closely related to one another beg less "selfishly."** The more likely it is that the nestlings in a brood are full siblings, the less "selfish" (i.e., less noisily) they beg. Data point 11 represents the begging loudness of a brood parasite, the brown-headed cowbird, which is unrelated to any of the other offspring in its host's nest. After Briskie et al. [156].

Illegitimate Receivers

The raccoon that listens in on the communication taking place between a brood of baby tree swallows and their parents is an **illegitimate receiver** in the sense that it uses information from the signal to the fitness detriment of the legitimate signalers and receivers. Illegitimate receivers can have powerful effects on the evolution of communication systems. For example, Mike Ryan wondered why male túngara frogs often give whining calls without chucks when females of their species prefer males who add the chucks. To address this question, Ryan and his associates asked whether the fitness *costs* of giving whine-chuck calls might not exceed the benefits to signal givers under some circumstances [1003]. The túngara frog has a special enemy, the predatory fringe-lipped bat, which sweeps down on calling males and hauls them away to eat (Figure 26). If males sometimes reduce the attractiveness of their calls to females, perhaps they do so to reduce the ease with which they can be located by fringe-lipped bats, thereby improving their chances to live and reproduce on another night.

This hypothesis generates a number of predictions, which Ryan and his colleagues tested. First, fringe-lipped bats should be attracted to the signals produced by their prey, and indeed, bat predators zero in on speakers broadcasting túngara frog calls while ignoring silent speakers. Second, fringe-lipped bats should be more attracted to whine-chuck calls than to whine calls alone; in fact, these bats were more than twice as likely to inspect, and even land upon, a speaker broadcasting a whine-chuck than a whine alone [998]. Third, the frogs should be less likely to give one-component whine calls when the risk of predation is lower. The chance of becoming a victim declines for frogs calling in large groups because of the dilution effect (see p. 204). As expected, males calling in large assemblages are much more likely to produce whine-chuck calls than are those in smaller groups [1003].

The risk of exploitation by an illegitimate receiver may also be responsible for the differences between the mobbing call and the "seet" alarm call of the great tit (Figure 27) [756]. These small European songbirds sometimes approach a *perched* hawk or owl and give a loud mobbing call whose dominant frequency

(A)

(B)

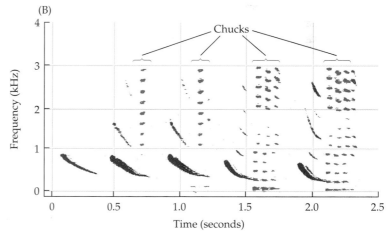

26 A calling male túngara frog (A) may inadvertently attract an illegitimate receiver of his signal: a deadly predator, the fringe-lipped bat. (B) The risk of attack is greater if the male's call includes one or more chucks (blue in the sonograms) as well as the introductory whine (purple in the sonograms). Sonograms courtesy of Mike Ryan. Photograph by Merlin D. Tuttle, Bat Conservation International.

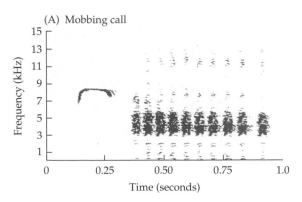

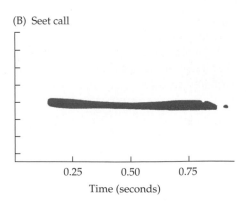

27 Great tit alarm calls. Sonograms of (A) the mobbing call and (B) the "seet" alarm call. Note the lower sound frequencies in the mobbing signal. A, courtesy of William Latimer; B, courtesy of Peter Marler.

Great tit

Sparrow hawk

is about 4.5 kHz. This easily located acoustical signal helps other birds find the mobbers and join in the harassment of their mutual enemy (see p. 184). If, however, a great tit spots a *flying* hawk, it gives a much quieter "seet" alarm call, which appears to warn mates and offspring of possible danger. This signal's dominant frequency lies within 7–8 kHz, so that the sound attenuates (weakens) after traveling a much shorter distance than the mobbing signal. The rapid attenuation of the "seet" call compromises its effectiveness in reaching distant legitimate receivers, but it also lowers the chance that a dangerous predator on the hunt will be able to tell where the caller is. Moreover, the frequencies of the "seet" call lie outside the range that hawks can hear best, while falling within the range of peak sensitivity of the great tit (Figure 28). As a result, a great tit can "seet" to a family member 40 meters away, but will not be heard by a sparrow hawk unless the predator is less than 10 meters distant [630].

If the "seet" call of the great tit has evolved properties that reduce the risk of detection by its enemies, then unrelated species should have convergently evolved alarm signals with similar properties. The remarkable convergence in the "seet" calls of many unrelated European songbirds suggests that selection by bird-eating hawks has favored the evolution of alarm calls that are hard for hawks to hear (Figure 29) [756].

28 Hearing abilities of a predator and its prey. A sparrow hawk can hear sounds in the 0.5–4 kHz range that are fainter (5–10 dB lower in intensity) than those that great tits can hear. But great tits can detect an 8 kHz sound (in the range of the "seet" call) that is fully 30 dB fainter than any 8 kHz sound that a sparrow hawk can detect. After Klump et al. [630].

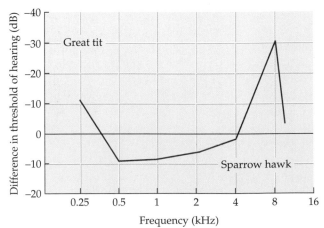

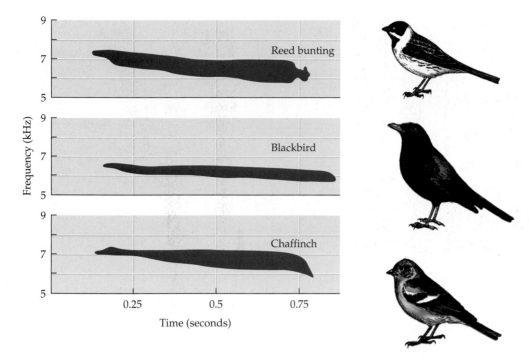

29 Convergent evolution in a signal. The great tit's high-pitched "seet" alarm call (see Figure 27B) is very similar to the alarm calls given by other unrelated songbirds when they spot an approaching hawk. After Marler [756].

Adaptive Signal Receiving

We have been using an adaptationist approach to examine cases in which signalers appear, at first glance, to be lowering their fitness. Let's now examine some instances in which *receivers* of signals seem to lose fitness by reacting to certain messages from others. First, why do so many animals resolve their disputes with highly ritualized, symbolic displays? Rather than bludgeoning a rival, males of many bird species settle their conflicts over a territory or a mate with much singing and feather fluffing but without ever touching one another. Likewise, caribou stags have magnificent antlers, but they almost never really attempt to harm one another. One team of observers recorded 1308 sparring matches, in which the males sedately clashed antlers, but only six real fights during the same period [74]. Even when genuine fighting does occur in the animal kingdom, the opponents often appear to be auditioning for a comic opera. After a body slam or two, a subordinate elephant seal generally lumbers off as fast as it can lumber, inchworming its blubbery body across the beach to the water, while the victor bellows in noisy, but generally harmless, pursuit.

Prior to the recognition that "for-the-good-of-the-species" hypotheses have serious logical problems (see p. 15), many persons proposed that animals resolved conflicts by threat displays and sham fights for the benefit of the species as a whole. The use of noncontact threats was thought to prevent injury to the supposedly superior individuals who were needed to father the next generation of offspring. However, if gentlemanly losers really were holding back for the welfare of all, then the special genes of these self-sacrificing individuals would disappear as other individuals propagated their genes more effectively without regard to the long-term survival or improvement of the species.

Therefore, adaptationists have hypothesized that even losers must gain fitness by resolving conflicts with little or no fighting [305]. Consider the European

30 Deep croaks deter rivals. Male European toads make fewer contacts with, and interact less with, silenced mating rivals when a tape of a low-frequency call is played than when a higher-frequency call is played. After Davies and Halliday [295].

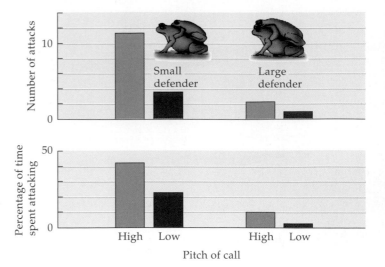

toad, *Bufo bufo*, whose males compete for receptive females. When a male finds another male mounted on a female, he may try to pull him from her back. The mounted male croaks as soon as he is touched, and often the other male immediately concedes defeat and goes away, leaving his croaking rival to fertilize the female's eggs. How can it be adaptive for the signal receiver in this case to give up a chance to leave descendants simply on the basis of hearing a croak?

Because European toad males come in different sizes, and body size influences the pitch of the croak produced by a male, Nick Davies and Tim Halliday proposed that males can judge the size of a rival by his croak. If a small male could tell, just by listening, that he is up against a larger opponent, then the small male ought to give up without getting involved in a fight he cannot win. If this hypothesis is correct, then deep-pitched croaks (made by larger males) should deter attackers more effectively than higher-pitched ones (made by smaller males) [295]. To test this prediction, the two researchers placed mating pairs of toads in tanks with a single male for 30 minutes. The paired male, which might be large or small, had been silenced by looping a rubber band under his arms and through his mouth. Whenever the second male touched the pair, a tape recorder supplied a 5-second call of either low or high pitch. Small paired males were much less frequently attacked if the interfering male heard a deep-pitched croak (Figure 30). Thus, deep croaks do deter rivals to some extent, although tactile cues also play a role in determining the frequency and persistence of an attack, as one can see from the higher overall attack rate on smaller toads.

Receivers May Require Honest Signals

So why don't small males appear to be large by giving low-pitched calls? Perhaps they would, if they could, but they can't. A small male toad apparently cannot produce a deep croak, given that body mass and the unbendable rules of physics determine the pitch of the signal that a male toad can generate. Thus, toads have evolved a warning signal that accurately announces their body size. By attending to this **honest signal**—one that conveys accurate information—a male can determine something about the size of his rival and thus his probability of winning an all-out fight with him, just as a parent bird can count on the begging signals of its chicks to indicate just how hungry and healthy they are.

When smaller rivals withdraw upon hearing an honest signal from a larger male, both parties gain: small males do not waste time and energy in a battle they are unlikely to win, and large males save time and energy that they would otherwise have to spend struggling with annoying smaller toads. Imagine two kinds of aggressive individuals in a population, one that fought with each opponent until physically defeated or victorious, the other that checked out the rival's aggressive potential and then withdrew as quickly as possible from superior fighters. The "fight no matter what" types would eventually encounter an opponent that would thrash them soundly. The "fight only when the odds are good" types would be far less likely to suffer an injurious defeat at the hands of an overwhelmingly superior opponent [773, 1213].

Further, imagine two kinds of superior fighters in a population, one that generated signals other lesser males could not produce, and another whose threat displays could be mimicked by smaller males. As mimics became more common in the population, natural selection would favor receivers that ignored the easily faked signals, reducing the value of producing them. This in turn would lead logically to the spread of the genetic basis for an honest signal that could not be devalued by deceitful signalers.

If this argument is correct, we can predict that the threat displays of many unrelated species should be expensive to produce and therefore difficult for small, weak, or unhealthy males to imitate, and that they should reliably inform rivals about the size of the displayer, given that body size, strength, and fighting capacity are linked in many animals [34]. The bizarre "antlers" and eye stalks of certain insects belonging to different families of flies meet these criteria beautifully (Figure 31). Males competing for access to mates confront each other,

31   Convergent threat displays. (Top) Males of this Australian antlered fly in the genus *Phytalmia* confront each other head to head, permitting each fly to assess his own size relative to the other's size. (Bottom) Male red deer clash aggressively, locking their horns and pushing in ways that enable opponents to judge the relative size and strength of their rivals. Top photograph by Gary Dodson; bottom photograph by J. H. Robinson, courtesy of Tim Clutton-Brock.

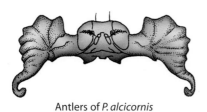

Antlers of *P. alcicornis*

32 Antler span in two New Guinean flies provides accurate information about body size, permitting males to make accurate decisions about an opponent's fighting ability. After Wilkinson and Dodson [1238].

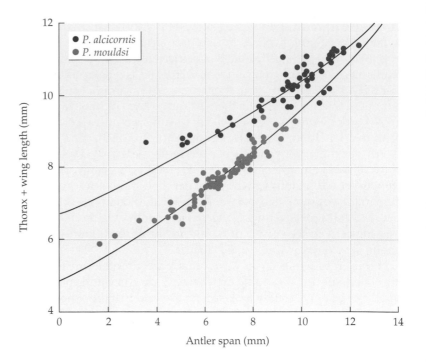

head to head, permitting rivals to compare the size of their large, and presumably expensive to produce, head projections. Antler span and eye span are highly correlated with body size (Figure 32), so that contestants can quickly and accurately assess their relative sizes by going head to head [1238]. Smaller males generally abandon the field of battle; the greater the difference between the width of their eye span and their opponent's, the more rapidly they depart, and the more energy they save [884].

Likewise, in the deer called red deer in Europe and elk in North America, antler-clashing and shoving matches permit males to evaluate the size and strength of an opponent, but these animals need not even come into contact in order to make these assessments. In fact, the typical clash between males occurs when the two competitors stand off at a considerable distance and roar loudly at each other (Figure 33) [231]. Only males in top condition can sustain roaring

33 An honest signal. Only red deer (elk) stags in top condition can roar for long periods. Photograph courtesy of Tim Clutton-Brock.

at a high rate for many minutes [230] because the activity is so costly in terms of energy expended, as is true of many other acoustical displays (e.g., [472]; but see [550]). Thus, just as in antlered flies and European toads, red deer opponents can quickly gain honest and accurate information about the fighting ability of a rival, and they use this information to terminate encounters they cannot win.

Why Does Deception Occur?

Although honest signalers provide useful information to each other for mutual benefit, deceptive signalers are not uncommon in nature. Recall the case of the orchids whose decoys lure male wasps into attempting to copulate with the flower (p. 95). Just as signal givers can lose fitness by providing information to exploitative receivers, receivers can lose fitness by responding to signals generated by **illegitimate signalers,** which use deception to reduce the fitness of a receiver. A famous example of this phenomenon involves the "femme fatale" fireflies studied by Jim Lloyd [710]. The females of some predatory fireflies in the genus *Photuris* answer the flashes given by males of certain other species in the genus *Photinus*. Each species of *Photinus* has its own code, with a female answering her male's distinctive flash pattern by giving a flash of her own after a precise time interval. Some *Photuris* females can respond at the correct interval to male signals of three *Photinus* species [711]. If a *Photuris* female succeeds in luring a male *Photinus* close enough, she will grab, kill, and eat him, reducing his reproductive potential to zero (Figure 34).

Deception of this sort poses a real puzzle for the adaptationist, not from the perspective of the predatory signaler, which gets a meal when its deception works, but from the standpoint of the male firefly that pays attention to the wrong "come hither" signal and gets killed. When an action is clearly disadvantageous to individuals, behavioral biologists generally propose one of the two following causes:

1. **Novel environment theory:** The maladaptive response is caused by a proximate mechanism that once was adaptive, but is no longer. The current maladaptation occurs because modern conditions are very different from those that shaped the mechanism in the past, and because there has not been sufficient time for advantageous mutations to occur that would "fix the problem."

2. **Exploitation theory:** The maladaptive response is caused by a proximate mechanism that is still adaptive in sum, but can be exploited by some indi-

34 A firefly femme fatale. This female *Photuris* firefly is feeding on a male *Photinus* firefly that she lured to his death by imitating the flashes given by females of his species. Photograph by Jim Lloyd.

viduals, leading to fitness losses that on average reduce, but do not eliminate, the net positive fitness gain associated with the mechanism and the behavior it controls.

Novel environments seem unlikely to account for the nature of interactions between *Photinus* males and their predatory relatives. This theory more often applies to cases in which very recent human modifications of the environment appear responsible for eliciting maladaptive behavior (see Figure 1, p. 116). Instead, the exploitation theory seems more likely to account for male *Photinus* behavior. The argument here is that *on average*, the response of male *Photinus* to certain light flashes increases his fitness, even though one of the costs of responding is the chance that he will be devoured by an exploitative *Photuris* signaler. Males that avoided these deceptive signals might live longer, but they would probably ignore females of their own species as well, and would leave few or no descendants to carry on their cautious behavior.

This hypothesis highlights the definition of adaptation employed by most behavioral biologists. As noted before, an adaptation need not be perfect, but it must contribute more to fitness on average than other possible alternative traits. The male firefly that responds to the signals of a predatory female of another species possesses a mechanism of mate location that clearly is not perfect, but is better than those alternatives that improve a male's survival chances at the cost of making him unlikely to reproduce.

If this adaptationist hypothesis is correct, then deception by an illegitimate signaler should exploit a response that has clear adaptive value under most circumstances [215]. We have discussed this idea in earlier chapters when explaining why deception works for orchids that take advantage of the generally adaptive sex drive of male wasps. Likewise, the anglerfish exploits smaller fish that possess a generally adaptive eagerness to attack visual stimuli associated with their prey. The predatory anglerfish provides the stimuli in question by waving a thin rod that projects out of the front of its head. On the end of the rod is a pale tip that looks like something to eat (Figure 35), luring victims close enough so that the anglerfish can engulf them in its massive mouth [919]. Although the deceived fish pay a heavy price for their interest in the lure, a complete failure to respond to these stimuli would probably doom them to starvation.

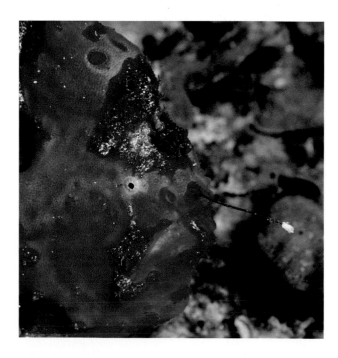

35   A deceptive signaler. This orange anglerfish has a pale tip on the thin appendage on the front of its head; by moving the lure, the predatory fish attracts prey within striking distance. Photograph by Ann Storrie.

Somewhat similarly, animals can be deceived into fleeing by a companion that produces an alarm call even though it has not actually seen an approaching predator. The trick works because animals that tend to ignore alarm signals will one day—probably sooner rather than later—find themselves out in the open with a hungry predator headed their way. When the penalty for not responding to a signal can be lethal, it pays to err on the side of caution [474]—which enables the occasional deceitful signaler to send its companions diving for cover so that it can catch a juicy insect that its departed companions had been pursuing [829]. So here too exploitation theory applies, with receivers sometimes fooled into behaving maladaptively when an illegitimate signaler activates a proximate mechanism that usually, but not always, produces an adaptive response.

Summary

1. A full understanding of the evolution of communication systems requires (1) information about the origins of the signal and the pattern of changes that took place in signalers and receivers and (2) information about the causal processes that made these changes occur.

2. The origins of signals and of responses to signals generally lie in small changes to behaviors and proximate sensory mechanisms that had a different function in the ancestors of the signal givers and receivers. The gradual accumulation of many such small adaptive changes through cumulative selection can result in the formation of traits that differ greatly from the original ancestral trait from which they are derived.

3. Each evolutionary change must be layered on that which has already evolved. In this sense, the traits already in place constrain or bias the pattern of evolution.

4. The sensory systems that have already evolved in a species make some signals much more likely to be detected by receivers than others. Mutant signalers that tap into preexisting sensory biases are more likely to cause evolutionary change in the communication system of their species than those whose signals are not readily perceived by other individuals.

5. Natural selection can cause signal changes to spread through the species only if both signalers and receivers derive net fitness benefits from their participation in the system. Applying the adaptationist approach to communication systems has resulted in the solution to some puzzling cases of signalers and receivers whose behavior seems to reduce, rather than increase, their fitness. In some instances, apparently self-sacrificing communication has been shown to actually raise an animal's chances of reproducing. In other cases, costly behavioral traits have been shown to occur because certain individuals (illegitimate signalers or illegitimate receivers) are able to exploit an otherwise adaptive communication system for their own benefit.

Discussion Questions

1. In studying the courtship behavior of the empid flies, E. L. Kessel was amazed to find a species, *Hilara sartor,* in which males gathered together to hover in swarms, carrying empty silken balloons [618]. Females flew to the swarm, approached a male, and received a balloon, which they held while mating occurred. In the overwhelming majority of fly species, including some other empid flies, courtship does not involve the transfer of any object from male to female. But in addition to (1) the empty balloon gifts of *H. sartor* and (2) the "no courtship gifts" of some empids, males of other species in the group courted by offering (3) gifts of an edible food item (a freshly killed dead insect), (4) gifts of a dried insect fragment wrapped in a silken covering, or (5) gifts of an edible prey

insect wrapped in silk. Construct a behavioral cladogram that minimizes the transitions from a possible ancestral pattern to the empty balloon gift-giving behavior. How would you test your hypothesis? (A fairly recent comparative analysis of empid courtship behavior appears in [271]; for a detailed study of a balloon-making species see [1007].)

2. Females of an African cichlid fish lay their eggs in depressions in the lake bottom made by males [1228]. The female picks up her large orange eggs in her mouth almost as quickly as she lays them. (She will protect the fertilized eggs and the fry, when they hatch, by holding them in her mouth.) As the female lays her eggs, the male who made the "nest" may move in front of her and spread his anal fin. The fin has a line of large orange spots. The female moves toward him and attempts to pick up the objects on the fin. As she does, the male releases his sperm, which are taken up in the female's mouth, where they fertilize her eggs. Use the theory of sensory exploitation to explain the evolutionary origins of the male's behavior. Was the first male to use this signal taking advantage of his mate?

3. Develop an honest signal hypothesis and a deception hypothesis to account for the fact that when copulating, males of some butterflies transfer a substance to their mates that makes these individuals sexually unappealing to other males [31]. What do you need to know to determine which explanation is correct?

4. In male cricket frogs, the larger the male, the lower the dominant frequency of his mate-attraction calls. But male cricket frogs will sometimes lower the dominant frequency of their calls when they hear a larger rival calling nearby [1180]. This ability may surprise you in light of our discussion of European toads. Why? William Wagner found that when males lowered their calls in this situation, they were significantly more likely to go on to attack the larger calling neighbor than when they did not do so. Does this finding help you evaluate the hypothesis that males who lower their calls are engaged in deceptive signaling?

5. Male fiddler crabs that lose their large claw regenerate one that is just as big, but weighs less and is therefore a less effective weapon. Even so, males that have their original, relatively heavy claw do not fight more often with males that have a regenerated claw [54]. How can you account for this puzzling response? Does the fact that more than 40 percent of the males in a population may have regenerated claws make it easier or more difficult to account for the failure of males with original claws to fight with these handicapped rivals? Why?

6. Among the modern stoneflies, species exist that swim on the water surface, that swim with wings raised for additional wind propulsion (skimming), that skim with all six legs on the surface, that skim with just four legs on the surface, that are hind-leg skimmers, and that jump from the water and fly freely. The existing species suggest an obvious historical sequence that goes from swimming to fly-

ing. What is it? A phylogeny of the stoneflies has been developed (see [1126] or http://www.bio.psu.edu/People/Faculty/Marden/skim.html). Use this phylogeny to test your historical scenario hypothesis. What conclusion is justified?

Suggested Reading

Animal communication has been reviewed by Jack Bradbury and Sandra Vehrencamp [143]. The weird and wonderful communication system of the spotted hyena has been studied independently by two teams of researchers [342, 401]. Cumulative selection is discussed by Richard Dawkins in his great book, *The Blind Watchmaker* [304]. Bernd Heinrich's *Ravens in Winter* is imbued with adaptationist thinking about communication [507]. For an adaptationist analysis of manipulation and deceit by communicating animals, read [305]; for the classic paper on honest signaling, see [1278].

11 *The Evolution of Reproductive Behavior*

I was thrilled when I first saw a male satin bowerbird fly down to his bower, which looked like the start of a child's miniature playhouse built with twigs. Had I been prepared to wait many hours near the bower, I might also have seen a female satin bowerbird come to the spot. Her arrival would have caused the male to bound about wildly while singing a bizarre medley of buzzes, screeches, and imitations of other bird songs. Yet despite the vigor of the male's courtship display, odds are that the female would have watched for a bit and then left without mating [128]. When a female does decide to mate with a male, she enters his bower and copulates quickly before flying away. She will have no further contact with her mate but will incubate her eggs and rear the young herself. The male bowerbird instead stays at or near his bower for most of the 8-month breeding season, ready to court any female that comes to inspect his handiwork. A few males with especially well decorated bowers get to copulate with more than one female (the record is 33 matings), but many hard-working males never mate, not even once, during a whole year [127].

Bowerbird behavior illustrates some fundamental questions about reproductive behavior that this chapter

*◀ **Males of this moth** have an astonishing inflatable apparatus at the tip of the abdomen from which they dispense a sex pheromone. Females of this species analyze the pheromonal message when making decisions about which male to choose as a mating partner. Photograph by Michael Boppré*

1 Bowerbird courtship. A male satin bowerbird uses a yellow flower in his courtship display. The female has entered his bower, a signal of her sexual receptivity. Photograph by Bert and Babs Wells.

will explore. Why are the reproductive roles of males and females so different? Why are male bowerbirds the ones that do the courting? Why are female bowerbirds so choosy when it comes to picking a mate? The patterns are widespread, and biologists ever since Darwin have tried to provide an evolutionary explanation for them.

The Evolution of Differences in Sex Roles

Female satin bowerbirds appear to reward males with well-decorated bowers (Figure 1). The importance of the bower for sexually motivated females raises an obvious evolutionary question: How did bower building originate and change over time [666, 667]? Bowerbirds are the only bird species that build display structures; indeed, some species in this group construct amazing "playhouses" that make the satin bowerbird's twig bower look positively amateurish (Figure 2, top). All 17 bower-building species have apparently evolved from a single bower-building ancestor of some sort (Figure 3). The current bower-building species can be divided into two groups, the avenue-bower builders (like the satin bowerbird) and the maypole-bower builders (like *Amblyornis inornatus* in Figure 2, bottom). An ancestral bower-building species split into these two lineages long ago, judging from molecular comparisons among the species. Within each lineage, bower evolution has been rapid and great, with large differences even among closely related species, especially among the maypole builders. In fact, two geographically separated populations of the same species, *A. inornatus,* build very different maypole bowers, even though the birds in the two areas are genetically very similar [1164]. The changes among species have been so extensive that the bowers of close relatives lack the kinds of shared features that would help us establish what the first bower looked like and what sequence of changes has taken place during the evolution of bower building.

2   Different bowers in different populations of the same bowerbird species. (Top) The huge playhouse bower of one population of the bowerbird *Amblyornis inornatus*. (Bottom) The same species builds a quite different maypole bower in another location. Photographs by Will Betz and Adrian Forsyth.

Why do the different species of bowerbirds currently build such distinctive bowers? One possibility is that females in different populations evolved different preferences for certain bower forms and ornaments, which then favored those males able to produce the preferred attributes. Indeed, females of the two populations of *A. inornatus* do not find the same bower characteristics attrac-

3 Evolutionary relationships among 14 of the 19 species of bowerbirds based on similarities in their mitochondrial cytochrome *b* gene. The icons on the right represent the shape of each species' bower. Note that two bowerbird species do not build bowers, having retained the ancestral trait of extinct species X. Extinct species Y presumably built a simple avenue bower or a cleared court; this species gave rise to all the current bower-building bowerbirds. After Kusmierski et al. [667].

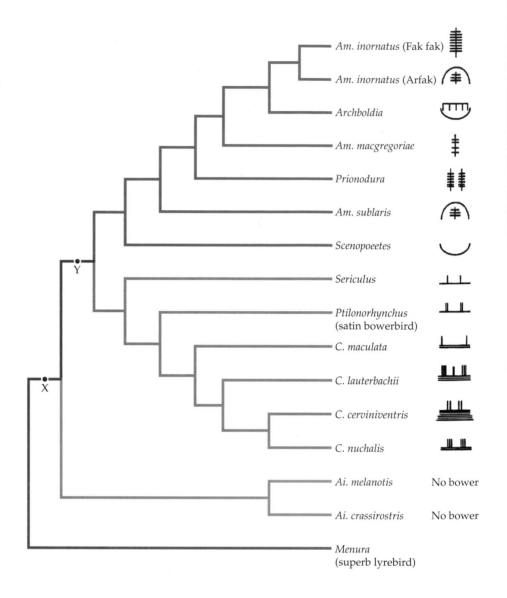

tive, but instead respond more positively to the kinds of bowers that their males currently produce [1164]. All of which raises a question: Why have female bowerbirds been able to dictate the evolution of male courtship behavior, and indeed, why is it that males of most animal species must persuade discriminating females to mate with them, rather than the other way around?

The reason why the sex roles of males and females of most species are so different almost certainly is related to the following fact of life: males produce sperm and females produce eggs. In all sexual species, males are (by definition) those individuals that produce sperm, the smaller gametes, usually no more than a set of genes in a package just large enough to contain the energy needed to drive them to an egg. Even in species in which males produce relatively gigantic sperm—such as a fruit fly whose males make sperm that are (when uncoiled) ten times longer than their bodies [924]—the mass of an egg is still vastly greater than that of a sperm. Indeed, females, are (by definition) the sex that produces eggs, the larger gametes, which in birds and most other animals are enormous relative to sperm (Figure 4). A single bird egg may constitute 15–20 percent of the female's body mass, and some go as high as 30 percent [671]. By way of contrast, a male splendid fairy-wren, a very small bird indeed, may have as many as 8 billion sperm in his testes at any given

4 Male and female gametes differ greatly in size. A hamster sperm fertilizing a hamster egg (magnified 4000 times) illustrates the trivial contribution of materials to the zygote by the male. Photograph by David M. Phillips.

moment, while his mate will lay no more than six eggs per nesting attempt [1160]. The same pattern applies to coho salmon, whose males shower about 100 billion sperm on a typical batch of 3500 eggs, according to a colleague of mine, Bob Montgomerie. Among mammals, the same correlation between gamete size and number of gametes produced holds true. A woman has only a few hundred cells that can ever develop into mature eggs [280]. In contrast, a single man could theoretically fertilize all the eggs of all the women in the world, given that each ejaculate contains on the order of 350 million sperm.

The point is that because sperm are relatively cheap to make, male animals almost always have gametes sufficient to inseminate many females. Therefore, a male's contribution of genes to the next generation depends on how many sexual partners he has. The more mates, the more eggs fertilized, the more descendants produced, and the greater the male's genetic success relative to less sexually active individuals. In contrast, the reproductive success of females is often limited by the number of eggs they can manufacture, and since these gametes are large, and often costly to produce, females have to take time out to secure the resources to make them. Moreover, females often expend additional time and energy caring for their offspring after the eggs have been fertilized. Thus, in the satin bowerbird, males are ready to mate on every day of the breeding season, whereas, for much of this period, females are likely to be occupied with foraging, or nest building, or caring for offspring. The pool of sexually receptive female bowerbirds is therefore smaller than the population of sexually eager males. In other words, because the *potential* rate of reproduction differs between the sexes [237], the **operational sex ratio** (the ratio of sexually receptive males to receptive females at any one time) [362] is heavily male-biased. Thus, each male must compete with many rivals for a limited number of receptive females, each of which donates more material resources to an offspring than her partner.

All of those expenditures of time, energy, and risks by a parent on one offspring that reduce the chance that the parent will have more offspring in the future are referred to as **parental investment** [1153]. Bob Trivers invented this term to emphasize the trade-offs for parents that make contributions to offspring. On the plus side, parental investment may increase the probability that an existing offspring will survive to reproduce. But this fitness benefit may come at the cost of the parent's ability to generate additional offspring down the road. In the animal world, the general rule is that females make a larger total parental investment per offspring than males. As we have seen, the differences start at the level of the gametes. When a female builds an egg, the many nutrients and cell organelles that she donates to that gamete constitute a much larger parental investment than is provided by the male to the single tiny sperm that will fertilize that egg. These differences between the sexes in gametic parental investment are often amplified by other kinds of parental investments, which include the food or protection parents sometimes give to their progeny (Figure 5) [1153] In mammals, for example, females nourish their embryos, then care for the babies

5 Parental investment takes many forms. (Clockwise from top left) A male frog carries his tadpole offspring on his back. A male crocodile transports babies to water just after they have hatched. A female eared grebe protecting her young by letting them ride on her back. A female cicada killer wasp drags a cicada she has paralyzed with a sting to a burrow she dug, where her offspring will feed upon the unlucky cicada. Photographs by Roy McDiarmid; Jeff Lang; Bruce Lyon; and the author.

when they are born and give them milk; the typical male mammal, in contrast, impregnates a female and departs, never to see or interact with his offspring.

But why is it that the sexes differ in the resources they donate to a fertilized egg? Geoffrey Parker and his colleagues have argued that the evolution of the two kinds of gametes, and thus the two sexes, stemmed from divergent selection that favored either (1) individuals whose gametes were good at fertilizing other gametes because of their size and mobility (a number-of-offspring strategy) or (2) individuals whose gametes were good at developing after being fertilized (a parental investment strategy) [889]. Sperm, small and highly active, are superbly designed to move rapidly toward an egg when released by a male. Bluegill sperm, for example, dart along at up to four sperm lengths per second. The sedentary egg, large and packed with nutrients, contains materials useful for the development of the zygote following fertilization. No single kind of gamete could be equally good at both tasks, which led to the separate evolution of effective fertilization devices (the sperm of males) that compete to unite with effective development devices (the eggs of females).

Testing the Evolutionary Theory of Sex Differences

We have suggested that males usually compete for mates because of inequalities between the sexes in parental investment, which lead to differences in potential reproductive rates and a male-biased operational sex ratio (Figure 6). These differences should favor males that compete aggressively with rivals for mates and copulate at every opportunity (Figure 7). In contrast, female fitness will rarely be advanced by receiving sperm from as many males as possible, so selection should favor instead those females that avoid the costs of additional matings after having chosen partners who have the most to offer in the way of resources or genes. Recall that female bowerbirds manage to rear

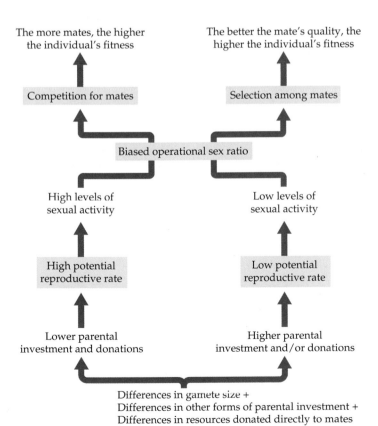

The more mates, the higher
the individual's fitness

The better the mate's quality, the
higher the individual's fitness

Competition for mates

Selection among mates

Biased operational sex ratio

High levels of
sexual activity

Low levels of
sexual activity

High potential
reproductive rate

Low potential
reproductive rate

Lower parental
investment and donations

Higher parental
investment and/or donations

Differences in gamete size +
Differences in other forms of parental investment +
Differences in resources donated directly to mates

6 Differences between the sexes in sexual behavior may arise from fundamental differences in parental investment that affect the rate at which individuals can produce offspring. The sex that can potentially leave more descendants gains from high levels of sexual activity, whereas the other sex does not. An inequality in the number of receptive individuals of the two sexes leads to competition for mates within one sex, while the opposite sex can be choosy about mating.

7 **Male copulatory drive is intense.** Scientists were able to weigh this male elephant seal without immobilizing him with a dart gun. Instead, they lured him to the area by playing a tape recording of the call of a copulating female elephant seal [315]. Once there, the animal saw a moving dummy female seal composed of urethane foam covered with fiberglass. The male climbed onto the scale while pursuing the moving model. Photograph by Chip Deutsch.

a brood after copulating just once, whereas male bowerbirds attempt to mate many times per breeding season.

We can test this theory by finding unusual cases in which males make the larger parental investment or have the lower potential reproductive rate. In species of this sort, males may make contributions other than sperm toward the welfare of their offspring or their mates because if they do not, females may refuse to accept or use their sperm to fertilize their eggs. For species in which males make the larger parental investment, the operational sex ratio should be biased toward females, leading to female competition for mates and careful mate choice by males—in other words, a **sex role reversal.** Such a reversal occurs in the mating swarms of certain empid flies, in which the operational sex ratio is heavily female-biased because most males are off hunting for insect prey to bring to the swarm as a mating inducement [1114]. When a male enters the swarm bearing his nuptial gift, he may find females advertising themselves with (depending on the species) unusually large and patterned wings or bizarre inflatable sacs on their abdomens (Figure 8) [412]. The male gets to choose from among many ornamented partners available to him.

Likewise, males of some fish species offer their mates something of real value—namely, a brood pouch into which the female can place her eggs. In these species, the potential reproductive capacity of males may be less than that of females [1169] (but see [765]). In the pipefish *Syngnathus typhle,* "pregnant" males provide nutrients and oxygen to a clutch of fertilized eggs for several weeks, during which time the average female produces enough eggs to fill the pouches of two males. Since the sex ratio is 1:1, male pouch space is in

8   Female advertisement of body size in the long-tailed dance fly, *Rhamphomyia longicauda*. When in a swarm, awaiting the arrival of a gift-bearing male, the female inflates her abdomen and holds her dark, hairy legs around her body, making herself appear as large as possible in the group of circling females. Photograph by David Funk, from [412].

short supply, and as expected, females compete for the opportunity to donate eggs to parental males. Males with free pouch space actively choose among mates, discriminating against small, plain females in favor of large, ornamented ones, which can provide selective males with larger clutches of eggs to fertilize [97, 990].

Another example of a species in which maximum reproductive capacity may be greater for some females than for males is the flightless Mormon cricket, which despite its common name has no religious affiliation and is a katydid, not a cricket. When male Mormon crickets mate, they transfer to their partners another kind of nuptial gift, an enormous edible spermatophore (Figure 9) [470]. Given that the spermatophore constitutes 25 percent of the male's body mass, most male Mormon crickets probably cannot mate more than once. In contrast, females may be able to produce several clutches of eggs, provided they can induce several males to mate with them.

Sometimes bands of Mormon crickets march across the countryside, eating farmers' crops and mating as they go. In these high density populations, when a male begins to stridulate from a perch, announcing his readiness to mate, females come quickly and jostle for the opportunity to mount him, the prelude to insertion of the male's genitalia and transfer of a spermatophore. In addition to this female sex role reversal, males often behave in a discriminating man-

9   An edible nuptial gift. A Mormon cricket female carries a large white spermatophore that her partner attached to her during copulation. Photograph by Darryl Gwynne.

10 Mate choice by males. Mormon cricket males prefer to mate with heavier females. As a result of their choosiness, they gain an average of 18 extra eggs to fertilize. After Gwynne [470].

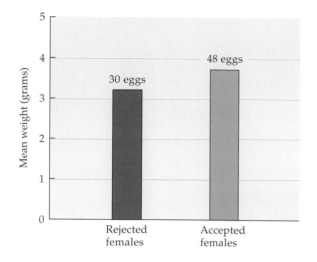

ner, refusing to transfer a spermatophore to lightweight females in favor of heavier, more fecund partners (Figure 10). A choosy male that rejects a 3.2-gram female in favor of one weighing in at 3.5 grams fertilizes about 50 percent more eggs as a result [471]. So here too the sex with the higher reproductive potential competes for access to members of the other sex.

Our theory also predicts that if the relative reproductive capacities of the two sexes within a species were to change, their reproductive roles should change as well. Data relevant to this point come from a study of a thin Australian katydid (Figure 11) whose food supply varies greatly over the course of the breeding season. When these katydids are limited to pollen-poor kangaroo paw flowers, the male's large spermatophore is both difficult to produce and valuable to

11 A costly gift. A female *Kawanaphila* katydid is consuming a nutritious gelatinous spermatophore that she has received from her mate. The female's two thin antennae and shorter, thicker ovipositor point downward. Photograph by Darryl Gwynne.

TABLE 1 Effect of the relative reproductive rates of males and females on sex roles in
***Kawanaphila* katydids**

Treatment[a]	Mean number of		Proportion of interactions with	
	Calling males	Matings per female	Male choice of mates	Female competition for mates
Food scarce: Males produce spermatophores slowly	0.4	1.3	0.4	0.2
Food abundant: Males produce spermatophores quickly	6.6	0.7	<0.1	0.0

[a]In each treatment, four cages containing 24 katydids each were observed for equal periods of time.
Source: Gwynne and Simmons [471]

females as a nutrient gift. Under these conditions, males are choosy about their mates, and females fight with one another for access to receptive, spermatophore-offering males. But when pollen-rich grass trees come into flower, the maximum reproductive rate for males increases (they can produce spermatophores more rapidly), whereas the females' reproductive rate is limited by the speed with which they can turn pollen into eggs. At this time, sex roles switch to the more typical pattern, with males competing for access to females and females rejecting some males. This same pattern of changes has been reproduced in the laboratory by controlling the food supplied to captive katydids (Table 1) [471].

Sexual Selection and Competition for Mates

Because sex roles differ, males and females affect each other's reproductive chances, creating a form of selection called **sexual selection.** Like almost everything else important in evolutionary theory, sexual selection was first recognized by Charles Darwin, who defined it as "the advantage which certain individuals have over others of the same sex and species, in exclusive relation to reproduction" [287]. Darwin thought natural selection resulted in the evolution of survival-enhancing traits, which promote the reproductive success of individuals. He devised sexual selection theory to account for the evolution of survival-*decreasing* traits, which nevertheless increase individual reproductive success by enabling individuals to acquire mates in competition with others (Figure 12). This special form of selection could help explain the bizarre sexual behavior of bowerbirds, whose males surely shorten their lives by spending their time and energy building bowers, gathering decorations, displaying to females, and fighting with other males. The survival costs of these traits, however, might be offset by the added reproductive success of aggressive males with elaborate bowers and dramatic displays.

Note that natural selection and sexual selection operate in fundamentally the same way. Both forms of selection can lead to evolutionary change if individuals vary in their hereditary attributes in ways that translate into differences in the number of surviving offspring produced. As noted already for bowerbirds, individual males vary in the quality of the bowers they build, which contributes to the great differences among males in their mating success. If these differences have a hereditary basis, the genetic makeup of the species will change over time. Although sexual selection is a subcategory of natural selection, the distinction Darwin made between the two processes usefully focuses attention on the selective consequences of sexual interactions *within* a species.

12 Sexually selected "ornaments" of males. Darwin believed that sexual selection via female choice was responsible for the evolution of elaborate plumage and remarkable displays in male birds such as the quetzal (left) and the sage grouse (top right). Darwin argued that the strange horns and snouts of certain beetles (bottom right) also arose via female choice, although males actually use these structures primarily as weapons when fighting for mates. Photographs by Bruce Lyon; Marc Dantzker; and the author.

Sexual selection consists of two components, one arising when the members of one sex compete for mates (often called intrasexual selection) and the other occurring when members of the choosier sex determine which members of the other sex will have a chance to mate (often called intersexual selection). The competition-for-mates component of sexual selection can be a powerful force for evolutionary change, as evidenced by the injurious, even life-threatening, battles among males that determine who gets to mate. For example, male *Cardiocondyla* ants use "their mandibles to pierce or crush the soft cuticle of young males, to sever their antennae or legs, to decapitate them or to cut their bodies in two" [510]. These ants have evolved large and powerful jaws to eliminate

13 **Males of many species fight,** using whatever weapons they have at their disposal. Here two giraffes slam each other with their heavy necks and clubbed heads. Photograph by Gregory Dimijian.

newly adult rivals for the several dozen virgin females that may emerge in their nest [508]. In vertebrates, lethal combat is not unknown, but males usually fight just long enough to determine a winner, with the loser then fleeing while still able to do so (see page 308). Living and extinct species [627] have fought with bites, kicks, antler locks, jaw grapples, head butts, even neck slams in the case of giraffes (Figure 13) [1064].

When two male elephant seals, each weighing 2000 kilograms, collide violently on a beach, the battle can decide which male remains in the company of dozens of receptive females and which male lumbers away to spend the rest of the breeding season elsewhere. From his records of winners and losers in a population of southern elephant seals on South Georgia Island in the Atlantic Ocean [776], T. S. McCann found that he could identify the dominant male, an individual who had from 14 to 157 encounters with each of nine other males, and won them all. The number two male elicited submission from all but the top male, and so on down a linear dominance hierarchy from top to bottom. If the ultimate significance of fighting is to acquire mates, then rank in a dominance hierarchy should correlate with male copulatory success over the breeding season. McCann's observations supported this prediction (Figure 14), as have similar findings for many other animals [320, 1185]. Moreover, recent genetic studies have confirmed that copulatory success in the southern elephant seal is tightly linked with actual paternity, so that alpha males do indeed achieve high reproductive success along with their social success [523].

14 Male dominance and mating success. High-ranking males of the southern elephant seal copulate many more times than low-ranking individuals. After McCann [776].

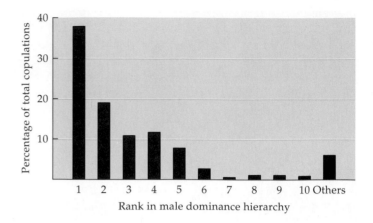

Dominance can also be a valuable commodity for females in those species in which females compete for social status, as is true for spotted hyenas (p. 285). Likewise, in African wild dogs, alpha females mate to the near exclusion of subordinates and produce at least three-quarters of all the litters [256]. But males do the bulk of the fighting in the animal kingdom, and when animals fight, larger individuals tend to have a competitive advantage that translates into greater mating success [34]. You might think, therefore, that sexual selection for the ability to dominate others would always favor large body size in males. Being large is not an unmitigated blessing, however, since it takes time and energy to grow into a big bruiser. Therefore, we can predict that costly investments in the growth and maintenance of large bodies will be produced by sexual selection only when truly exceptional reproductive rewards go to the largest individuals. If so, a strong **sexual dimorphism** in body size (in which the sexes differ greatly in length or weight) should occur only in those species in which very large males can monopolize many receptive females, while differences in male and female size should be small or absent in those species in which larger males cannot gain access to disproportionate numbers of females per breeding season [20].

This prediction is supported by data on body size in seals and their relatives (Figure 15). Thus, a massive male elephant seal can potentially mate with as

Harbor seal

Elephant seal

15 Sexual dimorphism is correlated with opportunities for extraordinary mating success in seals and other related mammals. In species in which some males can monopolize large numbers of females, sexual selection has resulted in extreme sexual dimorphism. In monogamous species, males and females weigh about the same. After Alexander et al. [20].

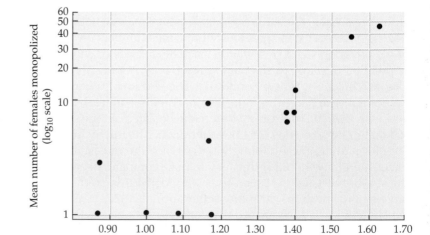

many as 100 (much smaller) females in one breeding season. In contrast, males and females of **monogamous** species, such as the harbor seal, whose males and females have only one mating partner, are more or less equal in length and weight. Essentially the same pattern applies to a group of birds that includes the gulls and sandpipers. The males of species that are potentially **polygynous** (those in which a male may mate with more than one female) generally have greater body mass and longer wings than females, whereas in **polyandrous** species (those in which a female may mate with more than one male) larger females pair off with several smaller males [1120].

Social Dominance and Male Fitness

Baboons are another example of a species in which males are much larger than females, and here too adult males compete violently for social dominance. Carlos Drews found that males were bitten about once every 6 weeks, an injury rate nearly four times greater than that of females [338]. In this species, however, the reproductive value of these costly struggles for dominance is problematic, since many researchers have found that high position in a dominance hierarchy does not necessarily result in extraordinary copulatory success. For example, when Glen Hausfater counted copulations in a troop of baboons with a clear dominance hierarchy, he found that males of low and high status were equally likely to copulate with females [499]. But Hausfater had initially assumed that the more times a male mated, the greater his chances of fathering offspring. If in reality females fertilize their eggs with sperm received during only a few days of their estrous cycle, shortly after they have ovulated, then males that copulate with them outside this time cannot produce offspring. It could be that high-ranking males are more likely to copulate with estrous females when it counts—that is, when their eggs can be fertilized. When Hausfater tested this possibility, he found that dominant males did outcopulate subordinates during the days of highest female fertility [499], and the same is true for other primates, including chimpanzees [770].

Although this result suggests that dominance is correlated with the number of *effective* copulations, the statistically significant relationship between rank and effective copulations in baboons arises only because young adult males have very low dominance and achieve few effective copulations [93]. For the fully adult male members of a baboon troop, the ability to defeat rivals in direct physical confrontations is only weakly related to the number of offspring produced. Indeed, among primates generally, the relationship between rank and reproductive success may be significant [247, 358], but it is not overwhelming, because socially subordinate males use alternative mating tactics to compete against stronger, larger rivals. Among baboons, for example, there are several paths to sexual success in addition to securing high status. Males can and do develop friendships with particular females, relationships that do not depend entirely on physical dominance, but rather on the willingness of a male to protect a given female's offspring, probably from other infanticidal males [883]. Once a male, even a moderately subordinate one, has demonstrated that he is willing and able to provide protection for a female and her infant, that female may seek him out when she enters estrus again [1106].

Male baboons also form friendships with other males. Through these alliances, they can sometimes collectively confront a stronger rival that has acquired a partner, forcing him to give her up, even though he is socially dominant in one-on-one encounters. Thus, for example, in one troop of yellow baboons that contained eight adult males, three low-ranking males (fifth through seventh in the hierarchy) regularly formed coalitions to oppose a single higher-ranking male who was in consort with a female. In 18 of 28 cases, the single male was forced to relinquish the female to the threatening gang of subordinates [842].

Alternative Mating Tactics

Baboons are far from the only species in which some males use special tactics in order to avoid getting thrashed by a stronger or more experienced opponent [34, 462]. Males of the gray seal, for example, compete to monopolize females on breeding beaches in much the same way as elephant seals, with a few socially dominant individuals gaining the large majority of matings observed by terrestrial behavioral ecologists. However, genetic tests of paternity revealed that more than half of the offspring were sired by males other than the dominant males that were so conspicuously successful in mating on land [1244]. Apparently, males swimming around the breeding colony meet females in the water and mate with them there, fathering a considerable proportion of the young produced in any one year.

The gray seals that intercept females at sea do not have to fight with powerful rivals on land. Likewise, small males of a Costa Rican rove beetle also adopt nonconfrontational tactics when they encounter larger rivals. The males perch on a pile of mammalian dung, which attracts flies as well as female beetles that come to feed on them (Figure 16). Instead of fighting for control of the dung, a smaller male may turn around and present the tip of his abdomen to his larger enemy, which may tap the abdomen of the little beetle as if he were a female. The female-mimicking male keeps moving slowly about the area, never permitting the larger courting male to consummate their affair. Occasionally, in the course of his meanderings, the courted male encounters a female, at which point he may court and copulate with her literally under the nose of his competitor [392].

Another alternative tactic suitable for smaller, competitively disadvantaged males has evolved in the famous marine iguana of the Galápagos Islands. These animals often live in dense aggregations, so when a small male runs over to mount a female, a larger male may arrive almost simultaneously to remove him forcibly from the object of his desire (Figure 17). Because it takes 3 minutes for a male iguana to ejaculate, one would think that small males promptly set packing would have no chance to inseminate their partners. However, the little iguanas have a solution to this problem, which involves having already released viable sperm into the cloaca *prior* to any copulation attempt. When, in the course of mounting a female, the small male everts his penis from the cloaca (at the

16   A conditional strategy in action. A male rove beetle perches on dung, which he will defend in order to mate with sexually receptive females drawn to this resource. If challenged by a much larger male, however, the territorial male can switch to female mimicry as an alternative mating tactic. Photograph by the author.

17 Sexual interference in the marine iguana. Two small males are mounted upon and trying to copulate with the same female; a much larger male is trying to pull both of the smaller iguanas away from the female. Photograph by Martin Wikelski, from Wikelski [1234].

base of his tail) to insert it into the female's cloaca, these "old" sperm begin to flow down a groove in his penis, so that even if he cannot ejaculate again while copulating, the small iguana can inseminate the female and father some of her offspring [1234].

More common alternative mating tactics (Figure 18) include the nonaggressive "satellite behavior" of males that simply remain near other males while their dominant companions call or display to females [188, 901], or defend a

18 Satellite male mating tactics. (Clockwise from top left) A satellite male Great Plains toad crouches by a caller, waiting to intercept females attracted by his signals. Six subordinate male bighorn sheep trail after a dominant male, who stands between them and the female with whom he will mate at intervals. Two satellite male horseshoe crabs grasp a female from either side; the female's primary mate is positioned directly behind her. Photographs by Brian Sullivan; Jack Hogg; Kim Abplanalp, courtesy of Jane Brockmann.

resource attractive to females [632], or defend females themselves [525]. In the horseshoe crab, for example, some males patrol the water off the beach, finding and grasping females heading toward the shore to lay their eggs. Other males are Johnny-come-latelies that crowd around a paired couple after they reach the beach [159]. As it turns out, the attached male fertilizes at least 10 percent more eggs than a single competing Johnny-come-lately [157, 158]. If being an attached male yields more offspring, why do any males practice the less successful satellite option? This question can be asked about almost any alternative male mating tactic, since these maneuvers typically result in lower fitness gains than the activities of dominant males.

A Conditional Strategy with Alternative Mating Tactics

When superior and inferior tactics coexist within a species, a possible explanation is that the tactic with lowered payoffs persists because it a component of a conditional strategy, as we discussed earlier (p. 264) when dealing with migrant and resident European blackbirds. In a conditional strategy, the "inferior" tactic enables socially disadvantaged individuals to gain more fitness than they would if they were to try to use the tactics of their dominant opponents. Let's apply the conditional strategy hypothesis to alternative mating tactics in a *Panorpa* scorpionfly in which (1) some males aggressively defend dead insects, a food resource highly attractive to receptive females, while (2) other males secrete saliva on leaves and wait for occasional females to come and consume this nutritional gift, and (3) still others offer females nothing at all, but instead grab them and force them to copulate (Figure 19) [1130]. In experiments with caged groups of ten male and ten female *Panorpa,* Randy Thornhill showed that the largest males monopolized the two dead crickets in the cage, which gave these males easy access to females and about six copulations on average per trial. Medium-sized males could not outmuscle the largest scorpionflies in the competition for the crickets, so they usually produced salivary gifts to attract females, but gained only about two copulations each. Small males were unable to claim crickets, and they could not generate salivary presents either, so these scorpionflies forced females to mate, but averaged only about one copulation per trial.

Since the three different tactics clearly yielded different fitness returns, each behavior could not be a separate hereditary strategy. (Remember that if the dif-

19 **A male *Panorpa* scorpionfly** with its strange scorpion-like abdomen tip, which it can use to grasp females in a prelude to forced copulation. Photograph by Jim Lloyd.

ferences between males using the three options were hereditary, then the one strategy that yielded the highest net benefits would inevitably replace the other two.) So Thornhill proposed instead that the three options were tactics specified by a single conditional strategy possessed by all the males of the species he was studying. This hypothesis predicts (1) that the differences between the phenotypes are environmentally caused and (2) that males should switch to a tactic yielding higher reproductive success if the social conditions they experience make the switch possible.

To test this prediction, Thornhill removed the large males that had been defending the dead crickets. When this change occurred, some males promptly abandoned their salivary mounds and claimed the more valuable crickets. Other males that had been relying on forced copulations hurried over to stand by the abandoned secretions of other males. Thus, male *Panorpa* can adopt whichever of the three tactics gives them the highest possible chance of mating, given their current competitive status. These results clinch the case in favor of a conditional strategy as the explanation for the coexistence of three mating tactics in *Panorpa* scorpionflies [1130].

Three Distinct Strategies: Three Mating Tactics

Almost all cases of alternative mating tactics represent examples of conditional strategies [462]. However, the male offspring of a sandpiper called the ruff generally behave like their territorial or satellite fathers [683], showing that two different, hereditarily distinct reproductive strategies can coexist within a species under some conditions (Figure 20) [1230]. Another unusual case of this sort is supplied by a marine isopod, *Paracerceis sculpta*, which vaguely resembles the more familiar terrestrial sowbugs and pillbugs that live in moist debris in suburban backyards. The marine isopod resides in sponges found in the intertidal zone of the Gulf of California. If you were to open up a sufficient number of sponges, you would find females, which all look more or less alike, and males, which come in three dramatically different sizes: large (alpha), medium (beta), and small (gamma) (Figure 21), each with its own behavior.

The big alpha males attempt to exclude other males from interior cavities of sponges that have one or more females living in them. If a resident alpha encounters another alpha male in a sponge, a battle ensues that may last hours before one male gives way. Should an alpha male find a tiny gamma male, how-

20 Two reproductive strategies in the ruff. The reddish male (left) with an erect ruff of feathers is an aggressive territorial resident at the display grounds where females come to choose a mate. The male with the white ruff is a nonaggressive, nonterritorial satellite male who displays in a portion of the resident's territory. (The third male is a "floater" with the potential capacity to become a territorial resident.) Photograph by Ola Jennersten, courtesy of Fredrik Widemo.

21   Three different forms of the sponge isopod: the large alpha male, the female-sized beta male, and the tiny gamma male. Each type not only has a different size and shape, but also uses a different tactic to acquire mates. After Shuster [1053].

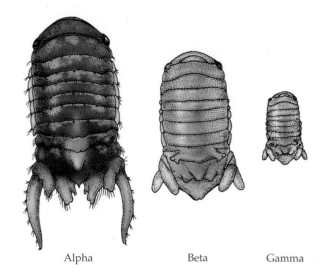

Alpha   Beta   Gamma

ever, the larger isopod simply grasps the smaller one and throws him out of the sponge. Gammas avoid alphas whenever possible, lurking about and trying to sneak matings from the females living with an alpha male [1052]. When an alpha and a medium-sized beta male meet inside a sponge cavity, the beta behaves like a female, and the male courts his rival ineffectually. Through female mimicry, the female-sized beta males coexist with much larger and stronger rivals and thereby gain access to the real females that the alpha male would otherwise monopolize, just as is true of the rove beetles we discussed earlier.

In this species, therefore, we have three different types of males, and one type has the potential to dominate others in male–male competition. If the three types represent three distinct strategies, then (1) the differences between them should be traceable to genetic differences and (2) the mean reproductive success of the three types should be equal. If, however, alpha, beta, and gamma males use three different tactics resulting from the same conditional strategy, then (1) the behavioral differences should be induced by different environmental conditions, not different genes, and (2) the mean reproductive success of males using the alternative tactics need not be equal.

Steve Shuster and his co-workers collected the information needed to check the predictions derived from these two hypotheses [1054, 1055]. First, they showed that the size and behavioral differences between the three types of male isopods are the hereditary result of differences in a single gene represented by three alleles. Second, they demonstrated that the three types of males probably have equal fitnesses by measuring their reproductive success in the laboratory. There they placed different numbers and types of males with different numbers of females in artificial sponges. The males used in this experiment had special genetic markers, distinctive characteristics that could be passed on to their offspring, enabling Shuster to identify which male had fathered which of the baby isopods that each female eventually produced. At the end of this experiment, Shuster found that the reproductive success of a male depended on how many females and rival males lived with him in a sponge. For example, when an alpha male and a beta male lived together with one female, the alpha isopod fathered most of the offspring. But when this male combo occupied a sponge with several females, the alpha male could not control them all, and the beta male outdid his rival, siring 60 percent of the resulting progeny. In still other combinations, gamma males outreproduced the others.

Shuster and Michael Wade then returned to the Gulf of California to collect a large random sample of sponges, each one of which they opened to count the isopods within [1055]. With these data, they could estimate the reproductive success of 555 males, given the laboratory results on male reproductive success in various mixes of cohabiting individuals. When the mathematical dust had settled, they estimated that alpha males had mated with 1.51 females on average, while betas checked in at 1.35 and gammas at 1.37 mates. Since these means were not significantly different, statistically speaking, Shuster and Wade concluded that the three genetically different types of males had essentially equal fitnesses in nature. The requirements for a three-distinct-strategies explanation had been met.

Sexual Selection and Sperm Competition

In most research on male–male reproductive competition, a male's fitness is measured by how many females he inseminates, since counting copulations is much easier than identifying the number of young actually fathered by a given male. But in many animal species, fertile females mate with more than one partner. When this happens, whose sperm fertilize the eggs? If some males' sperm have an advantage over those of other males, just counting a male's copulations will not measure his fitness accurately. And in many species, not all transferred sperm are equal in the sense of having an equal chance of fertilizing eggs [108]. Consider the sperm wars that occur within *Calopteryx maculata*, the black-winged damselfly of eastern North America [1177]. When a female flies to a stream to lay her eggs, she may visit and lay some eggs in two or three males' territories. The female's behavior creates competition among her partners to fertilize her eggs [888], and the resulting sexual selection has produced an extraordinary penis in males of this species.

To understand how the damselfly penis works, we need to describe the odd manner in which damselflies (and dragonflies) copulate. First, the male grasps the front of the female's thorax with specialized claspers at the tip of his abdomen. A receptive female then swings her abdomen under the male's body and places her genitalia over the male's sperm transfer device, which occupies a place on the underside of his abdomen near the thorax (Figure 22). The male damselfly then rhythmically pumps his abdomen up and down, during which time his spiky penis acts as a scrub brush (Figure 23), catching and drawing out any sperm already stored in the female's sperm storage organ. Jon Waage found that a copulating male *C. maculata* removes between 90 and 100 percent of any competing sperm before he releases his own gametes. Before interacting with a female, the male damselfly transfers sperm from his testes on the tip of his abdomen to a temporary storage chamber very near the penis. After

22 Copulation in the black-winged damselfly. The male (on the right) grasps the female with the tip of his abdomen; the female twists her abdomen forward to make contact with her partner's sperm-transferring organ. Photograph by the author.

23   Sperm competition in the black-winged damselfly. (Left) The male's penis has lateral horns and spines that enable him to scrub out a female's sperm storage organ before passing his own sperm to her. (Right) A close-up of a lateral horn reveals rival sperm caught in its spiny hairs. Photomicrographs by Jon Waage, from Waage [1177].

emptying the female's sperm storage organ, he lets his own sperm out of storage to enter the female's reproductive tract, where they remain for use when she fertilizes her eggs—unless she mates with another male before ovipositing, in which case his sperm will be removed by a rival male in turn [1177].

Although the male black-winged damselfly employs an unusually efficient mechanism for eliminating the sperm stored within his partner, this is not the only species in which males remove or destroy another male's stored sperm [108]—sometimes with the assistance of the female (Figure 24) [290]. Moreover, although sperm competition at first glance appears to be strictly the product of male–male competition, females of some species have ways of manipulating which of several males' sperm get to fertilize their eggs [349]. In a flour beetle, for example, when the same two males are paired with the same females, the second male's success in fathering offspring often varies greatly from female to female, showing that the females are not passive vessels within which males battle for egg fertilizations, but active players in the process [1252].

Mate Guarding

Competition among males for opportunities to fertilize a female's eggs has resulted in the evolution of mate guarding, a strategy in which a male attempts

24   Sperm competition in the dunnock involves female cooperation. A male pecks at the cloaca of his partner after finding another male near her; in response, she may eject a droplet of sperm-containing fluid. After Davies [290].

to prevent other males from gaining access to a potential or actual mate. By guarding a female, before or after copulating with her, the male improves the odds that his sperm will not have to compete with those of other males for the eggs of his partner. Guarding *after* insemination has occurred is sometimes achieved by sealing the female's genital opening with various secretions [326, 929], sometimes by keeping the female occupied after mating is finished [23], and sometimes by deceptively luring new suitors away from the female [385]. By far the most common postcopulatory guarding tactic, however, is simply to stay with a mate in order to repel other males should they dare approach her (Figure 25) [11].

The question is, does the time spent with an already inseminated female result in sufficient egg fertilizations to outweigh the costs of mate guarding, especially the loss of opportunities to seek out other females? Janis Dickinson measured this cost in the blue milkweed beetle, in which the male normally remains mounted on the female's back for some time after copulation. When Dickinson removed some males from their partners, about 25 percent of the separated males found new mates within 30 minutes. Thus, remaining mounted on a female after inseminating her carries a considerable cost for the guarding male, since he has a good chance of finding a new mate elsewhere if he would just leave his old one. On the other hand, nearly 50 percent of the females whose guarding partners were plucked from their backs also acquired new mates within 30 minutes. Since guarding males cannot easily be displaced from a female's back by rival males, they reduce the probability that an inseminated partner will mate again, giving their sperm a better chance to fertilize more of her eggs than otherwise. Dickinson calculated that if the last male to copulate with a female fertilizes even 40 percent of her eggs, he gains by giving up the search for new mates in order to guard a current one (Figure 26) [325].

In general, the *benefits* of mate guarding *increase* with the probability that unguarded females will mate again and use the sperm of later partners to fertilize their eggs. Therefore, we can predict that males should defend their mates only when there is a good chance that their partners, if left unguarded, would mate with another male and use his sperm to fertilize their eggs [84, 365]. Jan Komdeur and colleagues used an ingenious experiment to check this prediction for the Seychelles warbler. They placed a false egg in the nests of warbler pairs about 4 days before the female was due to deposit a real egg there. Females

25   Mate guarding occurs in many animals. (Left) A red male damselfly grasps his mate in the tandem position so that she cannot mate with another male. (Right) The male blueband goby, an Indonesian reef fish, accompanies his mate wherever she goes. Photographs by (left) the author and (right) Roger Steene.

26 **Adaptive mate guarding by the blue milkweed beetle.** The fitness of guarding males exceeds that of nonguarding individuals, provided that the last male to copulate with a female fertilizes a substantial proportion of her eggs. Guarding males are more likely than nonguarding males to gain the last male fertilization advantage. After Dickinson [325].

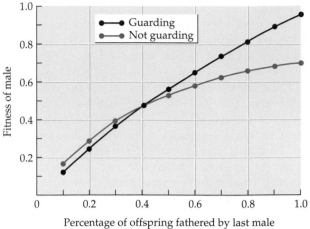

Blue milkweed beetle

of this species lay only a single egg per breeding attempt, after which the male devotes himself to protecting the egg against a local predator. Thus, when males saw the false egg, they immediately began guarding it by perching near the nest, leaving their mates unaccompanied. The fertile females attracted the attention of many neighboring males, some of whom succeeded in copulating with these unguarded individuals (Figure 27) [647]. This result demonstrates that mate guarding provides fitness benefits for male Seychelles warblers.

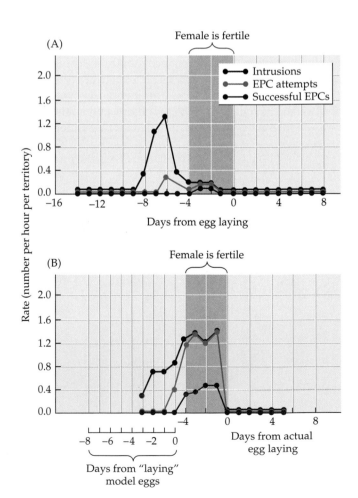

27 **Adaptive mate guarding by the Seychelles warbler.** The graphs show the rate of intrusions and extra-pair copulations (EPCs) by males other than the female's social mate in relation to the female's fertile period (shaded area). (A) Control pairs, in which the female's mate was present throughout her fertile period. (B) Pairs in which the female's mate was experimentally induced to leave her unaccompanied by the placement of a model egg in the nest. After Komdeur et al. [647].

Another way to test the hypothesis that mate guarding increases the guarding male's chance of paternity is to examine the following prediction: If males can be experimentally induced to *increase* the time spent in close association with their mates, they should fertilize a higher proportion of the eggs laid by those females than they would have otherwise. David Westneat enabled some male red-winged blackbirds to stay nearer their mates by giving them access to cracked corn, sunflower seeds, and mealworms at several feeding stations scattered about their territories. The food-supplemented males made significantly fewer trips off their territories than unfed control redwings, and so could monitor their mates' behavior more closely than the controls. These super-guarders sired nearly 90 percent of the nestlings produced on their territories, in contrast to about 70 percent for control males that had to leave their mates unguarded more of the time [1215].

Sexual Selection and Mate Choice

Mate guarding is one of the many dramatic evolutionary consequences of sexual selection arising from male–male mating competition. But although males of many species obviously affect one another's fitness, let's not forget about females. The mate choice component of sexual selection arises because females can influence the reproductive success of males, especially by mating with more than one individual (Table 2) [1178]. In addition, females sometimes regulate the size, chemical composition, or sex of their eggs [823]. When female zebra finches add more testosterone to an egg after mating with an exceptionally attractive male, they probably give this mating partner a fitness advantage. Eggs receiving the hormone supplement develop into chicks that beg more vigorously and so grow more rapidly, which should increase the odds that they will survive to reproduce [429].

As suggested earlier, females can affect which sperm get to fertilize their eggs, not only by actively choosing certain males as mating partners [34], but also in some cases by favoring the sperm of certain males over those of others (Figure 28) [925]. In the spotted sandpiper, for example, a species in which males care for the eggs (see p. 376), a female may produce a clutch of eggs for one male with whom she copulates frequently, then leave him to pair off with another male, with whom she also copulates. But when it comes to fertilizing her second clutch of eggs, she may use sperm she received from her first part-

---

TABLE 2 Ways in which females and males attempt to control reproduction while interacting with each other

Key reproductive decisions controlled primarily by females

Egg investment: What materials and how much of them to place in an egg

Mate choice: Which male or males will be granted the right to be sperm donors

Egg fertilization: Which sperm to use to fertilize each egg

Offspring investment: How much maintenance and care goes to each embryo and offspring

Ways in which males influence female reproductive decisions

Resources transferred to female: May influence egg investment, mate choice, or egg fertilization decisions by female

Elaborate courtship: May influence mate choice or egg fertilization decisions by female

Sexual coercion: May overcome female preferences for other males

Infanticide: May overcome female decisions about offspring investment

Source: Modified from Waage [1178]

28 Postcopulatory mate choice. Free-living hens eject the sperm of subdominant males, while retaining the sperm of preferred, dominant roosters. After Pizzari and Birkhead [925].

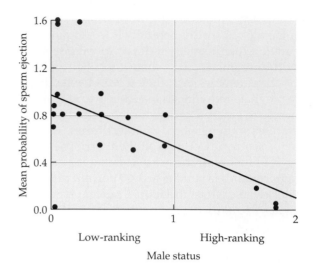

ner, which she has sustained internally in tiny sperm storage tubules in her uterine wall for days or weeks [864].

Likewise, females of the moth *Utetheisa ornatrix* mate more than once, but selectively use the sperm of males that provide them with large spermatophores, which contain large quantities of the defensive alkaloids that females use to protect their eggs [353] and themselves as well (see Figure 30, p. 208) [441]. Mated females somehow sense the mass of the spermatophores they have received, and they actively move only the sperm from large spermatophores to the sperm storage organ for later use in fertilizing their eggs [675, 676]. This moth, therefore, exhibits cryptic female choice *after* copulating with several males, a phenomenon that may be much more common than currently appreciated [348].

Female control over copulation and fertilization exerts selection on males in many different ways [1178]. Some females, for example, make it essential that males provide them with useful resources as a precondition for copulation (Figure 29). If some males refuse, they will not gain access to the females' eggs [1165].

29 Female mate choice based on nuptial gifts. Males of the blister beetle *Neopyrochroa flabellata* consume a substance called cantharidin and later transfer it, along with their sperm, to their mates. Before copulation occurs, the male must present his head to his potential partner (A), who samples secretions from a gland located there (B). If the secretions are cantharidin-rich, the female permits the male to mount and copulate with her (C), but if the male does not have cantharidin to transfer to her, she blocks intromission by extending and curling her abdomen under her body (D). Females add the cantharidin they receive to their eggs, making them unattractive to some egg-eating insects. Drawings based on photographs by Tom Eisner from [354].

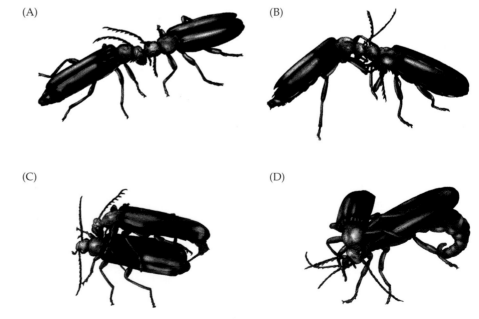

30 A potential nuptial gift. A male hangingfly has captured a moth, a material benefit to offer his copulatory partner. He advertises the availability of his gift by releasing a pheromone from abdominal glands. Photograph by the author.

A classic demonstration of this strategy comes from Randy Thornhill's study of the black-tipped hangingfly, whose females make copulation, and subsequent egg fertilizations, contingent upon receipt of an edible nuptial gift (Figure 30) [1129]. Females flatly refuse to mate with males that proffer unpalatable ladybird beetles and will let copulation begin only when the present is edible. If, however, the female can consume the nuptial gift in less than 5 minutes, she finishes it and separates from her partner without having accepted a single sperm. When the gift is large enough to keep the copulating female occupied for 20 minutes, she will depart with a full gut and a full complement of the gift-giver's sperm (Figure 31).

Males of many other animals provide food presents before or during copulation, including chimpanzees [1092], whose males supply chunks of monkey

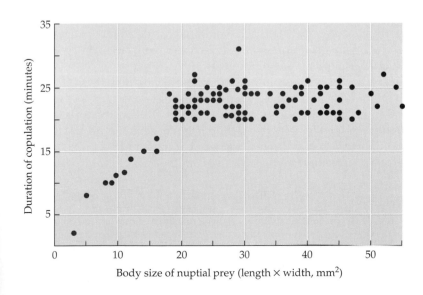

31   Sperm transfer and the size of nuptial gifts. In black-tipped hangingflies, the larger the gift, the longer the mating, and the more sperm the male is able to pass to the female. After Thornhill [1129].

32 Carotenoids in this katydid's spermatophore give the nuptial gift both its color and its special nutritional value to the female, which uses these vitamin-like chemicals in egg production. Photograph by Klaus-Gerhard Heller.

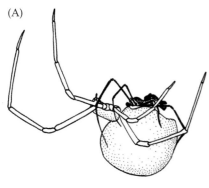

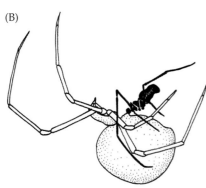

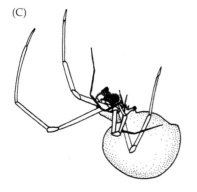

meat to potential mates, and assorted insects [123], such as the male katydids and crickets that attach edible spermatophores [512] to the genital openings of their mates (Figure 32; see also Figure 9). When male mantids and spiders are eaten by their mates (see Figure 2, p. 158), we can ask whether they have provided the ultimate in a nuptial gift [182, 937] or have been cannibalized only because hungry females have been able to take advantage of the sexually driven males' willingness to get close to them [707]. The second hypothesis gains support from the observation that male mantids exhibit great caution when approaching females; moreover, they apparently prefer well-fed females as mates rather than hungry ones [668, 771].

On the other hand, a male redback spider that is transferring sperm to a female performs a somersault that throws his body into his partner's jaws (Figure 33). In a majority of cases, the female obliges her sexual companion and devours him, but only after his invitation via the copulatory somersault [37]. Male redbacks are small, a mere 1 or 2 percent of the female's mass, barely more than a mouthful for a cannibalistic female. However, they may provide a useful snack, because females that are hungry are more likely to dine on a male [38]. When eaten, the deceased males derive substantial benefits from the experience, since they fertilize more of their partner's eggs posthumously than they would if they were to survive the mating. A cannibalistic female spider is less likely to mate again, and a male redback would have little chance of surviving the arduous search for a new partner [37]. Thus, the benefits to males from sexual suicide should usually outweigh the costs.

Males of some other animals offer parental care for the female's offspring, rather than a nuptial gift, and in these species we would expect females to exert sexual selection in favor of highly paternal males. To test this prediction, Katherine Noonan placed females of a cichlid fish whose males are paternal in the central compartment of a three-chambered aquarium [845]. The female cichlid could see a large male and a small male in the two end compartments

33 Sexual suicide in the redback spider. (A) The male first aligns himself facing forward on the underside of the female's abdomen while he inserts his sperm-transferring organ into her reproductive tract. (B) He then elevates his body and (C) somersaults backward into the jaws of his partner. She may oblige by consuming him while sperm transfer takes place. After Forster [391].

on either side of her, but the two males could not see each other because of two staggered black Plexiglas barriers in the central compartment. Females watched the visual courtship displays of both males, and in 16 of 20 trials, they spawned in a nest near the large male, which in nature would probably be the better father because of his greater capacity to repel egg-eating predators.

In the fifteen-spined stickleback, however, females do not favor large males in similar experiments. Instead, they associate with males that are able to produce frequent body shakes during courtship. Males capable of more body shaking also perform more nest fanning after courtship and egg laying; nest fanning sends oxygenated water flowing over the eggs in the nest, promoting gas exchange and egg hatching (see Figure 2, p. 398) [869].

The capacity to choose a paternal male extends to some birds, such as the pied flycatcher, whose males have either jet black and white plumage or a duller, browner appearance. In a three-chambered aviary (Figure 34) containing a female and two males, one naturally dull-plumaged male and one whose dull dark feathers have been artificially dyed dark black [1009], the female usually selects the dyed male. She demonstrates her preference by beginning to build a nest in the nest box next to the compartment containing the male of her choice.

If female flycatchers favor black-and-white males because of the direct parental care benefits they offer, then males with the preferred plumage should provide their offspring with more food than dull-plumaged males. In an experimental test of this prediction, the weight of nestlings fed by black-and-white males alone (after experimental removal of their mates) fell less than that of nestlings cared for by dull males alone, indicating that black-and-white males were indeed superior fathers [1010].

Bright plumage also carries the day with female house finches, who prefer males that have bright red feather patches as opposed to paler orange or yellow patches [518, 519]. Perhaps the redness of male plumage in this species signals the previous nutritional state of the male [520] or his physiological condition, especially his parasite load [1127]. If parasites do affect the color of a male's

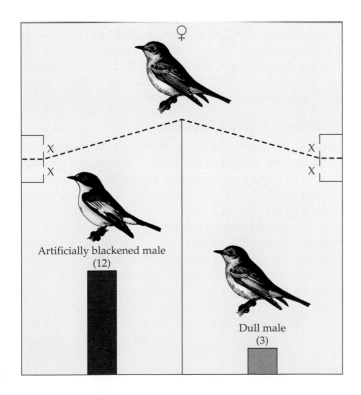

Artificially blackened male
(12)

Dull male
(3)

34 Mate choice by female pied flycatchers. Twelve of 15 captive females preferred to build a nest in a nest box advertised by a dull-plumaged male whose feathers had been artificially blackened. Females tended to ignore unaltered, dull-plumaged males. (The dashed lines indicate netting; the solid line indicates an opaque barrier; X's indicate nest box entrances.) After Sætre et al. [1009].

35 Mean plumage brightness of male house finches changed dramatically after a severe epidemic killed a large proportion of the population. The lower the score, the redder the plumage. The superior survival rate for brightly colored males (such as the bird on the right) suggests that they were in better physiological condition prior to the spread of the disease than their duller-colored companions. After Nolan et al. [843]; photograph by Geoff Hill.

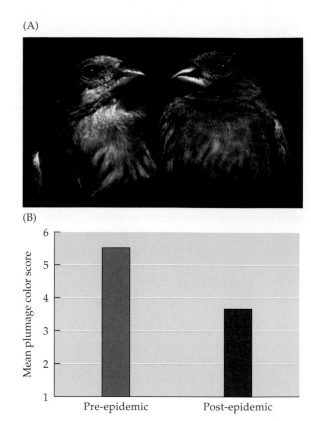

(A)

(B)

plumage, then if one were to experimentally infect house finches with a parasite, the infected birds' red plumage should gradually become duller—as it does when finches are dosed with a coccidia parasite [521]. Moreover, if bright colors and good physiological condition are correlated, then house finches that survive an epidemic ought to have brighter plumage on average than those that die. In the 1990s, the microorganism *Mycoplasma gallisepticum* spread from chickens to house finches, causing a massive die-off among East Coast finches. Samples of birds taken before and after the epidemic reveal that the birds with brighter plumage were more likely to have survived the disaster (Figure 35) [843]. Therefore, when house finch females choose brightly colored males as their partners, the chosen individuals are likely to be in better-than-average condition and thus better able to perform their parental duties.

Female Mate Choice without Material Benefits

The examples of female choice that we have just reviewed are consistent with **good parent theory,** which explains aspects of male color, ornamentation, and behavior as honest signals (p. 308) of a male's capacity to provide material benefits to females. Female choice based on these signals makes intuitive sense; female house finches or pied flycatchers that pair off with a competent paternal male seem likely to gain reproductively. Likewise, it is easy to see why females of some species might prefer males able to provide them with food or some other similar resource before copulating. For example, female black-tipped hangingflies that receive a large nuptial gift do not have to search for prey themselves, a task that involves the risk of flying into a spider's web [1128]. However, males of many species, including the satin bowerbird, do not provide food or any other material benefit to their mates. In these animals, females typically prefer males capable of more intense or more varied courtship stimulation [633].

The preference for intense courtship stimulation may even involve an evaluation by females of the tactile stimulation they receive during copulation (Fig-

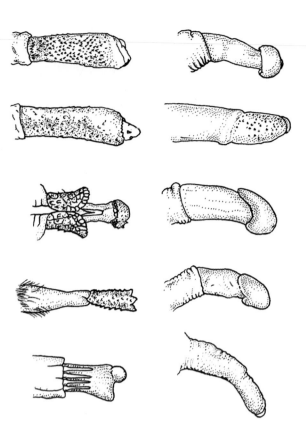

36 Female choice for stimulating male genitalia? The more elaborate penises on the left, which might provide more copulatory stimulation, are from primate species in which females regularly mate with more than one male, and so might evaluate the copulatory performance of several males. The simpler (less stimulating?) penises on the right are from species in which a single male usually copulates with a female during one estrous cycle. After Dixson [329] and Eberhard [346].

ure 36) [347]. If so, we would expect that male genitalia would be more complex and distinctive in polyandrous species than in monogamous species. Only polyandrous females have the chance to bias egg fertilization success toward the male or males that provide special genital stimulation. One way to test the female choice hypothesis for the evolution of stimulating male genitalia would be to compare genital differences within groups of closely related species that are either polyandrous or monogamous. Göran Arnqvist has performed the test in question by finding appropriate groups and then measuring certain components of genital shape for the species in each group. The group mean of each measurement was used as a standard against which any one species' value was compared. Polyandrous species generally differed far more from the mean than did members of a comparable group of monogamous species (Figure 37) [47].

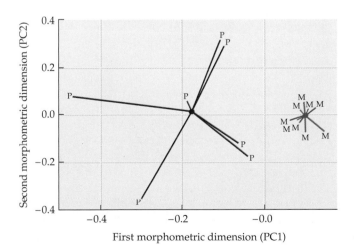

37 Mating system affects the evolution of male genitalia. Plots of two of the major components of genital shape against the mean value for a group of polygynous fly species in the genus *Dryomyza*, each represented by a "P," and a group of monogamous fly species in the genus *Lucilia*, each represented by an "M." The greater differences in the position of the "P" values reflect the evolution of greater differences in genital shape among these related species, whereas fewer differences exist among the genitalia of the monogamous flies. After Arnqvist [47].

38   Removal of eyespots from a peacock's tail reduces his attractiveness to females. After 20 eyespots had been cut from their tails, males averaged two fewer mates in the following breeding season compared with their performance in the previous year. After Petrie and Halliday [906].

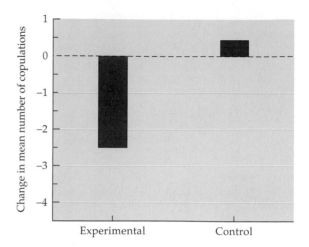

Although mate choice can evidently be based on aspects of copulation itself, visual courtship displays preceding mating also provide potential sensory cues for selective females. Consider that peahens prefer peacocks with relatively large numbers of eyespots in their immense tails, which the males spread and shake in front of potential mates [907]. The importance of these decorations was demonstrated when Marion Petrie and Tim Halliday captured some adult peacocks and removed 20 of the outermost eyespots from some males' tails. Birds so treated experienced a significant decline in mating success compared with their performance in the previous year. In contrast, control males, which were captured and handled but whose tails were left intact, were no less attractive to females after the treatment (Figure 38) [906].

If greater sensory stimulation from courting males does the trick, then experimentally augmenting a male's courtship ornaments should enhance his copulatory success. The relevant experiment was done first with the long-tailed widowbird, a species about the size of a red-winged blackbird that is endowed with a half-meter-long tail (Figure 39). The male flies around his grassland territory, displaying his magnificent tail to females passing by. Malte Andersson

39   A male long-tailed widowbird displays to potential mates while flying above his grassland territory. Photograph by Malte Andersson.

took advantage of the wonders of superglue to perform an ingenious experiment. He captured male widowbirds, then shortened the tails of some by removing a segment of tail feather, only to glue it onto another bird's tail, thereby lengthening that ornament [32]. The tail-lengthened males were much more attractive to females than those that suffered the loss of a portion of their ornaments. Moreover, the tail-lengthened males also did much better than controls, whose tails were cut and then put back together with applications of superglue.

On rare occasions, however, sexual selection can favor a *reduction* in tail size. In a small polygynous Australian bird called the golden-headed cisticola, males whose tails were artificially shortened acquired superior territories and induced more females to lay eggs in their territories than controls or cisticolas whose tails were elongated (Figure 40) [66]. In this species, whose tails are naturally rather short to begin with, males with even shorter tails experienced reduced aerodynamic drag, and so could fly faster, an ability that enabled them to chase rivals away more effectively and engage in faster display flights, thereby presumably providing greater sensory stimulation to potential mates. (On the other hand, tail-shortened males had more trouble maneuvering during slow-speed flights while foraging, which helps explain why male cisticolas have retained a tail of modest length over evolutionary time.)

The favorable response of females to heightened levels of sensory stimulation provided by a male's appearance or courtship displays says something about the proximate mechanisms of mate choice in these species, but does not explain the ultimate benefits, if any, of female mate choice when the chosen males have only their sperm to offer. What information do courting males and their elaborate ornaments provide to potential mates, and why are females sometimes persuaded by what they see, hear, smell, or feel? If males provide only sperm to their mates, females cannot use the physical attributes of those males, or their courtship behavior before, during, or after copulation, as indicators of the material benefits, such as help in rearing the young, that they will receive from their mates. But even though the good parent theory of female mate choice cannot apply to these species, females may still gain something by being choosy.

One of the major ideas about how females could benefit from their mate preferences is the **healthy mate theory** (Table 3), which states that male courtship displays and ornaments inform females of a potential sexual partner's health or parasite load [966]. If so, females could use this information to choose males that are less likely to transmit lice, mites, fleas, or bacterial pathogens, which could harm them or their future offspring. The finding that dominant jungle fowl roosters have larger red combs than subordinate males as well as a stronger

Golden-headed cisticola

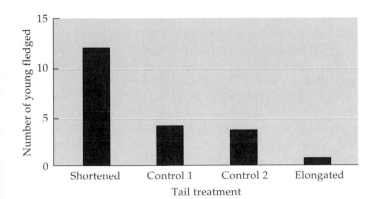

40 Tail length and reproductive success. Male cisticolas whose tails had been experimentally shortened fathered more fledglings than control males whose tails were not shortened or lengthened (controls 1 and 2) or males with experimentally elongated tails. After Balmford et al. [66].

TABLE 3	Three theories on why extreme male ornamentation and striking courtship displays have evolved in species in which males provide no parental care or other material benefits to their mates	
Theory	**Females prefer trait that is**	**Primary adaptive value to choosy females**
Healthy mate	Indicative of male health	Females (and offspring) avoid contagious diseases and parasites
Good genes	Indicative of male survival ability	Offspring may inherit the viability advantages of their father
Runaway selection	Sexually attractive	Sons inherit the trait that makes them sexually attractive; daughters inherit the majority mate preference

immune system [1289] suggests that female jungle fowl can identify healthier males on the basis of their comb ornaments [1290].

Of course, hens that mate with healthy roosters may also secure genes that will benefit their offspring. The **good genes theory** proposes that male courtship displays and ornaments provide information to females that enables them to choose males with viability-promoting genes. These genes might advance the survival chances of their offspring by conferring hereditary resistance to parasitic infection or disease or by promoting adaptive foraging skills or the ability to avoid predators. Parasite resistance took center stage in the version of good genes theory developed by W. D. Hamilton and Marlene Zuk, who realized that rapidly evolving parasites would exert continuous selection on hosts to combat their constantly changing enemies [486]. Considerable evidence indicates that parasites do quickly adapt to locally common host genotypes, which maintains selection for variants better able to escape parasitic infection [709]. In species subject to parasitic infection, individuals that can signal their relatively parasite-free state are desirable mates. Hamilton and Zuk predicted that selection for honest signals of noninfection would therefore lead bird species with numerous potential parasites to evolve strikingly colored plumage (which would be difficult to maintain in top condition by diseased individuals). The two researchers found a correlation between plumage brightness and the incidence of blood parasites in a large sample of bird species, supporting the view that males at special risk of parasitic infection engage in a competition that signals their condition to choosy females. Subsequent research has focused primarily on an examination of the connection between parasite levels, ornament condition, and female choice, and has provided some additional support for the Hamilton-Zuk hypothesis [34].

A third theory, **runaway selection theory** [45, 304], proposes that discriminating females acquire sperm with genes whose primary effect is to influence their daughters to prefer attractive males and to endow their sons with attributes that will be preferred by most females—even if these traits actually reduce the survival chances of individuals that possess them. Because the runaway selection alternative is the least intuitively obvious explanation for male courtship displays and female mate choice, let us briefly examine the argument underlying the mathematical models of Russell Lande [679] and Mark Kirkpatrick [625].

Imagine that a slight majority of the females in an ancestral population had a preference for a certain male characteristic, perhaps *initially* because the preferred trait was indicative of some survival advantage enjoyed by the male. Females that mated with preferred males would have produced offspring that inherited the genes for the mate preference from their mothers *and* the genes for the attractive male character from their fathers. Sons that expressed the preferred trait would have enjoyed higher fitness, in part simply because they possessed the key cues that females found attractive. In addition, daughters that responded positively to those male cues would have gained by producing sons with the trait that many females liked.

Thus, female mate choice genes and genes for the preferred male trait can be inherited together. This pattern could generate a runaway process in which ever more extreme female preferences and male traits spread together as new mutations occur. The runaway process would end only when natural selection against costly or risky displays balanced sexual selection in favor of traits that are appealing to females. Thus, if peahens originally preferred peacocks with larger than average tails because such males could forage efficiently, they might now favor males with extraordinary tails because this mating preference has taken on a life of its own, resulting in the production of sons that are exceptionally attractive to females and the production of daughters that will choose this kind of male for their own mates.

In fact, the Lande-Kirkpatrick models demonstrate that right from the start of the process, female preferences need not be directed at male traits that are utilitarian in the sense of improving survival, feeding ability, and the like. Any preexisting preference of females for certain kinds of sensory stimulation (see p. 292) could conceivably get the process under way. As a result, traits opposed by natural selection because they reduced viability could still spread through the population by runaway selection [625, 679]. Instead of mate choice based on genes that promote the development of useful characteristics in offspring, runaway selection could yield mate choice for *arbitrary* characters that are a burden to individuals in terms of survival, a disadvantage in every sense except that females mate preferentially with males that have them!

Testing the Healthy Mate, Good Genes, and Runaway Selection Theories

Discriminating among these three alternative explanations for elaborate male courtship displays and female choice in species in which males do not provide material benefits has proved very difficult, in part because the three hypotheses are not mutually exclusive. As just mentioned, female preferences and male traits that originated through a good genes process could then be caught up by runaway selection. Note too that males with hereditary resistance to certain parasites (good gene benefits) would also be less likely to infect their partners with these parasites (healthy mate benefits). Moreover, if at the end of a period of runaway selection, males had evolved extreme ornaments and elaborate displays, then only individuals in excellent physiological condition would be able to develop, maintain, and deploy their ornaments in effective displays. Males in such superb physiological condition would probably have to be highly effective foragers (good gene benefits) as well as parasite-free (healthy mate benefits), in which case females mating with such males would be unlikely to acquire parasites, and their daughters might well get some survival-benefiting genes while their sons received the pure attractiveness genes of their fathers.

Given the difficulty of separating these three possible processes, the best we can do for the moment is to look at predictions derived from one or another theory and test them by applying them to selected species. As an example, let's apply the good genes theory to the gray treefrog, a species in which males offer their mates nothing but sperm. Female gray treefrogs are attracted to males that give calls of relatively long duration. If the benefits of female choice are good genes, then if we were to take half of a female's clutch of eggs and fertilize them with sperm from a long-calling male, while the other half are doused with sperm from a short caller, then the tadpoles sired by the long-calling male should develop faster than their half-siblings whose father could manage only short calls [1204]. In three of four trials, the result matched this prediction, providing some support for the good genes argument.

An even more ambitious attempt to test the good genes theory was conducted by Marion Petrie, who applied it to peacocks and derived the following predictions: (1) males should differ genetically in ways related to their survival chances, (2) male behavior and ornamentation should provide accurate information on the survival value of the males' genes, (3) females should use this information to select mates, and (4) the offspring of the chosen males should benefit from their mothers' mate choice. In other words, males should signal their genetic quality in an accurate manner, and females should pay attention to those signals because their offspring will derive hereditary benefits as a result [634, 1278].

Petrie studied a captive but free-ranging population of peacocks in a large forested English park, where she found that males killed by foxes had significantly shorter tails than their surviving companions. Moreover, she observed that most of the males taken by predators had not mated in previous mating seasons, suggesting that females can discriminate between males with high and low survival potential, possibly on the basis of their ornamented tails [904]. The peahens' preferences translate into offspring with enhanced survival chances, as Petrie showed in a controlled breeding experiment. She took a series of males with different degrees of ornamentation from the park and paired them in large cages with four females chosen at random from the population. The young of all the males were reared under identical conditions, weighed at intervals, and then eventually released back into the park. The sons and daughters of males with more highly ornamented tails weighed more at day 84 and were more likely to be alive after 2 years in the park than the progeny of males with fewer eyespots (Figure 41) [905].

This combination of results is consistent with the view that peahen mate preferences create sexual selection that is *currently* maintaining or spreading the genetic basis for those preferences because offspring receive good "viability-enhancing" genes from their fathers. But it is also possible that healthy male benefits are involved as well; perhaps the peahens' mate choices lead them to avoid parasitized males and so reduce their risk of acquiring damaging para-

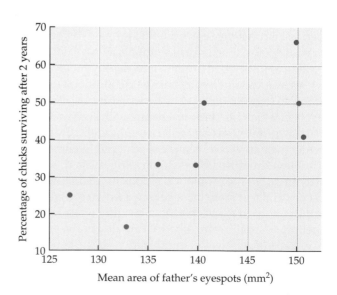

41 **Do male ornaments signal good genes?** Peacocks with more eyespots on their tails produced offspring that survived better when released from captivity into an English woodland park. After Petrie [905].

sites that they would pass on to their offspring, reducing their growth rates and survival chances. Moreover, a demonstration of current selective advantages associated with female preferences and male traits does not rule out the possibility that these attributes *originated* as side effects of runaway selection in the manner described above.

Yet another complication comes from the discovery that females of some species invest more resources in offspring that are sired by preferred males than in those fathered by less attractive males. As noted above, female zebra finches give eggs fertilized by favored males an extra dose of testosterone, which may give these offspring a developmental advantage. Likewise, female mallards make larger eggs after copulating with unusually attractive males [272], while black grouse produce and lay more eggs subsequent to mating with a top-ranked male [971]. All of these effects might easily be attributed to the "good genes" of the male when in reality they arise from the female's manipulation of her own parental investment, not the genetic contribution of her partner.

Sexual Conflict between Males and Females

Brett Holland and Bill Rice have proposed yet another scenario that could result in the evolution of extreme ornaments and displays in males [533]. According to their **chase-away selection theory,** exploitation and conflict between the sexes, rather than cooperation and mutual benefit, could drive the process in the following manner. Imagine that a male happened to have a mutation yielding a novel display trait, one that managed to tap into a preexisting sensory bias of the females of his species. A male that could activate that preexisting preference might induce females to mate with him even though he might not provide the material or genetic benefits offered by other males in the population [999]. The resulting spread of exploitative males over time would create selection on females favoring those that had psychological resistance to the purely attractive display trait. As females with a higher threshold for sexual responsiveness to the exploitative trait spread, selection would then favor males able to overcome female resistance, which might be achieved by mutations that exaggerated the original male advertising signals. A cycle of conflict between the sexes could ensue, involving increasing female resistance and male exaggeration of key characteristics, and leading gradually to the evolution of costly ornaments of no real value to the female and useful to the male only because without them, he has no chance of stimulating females to mate with him (Figure 42).

The approach underlying chase-away selection theory is an example of how the evolutionary view of sexual reproduction has changed in recent decades. Once considered a gloriously cooperative exercise between male and female, reproduction is now seen as an arena in which males and females battle for maximum genetic advantage. For example, until recently, termites were believed to be dedicated monogamists who worked together as pairs for years, but careful observation revealed that males of at least one species regularly abandon relatively lightweight females to search for heavier, more fecund partners, while some females invite male intruders into the nest (probably with pheromonal signals), where they may displace the female's previous mate [1042]. As it turns out, only a minority of individuals in this termite species actually remain with their first partner over the long haul.

In some sense, conflict between the sexes is inherent in the gametic strategies of males and females. Males essentially withhold useful resources from sperm, providing no developmental assistance for the female gamete, in order to maximize their production of sperm and thereby increase the number of eggs fertilized. Moreover, males can do more than simply withhold assistance. Con-

42 Chase-away selection theory. The evolution of extreme male ornaments and displays may originate with sensory exploitation of females. If sensory exploitation by males reduces female fitness, the stage is set for a cycle in which increased female resistance to male displays leads to ever greater exaggeration of those displays. After Holland and Rice [533].

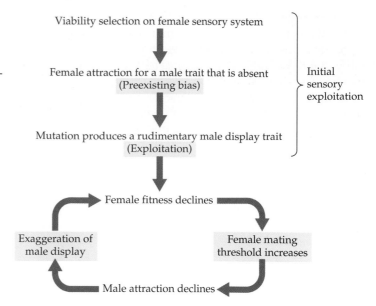

sider that infanticide by males (see Figure 11, p. 14) accounts for at least a quarter of all infant deaths among lions and several primate species [1077]. One possible evolutionary response by female primates to this kind of damaging conflict is to mate with several males, making the paternity of their offspring difficult to determine, and thereby perhaps reducing the risk of an infanticidal attack by any one of their partners, most of whom will not be the father of the infant [28, 854]. Thus, the females' actions could boost their own fitness but act against the genetic interests of most of their mates. Indeed, whenever a female mates with more than one male for whatever reason, an evolutionary conflict of interest is all but guaranteed. Polyandry means that some or all of a female's mates will fertilize fewer eggs. Likewise, when males that provide resources or parental care to their mates go on to copulate with several females, some of those females will usually get less than if their partners had been monogamous [197, 1158].

The evolutionary effects of the struggle to monopolize a mate's reproductive output or parental investment include a wide range of adaptations that strike most people as moderately unpleasant. For example, males of the polygynous fruit fly, *Drosophila melanogaster*, essentially poison their mates with toxic substances present in their seminal fluid. These damaging chemicals help the copulating male destroy rival sperm stored by their mates, but they also have negative effects on the females. The true nature of male–female relations in the fruit fly became apparent when many generations of flies were reared such that no male mated with more than one female. Under these conditions, selection favored males that did not harm their mates because their one partner's reproductive output constituted the male's only chance to contribute genes to the next generation. After more than 30 generations of selection in a monogamous environment, some females were mated once with "control" males taken from a typical polygynous population. These females received the standard toxic ejaculate from their mates, which resulted in fewer eggs laid and earlier female mortality compared with females who had evolved with polygamous males (Figure 43) [534]. In naturally polygynous populations, males can gain by supplying harmful spermicide to their sexual partners because they are unlikely to mate with the same female twice. Under these circumstances, males that increase the proportion of a current clutch of eggs fertilized by their own sperm can derive

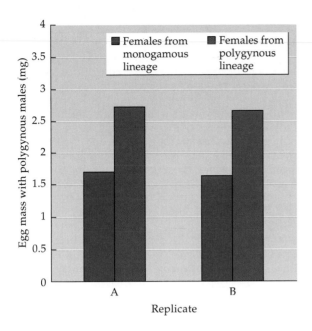

43 **Sexual selection and the evolution of male traits harmful to females.** Females from an experimental monogamous lineage of fruit flies have lost their biochemical resistance to the spermicidal toxins present in seminal fluids of polygynous male flies. They lay fewer eggs when mated with males from a control polygynous lineage than do control females that evolved with those males. After Holland and Rice [534].

fitness benefits even though their actions are detrimental to female fitness over the long haul.

Chemical warfare between the sexes lies beneath the surface, but other forms of sexual conflict are more apparent, such as the violent attempts some males make to force themselves on sexually resistant females [236] (see Figure 16, p. 477). Male water striders, for example, often grab and wrestle with females, which may or may not be able to somersault away (Figure 44). The struggles of the females cost them dearly, more than doubling their metabolic rate over that of solitary females cruising over the water surface [1187]. In addition, females fighting conspicuously with males attract the attention of predatory aquatic insects [993].

Damaging sexual harassment also occurs in a solitary bee whose males knock flower-visiting females to the ground. Although unreceptive females can usually break free, they often have to move to the less profitable flowers in concealed parts of the plant rather than foraging on the outer flowers, where they will be constantly assaulted [1102]. Males of our closest mammalian relative, the common chimpanzee, also sometimes viciously attack females, either to secure an immediate copulation or to intimidate the female into mating later, but sometimes injuring or even killing her in the process [1077].

Killing a potential mate obviously does little to advance the killer's reproductive success. However, despite the cost of wasted attempts to copulate, with or without female cooperation, a willingness to employ coercive tactics may still yield net fitness gains for males of some species. Female water striders that are unable to dislodge a mounted male will eventually mate with him [994], and persistent males secure more matings in species from Hawaiian fruit flies [117] to the savanna baboon [94]. Do females in these species give in because their harassers signal that they have the energy reserves to fuel their persistence, thus demonstrating their good physiological condition and their potential to supply their mates with either good genes or material benefits, such as protection from other males? Or do females mate with persistent males simply because the costs of continued resistance make it less expensive to mate with the unwanted male than to spend more energy on avoiding his advances [236, 1214]? We have much more to learn about the fitness consequences of the apparently hostile interactions that can occur between the sexes.

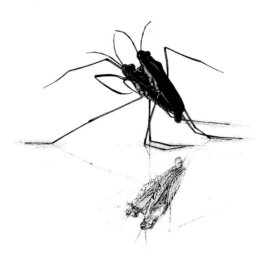

44 **Conflict between the sexes.** A male water strider struggles to maintain his grasp on a female, who is attempting to perform a backward somersault to dislodge her unwelcome partner. Drawing by G. Marklund, from Rowe et al. [994].

Summary

1. Sexual reproduction creates a social environment of conflict and competition among individuals as each strives to maximize its genetic contribution to subsequent generations. Males usually make many small gametes and usually try to fertilize as many eggs as possible, while providing little or no care for their offspring. Because females make fewer, larger gametes and often also provide parental care, they usually have a lower potential rate of reproduction than males. As a result, receptive females are scarce, and males typically compete for access to them, while females can choose among many potential partners. In a few species, however, these sex roles are reversed, providing an opportunity to test the theory that the differences between the sexes stem from differences in their relative parental investments (or in their potential reproductive rates).

2. Evolution by sexual selection occurs if genetically different individuals differ in their reproductive success (1) as a result of competition within one sex for mates or (2) because of their differential attractiveness to members of the other sex.

3. The competition-for-mates component of sexual selection is behind the evolution of many elements of male reproductive behavior, including competition for social dominance, alternative mating tactics, and mate guarding after copulation. Although in a typical species, males exert selection on one another in the competition for mates, females generally have the last word on reproduction because they control the production and fertilization of eggs. Interactions between the sexes can be viewed as a mix of cooperation and conflict as males seek to win fertilizations in a game whose rules are set by the reproductive mechanisms of females.

4. In a typical species, females often choose among potential mates, creating the mate choice component of sexual selection. Males of some species seek to win favor with females by offering them material benefits, including nuptial gifts, resources monopolized in territories, or parental care.

5. Mate choice by females occurs even in some species in which males provide no material benefits. Mate choice of this sort could arise as a result of females choosing males whose genes will enhance the viability of their offspring (good genes theory). Or female preferences for elaborate ornaments may arise because males with these attributes are healthy and parasite-free (healthy mates theory). On the other hand, extravagant male features could spread through a population in which even arbitrary elements of male appearance or behavior became the basis for female preferences. Exaggerated variants of these elements could be selected strictly because females preferred to mate with individuals that had them (runaway selection theory). A fourth possibility is that the extreme ornaments of males evolve as a result of an upward spiraling cycle of conflict between the sexes, with males selected for ever-improved ability to exploit female perceptual systems and females selected to resist ever more resolutely the ornament-based courtship of the opposite sex (chase-away selection theory). The relative importance of these various mechanisms of sexual selection remains to be determined.

6. Males often attempt to circumvent female mate choice by trying to inseminate unwilling females. Conflict between the sexes also takes other forms, including infanticide and sexual harassment, demonstrating that what is adaptive for one sex may be damaging to the other.

Discussion Questions

1. Male rats, sheep, cattle, rhesus monkeys, and humans that have copulated to satiation with one female are speedily rejuvenated if they gain access to a new female. This phenomenon is called the "Coolidge effect," supposedly because when Mrs. Calvin Coolidge learned that roosters copulate dozens of times each day, she said, "Please tell that to the President." When the President was told, he asked, "Same hen every time?" Upon learning that roosters select a new hen each time, he said, "Please tell that to Mrs. Coolidge." Provide a sexual selectionist hypothesis for the evolution of the Coolidge effect. Use your hypothesis to predict what kinds of male animals should lack the Coolidge effect.

2. Males of the barn swallow have long, thin outer tail feathers. When Anders Møller analyzed the effect of tail length on male mating success in the barn swallow in Europe, he did an experiment in which he made some males' tail feathers shorter by cutting them, and made other males' tail feathers longer by gluing feather sections onto their tails [804]. But he also created a group in which he cut off parts of the males' tail feathers and then simply glued the fragments back on to produce a tail of unchanged length. What was the point of this group? And why did he randomly assign his subjects to the shortened, lengthened, and unchanged tail groups? And why did a team of Canadian biologists repeat Møller's experiment on another continent [1072]?

(A)

3. We argued that mate guarding should be common in species in which females retain their receptivity after mating and are likely to use the sperm of their last mating partner when fertilizing their eggs. But there are species, including some crab spiders, in which males remain with immature, unreceptive females for long periods and fight with other males that approach these females [331]. How can "guarding" behavior of this sort be adaptive? Produce hypotheses and allied predictions.

4. In a pipefish species in which females compete for choosy males that care for the eggs they fertilize in a specialized brood pouch, males are more likely to accept eggs from courting females whose skin lacks the black spots that are caused by a parasitic trematode worm. The higher the trematode load, the lower the fecundity of the female. When Gunilla Rosenqvist and Kerstin Johansson tattooed black spots onto the skin of unparasitized females, these experimentally altered individuals experienced reduced success when courting males [991]. What proximate and ultimate explanations did the researchers examine in this study of mate choice, and what were the predictions they tested?

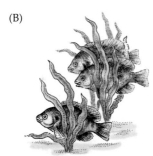
(B)

5. When male bluethroats (see Figure 28, p. 139) have their blue throats experimentally blackened, they become less attractive to females. Outline the costs and benefits to these males from guarding their mates, once they have secured a partner, and the costs and benefits of attempts to secure extra-pair matings with other females by visiting the territories of other pairs. Use conditional strategy theory to predict the tactics of experimental and control males. Discuss the significance of the finding that experimental males sired fewer of their social partner's offspring on average than did control males.

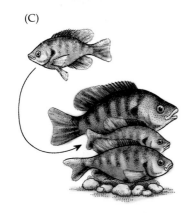
(C)

6. The figure to the right (after Gross [460]) shows three different tactics used by male bluegill sunfish for fertilizing eggs. (A) A territorial male guards a nest that may attract gravid females. (B) Little sneaker males wait for an opportunity to slip between a spawning pair, releasing their sperm when the territory holder does. (C) A slightly larger satellite male with the body coloration of a female hovers above a nest before slipping between the territorial male and his mate when the female spawns. How would you test the competing three-different-strategies versus one-conditional-strategy hypotheses to account for the existence of alternative mating tactics in this species? For test data, see [319].

Suggested Reading

To learn more about satin bowerbird sexual behavior, see [128, 129]. Malte Andersson's book on sexual selection is remarkably comprehensive [34], while Mart Gross focuses specifically on alternative mating tactics [462]. Sperm competition is the subject of a book edited by Tim Birkhead and Anders Møller [108], while Bill Eberhard examines how females might control which of several males' sperm will fertilize their eggs [348]. Finally, although this chapter did not explore the subject of how sexual reproduction evolved in the first place, you can learn about it from G. C. Williams [1240] and Bob Trivers [1155], with an update from Laurence Hurst and Joel Peck [561].

12

The Evolution of Mating Systems

As we saw in the previous chapter, male satin bower-birds are willing and able polygynists, capable of copulating with dozens of females in a single breeding season, although rarely having the good fortune to do so [127]. In contrast, males of the green catbird, another bowerbird, nest in alliance with a single female, and so are considered monogamous rather than polygynous. Polyandry comes into play in the spotted sandpiper, in which two or three monogamous males may share the same female in one breeding season [865]. Female spotted sandpipers are, how-ever, much less polyandrous than a queen honey bee, which may mate with more than 30 males during several nuptial flights away from the hive before she settles down to a life-time of egg-laying [408]. The males that copulate with a honey bee queen face certain monogamy; they violently pro-pel their genitalia into their one and only sexual partner, and so could not mate again even if they survived the traumatic loss of their genitalia, which they do not [1266].

In addition to defining mating systems on the basis of the number of mates that males and females have in a given breeding season, we can also distinguish between the social and genetic forms of the various mating systems.

When a male bluebird joins a single female in building a nest, defending a territory, and helping rear the female's brood (a case of social monogamy), their arrangement, as it turns out, does not necessarily translate into genetic monogamy for either partner. If, for example, the female bluebird copulates covertly with males other than her social partner and actually has offspring by several males, not just one, she has engaged in genetic polyandry. One of the revelations of the paternity analyses based on DNA fingerprinting techniques during the past decade or so is the extent to which apparently monogamous females actually acquire and use sperm from more than one male [1285]. The adaptive value of genetic polyandry by females is one of the central issues that this chapter, which will explore the evolutionary bases for the full range of mating systems.

Is Monogamy Adaptive?

Monogamy, whether social or genetic, offers a special challenge for the adaptationist. How can it possibly be advantageous for a honey bee drone (Figure 1), or a male of any other species, for that matter, to inseminate only one female per breeding period? Sexual selection theory (p. 327) suggests that males should generally try to mate with many females, not just one, because a male's reproductive success is usually related to the number of females he inseminates. As a result, males compete for mates, and the winners are polygynists that are likely to gain relatively high fitness. For example, male house wrens that attract two females produce about nine fledglings per year on average, whereas monogamous male wrens have fewer than six [1087]. Viewed in these terms, males that engage in genetic monogamy—and there are species with this mating system—are surprising.

Various explanations have been offered for why some males might stop hunting after having found one sexual partner. The **mate guarding hypothesis** argues that monogamy may be adaptive when a female left by one male would probably acquire another mate, whose sperm would then fertilize her eggs [362]. Mate guarding is especially likely to provide fitness payoffs when females remain receptive after mating and when they are widely scattered and difficult

1   The monogamous honey bee drone. (Left) An intact male and (right) a queen with the yellow genitalia of a deceased partner attached to the tip of her abdomen. Photographs by Christal Rau, courtesy of Nikolaus Koeniger.

2 A monogamous shrimp. When males of the clown shrimp encounter a potential mate, they remain with her, because receptive females are scarce and widely distributed. Here, a couple feeds on the severed arm of a starfish. Photograph by Roger Steene.

to locate—conditions that apply to many birds. Consider also the beautiful clown shrimp, *Hymenocera picta* (Figure 2), whose males spend weeks with one female rather than searching for more mates [1229]. As expected, the operational sex ratio (p. 321) is highly male-biased in this species because females are receptive for only a short period every 3 weeks or so. Because finding the widely dispersed receptive females is so difficult, males that encounter a potential mate guard her until she is willing to copulate. Here we have an example of what will be a recurrent theme in this chapter; namely, that ecological factors that determine female distributions and movement patterns have a major effect on how males compete for access to mates.

Alternatives to the mate guarding hypothesis for monogamy include the **mate assistance hypothesis,** which states that males remain with a single female to help rear their mutual offspring [362]. In some environments, the additional offspring that survive because of paternal care may more than compensate the monogamous male for giving up the chance to reproduce with other females. Males of the seahorse *Hippocampus whitei*, for example, take on the responsibility of pregnancy, carrying a clutch of eggs in a sealed brood pouch over 3 weeks or so (Figure 3). Each male has a durable relationship with one female, who provides him with a series of clutches. Pairs even greet each other each morning before moving apart to forage separately; they will ignore any others of the opposite sex they happen to meet during the day [1170]. Since a male's brood pouch can accommodate only one clutch of eggs, he gains nothing by courting more than one female at a time. Nor would he have much to gain by switching mates from one pregnancy to the next if his one mate can supply him with a new clutch of eggs as soon as he has reared the current brood. Many females apparently can keep their partners pregnant throughout the lengthy breeding season. Females also have an incentive to stick with one male, since the costs of attempted polyandry would probably be high, given the very low density of seahorses, their painfully slow swimming speed, and their vulnerability to sharp-eyed predators.

In other species, although males might gain by acquiring several mates, females attempt to block their partners' polygynous moves in order to monopolize their parental assistance, leading to **female-enforced monogamy.** In a

3 A pregnant male seahorse giving birth. Several youngsters are emerging, some tail-first. More are on the way (note the swollen belly of the pregnant male). Photograph by Rudie Kuiter.

Burying beetle

seabird called the razorbill, for example, which nests in clusters of a dozen or so pairs in crevices and under rock slabs on coastal cliffs, females attack their partners when they show an interest in a neighbor [1179]. The same is true for paired females of the burying beetle *Nicrophorus defodiens*. In this species, a mated male and female work together to bury a dead mouse or shrew, which will feed their offspring once they hatch from the eggs the female lays on the carcass. But once the carcass is buried, the male may release a sex pheromone to call a second female to the site. If this female also laid her clutch of eggs on the carcass, her larvae would compete for food with the larvae of the first female, reducing their survival or growth rate. Thus, when the paired female smells her mate's pheromone, she hurries to push him from his perch. These attacks reduce his ability to signal, as Anne-Katrin Eggert and Scott Sakaluk [351] showed by tethering the paired female so that she could not interact with her partner. Freed from a controlling spouse, the experimental males released scent for much longer than control males that had to cope with untethered mates (Figure 4). Thus, burying beetle males may often be monogamous not because it is in their genetic interest, but because females make it happen.

Monogamy in Mammals

Social monogamy is exceptionally rare in mammals, a group notable for the size of female parental investment in offspring. Given the nature of mammalian pregnancy and milk production, sexual selection theory suggests that mammalian males, which cannot become pregnant and do not offer milk to infants, should usually try to be polygynous—and they usually do. However, as always, exceptions to the rule can be instructive in helping to explain the evolution of a phenomenon.

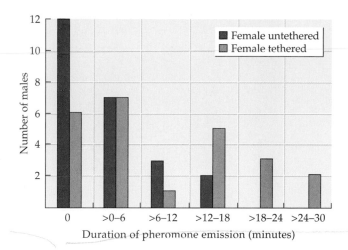

4 Female burying beetles combat polygyny. When a paired female beetle is experimentally tethered, the amount of time that her partner spends releasing sex pheromones to attract additional mates rises dramatically. After Eggert and Sakaluk [351].

If the mate assistance hypothesis for monogamy applies to mammals, then males of monogamous mammalian species should care for their offspring. Although fewer than 10 percent of all mammals exhibit male parental care [1263], some of these exceptions are indeed monogamous, including some notable rodents, among them the Djungarian hamster, whose males actually help deliver their partners' pups (Figure 5) [592]. Male parental care contributes to offspring survival in the hamster and also in the monogamous California mouse, pairs of which were consistently able to rear a litter of four pups under laboratory conditions, whereas single females always failed when they had to spend consid-

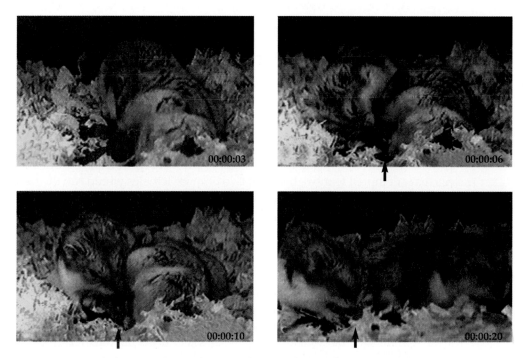

5 An exceptionally paternal rodent. Male Djungarian hamsters may pull newborns from their mate's birth canal and then clear the infants' airways by cleaning their nostrils, as shown in these photographs (the male is the hamster on the left; arrows point to the pink newborn). From Jones and Wynne-Edwards [592].

6 Male care of offspring affects female fitness. The mean number of offspring reared by female California mice falls sharply in the absence of a helpful male partner. After Gubernick and Teferi [465].

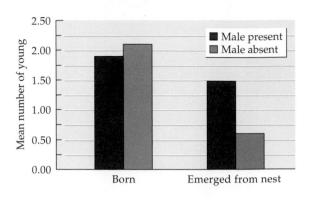

erable time away from their pups searching for food [192]. In addition, the number of young reared in natural nests of free-living California mice falls if a male is not present to help his mate keep the pups warm (Figure 6) [465].

In primates, one of the things a male can do for his offspring is to protect them against infanticidal male intruders, which seek to destroy the infant in order to mate with its mother (p. 14). If males and females form a social bond to reduce the risk of infanticide, then such monogamous associations should evolve in species in which new mothers tend to travel with their offspring (rather than leaving them in a nest), requiring that a protective male accompany the female if he is to defend her infant. In one group of primates, the prosimians, a nearly perfect correlation exists between mother–infant associations and year-round male–female pairings (Figure 7) [1166].

Although these results provide some support for the mate assistance hypothesis for mammalian monogamy, a systematic comparative test of the proposition produced completely negative results: male parental care and social monogamy are *not* tightly correlated in rodents, primates, or any other mammalian group [648]. For example, male parental care occurs in half of the 16 primate taxa (species or groups of related species) known to be monogamous, but male assistance also characterizes 35% of the 20 polygynous taxa. The difference between the two categories is not statistically significant, contrary to the expectation that paternal primate species should be monogamous while nonpaternal species should be polygynous.

In fact, the only pattern that survives comparative analysis is that social monogamy appears more often in mammals whose females live apart from one another in *small* territories. Under these conditions, males can effectively monitor the activities of one female at relatively modest energetic expense, but any male that attempted to move between two or more females would run the risk that intruder males would visit his unguarded female(s) while he was absent. These findings suggest that mate guarding probably played the primary role in the evolution of mammalian monogamy.

Monogamy in Birds

In about 90 percent of all bird species, males and females engage in social monogamy, with pairs forming long-term partnerships during a breeding season, thereby reducing or eliminating opportunities for males to live in association with several females (social polygyny) while also lowering the chances that a female will be able to bond with several social partners (social polyandry) [671]. In at least some birds, social monogamy does equate with genetic monogamy. In the common loon, for example, DNA fingerprinting of 58 young from 47 families revealed that all were the genetic offspring of the loons that raised them [922]. Similar techniques applied to Florida scrub jays

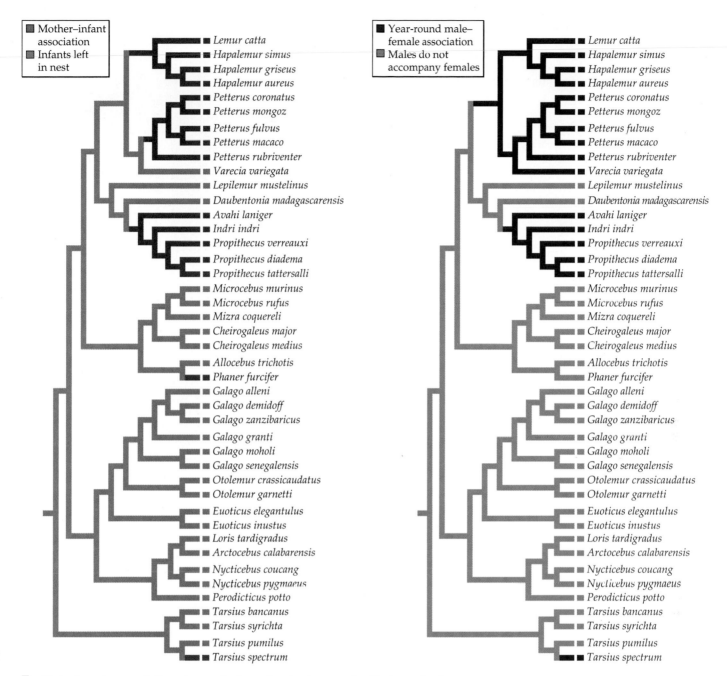

7 **Male–female associations in prosimian primates** have evolved independently four or five times in species or groups of species in which mothers carry their dependent infants with them. After van Schaik and Kappeler [1166].

also showed that nestlings were indeed the offspring of the adults that cared for them [950].

Why should social (and genetic) monogamy be so much more common in birds than in mammals? Gordon Orians argued that many male birds are monogamous because they, unlike most male mammals, can potentially increase their fitness substantially by incubating eggs and feeding nestlings [860]. One of the key predictions from the mate assistance hypothesis is that male parental care should in fact increase the number of surviving young reared by a female

8 Paternal assistance is valuable in the monogamous snow bunting. Control females often reared four or more offspring with the help of their male partners, whereas few females that had been experimentally "widowed" by removal of the male partner were able to fledge more than three chicks. After Lyon et al. [736].

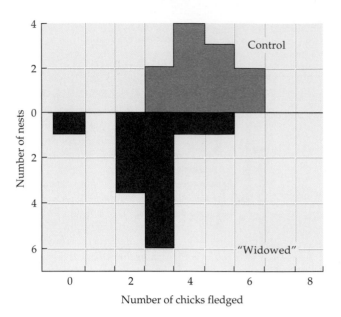

Snow bunting

bird, a proposition that has been checked in various ways. In the yellow-eyed junco, for example, observations of young fledglings reveal that these "bumblebeaks" are so inept at collecting and processing insect food that the male's food contributions to his fledglings are essential for their survival—and his fitness [1193]. Other studies have made the same point by experimentally removing males from pairs, leaving "widowed" females that sometimes reared fewer or lighter-weight young that did non-widowed controls (Figure 8) [736].

The mate assistance hypothesis has also been tested ingeniously by biochemically manipulating the parental behavior of male spotless starlings. The researchers treated some males with testosterone, which sharply reduced the starlings' willingness to feed their broods (see p. 176). Other males were given an anti-androgenic chemical, which blocked the effects of their naturally circulating testosterone and led them to feed their offspring about twice as often unmanipulated controls [818]. The mean number of fledged young per brood was highest for the starlings with inactivated testosterone and lowest for those with extra testosterone (Figure 9).

Spotless starling

9 Chemically altered paternal care alters the fledging rate for broods of the spotless starling. Males that received the anti-androgen cyproterone acetate (CA) provided more food for their broods and had the highest fledging rate per brood. Males given extra testosterone (T) provided less food and had the lowest fledging rate. Untreated controls were intermediate in feeding and fledging rates. After Moreno et al. [818].

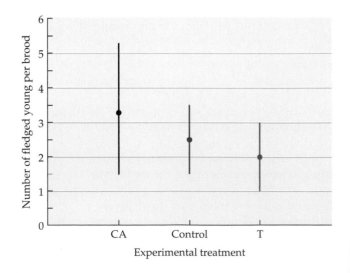

However, not every songbird study has found that male parental attention greatly increases the productivity of a breeding pair. In the savannah sparrow, some males are polygynous, and these sparrows provide little help to their secondary (later-arriving) mates. Even so, these secondary females have just as many fledglings, which weigh as much as those produced by females that have the full help of a male partner [1220]. Moreover, removal of the male has no appreciable effect on the number or weight of the young reared to fledging by the unassisted female sparrows [405]. In species of this sort, it is hard to argue that social monogamy evolved because the paternal male gained fitness by helping his mate care for their offspring.

Extra-Pair Copulations: The Male Perspective

An alternative hypothesis for the evolution of avian monogamy is that males form associations with females in order to prevent them from mating with other males. This mate guarding hypothesis demands that females of socially monogamous birds sometimes mate with other males, as is known to be true for many birds. Indeed, evidence has mounted that most socially monogamous birds regularly copulate outside the pair bond (e.g., [1217]). From the perspective of a male bird, these **extra-pair copulations,** or EPCs as they are known to acronym enthusiasts, have clear costs, including the risk just mentioned, namely, that a male's social partner, whose offspring he will assist, may fertilize her eggs with another male's sperm.

Whether extracurricular matings result in egg fertilizations is a matter that can be checked, thanks to what is popularly known as DNA fingerprinting. Several fingerprinting techniques exist, including what is now called multilocus DNA profiling [180] and microsatellite analysis [948]. Because the use of microsatellites has become increasingly popular for paternity assessment, let's briefly outline the technique. It takes advantage of the fact that scattered throughout the chromosomes are stretches of so-called "junk DNA" that consist of repeating triplets of nucleotides, such as

AAT AAT AAT AAT AAT
TTA TTA TTA TTA TTA

The number of times the base sequence AAT is repeated at a given location on a chromosome often varies greatly, so that one individual might carry five copies of AAT on one chromosome, whereas the other chromosome of the pair might have eight copies. Still other individuals might have other microsatellite alleles, such as $(AAT)_{10}$ or $(AAT)_{21}$. Microsatellite analysts employ procedures that enable them to identify a given nucleotide sequence, remove it from its chromosome, and copy the chromosome region over and over to make large numbers of the microsatellite (and the nucleotides immediately adjacent to it). The fluid containing the amplified DNA is then placed at one end of a gel (a thin strip of starch). When an electric field is run through the gel, microsatellites of different sizes will distribute themselves differently as they move up the gel.

Imagine that a female with the microsatellite genotype $(AAT)_5 (AAT)_8$ mates with two males, one of which has the genotype $(AAT)_{10} (AAT)_{21}$ while the other has the genotype $(AAT)_9 (AAT)_{33}$. Each offspring will carry a microsatellite allele from their mother, either $(AAT)_5$ or $(AAT)_8$, and a microsatellite allele present in the DNA of the sperm from their father. In order to determine the paternity of an offspring, one need only establish its microsatellite genotype; if it is, for example, $(AAT)_8 (AAT)_{33}$, we know who its father is (Figure 10). Studies of this sort have shown that EPCs often do result in reproductive success for philandering males. Indeed, extra-pair males fathered at least some offspring in 10 percent or more of the broods in 15 of the 32 bird species for which such data were available [107].

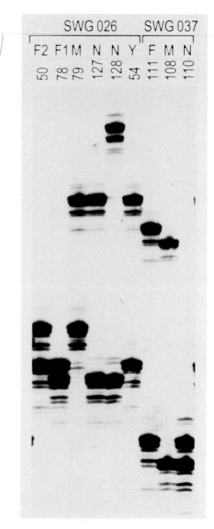

10 A DNA microsatellite gel showing the bands of individuals belonging to two family groups of the splendid fairy-wren, a socially monogamous Australian songbird. Each major black band (ignore the thin black lines) represents a microsatellite allele. All the birds are heterozygous (having two different alleles of this microsatellite). Family SWG037 consists of a breeding female (F111), a male (M108), and a nestling (N110), which has inherited one allele from each of its parents. In this instance, social monogamy has yielded genetic monogamy. Family SWG026 has an adult male (M79) paired socially with a female (F1-78); in addition, another adult female (F2-50) associates with the breeding pair and has helped feed nestlings (N127, N128, and Y54). The helping female and two of the nestlings are the genetic offspring of the socially bonded male and female, from whom they have inherited one microsatellite allele each. On the other hand, N128 has an allele from female F1-78 but no allele from male M79. His father is another male that mated with female F1-78, a genetically polyandrous individual. Photograph by Mike Webster.

Extra-Pair Copulations: The Female Perspective

Males obviously benefit by siring offspring that will be cared for by someone else (provided that their EPC attempts do not expose them to cuckoldry by their own social mate), but what might females gain by mating outside the pair bond? In some species, females probably do not benefit at all, judging from their attempts to avoid contact with EPC-eager males. In other species, however, females appear to accept, or even solicit, matings from two or more males [108]. As behavioral biologists have contemplated what females might gain from EPCs, they have come up with many possibilities (Table 1).

Extra-pair fertilizations could, for example, reduce the risk to a female of having an infertile partner as a social mate. This **fertility insurance hypothesis** is supported by the observation that the eggs of polyandrous female red-winged blackbirds are somewhat more likely to hatch than the eggs of monogamous females [453]. Similarly, in Gunnison's prairie dogs, polyandrous females became pregnant 100 percent of the time, compared with 92 percent of the time for monogamous females [546]. However, these results could also arise from very early embryonic failure caused by defective genes or the genetic incompatibility of the maternal and paternal genomes (see below).

The **good genes hypothesis** argues that females engage in EPCs in order to secure superior genes for their offspring—or genes that will at the very least promote normal embryonic development. The superb fairy-wren provides a possible example of this strategy. In this species, a socially bonded pair lives on a territory with a number of subordinate helpers, usually males. When the breeding female is fertile, she regularly leaves her territory before sunrise and travels to another territory, where she mates with a male before returning home [332]. Perhaps the female's choice of an extra mate to fertilize some of her eggs results in genetic benefits for her offspring. This possibility seems very likely in the blue tit. Female tits are especially likely to seek out extra-pair copulations if they have an unattractive social partner that secures relatively few extra-pair copulations himself (Figure 11) [609]. Genetically polyandrous females produce some offspring sired by attractive males whose survival chances are much higher than those of cuckolded males [608]. By mating with long-lived males,

Blue tit

TABLE 1 Why do females mate voluntarily with more than one male?	
Genetic benefits polyandry	
1. Fertility insurance hypothesis	Mating with several males reduces the risk that some of the female's eggs will remain unfertilized due to mating with a partially or completely infertile male.
2. Good genes hypothesis	Females mate with more than one male because their social partner is of lower genetic quality than other potential sperm donors whose genes will improve offspring viability or sexual attractiveness.
3. Genetic compatibility hypothesis	Mating with several males increases the genetic variety of the sperm available to the female, increasing the chance that some will be genetically compatible with her eggs.
Material benefits polyandry	
1. More resources hypothesis	More mates mean more resources or parental care received from the sexual partners of a female.
2. Better protection hypothesis	More mates mean more time with protectors who will keep other males from sexually harassing a female.
3. Infanticide reduction hypothesis	More mates mean greater confusion about paternity of a female's offspring and thus less likelihood of losing offspring to infanticidal males.

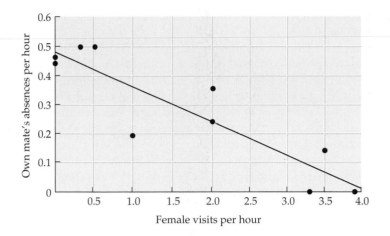

11   Selective polyandry by a songbird. In the blue tit, some females leave their home territories to visit and mate with males on other territories. Males that attract many visitors have mates that stay at home, suggesting that they are high-quality partners whose mates would gain little by extra-pair copulations elsewhere. In contrast, males whose mates often leave to visit other territories are rarely visited by other receptive females, suggesting that these males are considered inferior. After Kempenaers et al. [608].

females may acquire sperm endowed with genes that will promote the survival of their offspring [609]. Likewise, females of the familiar bird-feeder visitor, the black-capped chickadee, used the sperm of males dominant to their social partners in 11 of 15 cases of extra-pair paternity in which the identity of the extra-pair male was known [870].

The conclusion that genetically polyandrous females go out of their way to mate with males likely to possess good genes has also received support from studies of animals other than birds. Thus, the female sierra dome spider routinely mates with more than one male, preferring large and active partners capable of prolonged copulatory stimulation. When Paul Watson experimentally paired females with males of varying size and vigor, he found that the growth rates of the resulting offspring and their eventual size were correlated with their father's traits [1186]. It is also possible, however, that the genetic benefit of selective polyandry to females is the production of sons that have inherited the traits that made their fathers attractive. The sons of sexually successful field crickets are indeed more attractive to females than the sons of males that fail to acquire a mate when placed in a cage with a female and a competitor male (Figure 12) [1198]. This result suggests that females can gain superior qualities for their sons by choosing some potential fathers over others. Remember, however, that in all cases of this sort, it is possible that the improved performance of offspring stems from the willingness of females to invest more resources in the offspring of

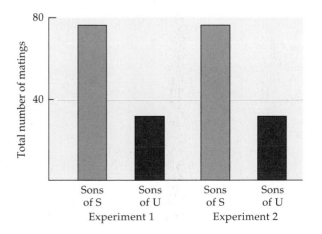

12   A father's mating success can be transmitted to his sons. In experiment 1, two male crickets were given an opportunity to compete for a female; one male (S) mated successfully leaving the other male (U) unsuccessful. When the sons of Male S were placed in competition with the sons of Male U (which had been given a female to mate with after failing to win the initial competition), the sons of S were about twice as likely to mate with the female as the sons of U. In experiment 2, a male that had won a mating competition was later allocated a female at random for breeding as was a male that had lost the competition. The sons of the two males were then placed in an arena with a female and as before, the sons of S were much more likely to mate with the female than the sons of U. After Wedell and Treganza [1198].

favored males, and not from those offspring having received genetic benefits from their fathers (p. 353).

Even if some males are capable of offering their mates sperm with good genes, an individual that provided "good genes" for one female might not be the ideal mate for another. Instead, the genetic benefits of mating with a male could depend on the compatibility of the sperm and egg genotypes, which could affect the viability of the resulting offspring. According to this **genetic compatibility hypothesis** [1283, 1284], the survival of an offspring might depend (for example) on its degree of heterozygosity, especially at the MHC locus (see p. 88). Offspring that are heterozygous for MHC genes may have immune responses superior to those of homozygotes, or they may develop more normally than homozygous individuals. This hypothesis predicts that the spontaneous abortion rate for fetuses that are heterozygous for MHC or other key genes will be less than that for homozygous embryos, which is true for an inbred human population [853]. Note that female mate choice in this case does not necessarily involve securing universally "good genes," but rather involves the acquisition of "different genes" that complement her own [169].

One way for a female to acquire a variety of sperm genotypes, at least some of which may perform well with her eggs' genotypes, is to mate with several partners. Jeanne Zeh showed that female pseudoscorpions (see Figure 20, p. 268) that mated with several males had lower rates of embryo failure (and more surviving offspring) than females that were experimentally paired with a single male (Figure 13) [1281]. As a result, polyandrous females increased their lifetime reproductive success [838]. Yet when Zeh compared the numbers of offspring different females produced from matings with the *same male*, she found no correlation. In other words, some males were not studs while others were duds. Instead, the effect of a male's sperm on a female's reproductive success depended on the match between their particular gametes, as predicted by the genetic compatibility hypothesis. Because a female's chances of securing genetically compatible sperm increase with the number of males she mates with, we can predict that females of this pseudoscorpion should avoid mating with the same male twice. Indeed, when a female was given an opportunity to mate with the same male 90 minutes after an initial copulation, she refused to permit him to pull her over his spermatophore in 85 percent of the trials. But if the partner was new to her, the female generally accepted his gametes [1282].

Our focus thus far on genetic gains for polyandrous females should not obscure the possibility that females sometimes mate with several males in order

Tropical pseudoscorpion

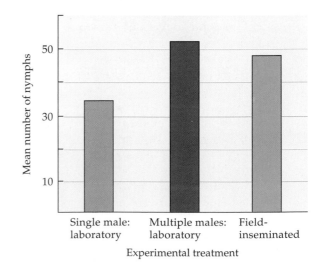

13 Female reproductive success in a polyandrous species. In laboratory experiments, female pseudoscorpions restricted to a single partner produced fewer nymphs than females that had mated with several males. Paternity tests of the offspring of wild-caught females confirm that females usually mate with several males under natural conditions. After Zeh [1281].

14 **Polyandry with material benefits.** By mating with many males, females of this megachilid bee gain access to pollen and nectar in the males' territories. Photograph by the author.

to secure certain material benefits. As is consistent with this **more resources hypothesis,** female red-winged blackbirds in some populations are allowed to forage for food on the territories of males with whom they have engaged in EPCs, whereas truly monogamous females are chased away [454]. Similarly, females of some bees must copulate with territorial males each time they enter a territory if they are to collect pollen and nectar there (Figure 14) [16]. Female hangingflies and butterflies also have an incentive to mate several times in order to receive a series of nuptial gifts or valuable spermatophores from their partners. The spermatophores of butterfly species in which females usually mate with a substantial number of males contain much more protein than those of species whose females are monogamous or mate with few males [109]. In other words, males of polyandrous butterfly species encourage females to mate with them by making it nutritionally worthwhile.

Among birds, females may solicit matings from more than one male in a breeding season to secure more parental assistance from their primary partner or to receive parental care from their several mates. Simply by being receptive to males other than her social mate, a female can encourage him to guard her more attentively and to invest more heavily in her offspring, since the fitness of attentive mate guarders will be largely a function of their social partner's fitness [364]. Males of many species do indeed associate closely with mates that have been seen near other males. Interestingly, in a sex-role-reversed bird species, the bronze-winged jacana (in which males perform all the egg incubation and chick care duties), females guard their several partners more closely when the males call loudly from within the female's territory. These signals, which announce that the males are ready to copulate, stimulate their polyandrous mate to hurry over to comply, especially with males that yell the loudest (Figure 15). The noisiest males often attract neighboring females, who may be able to lure them away from their current mate if she does not attend to the male and repel any intruding females that have responded to his calls [187].

In the dunnock, a small European songbird, a female that lives in a territory controlled by one male may actively encourage another, subordinate male to stay around by seeking him out and copulating with him when the alpha

15 A male bronze-winged jacana yelling that he is ready to mate. In this polyandrous species, a male's calls determine the extent to which his social partner guards him against other females. Photograph by Ian Newton, courtesy of Stuart Butchart.

male is elsewhere. Females dunnocks are prepared to mate as many as 12 times per hour, and hundreds of times before laying a complete clutch of eggs. The benefit to polyandrous females of distributing their copulations between two males is that both sexual partners will help them rear their brood—provided that they have copulated often enough. Female dunnocks ensure that this paternal threshold is reached by soliciting matings from whichever male, alpha or beta—usually the latter—has had less time in their company (Figure 16) [291]. Likewise, by mating with certain of their neighbors, female red-winged blackbirds ensure that they can count on these males for help when predators come close to their nests. Males who had mated with a neighboring female were much more vigorous in their attacks on a stuffed magpie placed in the territory of this extra-pair partner than when the magpie was placed in other nearby territories [454].

On the other hand, resident redwing males with unfaithful mates are *less* likely to provide nest defense, which can reduce the average number of fledglings produced by a polyandrous female compared with those who have been monogamous [1191]. Thus, extra-pair copulations carry costs for females, including loss of the primary partner's nest defense services, the time spent interacting with several males, and a heightened risk of venereal disease or parasitic

16 Copulations for paternal care. In territories occupied by two males, female dunnocks solicit copulations more often from males that spend relatively little time with them, whether these are the alpha or beta males. After Davies et al. [291].

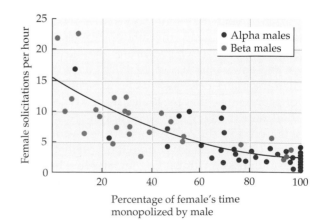

infection. On the other hand, as we have seen, any of several genetic or material benefits could conceivably outweigh these costs. The greater puzzle is why males ever tolerate social polyandry in their mates, which requires them to share, rather than monopolize, the reproductive output of a female [860]. However, because females can often control their mating choices for their own benefit, their decisions may leave a male the option of sharing or not mating at all.

Polyandry without Polygyny

A form of social polyandry that does challenge sexual selection theorists is practiced by the Galápagos hawk, in which as many as eight males may pair-bond for years with just one female, helping her rear a single youngster per breeding episode [374]. In this and another odd bird, the purple swamphen, all the males that associate with one female appear to have the same chance of fertilizing her eggs [575]. Thus, female mechanisms that regulate sperm competition may make it advantageous for several males to cooperate in keeping other males out of suitable breeding territories, which appear to be extremely scarce. The territory holders engage in an equal opportunity lottery that determines which male's sperm actually fertilizes the resident female's egg.

In some other polyandrous species, a more or less complete sex role reversal takes place, with males assuming all or most parental responsibilities and females competing for access to mates. In the wattled jacana, for example, an aggressive female whose territory encompasses those of several males gives each male a clutch of eggs that he incubates alone. Males that pair with a polyandrous female are likely to care for offspring sired by other males, since 75 percent of the clutches laid by polyandrous females are of mixed paternity [367]. Broods of mixed paternity also occur, but to a lesser degree, in the red phalarope, a shorebird in which females are both larger and more brightly colored than males (Figure 17). Female phalaropes fight for mates, which will take care of the eggs whether or not they have been fertilized by other males [277]. Cases of this sort illustrate the disadvantages of polyandry from the male fitness perspective, but male jacanas and phalaropes may have to accept the eggs their partners lay if they are to leave any descendants at all.

The same applies to the spotted sandpiper, whose females also behave like males in many ways [865]. In addition to taking the lead in courtship, females are larger and more combative than males, and they arrive on the breeding grounds first, whereas in most birds males precede females. Once on the breed-

17   **A polyandrous bird,** the red phalarope, whose females defend first one male and then another, donating a clutch of eggs to each male in turn. The more brightly colored bird (on the left) is the female. Photograph by Bruce Lyon.

18 **Resource defense polyandry.** Two female spotted sandpipers (left) about to fight and (right) in the act of fighting for possession of a territory that may attract several males to the winner. Photographs by Stephen Maxson.

ing grounds, females fight with other females for territories (Figure 18). A female's holdings may attract first one and then later another male. The first male mates with the female and gets a clutch of eggs to incubate and rear on his own in her territory, which she continues to defend while producing a new clutch for a second mate. As a result, a few females achieve higher reproductive success than the most successful males [863], an atypical result for animals generally (see p. 323).

Lew Oring believes that the understanding of **sex role reversal polyandry** in spotted sandpipers is advanced by recognizing a key historical constraint [862]. In all sandpiper species, females never lay more than four eggs in a clutch, and cannot adjust clutch size even if resources are especially abundant—perhaps because females that laid five eggs would actually have fewer *surviving* offspring than those that laid four eggs. A variety of egg addition experiments have shown that sandpipers given extra eggs sustain greater losses to predators or to egg damage caused by the incubator as it tries to brood the expanded clutch [46]. If spotted sandpipers are "locked into" an inflexible four-egg maximum, a trait inherited from an ancestral sandpiper, then they can capitalize on rich food resources only by laying more than one clutch, not by increasing the number of eggs laid in any one clutch. To do this, however, they must acquire more than one mate to care for their sequential clutches, making this a rare case in which female fitness is limited more by access to mates than by production of gametes.

Given the four-egg clutch of spotted sandpipers, social polyandry may be adaptive in this species because of the confluence of several unusual ecological features [682]. First, the adult sex ratio is slightly biased toward males. Second, spotted sandpipers nest in areas with immense mayfly hatches, which provide superabundant food for females and for the young when they hatch. Third, a single parent can care for a clutch about as well as two parents, in part because the young are precocial—that is, they can move about, feed themselves, and thermoregulate shortly after hatching. A comparative study has shown that female birds are more likely to evolve short pair bonds with males and seek a series of partners when their young are independent soon after hatching [1125]. This combination of excess males, abundant food, and precocial young means that female spotted sandpipers that desert their initial partners can find new ones without harming the survival chances of their first brood. Once he has been deserted, the male's options are limited. Were he also to desert the eggs, they would fail to develop, and he would have to start all over again. If all females are deserters, then a male that performs the role of sole parent presumably experiences greater reproductive success than he would otherwise, even if his partner acquires another mate to assist her with a second clutch. Furthermore, the *first* partner of a female spotted sandpiper may provide her with sperm that she stores and uses much later to fertilize some or all of her second clutch of eggs (p. 341). Thus, males that arrive on the breeding grounds relatively early and pair off quickly may not lose as much from their mate's acquisition of a second copulatory partner as was once believed [864].

Polygyny

Not all females achieve polyandry in the same manner, nor do males of every species employ the same tactics to become polygynous. Consider the behavior of bighorn sheep, black-winged damselflies, and satin bowerbirds. Even though males of all these species defend mating territories, those territories contain different things. Bighorn rams go where potential mates are, fighting with other males to monopolize females directly (**female defense polygyny**). Male black-winged damselflies wait for females to come to them, defending territories that contain the kind of aquatic vegetation in which females prefer to lay their eggs (**resource defense polygyny**). Male satin bowerbirds, on the other hand, defend territories containing only a display bower, not food or other resources that females might use to promote their reproductive success (**lek polygyny**).

How can we account for the variety of mating tactics among polygynous species? Steve Emlen and Lew Oring argued that the extent to which a male can monopolize mating opportunities depends on social and ecological factors that affect the distribution of females [362]. The degree to which receptive females are clumped in their environment varies as a function of such things as predation pressure and food distribution. Females that live together or aggregate at resource patches can be monopolized economically by males, whereas widely scattered females cannot, because as the size of a territory grows, the costs of defending it also increase.

The **female distribution theory** generates a key prediction: when receptive females occur in defensible clusters, males will compete for those clusters, and female defense polygyny will result. As predicted, social monogamy in mammals never occurs when females live in groups [648]. Instead, in animals as different as bighorn sheep and gorillas, clusters of females, which have formed in part for protection against predators, attract males that compete to control sexual access to the groups [492, 525]. Likewise, males of a tropical bat compete for roosting bands of females in their home cave (Figure 19); successful defenders may sire as many as 50 young with the females in the roost [780]. Similarly, males of the Montezuma oropendola, a tropical blackbird, try to control clusters of nesting females, which group their long, dangling nests in certain trees. The dominant male at a nest tree secures nearly 80 percent of all matings by excluding subordinates from the area [1194]. These dominant males shift from nest tree to nest tree, following females rather than defending a nesting resource, demonstrating that they deserve to be called female defense polygynists.

Groups of receptive females also occur in insects, and female defense polygyny has evolved here as well. For example, nests of *Cardiocondyla* ants produce large numbers of virgin queens that will mate with males inside the colony.

19 Female defense polygyny in a bat. The large male (on the bottom) guards a roosting cluster of smaller females. Many of the bats have been banded so that they can be recognized as individuals by the researcher. Photograph by Gary McCracken.

As we saw in Chapter 11, male nestmates stab and slice each other to death, and they may also induce the sterile workers in the nest to assist them in the destruction of rivals. The few males that survive the murderous melee within a colony copulate with dozens of freshly emerged females [508].

In still other species, males may actually create and control clusters of females by herding them together. Males of some tiny siphonoecetine amphipods construct elaborate cases composed of pebbles and fragments of mollusk shells found in shallow ocean bays. They then move about in these houses, capturing females by gluing the females' houses to their own. A male may succeed in constructing an apartment complex containing up to three potential mates (Figure 20) [595].

(A)   (B)

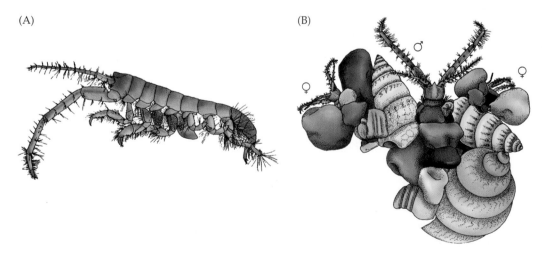

20 Female defense polygyny in a marine amphipod. (A) One individual is shown without his house. (B) A male has glued the houses of two females to his case. Drawings courtesy of Jean Just.

Female Defense Polygyny: The Female Perspective

Female defense polygyny has reproductive consequences for females as well as for males. In yellow-bellied marmots, females often remain near their birth site, so that clusters of mothers, daughters, and aunts end up living together, presumably for improved protection against coyotes and eagles [41]. Males search for locations with one or more resident females, and those males that are able to find and defend a meadow that is home to two or more females practice female defense polygyny. As one might expect, the reproductive success of a male increases as the number of females living in his territory increases. In contrast, females living in the *smallest* marmot groups have the greatest number of offspring per year (Figure 21) [336].

These observations suggest that females do not benefit from being part of a polygynous male's harem. One hypothesis to account for the failure of females to move from large to small groups is that their *lifetime* fitness is greater when they live with many others. If large groups thwart predators better, then females in large groups should live longer and reproduce during more years than monogamous females, compensating for their reduced output in any one year [356]. However, contrary to this prediction, one researcher found no correlation between the size of a group of females and female survivorship [41].

Why, then, do females stay together? Remember that what is adaptive for an individual depends on what its other options are [1261]. Perhaps female marmots that lived alone with a mate would indeed enjoy higher lifetime reproductive success than if they lived in a group. But in order to do this, they might have to drive off all other females, and this might be costly for a variety of reasons, particularly if those other females fought tenaciously to stay in high-quality habitat.

The idea that social interactions within a species have much to say about the evolution of that species' mating system can also be illustrated with lions. You will recall that lion females form groups whose members cooperatively hunt big game, but that debate exists on whether the caloric gains from cooperative hunting are sufficient to explain the formation of these prides (see p. 239). Female lions are long-lived animals that compete for permanent territories that are essential for successful reproduction. Therefore, lionesses may remain together in order to acquire critical living space and defend it against rival prides. If this hypothesis is correct, then clashes between prides over territories should be resolved in favor of the larger group. Craig Packer's team of lion watchers has observed 15 aggressive interactions between prides of known size, of which 13 were won by the larger group [879].

Yellow-bellied marmot

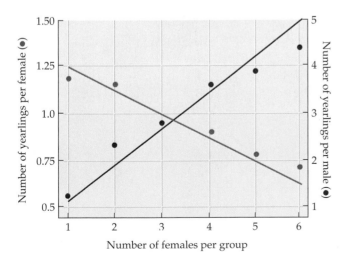

21   Annual reproductive success of male and female marmots. For a male, the number of offspring that survive their first year increases with the size of the group of females he controls, but for females, the number of offspring decreases with group size. After Armitage [41].

22   Resource defense polygyny in an Australian antlered fly. Males compete for possession of egg-laying sites that can be found only on certain species of recently fallen trees. Photograph by Gary Dodson.

Interactions between the sexes may also contribute to the maintenance of prides, even at the cost of short-term food shortages. By sticking together, females may have some chance of protecting their cubs after a pride takeover has occurred (see p. 17). When the new males encounter a single female parent, her cubs are doomed, whereas groups of two or more females occasionally prevent the infanticidal males from destroying all of their babies. Thus, the benefits to lionesses from social living may have more to do with coping with members of their own species than with maximizing food intake [879]. The various competitive and destructive interactions among lions may produce coalitions of females, which then benefit from association with one or more polygynous males that help protect their cubs.

Resource Defense Polygyny

In many species, females do not live permanently together, but a male may still become polygynous if he can control a rich patch of resources that females visit on occasion. Male black-winged damselflies, for example, defend floating vegetation that may attract a series of sexually receptive females, which mate with the male and lay their eggs in the vegetation he controls. Likewise, male antlered flies fight for small rot spots on fallen logs or branches of certain tropical trees because these locations attract receptive, gravid females, which mate with successful territory defenders before laying their eggs (Figure 22) [330].

A safe location for eggs constitutes a defensible resource for a host of animal species; the more of this resource a territorial male holds, the more likely he is to acquire several mates. In an African cichlid fish, *Lamprologus callipterus*, a female deposits a clutch of eggs in an empty snail shell, then pops inside to remain with her eggs and hatchlings until they are ready to leave the nest. Territorial males of this species, which are much larger than their tiny mates, not only defend suitable nest sites, but collect shells from the lake bottom and steal them from the nests of rival males, creating middens of up to 86 shells (Figure 23). Because as many as 14 females may nest simultaneously in different

23   Resource defense polygyny in an African cichlid fish. (Left) A territorial male bringing a shell to his midden. (Right) The tail of one of his very small mates can be seen in a close-up of the shell that serves as her nest. Photographs by Tetsu Sato.

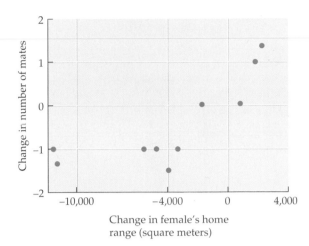

24 **A test of the female distribution theory of mating systems.** When food supplements reduce the size of a female dunnock's home range greatly, a male can monopolize access to that female and reduce the number of males with whom she interacts. After Davies and Lundberg [299].

shells in one male's midden, the owner of a rich territory can enjoy extraordinary reproductive success [1016].

If the distribution of females is controlled by the distribution of key resources, and if male mating tactics are in part dictated by this fact of life, then it ought to be possible to alter the mating system of a species by moving resources around, thereby altering where females are located. This prediction was tested by Nick Davies and Arne Lundberg, who first measured male territories and female foraging ranges in the dunnock. In this drab little songbird, females normally roam over such large areas hunting for widely dispersed food items that the ranges of two males may overlap with those of several females. However, when Davies and Lundberg gave some females supplemental oats and mealworms for months at a time, female home ranges contracted substantially [299]. Those females with the most reduced ranges had, as predicted, fewer social mates than other generally polyandrous females whose ranges remained the same size or increased slightly (Figure 24). These results support the general theory that males attempt to monopolize females within the constraints imposed on them by the spatial distribution of potential mates, which in turn may be related to the spatial distribution of key resources.

Dunnock

Resource Defense Polygyny: The Female Perspective

A female that pairs with a male that already has a partner is forced to share the resources under his control with his first mate. How can such resource sharing be reproductively advantageous to a female? According to the **polygyny threshold model,** when the quantity or quality of resources controlled by males varies greatly, females that join already paired males on very rich, or very safe, territories may have more surviving offspring than if they were to pair off with single males on resource-poor, or predator-vulnerable, territories [860].

The polygyny threshold hypothesis has been tested in a study of the polygynous great reed warbler. If females prefer some males over others because their territories are safer, then eggs in nests in attractive territories (those that are quickly occupied by nesting females) should be less often destroyed by egg-eating predators than eggs in less attractive territories. The data support this hypothesis (Figure 25) [490], suggesting that females may accept polygynous males in order to secure a safer nest site.

Of course, a direct way to test the polygyny threshold model is to count the number of young fledged in territories controlled by monogamous versus polygynous males. Michael Carey and Val Nolan did the counting for a popu-

Great reed warbler

25 The risk of predation is related to the attractiveness of a territory to nesting female great reed warblers. Females that nest in territories that are quickly occupied, which are usually held by polygynous males, are more likely to have a nest that survives. Females that nest in less attractive territories, which are more likely to be held by monogamous males, often lose their nests to predators. After Hansson et al. [490].

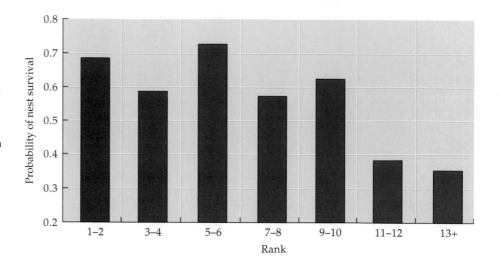

lation of indigo buntings, a small songbird whose parental males may attract two, one, or no mates to their territories in overgrown fields [195]. Monogamous females whose relationship with a male lasted the whole breeding season had only slightly greater reproductive success (1.6 young fledged) than females who participated in a polygynous arrangement (1.3 young fledged). Therefore, females choosing polygynous mates were not penalized heavily, if at all, by moving into an already occupied territory.

In the pied flycatcher, however, monogamous females do better than those in polygynous relationships. In this species, polygynous males manage to attract two mates by maintaining two separate territories, each with a nest cavity in a tree (a vital resource for the female), so that a second female could pair off with a male without necessarily learning that he already has a partner somewhere else. The second female then suffers substantially because her polyterritorial partner helps his first mate's brood at the expense of his second mate's offspring (Figure 26). Thus, the second female might be taken in by a "deceitful" polygynist male [9].

This deception hypothesis predicts that females should not be able to determine the pairing status of a potential mate, but Svein Dale and Tore Slagsvold have shown that mated male flycatchers spend less total time singing than unmated individuals, thus providing an important potential cue for perceptive females [279], which may visit as many as seven males in a single hour [278], returning repeatedly to visit those they consider good prospects. As a result, a male's mating status should usually become apparent to a female when she finds him absent from what is his secondary territory (because he is away checking on his primary territory and primary mate). Therefore, females may not be deceived when they accept an already mated male [1124]. Perhaps they choose him anyway because of the high costs of finding or evaluating other potential partners [1097], or because the remaining unmated males have extremely poor territories [1124], in which case the real options for some females may be to accept a secondary territory or not to breed at all.

Scramble Competition Polygyny

Although female defense and resource defense tactics by competitive males make intuitive sense when females or resources are clumped in small, defensible areas, in many other species receptive females and the resources they need are widely dispersed. Under these conditions, the cost–benefit ratio of mating territoriality falls, and males compete instead by trying to outrace rivals to recep-

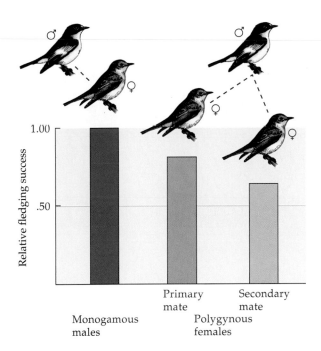

26   Does polygyny reduce female fitness? Female pied flycatchers that mate with monogamous males enjoy higher reproductive success than either primary or secondary mates of polygynous males. After Alatalo and Lundberg [9].

tive females in a mating system that can be called **scramble competition polygyny.** The scarce receptive females of a flightless *Photinus* firefly, for example, can appear almost anywhere over wide swaths of Florida woodland. Males of these species make no effort to be territorial. Instead they search, and search, and search some more. When Jim Lloyd tracked flashing males, he walked 10.9 miles in total, following 199 signaling males, and saw exactly two matings. Whenever Lloyd spotted a signaling female, a firefly male found her in minutes [712]. Mating success in this species probably goes to the searchers that are the most persistent, durable, and perceptive, not the most aggressive.

Male thirteen-lined ground squirrels behave like fireflies, searching widely for females, which become receptive for a mere 4 to 5 hours during the breeding season. The first male to find an estrous female and copulate with her will fertilize about 75 percent of her eggs, even if she mates again [389]. Given the widely scattered distribution of females and the first-male fertilization advantage, the ability to keep hunting should greatly affect a male's reproductive success. In addition, however, male fitness may also depend on a special kind of intelligence [389], namely, the ability to remember where potential mates can be located. After visiting a number of females near their scattered burrows, searching males often return to these places on the following day. When researchers experimentally removed several females from their home sites, returning males spent more time searching for the missing females that had been on the verge of estrus. Moreover, the males did not simply inspect places that females used heavily, such as their burrows, but instead biased their search in favor of the spot where they had actually interacted with an about-to-become-receptive female [1028, 1029]. Individuals with superior spatial memory can probably relocate females more efficiently (Figure 27).

A quite different form of scramble competition polygyny, the **explosive breeding assemblage,** occurs in species with still more highly compressed breeding seasons. One such species is the horseshoe crab, whose females lay their eggs on just a few nights each spring and summer. Males are under the gun to be near the egg-laying beaches at the right times and to accompany a

Thirteen-lined ground squirrel

27 Scramble competition polygyny selects for spatial learning ability. Male thirteen-lined ground squirrels remember the locations of females that are about to become sexually receptive. When males returned to an area where such a female had been the previous day, but from which she had been experimentally removed, they spent more time searching for her, and returned to her home range more often, than in the case of removed females that had not been about to enter estrus when the males visited them the previous day. After Schwagmeyer [1029].

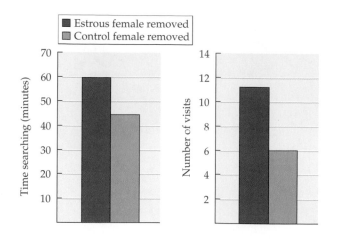

female ashore, where egg laying and fertilization occur [159] (see Figure 18, p. 333). A race to find and pair off with a female also occurs in the wood frog, another species in which the opportunity to acquire a mate is restricted to one or a few nights each year. On that night or nights, most of the adult males in the population are present at the ponds that females visit to mate and lay their eggs. Just as in horseshoe crabs, the high density of rivals raises the cost of repelling them from a defended area. And because females are available only on this one night, a few highly aggressive territorial males cannot monopolize a disproportionate number of them. Therefore, male wood frogs eschew territorial behavior and instead hurry about trying to encounter one or more egg-laden females before the one-night orgy ends (Figure 28) [153].

28 An explosive breeding assemblage. A male wood frog grasps a female (upper left) that he has found before rival males, two of whom are near the mating pair. Numerous fertilized egg masses float in the water around the frogs. Photograph by Rick Howard.

29 **Lek polygyny in the white-beard-ed manakin.** The black and white males defend tiny display sites that are clustered together; the greenish females visit leks to select a mate from the many males displaying there. Photograph by J. Dunning.

Lek Polygyny

An especially intriguing polygynous mating system goes by the name of lek polygyny. In lekking species, males do not hunt for mates. Instead, females come to watch males display at territories that do not contain food, nesting sites, or anything useful to the visitors [141]. The males' little territories may be dispersed, as is true for satin bowerbirds, or clustered in a traditional display area, or lek. For example, males of the South American white-bearded manakin [700, 701], a sparrow-sized bird, defend a single sapling or two rising out of a barren, resource-free display court (Figure 29). As many as 70 males may pack their display courts into an area only 150 meters square. Each male begins his display routine with a series of rapid jumps between perches, loudly snapping certain of his wing feathers together, before pausing with body tensed. The bird then jumps to the ground with a snap and immediately back to the perch with a buzz, and then back and forth "so fast he seems to be bouncing and exploding like a firecracker" [1079]. The arrival of a female at the lek encourages many males to display simultaneously, producing an uproar. If the female is receptive and chooses a partner, she flies to his perch for a series of mutual displays, followed by copulation. Afterward, she leaves to begin nesting, and the male remains at the lek to court newcomers.

Alan Lill found that in a lek with ten manakins, where he recorded 438 copulations, one male contributed nearly 75 percent of the total; a second male mated 56 times (13 percent), while six other males together accounted for a mere 10 matings [701]. You will recall similar inequalities in mating success among males of the satin bowerbird [127]. Likewise, 6 percent of the males in a lek of the bizarre West African hammer-headed bat (Figure 30) were responsible for 80 percent of the matings. In this species, males gather in groups along riverbanks, each bat defending a display territory high in a tree, where he produces loud cries that sound like "a glass being rapped hard on a porcelain sink" [140]. Receptive females fly to the lek and visit several males, each of which responds with a paroxysm of wing-flapping displays and strange vocalizations (note the behavioral convergence with manakins and satin bowerbirds).

30   A lek-polygynous mammal: the hammer-headed bat. Photograph by A. R. Devez.

Why Do Males Aggregate in Leks?

Why do male manakins and hammer-headed bats behave the way they do? Jack Bradbury has argued that lekking evolves only when other mating tactics do not pay off for males, thanks to a wide and even distribution of females [141]. Female manakins and hammer-headed bats do not live in permanent groups, but instead travel great distances in search of widely scattered sources of food of unpredictable availability, especially figs and other tropical fruits. A male that tried to defend one tree might have a long wait before it began to bear attractive fruit, and when it did, the fruit would attract hordes of consumers, which would overwhelm the territorial capacity of a single defender. Thus, the feeding ecology of females of these species makes it hard for males to monopolize them, directly or indirectly. Instead, males display their merits to choosy females that come to leks to inspect them.

But why, in most lekking species, do many males congregate in a small area instead of displaying on their own isolated territories? Of the many hypotheses proposed for male clustering on mini-territories, we shall review three: (1) the **hotspot hypothesis**, according to which males cluster because females tend to travel along certain routes that intersect at particular points, or "hotspots" [144]; (2) the **hotshot hypothesis**, according to which males cluster because subordinate males gather around highly attractive males in order to have a chance to interact with females drawn to these "hotshots" [142]; and (3) the **female preference hypothesis**, according to which males cluster because females prefer sites with large groups of males, where they can more quickly, or more safely, compare the quality of many potential mates [141].

One way of testing the hotspot versus hotshot hypotheses is by temporarily removing males that have been successful in attracting females. If the hotspot hypothesis is correct, removal of these successful males from their territories will enable others to move into the favored sites. But if the hotshot hypothesis is correct, removal of attractive males will cause the cluster of subordinates to disperse to other popular males or to leave the site altogether.

In a study of the great snipe, a European sandpiper that displays at night, removal of central dominant males caused their neighboring subordinates to leave their territories. In contrast, removal of a subordinate while the alpha snipe was in place resulted in quick replacement of the vacant territory by another subordinate. At least in this species, the position taken by attractive males, not the real estate per se, determines where clusters of males form [527, 528]. Likewise, in the unrelated black grouse, the exact location of the most successful territory changed slightly from year to year within a long-lasting lek, suggesting that a popular male, rather than a particular spot, most influences female behavior (Figure 31). In black grouse leks, the top males, which have relatively high testosterone levels [8], fight often with neighbors over the central territories. Males that occupy sites next to popular males gain slightly improved mating success, as expected from the hotshot hypothesis [970].

If the hotshot hypothesis is correct, we can also predict that females should be attracted primarily by particular lekking males, rather than by the location of their display sites. In fallow deer, dominant, attractive bucks can be induced to abandon their central territories by placing sheets of black polyethylene on these sites. After moving to a new site, these hotshots still continue to attract a disproportionately large number of females, despite their shift in position [233].

Although the hotshot hypothesis seems likely in some cases, the hotspot hypothesis has received support in others (Figure 32). For example, when logging activity altered the paths regularly followed by female fallow deer, some bucks eventually abandoned one lek and moved to a neighboring site on a travel route still used by the does [40]. That the tracks followed by females can determine where leks form seems likely, given that prominent topographic features, such as mountain ridgelines or forest streams, serve as lekking areas for a diverse array of insects and tropical birds. These observations suggest that topography determines where females go, and males follow [10, 1210].

On the other hand, the hotspot hypothesis has been found wanting for a number of species whose males do *not* form leks in the places where females are concentrated (Figure 33) [65]. Indeed, estrous females of several lekking

Black grouse

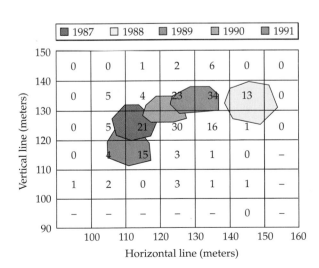

31 Hotspot or hotshot? The total number of copulations recorded in 100-square-meter sectors of a black grouse lek from 1987 to 1991. The irregular polygons show the location of the top territory for each of the five years. The shifts in preferred territory suggest that male attractiveness, rather than the territory itself, plays the key role in lek polygyny in this species, as required by the hotshot hypothesis. After Rintamäki et al. [970].

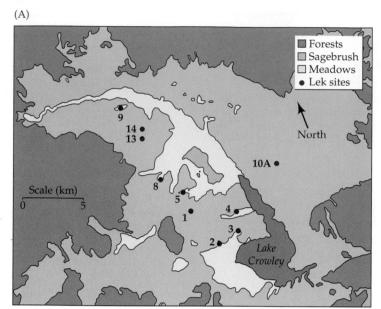

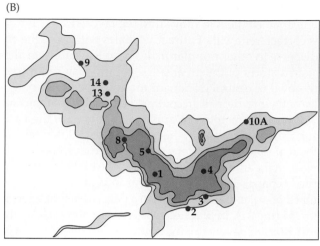

32   A test of the hotspot hypothesis. (A) The position of sage grouse leks (numbered red circles) in relation to sagebrush (shown in white), meadows (pale green), water (blue), and forests (dark green). (B) The distribution of nesting females, in relation to the leks (red circles). The darker the shading, the more females present. Males form leks near concentrations of females. After Gibson [428].

Kafue lechwe

ungulate species leave their customary ranges to visit groups of males elsewhere, perhaps to compare the performance of many males simultaneously. James Deutsch tested whether female preferences push male Uganda kob, an African antelope, into large groups. If so, then those leks with a relatively large number of males should attract proportionately more females than leks with few males. However, the operational sex ratio was the same for leks across a spectrum of sizes (Figure 34), so that males were no better off in large groups

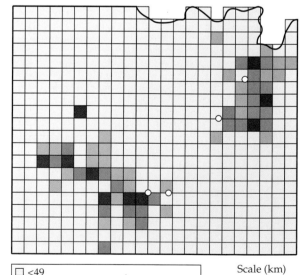

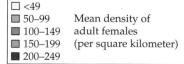

33   Female density is not correlated with lek formation in an African antelope, the Kafue lechwe. The four leks in this region (open circles) are not located in the areas of highest female density, evidence against the hotspot hypothesis in this species. After Balmford et al. [65].

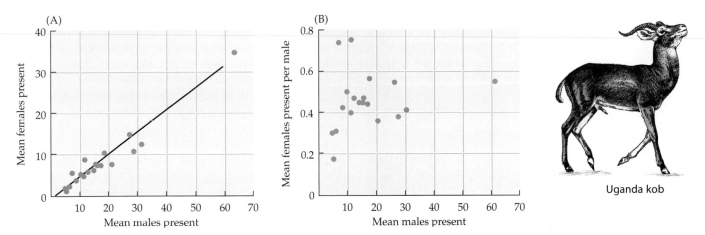

34   Female Uganda kob do not prefer to visit leks with large numbers of males. (A) Female attendance at leks is simply proportional to the number of males displaying there. (B) As a result, the female-to-male ratio does not increase as lek size increases. After Deutsch [316].

than in small ones [316]. For this species at least, the female preference hypothesis can be rejected. The same is true for the ochre-bellied flycatcher, because males in leks were not significantly more likely to be visited by females than males that displayed alone [1211]. In an Australian fruit fly, the female encounter rate for males does increase as the number of males in a lek rises to 20, but then falls off [48], and the same pattern applies to the ruff (Figure 35) [1231].

Why Do Many Females Mate with the Same Males at Leks?

Whatever the reason(s) for male aggregation at leks, the typical rule is that only 10 to 20 percent of the males at a given lek secure more than half the matings [739]. Evidently, most females choose the same individuals that happen to have the most striking body ornaments and the ability to display long and often [387]. What do females gain by selecting these males?

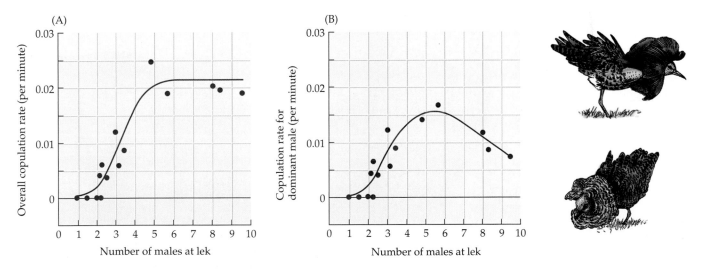

35   Lek size and copulation rate in the ruff, a lekking sandpiper. (A) Up to a point, larger leks attract more receptive females than small leks. (B) At leks of six or more males, however, the number of attracted females levels off, lowering the average reproductive success of the dominant males in attendance because some lower-ranking individuals copulate with some of the visiting females. After Widemo and Owens [1231].

In the lekking black grouse, the males that females mate with have survival rates twice those of the least attractive individuals, suggesting that females gain "good genes" from their choices [7], as is also possible in peacocks (see p. 352). However, in order to derive this genetic benefit, females must choose among males that differ in their genetic quality. In theory, if all females in a population chose the same few males, any genetic differences among males would disappear in a relatively few generations, eliminating any genetic benefits for selective females. The paradoxical persistence of strong female preferences in lekking species creates what has been called the **lek paradox.**

The lek paradox has one solution that is something of a letdown, namely, that females are not as biased in their mate choice as they seem to be. In the buff-breasted sandpiper, for example, most females watched at leks mated with a handful of males, and none was seen to mate with two or more partners. But when paternity tests were done on 109 offspring from 30 broods, at least 39 different fathers were represented [677, 678]. Some females obviously did accept and use sperm from more than one male, presumably maintaining genetic variation in the population, which decreases the extent to which the lek paradox applies to this lekking sandpiper.

Indeed, as it turns out, lekking species often exhibit large amounts of genetic variation among males [931]. Evidence for this conclusion comes from various sources. One can, for example, artificially select for increased or decreased acoustical signaling in the lesser waxmoth, a lekking insect whose males call mates to them with wing vibrations (Figure 36) [582]. Thus, hereditary variation still exists in this trait, despite strong sexual selection for high rates of calling by males over millennia. The reasons for the maintenance of genetic variation in sexually selected traits are doubtless varied, but the development of some of these attributes is strongly affected by variable environmental factors, such as temperature or food availability. In different environments, different alleles may be more likely to yield adaptive developmental outcomes, and under these circumstances, multiple alleles may persist in the population, making it possible for females to benefit year after year from discriminating mate choice.

Finally, John Reynolds and Mart Gross have pointed to several nongenetic gains that females might secure by choosing males carefully from a lek [966]. For one thing, by picking dominant males as copulatory partners, females may avoid the injuries and harassment that come from dealing with individuals that

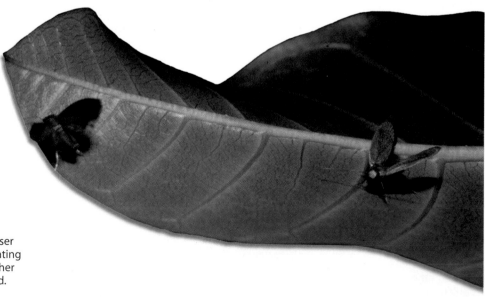

36 **A lekking insect.** Males of the lesser waxmoth "call" to attract females by vibrating their wings while perched on a leaf or other object. Photograph by Michael Greenfield.

have not fully intimidated their rivals. Among fallow deer, for example, females that are being pursued move to the territories of top-ranking males, which effectively repel sexually aggressive rivals [233]. (However, in the lekking great snipe, females chased away by neighbors of the males they are visiting are as likely to return to the original male for a second look as to a male whose courtship display has not been interrupted by a rival, showing that female mate choice, in this species at least, is not related to harassment avoidance [1008].)

If nothing else, the time costs of searching for and evaluating mates are reduced by systems in which groups of males conspicuously advertise their presence, which may make it advantageous for males to cooperate by lekking and for females to take advantage of the opportunity for rapid assessment of potential mates. Perhaps selection in favor of efficient mate choice by females has contributed to lek evolution, but behavioral biologists will have to spend many more hours watching males strut their stuff at leks before this issue can be fully resolved.

Summary

1. Mating systems can be defined in terms of the number of mates an individual forms an alliance with during a breeding season: one male may live with several females (social polygyny), one male may ally himself with a single female (social monogamy), and one female may associate with several males (social polyandry). However, the genetic consequences of these associations may not match their titles; a socially monogamous female may be genetically polyandrous if she mates outside the pair bond, and a socially polygynous male could be genetically monogamous if only one of his social partners actually uses his sperm to fertilize her eggs. Males of most species attempt to engage in genetic polygyny, which is not surprising because male fitness usually increases with the number of females inseminated.

2. Understanding variation in mating systems requires consideration of the different social factors that influence male and female reproductive success. For example, male genetic monogamy may be adaptive when there are large payoffs from male parental care or from guarding a single mate. Alternatively, conflict between females and males may thwart male attempts to be genetically polygynous, even if it would be advantageous from the male perspective.

3. Likewise, although females of most birds and other animals can secure all the sperm they need to fertilize their eggs from a single mate, female genetic polyandry is common because of various genetic and material benefits that females can gain by mating with several partners. Because females usually control how many males have sexual access to them, both female choice and conflict between the sexes play major roles in the evolution of mating systems.

4. Ecological factors that affect males' ability to monopolize receptive females are also important influences on mating system diversity. In particular, differences in the distribution of females may affect the profitability of different kinds of territorial mating tactics for males. When females or the resources they need are clumped in space, female defense or resource defense polygyny becomes more likely. If, however, females are widely dispersed or male density is high, males may engage in nonterritorial scramble competition for mates, or they may acquire mates by displaying at a lek.

5. The evolutionary analysis of lekking behavior provides a good illustration of the productivity of an adaptationist approach to animal behavior, which has enabled researchers to identify and investigate several major puzzles associated with this mating system, such as why males compete in dense display aggregations, and why many females so often choose the same partner from among the many males displaying at a lek.

Discussion Questions

1. We began this chapter with a mention of male monogamy in the honey bee. Try to explain its suicidal mating behavior. Include in your list of hypotheses one based on group selection theory. What predictions follow from the alternative hypotheses you have considered? What data are required to resolve the issue?

2. Construct a list of at least three hypotheses on the adaptive value of extra-pair copulations to female birds, and then make a table of predictions that follow from each hypothesis. Identify the key predictions that would enable you to discriminate among the hypotheses for a particular case. Now, what about those species in which females mate repeatedly with the *same* male, often their social partner? Do the same hypotheses apply for this kind of multiple mating as for polyandry? Check [558] after completing your comparison.

3. Apply optimal foraging theory (see p. 233) to the problem of female mate choice. Contrast the following two options (and any others you choose to consider) for females that encounter one male after another: (1) the best-of-*n*-males tactic, in which females inspect a pool of males and then pick the top-quality individual, or (2) the threshold tactic, in which females have a minimum standard and mate whenever they encounter a male that meets that standard [576]. Under what conditions would tactic 1 be superior to tactic 2, and vice versa? How does cost–benefit analysis come into play here? How might male clustering at leks, for example, affect the payoff to females using each tactic?

4. In a small African antelope called Kirk's dik-dik, most males and females engage in social monogamy [160]. Evaluate alternative hypotheses for monogamy in light of the following evidence: male presence does not affect the survival of their offspring; males conceal the female's estrous condition by scent-marking over all odors deposited by their mates in their territory; males sire their social partner's offspring; females left unaccompanied wander from the pair's territory; some territories contain five times the food resources of others; the few polygynous associations observed do not occupy larger, or richer, territories than monogamous pairs of dik-diks.

5. The average proportion of extra-pair offspring in a brood varies among bird species from 0.0 to almost 0.8, as shown in the figure below [908]. Consider how variation in the benefits and costs of extra-pair paternity (EPP) might be responsible for variation in the willingness of females to engage in EPCs. For example, how might low variation in male genetic quality in a species affect the EPP figure? What about differences among species in the risk of venereal disease, or in the likelihood that partners can detect and punish the sexual infidelity of a mate? Also, if mating with other males provides females of a given species primarily with the opportunity to trade up to a genetically or materially superior partner [209], what is the predicted relationship between extra-pair copulations and divorce among songbirds?

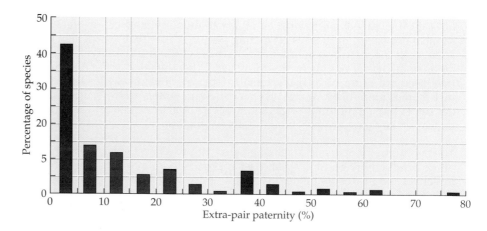

6. The alpine accentor is a songbird with a mating system rather like that of its close relative, the dunnock, in that an alpha and beta male may both copulate with the same female(s) living in the same area. However, unlike the dunnock, alpha male accentors reduce the amount of help they give to females as their mating share decreases, whereas the beta male does not make paternal care donations dependent on past mating frequency [296]. If female accentors have evolved a mating strategy designed to increase the parental care they receive from males, how should their social polyandry differ from that of female dunnocks?

Suggested Reading

Steve Emlen and Lew Oring's now classic paper on the ecology of mating systems changed the analysis of mating systems [362]; see an article by John Reynolds [965] for an update. For an especially thorough examination of a single species with multiple mating systems, read *Dunnock Behaviour and Social Evolution* [292]. Jacob Höglund and Rauno Alatalo [526] have written a useful book on leks. Malte Andersson's book on sexual selection explores mating system evolution in depth [34], while Dave Ligon focuses exclusively on avian mating systems [697].

13 The Evolution of Parental Care

Although most human beings consider care by both parents to be normal and desirable, in a great many species neither the mother nor the father lifts a finger for its brood, while in still others only the female, or much more rarely, only the male, lends a helping hand. This diversity in parental behavior has been hinted at in previous chapters where mention was made of the biparental care provided by the California mouse, the strictly maternal care provided by most other mammals, and the fathers-take-charge approach of the spotted sandpiper and giant water bug. Variety of this sort, as my readers know by now, always interests evolutionary biologists. The key to explaining the variety in parental behavior lies with the cost–benefit approach used by behavioral ecologists. The benefits of parental care are generally pretty obvious, having to do with the improved survival of assisted offspring. But the costs of this parental investment must also be considered if one is to deal with the major evolutionary questions that surround caring for offspring. We will look at several of these questions in this chapter, starting with the issue of why in those species in which either parent could conceivably help its offspring survive, mothers are much more likely to oblige than fathers.

Why Is Parental Care More Often Maternal than Paternal?

The first step toward explaining why parental behavior is not the same in every species is to recognize the physiological expense of the behavior, which can be truly impressive. Think of the male emperor penguin standing in the blackness of the frigid Antarctic winter with a blizzard in progress while he cradles an egg on his feet. The penguin will not eat for more than 2 months while he incubates the egg, and then the chick, under these miserable conditions until his mate finally returns to take care of the chick [27]. Or consider that male mallee fowl, although living in a more hospitable environment, nevertheless invest 5 hours a day for 6 months each year in building and maintaining a nest—a huge mound of sand and decomposing organic material (Figure 1). Each time a female lays an egg in the mound, the male digs a deep pit to receive the egg, then covers it up, which requires moving about 850 kilograms of compost and sand—nearly 500 times his own weight—clawful by clawful [1192]. Between eggs, he regulates the internal temperature of the nest by moving material around so as to ensure that the compost produces just the right amount of heat.

The time and energy given over to being a parent penguin or mallee fowl surely create fitness *disadvantages* for males, which may be less able to survive another year to reproduce again. Can the hard-working mallee fowl increase the survival chances of the eggs in his nest enough to overcome the fitness costs of his exhausting activities? And having worked so diligently to provide an appropriate environment for the eggs, why then does he completely ignore the chicks when they hatch and leave the nest? Moreover, why is it that emperor penguins and mallee fowl are exceptions to the general rule that mothers do the majority of parenting?

One of several intuitive explanations for the female domination of parental caregiving takes the following form: "Because females (unlike males) have already invested so much energy in making eggs, they have a special incentive to make sure their large initial gametic investment is not wasted. Females therefore continue to provide parental care after the eggs have been produced and fertilized." This hypothesis comes to grief, however, when we observe the mallee fowl, or the spotted sandpiper, or the many frogs and fishes whose

1 A highly paternal bird. The male Australian mallee fowl provides most of the parental care his offspring will receive. Among his duties is the construction of a huge nest of compost and sand, which he modifies each time a female lays an egg in the nest. Photograph by the author.

females abruptly terminate their parental investment after laying their large and expensive-to-produce eggs, leaving them in the care of parental males. These species show that a considerable initial investment in offspring does not automatically make it advantageous for females to invest still *more* in their brood. Instead, each increment of maternal care is subject to selection. When the costs of providing an additional unit of care exceed the fitness benefits gained from the act, females that do not make the added investment should leave more surviving descendants on average than females that provide the extra care.

So we must find another explanation for the general pattern of female-only parental care. David Queller comes to the rescue by pointing out that if the cost–benefit ratio for parental care were usually lower for females than for males, as it may well be, then we would expect females to provide care more often than males [946]. Let's assume for the sake of simplicity that one standard unit of parental care invested in a current offspring reduces the future reproductive output of a male and a female by the same amount. Let us also assume that we are looking at a species in which females sometimes mate with more than one male in a breeding season. In this case, the average benefit to a male from caring for a brood of offspring will be reduced to the extent that some of those offspring were fathered by other males. For example, if paternity is on average 80 percent, then for every five offspring assisted, the male's investment can yield at best only four descendants, whereas all five youngsters could advance the female's genetic success. In other words, when cuckolded paternal males waste some of their costly parental care on nonrelatives, they become less likely than their mates to experience a favorable cost–benefit ratio for parenting.

Not only are the benefits from paternal care likely to be less than the benefits from comparable amounts of maternal care, but the costs are likely to be greater for males than females. As we noted when discussing sexual selection theory (p. 327), males generally attempt to acquire many mates because those that succeed leave many descendants. Time and energy spent in caring for offspring cannot be invested in mating effort, and therefore, sexual selection should often lead to the evolution of males that provide less parental care than females. Imagine a lek of black grouse in which the top male fertilizes most of the eggs of the 20 or so females that come to the lek to mate. As noted earlier, regular attendance at the lek is one of the main correlates of male mating success; grouse with the potential to become alpha males with extraordinary access to females would pay a heavy price in terms of lost offspring if they were to take time away from lekking in order to conserve the energy needed to help care for offspring [946]. In most species, the *potential* reproductive rate of males exceeds that of females (p. 321), making the cost of parental care greater for males than for females, and thereby decreasing the likelihood of its evolution.

Exceptions to the Rule

Nonetheless, the general rule that males are not parental has many exceptions (see Figure 5, p. 322). If evolutionary theory is going to help explain these unusual cases, then paternal care systems have to make sense in terms of the relative fitness costs and benefits to males versus females in these species. Exceptions to the rule of female-only parental care are especially numerous among fishes (Figure 2). However, male fish are no different from males of other animals in that they produce vast quantities of sperm and could potentially have many more offspring than the most fecund female. Therefore, male fish would seem to have much to lose by diverting time and energy from mating effort to parental effort. Indeed, parental male fish often do lose some chances to mate as a result of protecting their brood. For example, when a nest-

Why Do Male Water Bugs Do All the Work?

Under certain circumstances, a good many male fishes will take care of their young. In contrast, male parental care is infrequent among the insects [1279], although the termites have paternal males that cooperate with their mates in rearing young [1041], and some male water bugs rear their offspring single-handedly. Male water bugs in the genus *Lethocerus*, for example, guard and moisten clutches of eggs that females glue onto the stems of aquatic vegetation above the water line (see Figure 13, p. 18) [563]. Males in some other genera of water bugs (e. g., *Abedus* and *Belostoma*) permit their mates to lay a clutch of eggs on their backs (Figure 5). The males then spend hours perched near the water surface, pumping their bodies to keep well-aerated water moving over the eggs. Clutches that are experimentally separated from a male attendant do not develop, demonstrating that male parental care is essential.

Robert Smith has explored both the history and adaptive value of these unusual paternal behaviors [1075]. Since the closest relatives of the Belostomatidae, the family that contains these paternal water bugs, are the Nepidae, a typical family of insects without male parental care, we can be confident that the brooding species evolved from nonpaternal ancestors (Figure 6). Whether out-of-water brooding and back brooding evolved independently from such an ancestor, or whether one preceded the other, is not known, although some evidence suggests that back brooding came later. In particular, female *Lethocerus* sometimes lay their eggs on the backs of other individuals, male or female, when they cannot find suitable vegetation for the purpose. This unusual behavior indicates how the transition from out-of-water to back brooding might have occurred. Females with the tendency to lay their eggs on the backs of their mates could have reproduced in temporary ponds and pools where emergent aquatic vegetation was scarce or absent.

But why do the eggs of water bugs require brooding? Huge numbers of aquatic insects lay eggs that do perfectly well without a caretaker of either sex. However, Smith notes that belostomatid eggs are much larger than the standard aquatic insect egg. A large sphere has a smaller surface area relative to its volume than a small sphere. Thus, as a large belostomatid egg develops, it has

5 Male water bugs provide uniparental care. A male belostomatid broods eggs glued onto his back by his mate. Photograph by Bob Smith.

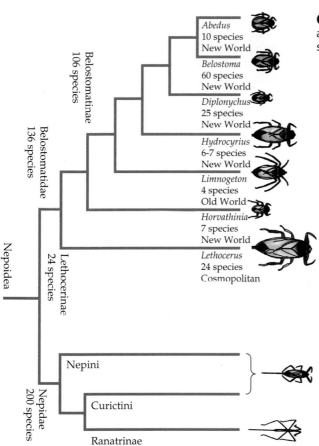

Abedus
10 species
New World

Belostoma
60 species
New World

Diplonychus
25 species
New World

Hydrocyrius
6-7 species
New World

Limnogeton
4 species
Old World

Horvathinia
7 species
New World

Lethocerus
24 species
Cosmopolitan

Belostomatinae
106 species

Belostomatidae
136 species

Lethocerinae
24 species

Nepoidea

Nepini

Curictini

Ranatrinae

Nepidae
200 species

6   Cladogram of the Nepoidea, the group that includes the belostomatid water bugs. The illustrations, drawn to scale, show the largest representatives of each group. After Smith [1075].

a special problem—how to move carbon dioxide out and oxygen in rapidly enough to sustain the high metabolic rate required for embryonic development. Since oxygen diffuses through air much more easily than through water, laying eggs out of water can solve that problem. But this solution creates another problem, which is the risk of desiccation that the eggs face when they are exposed to the atmosphere. The solution, brooding by males that moisten the eggs repeatedly to prevent their drying out, sets the stage for the evolutionary transition to back brooding at the air–water interface.

Wouldn't things be simpler if belostomatids simply laid small eggs with large surface-to-volume ratios? To explain why water bugs produce eggs so large that they need to be brooded, Smith points out that water bugs are among the world's largest insects, almost certainly because they specialize in grasping and stabbing large vertebrates, including fish, frogs, and tadpoles. Water bugs, like all other insects, grow in size only during the immature stages. After the final molt to adulthood, no additional growth occurs. As an immature insect molts from one stage to the next, it acquires a new flexible cuticular skin that permits an expansion of size, but no immature insect grows more than 50 or 60 percent per molt. One way for an insect to grow large, therefore, would be to increase the number of molts before making the final transition to adulthood. However, no member of the belostomatid family molts more than six times, suggesting that they are locked into a five- or six-molt sequence, just as spotted sandpipers evidently cannot exceed a four-egg clutch. If a water bug is to grow large enough to kill a frog in just five or six molts, then the first instar (the nymph that hatches from the egg) must be large, because it will get to undergo only five or so 50-percent expansions. In order for the first-instar nymph to be large, the egg must be large. In order to overcome the upper size limit for egg survival under water, large eggs must have access to the air, which is where male brooding comes into play, an ancillary evolutionary development whose foundation lies in selection for large body size in conjunction with prey specialization [1075].

However, females water bugs could conceivably provide care for their own eggs after laying them on exposed aquatic vegetation. Why is it that males do the brooding, never the females? Here the situation parallels the fish story closely. First, male water bugs with one clutch of eggs can sometimes attract a second female, which means that remaining with one clutch is less costly than it appears at first glance. Second, just as is true for some fishes, the costs of parental care may be disproportionately great for females. In order to produce large clutches of large eggs, female belostomatids require far more prey than do males, which need only maintain themselves. Because brooding limits mobility and thus access to prey, parental care probably has greater fitness costs for females than for males, biasing selection in favor of male parental care.

Discriminating Parental Care

No matter which parent provides care, an adaptationist would predict that a parent or parents should avoid providing valuable assistance to young animals that are not their own offspring. But can parental animals always identify their own progeny? Consider the Mexican free-tailed bat, which migrates to certain caves in the American Southwest, where pregnant females form colonies in the millions. After giving birth to a single pup, a mother bat leaves her offspring clinging to the roof of the cave in a crèche that may contain 4000 pups per square meter (Figure 7). When the female returns to nurse her infant, she flies back to the spot where she last nursed her pup and is promptly besieged by a host of hungry bat babies [779]. Given the swarms of pups, early

7 **A crèche of Mexican free-tailed bat pups** left together by their mothers, who are foraging outside the cave but will return to nurse them. Photograph by Gary McCracken.

observers believed that mothers could not possibly identify their own offspring and instead had to provide milk on a first-come, first-served basis.

But do Mexican free-tailed bat mothers really nurse indiscriminately? To find out, Gary McCracken captured female bats and the pups nursing from them and took blood samples from both [778]. He then analyzed the samples using starch-gel electrophoresis, a technique that can be used to determine if two individuals have the same variant form of a given enzyme, and thus the same allele of the gene that codes for that enzyme. If female bats are indiscriminate care providers, the enzyme variants of females and the pups they nurse should often be different. But if females tend to nurse their own offspring, then females and pups should tend to share the same alleles. McCracken focused on the gene for superoxide dismutase, an enzyme represented by six different forms in the population he sampled. Despite the chaos of the bat colony, the enzyme data indicated that females found their own pups at least 80 percent of the time [778]. More recent direct observational studies indicate that females probably do better than that, almost always recognizing their own pups through vocal and olfactory signals [61, 781]. Thus, mother free-tailed bats clearly deliver their parental care primarily to their own pups.

Offspring Recognition: Comparative Studies

The hypothesis that offspring recognition functions to prevent misdirected parental care leads to the prediction that parents should be especially good at identifying their own young in colonial species, such as the free-tailed bat, in which parents have ample opportunity to encounter youngsters that are not their own. In contrast, offspring recognition is expected to be much less well developed in solitary species whose parental adults are extremely unlikely to encounter babies other than their own.

Let's check this prediction by comparing offspring recognition in the bank swallow and rough-winged swallow. Although both species nest in clay banks, one is colonial and the other is solitary. The mobile fledglings of the colonial bank swallow produce highly distinctive vocalizations, giving their parents a potential cue to use when making decisions about which individuals to feed, allowing them to distinguish between their own offspring and other fledglings that wind up in the wrong nests begging for food [787]. Bank swallow parents rarely make mistakes, despite the high density of nests in their colonies. The solitary rough-winged swallow, on the other hand, never has a chance in nature to feed another's fledglings, and so is not expected to show the same skillful recognition of offspring as its cousin, the bank swallow. Indeed, when Michael and Inger Beecher transferred fledglings between distant rough-wing burrows, they found that the parents readily accepted the transplants. Rough-winged swallows will even act as foster parents for fledgling bank swallows [83].

Cliff swallow

Two other swallow species, the highly colonial cliff swallow and the less social barn swallow, also differ in the degree to which they are exposed to the risk of misdirected parental care. We would expect that cliff swallow chicks should have more distinct characteristics that make it easier for their parents to recognize them, and they do. Cliff swallow chicks utter more variable and complex begging calls than barn swallow nestlings. By measuring several elements of begging calls, such as the duration and range of frequencies in these vocalizations, Mandy Medvin and her colleagues demonstrated that cliff swallow chicks produce signals containing about sixteen times as much variation as the corresponding calls of barn swallow chicks (Figure 8) [788].

If cliff swallow adults have experienced stronger selection to avoid feeding genetic strangers, then they should be better discriminators among chick calls than barn swallow adults. They are. In operant conditioning experiments that required adults of both species to discriminate between pairs of chick calls, cliff

Barn swallow

8   Call distinctiveness and offspring recognition.
The chicks of cliff swallows, a colonial species, produce highly structured and distinctive calls, enabling their parents to recognize them as individuals. The calls of barn swallow chicks, a less colonial species, are much less well defined and more similar. After Medvin et al. [788].

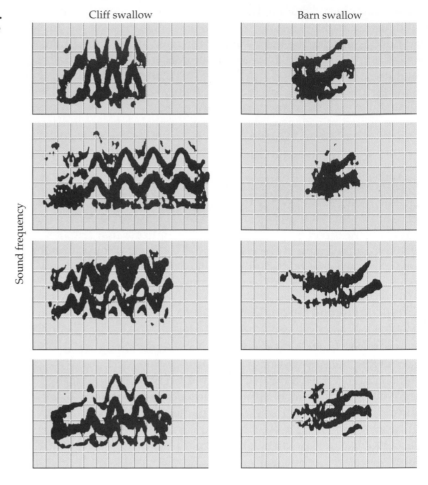

swallows reached criterion (85 percent accuracy) faster than barn swallows, suggesting that the perceptual systems of cliff swallows have evolved to promote accurate offspring recognition [715].

Why Adopt Genetic Strangers?

The cases described thus far support the prediction that parents should recognize their own young and discriminate against others when the probability of being exploited by someone else's hungry offspring is high. And yet both the colonial herring gull and ring-billed gull occasionally adopt unrelated chicks. Although researchers initially reported that adults rejected older, mobile chicks when they were experimentally transplanted between nests [791], attacks by adults on these transferred youngsters apparently occurred because of the frightened behavior of the displaced chicks [452]. When juveniles *voluntarily* leave their natal nest—which they sometimes do if they have been poorly fed by their parents (Figure 9)—they do not flee from potential adopters, but instead beg for food and crouch submissively when threatened. They have a good chance of being adopted, even at the advanced age of 35 days [538], and if they are taken in, they are more likely to survive than if they had remained with the genetic parents that were failing to supply them with enough food [174].

The adoption of strangers qualifies as a Darwinian puzzle, for here we have parents apparently failing to act in the best interests of their genes. As usual, resolving the paradox requires thinking about the costs, not just the benefits, of

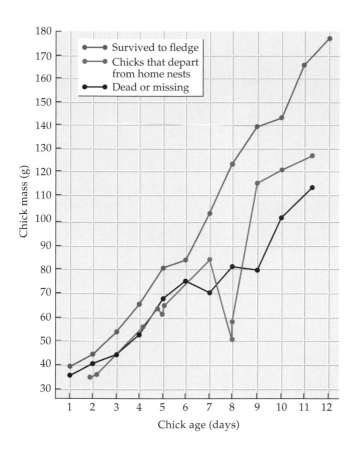

9 Why seek adoptive parents? Gull chicks that abandoned their natal nest in search of foster parents weighed much less than average for chicks their age. After Brown [174].

a trait of interest. And learned recognition of offspring carries costs as well as benefits, notably the risk of making a mistake and not feeding, or even attacking and killing, your own offspring. Rather than erring on the side of harming their genetic offspring, gulls have evolved a readiness to feed any chicks in their nest that beg confidently when approached by an adult [916]. Sometimes this rule of thumb permits a genetic outsider to steal food from a set of foster parents by slipping into a nest with other youngsters of its age and size [631]. When adoption occurs, the adoptive parents lose about 0.5 chicks of their own on average; however, adoption is rare, with fewer than 10 percent of adult ring-billed gulls taking in a stranger in any year [174]. The modest average annual fitness cost of a rule of thumb that results in occasional adoptions has to be weighed against the cost of rejecting one's own genetic offspring that would arise if parent gulls were more cautious about feeding chicks in their nests.

The argument here is that a less than perfect "decision rule" or "rule of thumb" can be selectively advantageous if the proximate mechanism responsible for the rule has a more favorable cost–benefit ratio than that of alternative psychological mechanisms that control different decision rules. This hypothesis has been used to explain why male western bluebirds adopt broods under some circumstances. When Janis Dickinson and Wes Weathers removed some nesting males, most of the experimentally widowed females soon attracted replacement partners, about half of whom fed the chicks of the females they had joined, even though at least some of those chicks had surely been fathered by the original male [327]. The observation that the paternal replacement males were those that found a female while she was still laying eggs (and thus both sexually receptive and potentially fertile) suggests that male western bluebirds have a proximate all-or-none decision-making mechanism that regulates their

10   Effective begging depends on how high a baby bird can reach relative to its nestmates. Brown-headed cowbird nestlings generally can reach higher than their hosts' offspring, which is why they have a higher probability of getting fed. After Lichtenstein and Sealy [695].

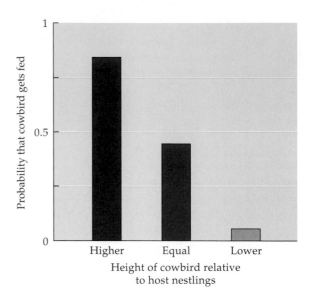

parental caregiving. If the male joins a female during her fertile phase, he exhibits all-out parental care; if he joins her after her fertile period is over, then the rule is "no parental care at all." Likewise, in the white-browed scrub-wren, a polyandrous species with alpha and beta males, the rule for the subordinate is to be paternal only if he has copulated with the female [1226]. For both bluebirds and beta scrub-wrens, males apparently use proximate decision rules that motivate them to help broods that probably contain some of their genetic offspring, as do male dunnocks (p. 374).

The use of another generally, but not universally, adaptive rule of thumb may explain why breeding adults of various species devote themselves to caring for the young of specialized brood parasites, such as certain cuckoos and cowbirds, that place their eggs in the nests of other birds. Parental songbirds typically distribute the food they bring to the nest in a biased fashion, preferentially feeding their larger, more actively begging nestlings (p. 302). This kind of bias enables parents to avoid wasting hard-earned resources on weakened or sickly offspring unlikely to reach reproductive age, but the decision rule can be exploited by brood parasites. Exploitation of the host is made more likely because the young cuckoo or cowbird tends to hatch a day or two sooner than the young of the host species, which enables the parasite to start growing sooner and get larger than its competitors in the nest. Because adult brood parasites are generally larger birds than the species they victimize (see Figure 7, p. 120), their babies soon become much larger than the hosts' own youngsters. When a strapping young cowbird stretches up higher than its nestmates to greet a returning parent bird, odds are that the parasite will get fed (Figure 10) [695].

The History of Interspecific Brood Parasitism

But how did cowbirds, cuckoos, and the like come to specialize in parasitizing other bird species? Two alternative historical scenarios for the origins of specialized brood parasitism have been proposed, involving (1) an initial period in which the parasites targeted nesting adults of their own species, with the shift to members of one or more other species occurring later, or (2) the exploitation of adults of another species right from the start. The first, gradualist scenario leads to the prediction that females of some of today's birds should lay their eggs in nests of their *own* species, and indeed, parasitism of this sort has been documented in about 200 species and may occur in many others as well

[293, 935]. A representative *intraspecific* brood parasite is the coot, a common black water bird, in which "floater" females that lack nests or territories of their own will lay their eggs in the nests of other coots, apparently in an effort to make the best of a bad situation, since they cannot brood their own eggs themselves. In addition, territorial females with nests of their own regularly pop some surplus eggs into the nests of unwitting neighbors. Since there are limits to how many young one female coot can rear with her partner, even a territorial female can boost her fitness a little by surreptitiously enlisting the parental care of other pairs [733].

The gradual shift hypothesis also yields the prediction that intraspecific parasites should probably branch out by first exploiting other related species with similar nestling food requirements. Currently, most specialized brood parasites take advantage of species that are not closely related to them, but this observation could simply reflect the fact that most brood parasites have been evolving for many millions of years since the onset of their *interspecific* parasitic behavior. Thus, to check this prediction, we need to find brood parasites that have a relatively recent origin. The familiar cowbirds of the Americas are one such group, with the parasitic species having originated "only" 3 to 4 million years ago, whereas cuckoos evolved at least 60 million years before the present [293]. The living cowbird species believed to be closest to the ancestral brood parasite does indeed parasitize a single host species that belongs to the same genus that it does; the next closest species to the ancestral parasite parasitizes other birds belonging to its own family (Figure 11) [684]. These data, if they have been interpreted properly, provide support for the gradual shift hypothesis.

On the other hand, the very large majority of living brood parasites take advantage of unrelated species much smaller than they are [1068]. Some persons interpret this finding to mean that the original brood parasites may have made an abrupt shift to one or more unrelated smaller host species because such a shift was more likely to succeed given that, as noted above, parasite nestlings that become larger than their hosts' offspring are more likely to be fed.

The importance of the size disparity between host and parasite has been demonstrated by creating some experimental brood parasites. When Tore

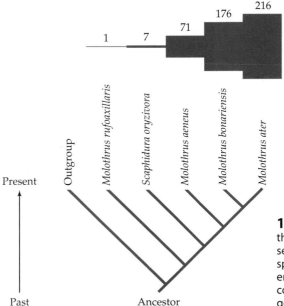

11   Evolution of brood parasitism among cowbirds. The cladogram depicts the evolutionary relationships among cowbirds as determined by genetic analyses. Above the cladogram, the number of hosts parasitized by each current living species of cowbird is shown. (The "outgroup" is a species of cowbird that does not engage in brood parasitism.) The pattern suggests that the first brood parasitic cowbird victimized only a single closely related host species, with increasingly generalized brood parasitism probably evolving subsequently. After Lanyon [684].

12   **The size of an experimental "parasite" relative to its host species** determines the survival chances of nestlings transferred from their parents' nest to a host nest. Larger great tits survived well in the nests of smaller blue tits, whereas blue tits did poorly in great tit nests. After Slagsvold [1068].

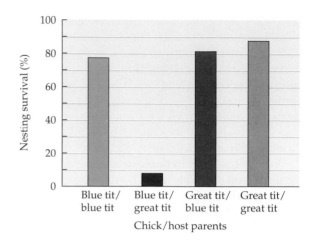

Blue tit

Great tit

Slagsvold shifted blue tit eggs into great tit nests, the "parasitic" blue tit nestlings, which are smaller than those of great tits, did very poorly. In the reciprocal experiment, however, most of the great tit chicks cared for by blue tit parents survived to fledge (Figure 12) [1068]. These findings indicate that unless the original mutant interspecific parasites happened to deposit their eggs in the nests of a smaller host species, the likelihood of success (from the parasite's perspective) was not great. Therefore, proponents of both the gradual and abrupt shift scenarios for the evolution of interspecific brood parasitism have at least some supportive evidence to which they can point, which is why the issue is still unresolved.

Why Accept a Parasite's Egg?

Whatever the origin and subsequent history of interspecific brood parasitism, it is true that in order for a parasite nestling to take advantage of a host species' parental decision rules, the egg containing the parasite has to hatch. Why, then, don't host species take immediate action against the parasite when it is in the egg stage? Some birds do, by recognizing and removing the foreign egg from the nest or abandoning the nest and the parasite's egg altogether. However, each of these options has its disadvantages [1254]. Parent birds that made incubation of eggs dependent on their learned recognition of the eggs they had laid might sometimes abandon or destroy some of their own eggs by mistake, treating them as if they were foreign eggs. Indeed, reed warblers sometimes do throw out some of their own eggs while trying to rid themselves of a cuckoo's egg [294]. If brood parasites victimize only a very small minority of the host population, then even a small risk of costly recognition errors by the host can make accepting parasite eggs the adaptive option [724].

Acceptance of the parasite egg is even more likely to be adaptive when the host is a small species unable to grasp and remove large cowbird or cuckoo eggs [981]. For these small-billed birds, the only options are to abandon the clutch, either by leaving the site or by building a new nest on top of the old one, or to stay put and continue brooding the clutch along with the parasite egg. The abandonment option imposes heavy penalties on the host, which must find a new nest site, build a new nest, and lay a new clutch of eggs. The costs of this option are especially high for hole-nesting species because suitable tree holes are generally rare. Lisa Petit tested the prediction that the availability of artificial nest boxes and natural nest holes would determine the likelihood that a prothonotary warbler with a parasitized nest would start all over again elsewhere [903]. As expected, prothonotary warblers with several nearby replacement nest sites more often abandoned their parasitized nests than warblers that had few alter-

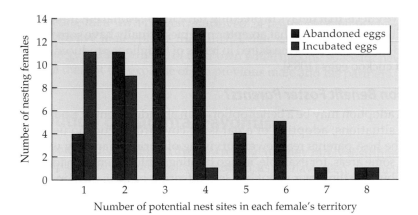

13 The cost of abandoning a parasitized nest depends on the availability of new nest sites. If few other potential nest sites are available, prothonotary warblers often continue to incubate the eggs in their parasitized nest. However, warblers that have access to several other nest cavities in their territories are more likely to abandon a nest with a cowbird egg. After Petit [903].

native sites (Figure 13). Likewise, among yellow warblers nesting in southern Canada, individuals parasitized by cowbirds near the end of the short breeding season tended to accept the foreign egg, probably because they had too little time left to rear a new brood from scratch [1030].

Even if host birds could throw out or cover up a parasite's eggs without risk of error, the parasite might make this option unprofitable by returning to the nest to destroy or consume the host's eggs or young if it found that its offspring had been harmed (Figure 14). This "Mafia hypothesis" has been tested by examining the interactions between European magpies and parasitic great spotted cuckoos [1083]. As predicted by the hypothesis, magpie nests from which cuckoo eggs had been ejected suffered a significantly higher rate of predation than nests with accepted cuckoo eggs (87 percent versus 12 percent in one sample). Furthermore, when researchers removed the cuckoo egg from a nest that was apparently being checked by a cuckoo, and replaced the magpie's eggs as well with plasticine imitations, the cuckoo approached the nest after the researchers had finished and violently pecked the magpie pseudo-eggs. In nature, magpies that lose their clutch have to renest, which exposes them to

Prothonotary warbler

14 An avian mafioso. European great spotted cuckoos are large birds capable of removing or destroying the eggs of potential hosts. Photograph by Ian Wyllie.

17 Weapons of siblicide. Very young spotted hyenas sometimes use their sharp, fully developed canines to slash a sibling to death. Photograph by Laurence Frank.

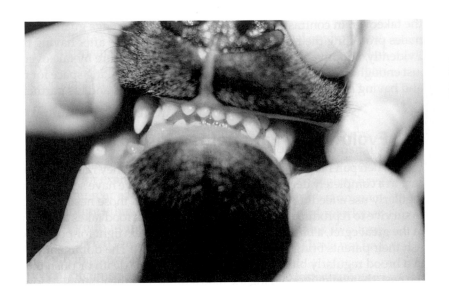

siblings. Because even full siblings typically share only half their genes on average, members of broods are expected to compete for parentally donated resources whenever these are in short supply [800]. The fitness gains to the selfish monopolizer can potentially exceed the costs it incurs in terms of lost siblings whose deaths eliminate some of the genes they share as a result of having the same parents (see p. 55). Under some circumstances, selection on parents might favor adults able to thwart their offspring's genetically selfish behavior. As Bob Trivers has explained, **parent–offspring conflict** is a theoretical possibility because some actions that advance the fitness of an offspring can potentially reduce the lifetime reproductive success of the parent, and vice versa [1154].

At first glance, parent–offspring conflict seems evident in spotted hyenas, whose females deposit their newborns, usually twins, at the entrance to a small underground burrow, usually an old aardvark den. When the babies crawl in, they are safe from most predators, and free from maternal supervision as well. Inside the burrow, the young hyenas often employ their fully functional canines and incisors in slashing attacks on each other (Figure 17) [402]. The pups also fight when attempting to suckle from their mother. The violence between siblings sometimes results in the death of one pup, generally before it is 2 months old [440].

Sibling aggression in hyenas appears to occur more often when the two sibs are of the same sex (Figure 18) [402, 440], which may explain why surviving brother–brother and sister–sister pairs are scarce compared with surviving brother–sister twins (but see [1069]). The consistent ability of females to rear two offspring (provided they are a son and a daughter) suggests that the **siblicide** that occurs within same-sex pairs may reduce parental fitness. On the other hand, perhaps mother hyenas gain fitness when an extremely healthy, strong, siblicidal offspring removes a less healthy rival of the same sex, especially in seasons when food is in short supply. The surviving pup will have more milk and a better chance to develop into a powerful adult that will not have to face potentially costly competition with a same-sex sibling. These pampered offspring may have a better chance of becoming reproductively successful, high-ranking individuals in this highly socially stratified species (see p. 285). Perhaps this is why adult spotted hyenas appear to tolerate vicious attacks by one twin on another [403].

In some other animals in which siblicide occurs, parents apparently can and do resist their progeny's siblicidal behavior. Evidence for this claim comes from

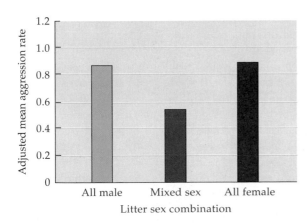

18   Aggression between young spotted hyenas is more pronounced in same-sex litters. Shown here are measures of aggression that have been adjusted to control for other factors, such as the age of the pups. After Golla et al. [440].

studies of colonially nesting seabirds called boobies. Only some of these species exhibit "early siblicide," in which an older "A" chick disposes of a younger "B" chick in the first few days of its sibling's short and unhappy life [725]. The A chick's ability to kill its brother or sister stems in part from the pattern of egg laying and incubation in boobies. In these birds, females lay one egg, begin incubating it at once, and then some days later lay a second egg. Because the first egg hatches sooner than the second, the A chick is relatively large by the time the B chick comes on the scene. In early siblicidal species, the A chick immediately begins to manhandle the B chick, soon forcing it out of the nest scrape, where it dies of exposure and starvation (Figure 19).

19   Early siblicide in the brown booby. A very young chick is dying in front of its parent, which continues to incubate the other, larger chick that has forced its sibling out of the nest and into the sun. Photograph by the author.

20 Parent boobies can control siblicide to some extent. The rate of early siblicide declines when masked booby (MB) chicks are placed in nests with intervention-prone blue-footed booby (BFB) foster parents. Conversely, the rate of early siblicide is higher when blue-footed booby chicks are given laissez-faire masked boobies as foster parents. After Lougheed and Anderson [725].

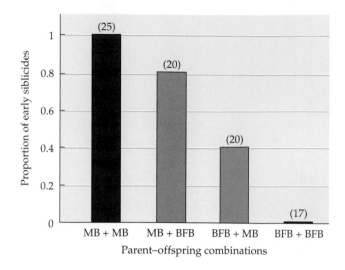

Early siblicide is standard practice for the masked booby, but not the blue-footed booby, whose A chicks engage in siblicide less often and generally later in the nesting period. If, however, you give a pair of blue-footed booby chicks the chance to be cared for by masked booby parents, which tolerate early sibling aggression, the A chick often quickly kills the B chick under the vacant gaze of its substitute parents. In contrast, blue-footed booby parents are believed to keep their A chick under control during its initial days with its sibling. If so, when masked booby chicks are given to blue-footed booby parents, the foster parents should sometimes be able to prevent them from immediately killing their siblings. The actual results match the predicted ones (Figure 20) [725], providing evidence that parents can evolve the ability to interfere with lethal sibling rivalries, should it be in their interest to do so.

No such intervention occurs in the case of siblicidal juvenile egrets. Indeed, lethal sibling interactions are actually promoted by two earlier parental decisions about what to put in the eggs and when to begin incubating them. First, when female cattle egrets (a siblicidal species related to the great egret) produce eggs, they vary the hormones that they incorporate into the yolk. Relatively large quantities of androgens go into the first two eggs of a three-egg clutch [1027]. The more androgen contained in an egg, the more aggressive the nestling will probably be, another example of the female exercising her ability to influence the attributes of her offspring (see p. 341).

Second, as soon as a female egret lays her first egg, incubation begins, just as is true for boobies. Because 1 or 2 days separate the laying of each egg in a three-egg clutch, the young hatch out asynchronously, with the firstborn getting a head start in growth. As a result, this chick will not only be more aggressive, but will also be much larger than the third-born chick, ensuring that the senior chick monopolizes the small fish its parents bring to the nest. Unequal feeding rates further exaggerate the size differences among siblings, creating a runt of the litter that often dies from the combined effects of starvation and assault.

Thus, parent egrets not only tolerate siblicide, but actually promote it. Why? Perhaps because parental interests are served by having the chicks eliminate those members of the brood that are especially unlikely to survive to reproduce. In most years, food will be moderately scarce, making it impossible for the adults to find enough to feed all three offspring. In good years, they can, and it is then that the ability to lay three eggs sometimes pays off. Under conditions of food scarcity, however, a reduction in the brood accomplished by siblicide saves the parents the time and energy that would otherwise be wasted on offspring with little or no chance of reaching adulthood, even if their siblings left them alone.

One way to test this hypothesis is to create unnaturally synchronous broods of cattle egrets [801]. In one experiment, synchronous broods were formed by putting chicks that had hatched on the same day in the same nest; normal asynchronous broods were assembled by bringing chicks together that differed in age by the typical 1.5 day interval. A category of exaggeratedly asynchronous broods was also created by putting chicks that had hatched 3 days apart into the same nests. If the normal hatching interval is optimal in promoting efficient brood reduction, then the number of offspring fledged per unit of parental effort should be highest for the normal asynchronous broods. This prediction was confirmed. Members of synchronous broods not only fought more and survived less well, but required more food per day than normal broods, resulting in low parental efficiency (Table 1). The same result has been recorded for the blue-footed booby, in which experimental synchronous broods of two fought more and required much more food than control broods composed of an asynchronously hatched pair of chicks [868].

Thus, cattle egret parents and others like them know (unconsciously) what they are doing when they manipulate the hormone content of their eggs and incubate them in ways that ensure differences in size and fighting ability among their chicks. Sibling rivalry and siblicide actually help parents deliver their care only to offspring that have a good chance of eventually reproducing while enabling the parents to keep their food delivery costs to a minimum. Although cases of this sort represent extreme examples of parental favoritism, even those nesting birds that bias their allocation of food resources toward vigorous offspring are really practicing infanticide, speeding the demise of those progeny unlikely to reproduce even if they were well fed.

Parents can bias food delivery to their offspring based on signals provided by their infants, such as the bright red lining of the mouth, which is conspicuously displayed by many baby songbirds as they reach up to solicit food from a returning parent (Figure 21). A research team led by Nicola Saino proposed that parental food allocations might be based in part on this cue [1012]. Because the red color of the mouth lining is generated by carotenoid pigments in the blood, and because carotenoids are believed to contribute to immune function, a bright red gape could signal a healthy nestling deserving of extra food. Parents that preferentially fed those members of their brood that had bright red gapes would be investing in healthy nestlings that were more likely to fledge and eventually become reproducing adults than their sickly nestmates. If parents do indeed make adaptive parental decisions of this sort, then nestlings made ill by injection of a foreign material should have paler mouth linings than those that have not had their immune systems challenged. In addition, parents should feed offspring with artificially reddened gapes more than they feed

---

TABLE 1 The effect of hatching asynchrony on parental efficiency in cattle egrets

	Mean survivors per nest	Food brought to nest per day (ml)	Parental efficiency[a]
Synchronous	1.9	68.3	2.8
Normal asynchronous	2.3	53.1	4.4
Exaggerated asynchronous	2.3	65.1	3.5

[a]The number of surviving chicks divided by the volume of food brought to the nest per day × 100.

Source: Mock and Ploger [801]

21   A honest signal of condition? The red mouth gape of nestling lark buntings is exposed when the birds beg for food from their parents. Photograph by Bruce Lyon.

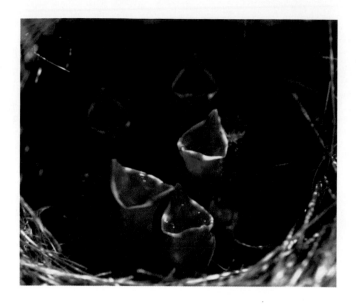

offspring with unaltered mouth coloration. When Saino's research team tested both predictions in the barn swallow, the results were positive (Figure 22).

Carotenoids may also play a role in the system that parent coots use to establish favorites among their numerous offspring. Baby coots emerge from the egg with long, orange-tipped plumes on their backs and throats. Bruce Lyon and his colleagues suspected that these purely ornamental plumes provided signals to parents that they might use to determine which individuals to feed preferentially. To test the link between chick ornaments and parental feeding decisions, Lyon and company trimmed the thin orange tips from these special feathers on half of the chicks in a brood, while leaving the other members of the brood untouched. The unaltered orange-plumed chicks were fed more frequently by their parents, and they grew more rapidly as well (Figure 23) [735]. In control

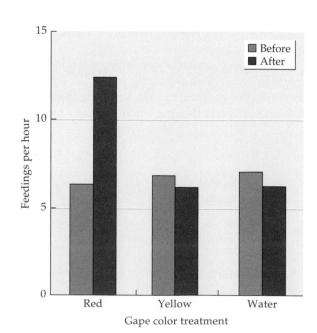

22   The color of the mouth gape affects the amount of food nestlings are given by their parents. Nestling barn swallows were fed more after the experimenters had colored their gapes with two drops of red food coloring. Nestlings that received two drops of yellow food coloring or water were not fed more after the treatment. After Saino et al. [1012].

broods, in which *all* of the chicks had their orange feathers trimmed, the youngsters were fed as often, and survived as well, as broods consisting only of untouched orange-feathered chicks. This result shows that the parents of the experimental mixed broods discriminated against the black chicks because they were not as strongly ornamented as their feather-intact broodmates, not because the parents failed to recognize them as their offspring [735].

One wonders whether the orange ornaments of baby coots signal something about offspring health, thereby enabling parents to make wise decisions about parental care. The general message provided by coots, barn swallows, and cattle egrets is that parents do not always treat each offspring the same, but instead often help some survive at the expense of others. Cases of this sort remind us that selection acts not on the number of babies produced, but on the number that survive to reproduce and pass on the hereditary traits of their parents.

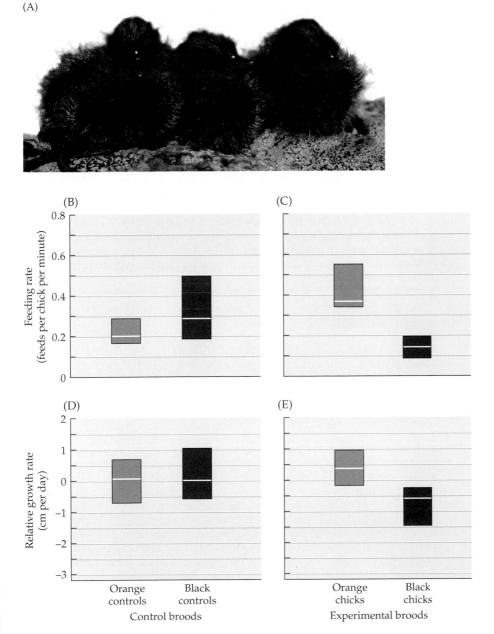

23 **Orange feather ornaments** are the basis for parental favoritism in coots. (A) Baby coots have unusually colorful feathers near the head. (B) Control groups composed *entirely* of either unaltered (orange) chicks or chicks that had had the orange tips trimmed from their ornamental feathers (purple) were fed at the same mean rate. (C) In experimental broods composed of half orange and half black chicks, the ornamented individuals received more frequent feedings from parent birds. (D) The relative growth rate of chicks in both control groups was the same, but (E) ornamented chicks grew faster in mixed broods compared with the experimentally altered chicks. After Lyon et al. [735]; photograph by Bruce Lyon.

Summary

1. Parental care is a form of parental investment, in which time, energy, and resources donated to an existing offspring have costs, including reduced fecundity in the future and fewer opportunities to mate in the present, and potential benefits, especially improved survival by the assisted progeny.

2. When the circumstances are such that the benefits of parental care exceed the costs, females are more likely than males to provide it, perhaps because fathers that help broods of mixed paternity gain less from their paternal care than do mothers, which are usually genetically related to all the offspring in their broods.

3. Male fish are unusual in that they often provide uniparental care. The cost–benefit ratio of paternal care in fishes may be favorable because males caring for eggs laid in their territories are often more attractive to potential mates than are males without eggs to brood. In contrast, the cost of parental care to females may be large reductions in growth rate and consequent losses of fecundity.

4. An evolutionary approach to parental care yields the expectation that parents will be able to identify their offspring when costly investments in genetic strangers are possible. Although offspring recognition is widespread, particularly in colonial species in which the potential for misdirected parental care is high, adults of many species adopt nongenetic offspring, even brood parasites. Multiple hypotheses exist to account for these puzzling cases, including the possibility that highly discriminating host adults could lose fitness by sometimes mistakenly refusing to care for their own offspring.

5. Another Darwinian puzzle is the indifference shown by some animals to lethal aggression among their young offspring. Cases like this may be explained as part of a parental strategy to let the offspring themselves identify which individuals are most likely to survive, and therefore which youngsters will provide a payoff for continuing parental investment. The more general principle is that selection rarely favors completely even treatment of offspring because some youngsters are more likely than others to survive to reproduce.

Discussion Questions

1. Gary McCracken found that although female Mexican free-tailed bats usually feed their own pups, they do make some "mistakes," which they could have avoided by leaving the pup in a spot by itself instead of in a crêche with hundreds of other babies [778]. Does this mean that the parental behavior of this species is not adaptive? Use a cost–benefit approach to develop alternative hypotheses to account for these "mistakes."

2. A hawk called the lesser kestrel nests in colonies. In this species, young birds sometimes move to the nests of adults other than their parents. Here are some data on the adoptions that result from the young kestrels' behavior: (1) the nest-switching nestlings were not in poor condition before moving, (2) in the new nest, the adopted nestlings were not fed at a higher rate than they had been prior to the switch, and (3) the adoptees did not achieve higher rank in their new brood. What hypothesis do you suppose these data were designed to test? What conclusion must the researchers of this study have reached on the basis of their tests? Of what possible relevance is the information that the lesser kestrel once nested primarily on small cliff ledges but now often nests on rooftops [1123]?

3. You observe a bird in which females typically exhibit stronger defense of their eggs and nestlings than do their male partners. Use an optimality approach to

develop testable hypotheses on why females invest more in defense than males in this species. Consider the costs and benefits of parental defense to members of the two sexes, keeping in mind that the sexes might differ in size, color, certainty that the offspring were indeed their own, and so on [813].

4. Many birds parasitized by cuckoos can recognize the parasite's eggs, and they either remove them or abandon the nest. These egg discriminators learn the key visual features of their own eggs and then reject those that do not properly match. But the same species that are extremely good at learning to recognize egg features usually completely fail to recognize a cuckoo chick [723]. This failure appears to be maladaptive, but consider the consequences of learned chick recognition to birds that were successfully parasitized in their first year of breeding by a single cuckoo chick that took over the nest and eliminated all the host's youngsters, as is their habit. How would the host adults respond to their own chicks in a second breeding attempt? How does this case illustrate the importance of cost–benefit analyses in behavioral ecology as well as the value of considering the proximate mechanisms by which individuals achieve behavioral goals?

5. Males of the Seychelles warbler sometimes remain by their nests in the absence of the female to protect the nest's contents, a single egg, from the Seychelles fody, an egg-eating bird. When some warblers were moved to an island where fodies were absent, the translocated males spent almost no time guarding the egg, and instead foraged for food. What do these results tell us about the proximate mechanism that controls nest-guarding behavior in the warbler? What do these results tell us about the fitness costs of parental care in this species? What predictions can you make about the consequences of leaving nests unguarded in the two locations? You can check your predictions against data in [646].

6. Among certain monkeys and apes with prolonged parental care, females live longer than males in species in which females provide most or all of the parenting, but males live longer than females in species in which males make the major contribution to offspring care [26]. In other words, the adults of the parental sex tend to live longer than the nonparental sex. Does this finding indicate that parental care provides a fitness benefit for caretakers in the form of improved survival? Someone claims that the longer life of the parental sex has been selected for because primate young are very slow to develop and therefore parents must live long enough to get their offspring to the age of independence in order to maintain a stable population. What do you have to say about that hypothesis? Do you have an alternative explanation for the observed pattern?

7. Employ a cost–benefit analysis to help solve the problem of why interspecific brood parasitism is so rare. Make a prediction about which group of birds, those with precocial young or those with altricial young, are more likely to become specialist brood parasites. (In altricial species, the eggs are small in relation to parental body weight, but the hatchlings are initially completely dependent on food supplied to them by a parent. In precocial species, the eggs are relatively large, but the youngsters can move about and feed themselves shortly after hatching.) Check your prediction against [734].

Suggested Reading

Good books on parental care include *The Evolution of Parental Care* by Timothy Clutton-Brock [228] and *Mother Nature* by Sarah Hrdy [555]. Different approaches to the question of why females are more likely to be maternal than males are to be paternal can be found in [946, 1207, 1216]. A vast literature exists on brood parasites and their interactions with their hosts; for a superb review see *Cuckoos, Cowbirds and Other Cheats* by Nick Davies [293]. Avian siblicide is the subject of an excellent article by Doug Mock and his co-workers [798]; see also [799].

14 The Evolution of Social Behavior

D o not reach for the Raid the next time you encounter a paper wasp colony (see Figure 11, p. 86). The wasps will not sting you, provided you approach cautiously—not that most people need much encouragement to be careful around wasp nests. When I watch paper wasp nests around my home, I occasionally see a female insert her abdomen into an empty cell and lay an egg there. Or a wasp flying in with food in her mouth will be greeted by several of the five or six wasps on the nest, who walk over to take chunks of caterpillar flesh from the newcomer. At other times, a returning wasp goes to a cell nearly filled by a shiny grub. The adult places her head into the cell, regurgitating a droplet of fluid that the grub greedily consumes.

If I were to tell you that only one of the wasps at the nest might be the mother of all the grubs housed within the paper cells, and that these youngsters are fed with food collected by females other than their mother, I hope you would be at least mildly surprised. Although parental care can sometimes evolve by natural selection when the benefits of the behavior exceed its costs (see Chapter 13), it is hard to imagine how an adult's fitness could be increased by behaving parentally toward someone else's youngsters.

◀ *Weaver ants form superbly cooperative societies based on a sterile worker caste. One of the many products of their altruism is a magnificent leaf nest woven together with silk provided by the colony's larvae. Photograph by the author*

Yet helpers at nests are found not just in paper wasps but in a host of other insects, as well as in birds and mammals. These self-sacrificing altruists pose a wonderful Darwinian puzzle whose solution has engaged some of the best evolutionary biologists in the world, including Darwin himself.

This chapter focuses on how altruism and other helpful acts can be analyzed from an adaptationist perspective. But first we must ask a more basic question: Why should animals live together at all?

The Benefits and Costs of Social Life

Before examining the evolution of helpful behavior in detail, we first need to erase the misconception that species with elaborate social behavior are somehow "better adapted" than those whose members lead largely solitary lives. Because humans are highly social, we like to think that complex societies represent the crowning achievement of evolution. But just as a cost–benefit approach demonstrates that parental care is not always adaptive, "advanced" social behavior can be less adaptive than so-called "primitive" solitary behavior under some circumstances [17].

Indeed, if group living were the universally superior lifestyle, we would expect social species to outnumber solitary ones, but exactly the opposite is true. Why? Almost certainly because in many environments, the costs of living with others are prohibitively high. Just having to signal one's submissive state to more dominant members of the group can take up a major share of a social animal's time and energy (Figure 1) [1122]. Consider also the costs of reproductive interference, which can more easily occur when individuals live in close association with one another. Males in groups may run a higher risk of being cuckolded, and females may more often incubate eggs laid in their nests by parasites of their own species [551, 805].

(A)

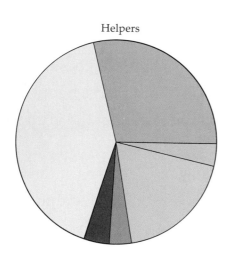

Helpers

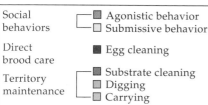

Social behaviors	☐ Agonistic behavior ☐ Submissive behavior
Direct brood care	☐ Egg cleaning
Territory maintenance	☐ Substrate cleaning ☐ Digging ☐ Carrying

(B)

1   The energy budget of subordinate, nonbreeding "helpers" that associate with breeding pairs in the fish *Neolamprologus pulcher*. (A) The largest proportion of the subordinate fish's energy is expended on the performance of submissive behaviors—specifically, a tail-quivering display. Most of the remainder of the subordinate fish's energy budget is spent attacking intruders and removing sand and debris from the area defended by the breeding pair and their helpers. (B) A subordinate helper quivers his tail as the dominant fish approaches from behind. A, after Taborsky and Grantner [1122]; B, photograph by Michael Taborsky.

2 Reproductive interference in a social animal. One member of a breeding group of acorn woodpeckers removes an egg laid by a companion female from their communal nest. Photograph courtesy of Walt Koenig.

Such costly reproductive competition occurs in the acorn woodpecker, a bird that forms breeding groups containing as many as three females and four males. The several females all lay their eggs in the same tree hole nest, perhaps because any female that tries to nest alone will have all her eggs destroyed by her companions [637]. Even when several females settle on the same nest, the first eggs laid are almost always removed by another female member of the band (Figure 2) [827]. Eventually these "cooperatively breeding" females all lay eggs on the same day, at which time they finally stop tossing each other's eggs out and incubate the clutch. By this time, however, more than a third of the eggs laid by the woodpeckers may have been destroyed.

In addition to these direct reproductive costs, sociality has two other nearly inescapable costs. The first is heightened competition for food, which occurs in animals as different as colonial fieldfares (Figure 3) [1236] and lions, whose females are often pushed from their kills by hungry males [1020]. The second is increased vulnerability to parasites and pathogens [17] (but see [989]). As a general rule, the larger the group, the greater these costs, and the lower the net

Fieldfare

3 Competition for food is a cost of sociality in the fieldfare, a songbird that nests in loose colonies in woodlands. The larger the colony, the lower the survival rate of nestlings, due to increased mortality caused largely by starvation. After Wiklund and Andersson [1236].

4 Reproductive success and group size in a social spider. The spider *Anelosimus eximius* lives in groups of up to several hundred spiders on a single web. Initially, individual reproductive success increases with increasing group size, but the net benefits of associating with others soon level off, then decline as the number of communally living spiders grows still larger. After Avilés and Tufiño [50].

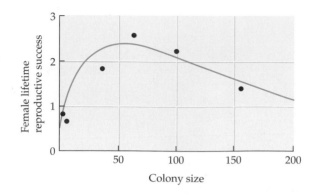

fitness of the members of the social unit (Figure 4) [50]. For example, the larger a nesting colony of cliff swallows, the greater the chance that at least one bird will be infected with blood-sucking swallow bugs, which can spread from brood to brood because the birds build their nests side by side. Charles and Mary Brown found that nestlings in large colonies were heavily assaulted by blood-sucking, growth-stunting swallow bugs [164]. The Browns demonstrated that the bugs were guilty of harming the nestlings by fumigating a sample of nests in an infested colony while leaving other control nests untreated. The nestlings doused with insecticide weighed much more, and were more likely to survive, than those plagued by many parasites (Figure 5).

The parasites that target swallow chicks and other social creatures demonstrate that if sociality is to evolve, costs of this sort must be outweighed by benefits to social individuals. As noted in Chapter 8, some social animals gain by following companions to good feeding sites [165, 455], while others, like egg-brooding male emperor penguins, save thermal energy by huddling together during the brutal Antarctic winter [27]. Still others join forces to fend off enemies of their own species, such as infanticidal males [1098].

Perhaps the most widespread fitness benefit for social animals, however, may be improved protection against predators (Figure 6) [17]. Many studies have

5 Effect of parasites on cliff swallow nestlings. The much larger nestling on the right came from an insecticide-treated nest; the stunted baby of the same age on the left occupied a parasite-infested nest. From Brown and Brown [164].

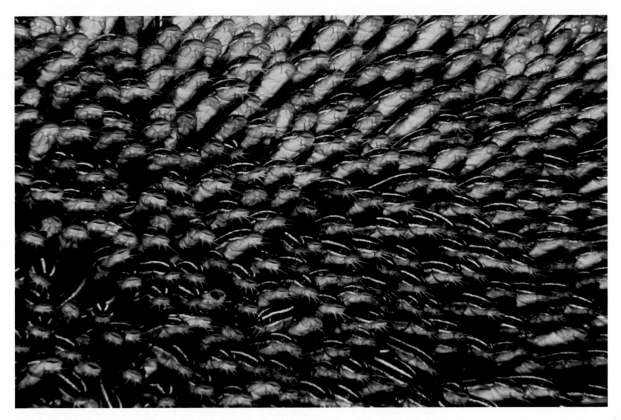

6   Social living with defensive benefits? The members of this dense school of small (5 cm long) striped catfish living on a coral reef near Sulawesi have almost certainly joined forces to improve their chances of survival, either by intimidating some predators through the collective size of the massed school or to amplify their defenses, if the fish are protected by spines or chemical repellents to some extent. Photograph by Roger Steene.

shown that animals in groups do gain by diluting the risk of being captured, or by spotting danger sooner, or by repelling their enemies more effectively (p. 202). In the maomao, a reef fish with a wonderful Latin name (*Abudefduf abdominalis*), individual males in small nesting groups are forced to chase other egg-eating fish about four times as often as males in large nesting aggregations. And when a nest-defending male is removed from a small group, his eggs are attacked by a predator sooner than the eggs of a male removed from a large group, indicating again that males derive mutual antipredator benefits by nesting together [1161].

Likewise, males in nesting colonies of the bluegill sunfish cooperate in driving egg-eating catfish away from their nests at the bottom of freshwater lakes (Figure 7) [463]. If the social behavior of the bluegill has indeed evolved in response to predation, then closely related species that nest alone should suffer less from predation. As predicted, the solitary pumpkinseed sunfish, a member of the same genus as the bluegill, has powerful biting jaws and so can repel egg-eating enemies on its own, whereas bluegills have small, delicate mouths good only for inhaling small, soft-bodied insect larvae [463]. The pumpkinseed sunfish are in no way inferior to or less well adapted than bluegills because they are solitary; they simply gain less through social living, which makes solitary nesting the adaptive tactic for them.

Although the comparative study of the two sunfishes supports the antipredator hypothesis for the evolution of sociality, one comparative analysis of colonial

7 Mutual defense in a society of bluegills. Each colonial male defends a territory bordered by the nest sites of other males, while bass (above), bullhead catfish (left), snails, and pumpkinseed sunfish (right foreground) roam the colony in search of eggs. Drawing courtesy of Mart Gross.

nesting in marine birds provided no definitive support for the hypothesis [982]. Today's seabirds that nest on the ground in vast colonies (Figure 8) almost certainly evolved from a solitary ancestral species whose nests were not so exposed and conspicuous. But whether colonial nesting evolved as a direct response to predation on dispersed ancestors or whether it evolved for some other reason (such as the benefits to nesting birds of finding food by observing other foragers) is an unresolved issue. Thus, we have more to learn about the precise causes of sociality even for some of the most conspicuous of all social animals.

8 A seabird society. A breeding colony of gannets is composed of nesting pairs that defend tiny territories packed close together. The evolutionary history and adaptive value of these social aggregations are not fully understood. Photograph by Bruce Lyon.

The Evolution of Helpful Behavior

Once animals live together, no matter what the origin of their social arrangements, they have the potential to assist one another, and they often do. Until the mid-1960s, biologists took helpful behavior for granted because they assumed that animals should assist one another for the benefit of the species as a whole. But with the recognition that group selection was unlikely to have played a major role in shaping behavioral evolution (see p. 15), helpful actions become considerably more interesting. Here, as elsewhere, the trick is to consider the costs and benefits of a helpful action (Table 1).

Helping sometimes generates immediate payoffs for both helper and helpee. When one lioness drives a wildebeest into a lethal ambush set by her fellow pride members [1091], the driver will get some meat, even if she did not personally pull the antelope down and strangle it. Likewise, if several male bluegills succeed in fending off a bullhead catfish that has entered their part of the nesting colony, the eggs in all the males' nests are more likely to survive to hatch. When both helper and recipient enjoy reproductive gains from their interaction, they have engaged in **mutualism,** or cooperation, which requires no special evolutionary explanation.

This is not to say that mutualism is boring. Consider the coalitions of male lions that form to oust rival males living with a pride of females. When cooperating males are successful, they may gain sexual access to a large group of receptive females. When Craig Packer and his associates analyzed the genetic relationships among male lion coalitions, they found that partnerships of two or three males were generally composed of unrelated individuals, and that they shared the females fairly evenly [878]. Nevertheless, some males in these coalitions do not do as well as others. Why do the disadvantaged males tolerate their situation? Probably because they could do no better alone. If they went solo, their chance of acquiring and defending a pride would be next to zero because one male has little chance against two or three rivals. Thus, some males are forced to cooperate with domineering companions if they are to have any chance of mating [635].

Likewise, subordinate yearling male lazuli buntings, which have dull brown plumage, engage in a special mutualism with brightly colored yearling males (Figure 9). The dominant bright-plumaged males aggressively drive other males with bright or intermediate plumage away from top-quality territories with good shrub cover. However, they tolerate the presence of dull-plumaged males, which are therefore able to claim territories in good habitat next to brightly colored companions. One hypothesis for the bright males' surprising acceptance of their brownish neighbors is that bright males are unlikely to lose paternity to dull neighbors, which probably cannot induce females to engage in extra-pair copulations. If this hypothesis is true, then extra-pair paternity should be

TABLE 1 **The reproductive success of individuals that engage in different kinds of social interactions**

Type of interaction	Effect on reproductive success of	
	Social donor	Social recipient
Mutualism (Cooperation)	+	+
Reciprocity	+ (delayed)	+
Altruism	−	−
Selfish behavior	+	−
Spiteful behavior	−	−

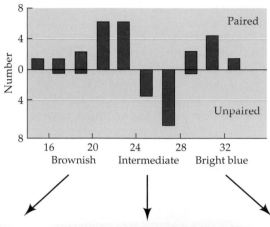

9 Cooperation among competitors. Young male lazuli buntings range in color from dull brownish to bright blue and orange. Bright yearling males permit dull males, but not males of intermediate brightness, to settle on neighboring territories. Reproductive success ranges from slight for the dull males to nil for the intermediate males, most of which are unpaired, to high for brightly colored yearlings. After Greene et al. [456]; photographs courtesy of Erick Greene.

high in the broods of the lazuli bunting, but less so in the nests of brightly colored males and their mates. In keeping with these predictions, about 50 percent of all nests sampled by Erick Greene and his co-workers had at least one chick fathered by a male other than the resident male, but the risk of cuckoldry for bright males was only half as great as for dull ones, which cared for an average of 1.5 extra-pair young. Most of these baby birds were probably fathered by the dull males' next-door neighbors [456].

Given these findings, what do dull males gain by accepting territories near bright males that are likely to cuckold them? Because dull subordinate buntings are permitted to hold high-quality territories, they are more likely to acquire a social mate than are males of intermediate brightness, which are often forced by dominant rivals into habitat so poor that no female will join them. Dull yearling males may often rear the chicks of other males, but they also produce their own on occasion, achieving at least some reproductive success, unlike most males of intermediate plumage and in-between social status, which have to wait another year to breed [456].

The fact that both dull and bright yearling neighbors gain fitness from their interactions means that their social arrangement constitutes a mutualism. But what about the male coalitions of the long-tailed manakin studied by David McDonald, in which only one of two cooperative males appears to reproduce? In this bird, males form pairs and sing loud duets over and over to attract females to a display court [393, 782]. Visiting females land on the pair's display perch, often a horizontal section of liana that lies a foot or so above the ground. In response, the two males dart in and land close to the prospective mate before performing an astonishing cartwheel display (Figure 10). After a series of these moves, the males may perform the "butterfly flight," in which they flutter slowly back and forth in front of the female, displaying their beautiful plumage and coordinated flight capacity. Should a female visitor start

10   Cooperative courtship of the long-tailed manakin. The two males are in the cartwheeling portion of their dual display to a female, who is perched on the vine to the right.

jumping on the perch in response to these displays, one member of the duo quickly leaves, while the remaining male stays to copulate with her. The female then flies off, after which the mated male calls for his display partner, who flies back to resume his duties.

By marking the males at display perches, McDonald and his manakin watchers found that each site has only one mating male. This alpha male is unrelated to his display companions, of which he has a favorite, a beta male, who in turn is dominant to several other part-time cooperators [782]. How can it be adaptive for the subordinates to work so hard on behalf of the sexually monopolistic alpha male? By patiently following males year after year, McDonald established that whenever an alpha male disappeared, the beta male took over, after which a lower-ranking individual moved up to become the ex-beta's main nonbreeding display partner. Therefore, by cooperating with the alpha male, a beta individual establishes his claim to be next in line, keeping other (younger?) birds at bay for years. When a beta male becomes an alpha, he usually gets to mate with many of the same females that copulated with the previous alpha (Figure 11) [782]. Thus, beta males form a mutualism with their exclusionary partners because this is the only way to become a reproducing alpha male (eventually).

Reciprocal Altruism or Personal Gain?

The study of long-tailed manakins shows us that some superficially self-sacrificing actions actually advance the reproductive chances of helpful individuals. Another possible case of this sort involves the meerkat, a small African mammal that forages in groups. From time to time, one meerkat will stop digging for insects in the soil and climb a tree or a termite mound to look around for approaching predators (Figure 12) [234]. Should a goshawk come swoop-

11 Cooperation with an eventual payoff. After the death of his alpha male partner, the beta male long-tailed manakin (now an alpha) copulates about as frequently as his predecessor did, presumably because the females attracted to the duo in the past continue to visit the display arena when receptive. After McDonald and Potts [782].

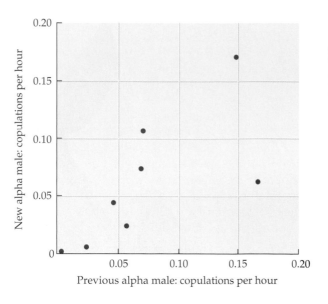

ing in, the elevated sentinel is usually the first to give an alarm, which sends all the still-foraging meerkats dashing for cover. One explanation for this behavior is that sentinels help others at personal cost now because they will be repaid later by their teammates when they take their turns at being lookouts. Bob Trivers gave this kind of social relationship the label **"reciprocal altruism"** [1152] because the helped individuals return the favor at a later date.

However, consider an alternative explanation for sentinel behavior. Perhaps the lookouts have full bellies and choose to climb a tree in order to better spot

12 A meerkat sentinel on the alert for approaching predators. Photograph by Nigel J. Dennis.

danger to themselves. Rather than offering costly assistance to others in their band, the "sentinels" could be securing personal fitness benefits, especially if an approaching goshawk is more likely to chase one of the lookout's fleeing companions rather than the alert sentinel. (Such an argument requires that the signaler be safer if it can get its companions to run for cover than if it slips off silently by itself; the hypothesis also requires that the signaler's companions gain more by dashing for a burrow than by remaining frozen in place in an effort to avoid detection by the onrushing predator.)

How can we test the reciprocal altruism hypothesis against the personal safety alternative? The reciprocal altruism hypothesis predicts that individuals should follow a regular rotation of sentinel duty and that sentinels should be especially exposed to predators, but apparently neither condition applies to meerkat bands. Sentinel duty is established haphazardly, and sentinels are usually closer to an escape burrow than are their fellows, suggesting that lookouts do not put themselves in special danger. The personal safety hypothesis receives support from the finding that solitary meerkats spend about the same proportion of each day in sentinel behavior as do the members of a band. Moreover, when meerkats are given supplemental food, which reduces the cost of taking time out to look around for predators, they increase the amount of time spent on the lookout perch. Thus, what initially appears to be a rotation of lookouts may actually be the product of individuals spending as much time as possible during the day in the safest possible position [234].

This is not to say that reciprocity is absent from nature [875, 1237]. In many primates, for example, individuals spend considerable time carefully grooming the fur of a companion (Figure 13). The groomer helps the groomee by removing parasites and debris, but gains no immediate benefit from its actions, cashing in only when the animal it helped returns the favor. In bands of baboons, pairs of females do take turns in grooming and being groomed, as predicted from the reciprocal altruism hypothesis [1058].

Reciprocity is not particularly common, however, perhaps because a population composed of reciprocal altruists may be vulnerable to invasion by individuals who accept help but later neglect to return the favor. "Cheaters" reduce the fitness of "noncheaters" in such a system, which ought to make reciprocal altruism less likely to evolve. The problem can be illustrated with a game theory model called the **prisoner's dilemma** (Figure 14), which is based on a human situation. Imagine that a crime has been committed by two persons, who agree

13 Reciprocity in a social primate. A vervet monkey grooms a companion. The groomee will return the favor at a later date. Photograph by Dorothy Cheney.

14 **The prisoner's dilemma.** The diagram lays out the payoffs for player A associated with the different options available to two potential cooperators.

not to squeal on each other if caught. The police have brought them in for interrogation and have put them in separate rooms. The cops offer each suspect his freedom if he will implicate his pal in the crime. If suspect A accepts the tempting offer ("Defect") while B maintains their agreed-upon story ("Cooperate"), A gets his freedom (the maximum reward) while B gets hit with the maximum punishment (the "sucker's payoff"). If together they maintain their agreement ("Cooperate + Cooperate"), they make it harder, though not impossible, for the police to convict either one. And if each one fingers the other, the police will use this evidence against both and renege on their offer of freedom for the snitch, so that both A and B will be punished for their mutual defection.

In a setting in which the payoffs for the various responses are ranked "Defect while other player cooperates" > "Both cooperate" > "Both defect" > "Cooperate while other player defects," the optimal response for suspect A is always to defect, never to cooperate. Under these circumstances, if suspect B maintains their joint innocence, A gets a payoff that exceeds the reward he could achieve by cooperating with a cooperative B; if suspect B squeals on A, defection is still the superior tactic for A, because his payoff when both players defect is greater than that from cooperating when his companion squeals on him. By the same token, suspect B will come out ahead on average if he defects and points the finger at his buddy.

This model predicts, therefore, that reciprocal cooperation should never evolve. How, then, can we account for the cases of reciprocity that have been observed in nature? One answer comes from examining scenarios in which two players interact repeatedly, not just once. Robert Axelrod and W. D. Hamilton have shown that when this condition applies, individuals that use the simple behavioral rule "Do unto individual X as he did unto you the last time you met" can reap greater gains than cheaters who accept assistance but do not return the favor [52]. When multiple interactions are possible, the rewards for back-and-forth cooperation add up, exceeding the short-term gain from a single defection.

More sophisticated tactics than the simple tit-for-tat just described are possible [848, 849]. These game-playing rules can yield even higher returns for cooperators who interact repeatedly. Individuals using a "generous tit-for-tat" strategy, for example, can avoid getting stuck in a long sequence of low-return mutual defections by "generously" being cooperative on occasion after a companion has defected on the previous round. Such a response can reinstate cooperative responses from the companion, leading to a series of mutually beneficial interactions. If alternating rounds of cooperation yield higher payoffs than rounds of mutual defection, then the individual who can sometimes "forgive" his fellow player for a defection or two will come out ahead after many plays of the prisoner's dilemma game [1196]. In other words, certain conditions make reciprocity quite likely to evolve.

Altruism and Indirect Selection

Reciprocal altruism is really a special kind of mutualism in which the helpful individual endures a short-term loss until its help is reciprocated, at which time it earns a net increase in fitness. In evolutionary biology, the term **altruism** is restricted to cases in which the donor really does permanently lose opportunities to reproduce as a result of helping another produce more surviving offspring. Altruistic actions, if they exist, are an especially exciting puzzle for adaptationists because they violate the "rule" that traits cannot spread over evolutionary time if they reduce an individual's personal reproductive success (see p. 15).

In order to explain how altruism could evolve, W. D. Hamilton developed a special explanation that did not rest on for-the-good-of-the-group arguments [484]. Instead, Hamilton's theory is based on the premise that individuals reproduce with the unconscious goal of propagating their special alleles more successfully than other individuals. Personal reproduction contributes to this ultimate goal in a direct fashion. But helping genetically similar individuals— that is, one's relatives—survive to reproduce can provide an indirect route to the very same end.

To understand why, we must return to the concept of the coefficient of relatedness, the probability that two individuals both possess the same rare allele by virtue of inheriting it from a recent common ancestor (see p. 55). Imagine, for example, that a parent has the genotype amy^1/amy^2, and that amy^2 is a rare form of the gene. Any offspring of this parent will have a 50 percent chance of inheriting the amy^2 allele because any egg (or sperm) that the parent donates to the production of an offspring has one chance in two of bearing the amy^2 allele. The coefficient of relatedness (r) between parent and offspring is therefore ½, or 0.5. In contrast, r is close to 0 between this adult and other unrelated individuals, which almost surely lack the allele in question.

The coefficient of relatedness varies for different categories of relatives. For example, an uncle and his sister's son have one chance in four of sharing a rare allele by descent because the man and his sister have 1 chance in 2 of sharing the allele in question, and the sister has 1 chance in 2 of passing that allele on to any of her offspring. Therefore, the coefficient of relatedness for an uncle and his nephew is ½ × ½ = ¼, or 0.25. For two cousins, the r value falls to ⅛, or 0.125.

With knowledge of the coefficient of relatedness between altruists and recipients of their help, we can determine the fate of a rare "altruistic" allele that is in competition with a common "selfish" allele. The key question is whether the altruistic allele becomes more abundant if its carriers forgo reproduction and instead help relatives reproduce. Imagine that an animal could potentially have one offspring of its own, or invest its efforts in the offspring of its siblings, helping three nephews or nieces survive that would have otherwise died. A parent shares half its genes with an offspring; the same individual shares one-quarter of its genes with each nephew or niece. Therefore, in this example, personal reproduction yields $r \times 1 = 0.5 \times 1 = 0.5$ genetic units contributed directly to the next generation, whereas altruism yields $r \times 3 = 0.25 \times 3 = 0.75$ genetic units passed on indirectly in the bodies of relatives. In this example, the altruistic tactic results in more shared alleles being transmitted to the next generation.

Another way of looking at this matter is to compare the genetic consequences to individuals who aid others at random versus those who direct their aid to close relatives. If aid is delivered randomly, then no one form of a gene is likely to benefit more than any other. But if close relatives aid one another selec-

15 The components of selection and fitness. (A) Direct selection acts on variation in individual reproductive success. Indirect selection acts on variation in the effects individuals have on their relatives' reproductive success. (B) Direct fitness is measured in terms of personal reproductive output; indirect fitness is measured in terms of genetic gains derived by helping relatives reproduce. Inclusive fitness can be considered the sum of the two measures and represents the total genetic contribution of an individual to the next generation. After Brown [168].

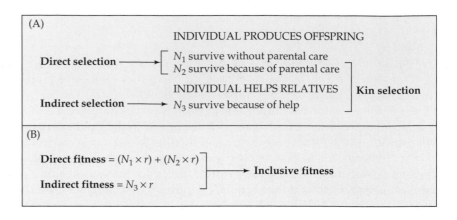

(A)

INDIVIDUAL PRODUCES OFFSPRING

Direct selection ⟶ $\begin{bmatrix} N_1 \text{ survive without parental care} \\ N_2 \text{ survive because of parental care} \end{bmatrix}$

INDIVIDUAL HELPS RELATIVES   Kin selection

Indirect selection ⟶ N_3 survive because of help

(B)

Direct fitness = $(N_1 \times r) + (N_2 \times r)$

Indirect fitness = $N_3 \times r$

⟶ Inclusive fitness

tively, then any distinctive alleles they possess may survive better, helping those alleles increase in frequency compared with other forms of the gene in the population at large. When one thinks in these terms, it becomes clear that a form of natural selection can occur when genetically different individuals differ in their effects on the reproductive success of close relatives. Jerry Brown calls this form of selection **indirect selection,** which he contrasts with **direct selection** for traits that promote success in personal reproduction (Figure 15) [168].

A brief digression is necessary here to deal with yet another term, **kin selection,** which was originally defined by John Maynard Smith to embrace the evolutionary effects of *both* parental aid given to descendant kin (offspring) and altruism directed to **nondescendant kin** (relatives other than offspring). Biologists have long recognized that parents can affect the evolutionary process by improving the survival chances of their offspring, and that parental care spreads when the resulting increase in the survival of the aided offspring more than compensates a parent for the loss of opportunities to produce additional offspring in the future (see p. 397). In genetic terms, parents can gain via parental investment because they share 50 percent of their genes with each offspring. By the same token, however, individuals can promote the survival of certain of their genes by helping relatives other than offspring. Altruism can be favored by the component of kin selection that Brown calls *indirect selection,* and the use of this term keeps the focus clearly on the distinction between parental effects on offspring and an aid-giver's effects on nondescendant kin [168]. The term *kin selection* is, however, currently used very widely by evolutionary biologists as a synonym for indirect selection, and readers should be aware that when they see the term, odds are that it is being used to refer to selection for altruism directed to relatives other than offspring.

The Alarm Call of Belding's Ground Squirrel

Having laid the necessary theoretical groundwork, let's use it in analyzing a possible case of altruism. African meerkats are not the only animals to call when they detect a predator. In North America, the Belding's ground squirrel gives a staccato alarm whistle (Figure 16) when a coyote or badger approaches; other ground squirrels that hear the signal rush for safety. Is this behavior a product of direct selection, as is apparently true for meerkat sentinel behavior, or has indirect selection shaped the willingness of the squirrels to give the alarm?

Paul Sherman collected the evidence needed to evaluate these hypotheses [1044]. He found that alarm-calling ground squirrels are tracked down and killed by weasels, badgers, and coyotes at a higher rate than noncallers, a dis-

16 A Belding's ground squirrel gives an alarm call after spotting a terrestrial predator. Photograph by George Lepp, courtesy of Paul Sherman.

covery that eliminates the direct selection hypothesis that alarm calling benefits the signaler by confusing or deterring predators. Moreover, the possibility that alarm calling evolved as a form of reciprocity is also unlikely because the probability that an individual will give an alarm call is not correlated with familiarity or length of association between the caller and the animals that benefit from its signal [1045]. Remember that tit-for-tat reciprocity is more likely to evolve when reciprocators form long-term associations.

Sherman's observation that adult female squirrels with relatives nearby are more than twice as likely as males to give costly alarm calls is consistent with both a parental care hypothesis (based on direct selection) and an altruism hypothesis (based on indirect selection). If the parental care hypothesis is correct, we would expect females to give more alarm calls than males because only female squirrels live near their offspring, whereas males move to new locations away from their offspring. If the altruism hypothesis is correct, we can also predict a female bias in alarm calling because females not only live near their genetic offspring, but are also surrounded by other female relatives, such as sisters, aunts, and female cousins. When self-sacrificing females warn their nondescendant kin, they could be compensated for the personal risks they take by an increased probability that kin *other than offspring* will survive to pass on shared genes, resulting in indirect fitness gains for the altruists. Females with offspring living nearby as well as females with *only* nondescendant kin as neighbors are more likely to call upon detecting a predator than are females who lack relatives in their neighborhood. These findings suggest that both direct and indirect selection contribute to the maintenance of alarm calling behavior in this species [1044].

The Concept of Inclusive Fitness

Because fitness gained through personal reproduction (**direct fitness**) and through increased production of surviving nondescendant kin (**indirect fitness**) can both be expressed in identical genetic units, we can sum up an individual's total contribution of genes to the next generation, creating a quantitative measure that can be called **inclusive fitness** (see Figure 15B). Note that an individual's inclusive fitness is not calculated by adding up that animal's genetic representation in its offspring plus that in all its other relatives. Instead, what counts is an individual's own effects on gene propagation (1) directly in the bodies of its surviving offspring *that owe their existence to the parent's actions, not to the efforts of others*, and (2) indirectly via nondescendant kin that would not have existed except for the individual's assistance. For example, if the animal we mentioned earlier successfully reared one of its own offspring and also adopted three of its sibling's progeny, then its direct fitness would be $1 \times 0.5 = 0.5$ and its indirect fitness would be $3 \times 0.25 = 0.75$; the union of these two figures provides a measure of the animal's inclusive fitness ($0.5 + 0.75 = 1.25$).

Inclusive fitness, however, is not used to secure absolute measures of the lifetime genetic contribution of individuals, but rather to help us compare the evolutionary (genetic) consequences of two alternative hereditary traits [945]. In other words, inclusive fitness becomes important only as a means to determine the relative genetic success of two or more behavioral strategies. For example, if we wish to know whether an altruistic strategy is superior to one that promotes personal reproduction, we can compare the inclusive fitness consequences of the two traits. In order for an altruistic act to be adaptive, the inclusive fitness of altruistic individuals has to be greater than it would have been if those individuals had tried to reproduce personally. As W. D. Hamilton showed, in what is now called **Hamilton's rule**, a gene "for" altruism will spread only if the loss of direct fitness for the altruist (the number of offspring not produced, *c*, times the coefficient of relatedness between parent and offspring, r_c) is less than the indirect fitness gained by the altruist (the *extra* number of relatives that exist thanks to the altruist's action, *b*, times the mean coefficient of relatedness between altruist and recipients, r_b). For example, if the genetic cost of an altruistic act were the loss of one offspring ($1 \times r = 1 \times 0.5 = 0.5$ genetic units), but the altruistic act led to the survival of three nephews that would have otherwise perished ($3 \times r = 3 \times 0.25 = 0.75$ genetic units), the altruist would experience a net gain in inclusive fitness, thereby increasing the frequency of any distinctive allele associated with its altruistic behavior.

The value of Hamilton's rule can be illustrated by Uli Reyer's study of the pied kingfisher [964]. These attractive African birds nest colonially in tunnels in banks by large lakes. Some year-old males are unable to find a mate, and instead become *primary helpers* that bring food to their mother and her nestlings while attacking predatory snakes and mongooses. The key question is, are these males propagating their genes as effectively as possible by helping to raise their siblings? They do have other options: they could help unrelated nesting pairs in the manner of *secondary helpers,* or they could simply sit out the breeding season, waiting for next year in the manner of *delayers.*

To learn why primary helpers help, we need to know the costs and benefits of their actions. Primary helpers work harder than delayers and the more laid-back secondary helpers (Figure 17). The greater sacrifices of primary helpers translate into a lower probability of returning to the breeding grounds the next year (just 54 percent return) compared with secondary helpers (74 percent return) or delayers (70 percent return). Furthermore, only two in three surviving primary helpers find a mate in their second year and reproduce personally, whereas 91 percent of returning secondary helpers succeed in breeding. Many one-time secondary helpers breed with the female they helped the preceding year (10 of 27 in Reyer's sample), suggesting that improved access to a potential mate is the ultimate payoff for their initial altruism.

These data enable us to calculate the direct fitness cost to the altruistic primary helpers in terms of reduced personal reproduction in their second year of life. For simplicity's sake, we shall restrict our comparison to solo primary helpers that help their parents rear siblings in the first year and then breed on their own in the second year, if they survive and find a mate, versus secondary helpers

Pied kingfisher

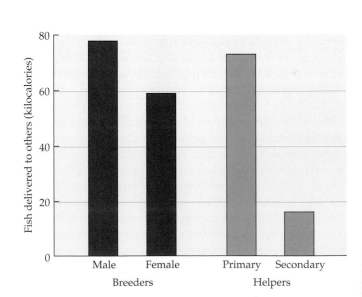

17   Altruism and relatedness in pied kingfishers. Primary helpers deliver more calories per day in fish to a nesting female and her offspring than do secondary helpers, which are not related to the breeders they assist. After Reyer [964].

TABLE 2 Calculations of inclusive fitness for male pied kingfishers

Behavioral tactic	First year			Second year				
	y	r	f_1	o	r	s	m	f_2
Primary helper	$1.8 \times 0.32 = 0.58$			$2.5 \times 0.50 \times 0.54 \times 0.60 = 0.41$				
Secondary helper	$1.3 \times 0.00 = 0.00$			$2.5 \times 0.50 \times 0.74 \times 0.91 = 0.84$				
Delayer	$0.0 \times 0.00 = 0.00$			$2.5 \times 0.50 \times 0.70 \times 0.33 = 0.29$				

Symbols: y = extra young produced by helped parents; o = offspring produced by breeding ex-helpers and delayers; r = coefficient of relatedness between the male and y and o; f_1 = fitness in first year (indirect fitness for the primary helper); f_2 = direct fitness in second year; s = probability of surviving into the second year; m = probability of finding a mate in the second year.

Source: Reyer [964]

that help nonrelatives with no other helpers present in the first year and then reproduce on their own in the second year, if they survive and find a mate.

Primary helpers throw themselves into helping their parents produce offspring *at the cost of having less chance of reproducing personally in the next year.* Although primary helpers do better than delayers in the second year (0.41 versus 0.29 units of direct fitness), secondary helpers do better still (0.84 units of direct fitness) because they have a higher survival rate and greater probability of securing a partner (Table 2).

But is the cost to primary helpers of 0.43 lost units of direct fitness (0.84 − 0.41 = 0.43) in the second year offset by a gain in indirect fitness during the first year? To the extent that these males increase their parents' reproductive success, they create siblings that would not otherwise exist, indirectly propagating their genes in this fashion. In Reyer's study, the parents of a primary helper gained an *extra* 1.8 offspring on average when their son was present. Some primary helpers assisted their genetic mother and father, in which case the extra 1.8 siblings were full brothers and sisters, with a coefficient of relatedness of 0.5. But in other cases, one parent had died and the other had remated, so that the offspring produced were only half-siblings ($r = 0.25$). The average coefficient of relatedness for sons helping a breeding pair was between one-quarter and one-half ($r = 0.32$). Therefore, the average gain for helper sons was 1.8 extra sibs × 0.32 = 0.58 units of indirect fitness, a figure higher than the mean direct fitness loss experienced in their second year of life.

Reyer used Hamilton's rule to establish that primary helpers sacrifice future personal reproduction in year 2 in exchange for *increased* numbers of nondescendant kin in year 1 [964]. Because these added siblings carry some of the helpers' alleles, they provide indirect fitness gains that more than offset the loss in direct fitness that primary helpers experience in their next year relative to secondary helpers.

Inclusive Fitness and Helpers at the Nest

In the pied kingfisher, primary helpers raise their fitness indirectly through the increased production of nondescendant kin, whereas secondary helpers raise their fitness directly by increasing their future chances of reproducing personally. Primary helpers demonstrate that altruism can be adaptive; secondary helpers show that helping need not be altruistic, but instead may generate direct fitness benefits to helpers. Thus, this one species offers support for two very different adaptationist hypotheses on the evolution of helping behavior. These hypotheses can be tested for other cases of helpers at the nest, which are found in a variety of other birds, as well as fishes, mammals, and insects [168, 581, 949, 1121].

Each case of helping at the nest presents a separate puzzle that deserves to be analyzed in light of a full range of hypotheses, including the possibility that caring for another's offspring is a *nonadaptive* side effect of other adaptive traits. As Ian Jamieson has pointed out, helping may have *originated* in some bird species as an incidental side effect of genetic or ecological changes that made it adaptive for young adults to delay their dispersal from their natal territory [573, 574]. If these stay-at-home birds were exposed to the nestlings being cared for by their parents, the begging behavior of the baby birds might have activated parental behavior in the young nonbreeding adults. How the helping trait first appeared is an issue to be resolved by testing alternative historical scenarios, but once it did, the question becomes what might maintain the trait in a population. One nonadaptationist hypothesis is that the behavior could be maintained as a by-product of two adaptive traits, delayed dispersal and the ability to care for one's own offspring, even if feeding someone else's young reduced the fitness of nonbreeding helpers.

This nonadaptationist by-product hypothesis assumes that selection could not eliminate costly feeding of its parents' offspring without also destroying the capacity of the stay-at-home bird to invest in its own nestlings at a later date. This assumption is questionable, but we can test the hypothesis anyway. One of its key predictions is that the underlying mechanisms of parental care should be no different in species with helpers than in species without helpers. In the Mexican jay (*Aphelocoma ultramarina*) and Florida scrub jay (*Aphelocoma coerulescens*), some nonbreeding birds help their parents rear additional siblings (Figure 18). The western scrub jay (*Aphelocoma californica*) is a member of the same genus, but lacks helpers at the nest. Helpers in the Mexican jay and Florida scrub jay have prolactin levels that match those of their breeding parents and are much higher than those in nonbreeding western scrub jays [1025, 1172]. (Prolactin is a hormone thought to play a critical role in regulating parental care in birds.) Moreover, the prolactin levels of nonbreeding Mexican jays rise to peak levels *before* there are young to feed in the nest (Figure 19), suggesting that selection has favored nonbreeding individuals of this species that are capable of becoming hormonally primed to rear their siblings [1172].

18   Cooperation among scrub jay relatives. Helpers at the nest in the Florida scrub jay provide food for the young, defense for the territory, and protection against predators. Based on a drawing by Sarah Landry, from Wilson [1247].

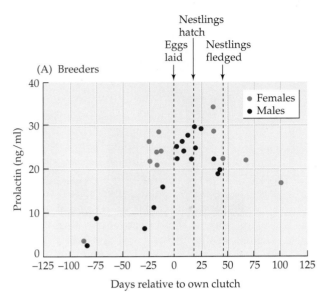
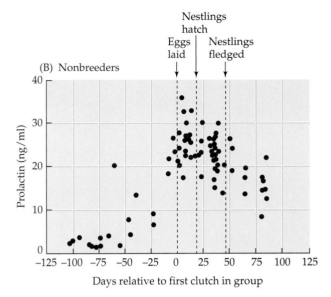

19   Seasonal changes in prolactin concentrations in helpers at the nest in the Mexican jay. Nonbreeding birds in a group exhibit the same pattern of increased prolactin production prior to the hatching of eggs as do breeding adults. After Brown and Vleck [172].

These results are at odds with the nonadaptationist by-product explanation for helping.

Another way to test alternative hypotheses for helping requires information on whether helpers derive inclusive fitness gains via the direct or indirect routes. In both the Mexican jay and Florida scrub jay, for example, some stay-at-home helpers inherit their natal territories from their parents—a direct fitness benefit. Furthermore, parents with helpers rear more offspring than parents without helpers, which generates an indirect fitness benefit for the helpers as well (Table 3) [168, 1265]. However, the apparent increase in the number of offspring fledged by pairs with helpers might arise strictly because helpers live with parents on better territories, which provide more food or superior nesting sites. The hypothesis that territory quality, not helpers at the nest, is truly responsible for differences among breeding pairs in the number of fledglings produced has been tested by Ronald Mumme. He captured and removed the nonbreeding helpers from some randomly selected breeding pairs, while leaving other

TABLE 3   Effect of Florida scrub jay helpers at the nest on the reproductive success of their parents and on their own inclusive fitness

	Parents do not have breeding experience	Parents do have breeding experience
Average number of fledglings produced with no helpers	1.03	1.62
Average number of fledglings produced with helpers	2.06	2.20
Increased reproductive success due to help	1.03	0.58
Average number of helpers	1.70	1.90
Indirect fitness gained per helper	0.60	0.30

Source: Emlen [359]

20 Helpers at the nest help parents raise more siblings in the
Florida scrub jay. The graph shows numbers of offspring alive after 60 days
in experimental nests that lost their helpers and in unmanipulated control
nests during a 2-year experiment. After Mumme [826].

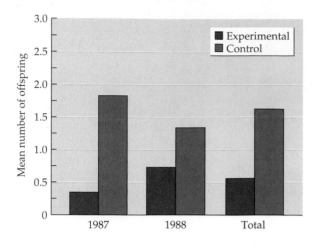

helpers untouched. The experimental removal of helpers reduced the reproductive success of the experimental pairs by about 50 percent, as measured by
the number of offspring known to be alive 60 days after hatching (Figure 20).
Helpers apparently really do help in this species [826].

In fact, helper scrub jays also improve the chances that their parents will live
to breed again another year, as do helpers in the pied kingfisher. Improved
parental survival means that helpers are responsible for still more siblings in
the future; these extra siblings yield an average of about 0.30 additional indirect fitness units for helpers [828]. Thus, the total indirect fitness gains from
altruistic helping can potentially exceed its costs in terms of lost direct fitness,
particularly if the young birds have almost no chance of reproducing personally. When very few openings are available for dispersing young adults, helping is more likely to be the adaptive option.

Whether saturated nesting habitats are responsible for the maintenance of
helping at the nest is a testable proposition. If young birds remain on their natal
territories because they cannot find suitable nesting habitat, then yearlings given
an opportunity to claim good open territories should promptly become breeders. Jan Komdeur did the necessary experiment with the Seychelles warbler, a
drab little brown bird that has been very helpful in supplying evidence on the
evolution of helping at the nest. When Komdeur transplanted 58 birds from one
island (Cousin) to two other nearby islands with no warblers, he created vacant
territories on Cousin, and helpers at the nest there immediately stopped helping in order to move into open spots and begin breeding. Since the islands
that received the transplants initially had many more suitable territorial sites
than warblers, Komdeur expected that the offspring of the transplanted adults
would also leave home promptly in order to breed elsewhere on their own. They
did, providing further evidence that young birds help only when they have
little chance of making direct fitness gains by dispersing [640].

Moreover, the sophisticated conditional strategy that controls the dispersal
decisions made by youngsters is sensitive to the quality of their natal territory.
Breeding birds occupy sites that vary in size, vegetational cover, and insect supplies. By using these variables to divide warbler territories into categories of
low, medium, and high quality, Komdeur showed that young helpers on good
territories were likely to survive there while also increasing the odds that their
parents would reproduce successfully compared to adults without helpers on
comparable territories. Young birds whose parents had prime sites often stayed
put, securing both direct and indirect fitness gains in the process. In contrast,

Seychelles warbler

TABLE 4 The effect of natal territory quality on the fitness consequences of helping at the nest by young Seychelles warblers

	Territory quality		
	Low	Medium	High
Increase in number of fledglings produced by pairs with a helper	0.03	0.34	0.42
Proportion of helpers that survive the year	0.30	0.67	0.86
Proportion of young birds that remain in natal territory to help	0.29	0.69	0.93

Sources: Komdeur [639] and Mumme [825]

young birds on poor natal territories had little chance of making it to the next year, nor could they have a positive effect on the reproductive success of their parents (Table 4). They left home and tried to find a breeding opportunity of their own.

On the once vacant island of Aride, the transplanted warblers bred successfully, and they and their offspring began to fill up appropriate breeding habitat. Komdeur and colleagues predicted that helpers would appear first in the best territories, later on in medium-quality sites, and last in low-quality territories on Aride [645]. The birds behaved as predicted, making decisions that maximized their inclusive fitness, whether achieved directly by reproducing personally or indirectly by helping their parents rear more siblings (Figure 21).

In the Seychelles warbler, daughters are more likely than sons to be helpers, all other things being equal, perhaps because helper females gain experience that substantially increases their chances of successful reproduction when they breed on their own for the first time [642]. Because Komdeur was aware of this sex bias in helping tendency, he realized that adults holding superior territories would gain if they could produce a daughter who would be predisposed to help them feed their next offspring. In contrast, breeding pairs on poor territories would do better to produce a son, who would usually leave home rather than remain on site to become a burden on the limited food supplies there.

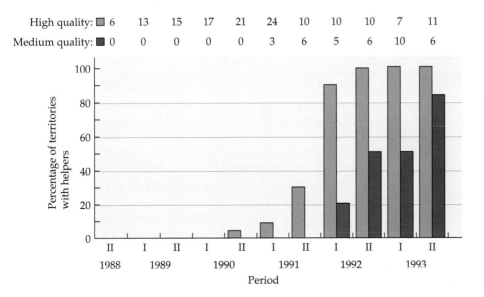

High quality: ▨ 6   13   15   17   21   24   10   10   10   7   11
Medium quality: ■ 0   0   0   0   0   3   6   5   6   10   6

21   Territory quality influences the decision to become a helper at the nest. After the introduction of the Seychelles warbler to Aride Island, the population grew until all the best territories were taken, after which some birds were forced to settle for medium-quality sites. Subsequently, helpers at the nest first appeared in the high-quality territories, and only later in some medium-quality territories. Numbers at the top are the numbers of established territories that could have had helpers. After Komdeur et al. [645].

22 Adaptive sex ratio manipulation in the Seychelles warbler. The eggs laid by females nesting on low-quality territories tend to produce male nestlings, whereas the eggs of females in possession of high-quality territories usually produce daughters, whose assistance as helpers will boost parental fitness. After Komdeur et al. [643].

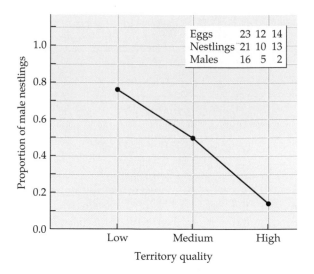

	Low	Medium	High
Eggs	23	12	14
Nestlings	21	10	13
Males	16	5	2

Remarkably, female Seychelles warblers do have the ability to bias the sex ratio of their offspring in the predicted, adaptive direction (Figure 22) [643]. In some way, the female warblers use information gathered about territory quality to manipulate meiosis. Females on good territories are able to bias gamete production toward eggs that carry the W sex chromosome; when an egg cell of this sort unites with a Z-bearing sperm cell, it will yield a WZ zygote that is female. (In birds, females are the sex with two different sex chromosomes, unlike mammals, in which females have two copies of the X sex chromosome while XY-bearing individuals are males.)

Insect Helpers at the Nest

The parental ability to manufacture helpers of the appropriate sex has also evolved in certain insects. For example, the paper wasp colonies mentioned at the beginning of this chapter usually consist of one or more reproductively active females and a number of helpers at the nest, or workers, which are always females, never males. In order to generate a set of female helpers early in the nesting cycle, egg-laying female wasps fertilize some eggs by releasing sperm from a sperm storage organ as the eggs pass down the oviduct. In order to produce sons, the females need only lay unfertilized eggs. Sex determination in the Hymenoptera (the wasps, bees, and ants) is usually achieved in this fashion, with fertilized eggs producing diploid females and unfertilized eggs giving rise to haploid males.

The degree to which hymenopteran queens can control the fertilization of their eggs has been measured in the honey bee. In this species, worker (female) larvae are reared in smaller brood cells than are drone (male) larvae. Direct genetic analysis of over 100 eggs laid in small cells revealed that they were all diploid and so had been fertilized before oviposition occurred, whereas all the eggs removed from large cells were haploid and so had not been fertilized. In this hymenopteran species at least, females have essentially perfect control over the sex of their progeny [955].

In some social hymenopterans, including paper wasps, the daughters produced early in the season have functional ovaries, but they may not use them if they stay at their natal nest to help care for the additional offspring that their mother continues to manufacture. (Some potential workers, however, do not stay on as helpers, but retire to wait for the next season, at which time they attempt to found nests of their own [959].) Daughter workers help to rear more offspring, some of which will be brothers and sisters capable of reproducing late in the season. Their brothers die soon after mating, but their mated sisters overwinter in a shelter of some sort, storing the sperm they received for use in the next spring, when they become active again and build a paper nest to start the cycle anew.

Those daughters that forgo reproduction in order to help out at their natal nest gain indirect fitness to the extent that their actions increase the number of their surviving, reproducing brothers and sisters. But what about those helpers that are not the queen's daughters, but instead are females of her own generation that have joined a foundress at her nest? All the members of these foundress associations are physiologically capable of reproducing, but often only one female monopolizes the egg laying while the others assist the queen by caring for her brood. For example, the dominant female at the nests of the paper wasp *Polistes dominulus* usually lays over 95 percent of all the eggs, granting the others almost no reproductive rights at all. Why do the subordinates accept their status?

Researchers have used molecular techniques to establish that about half of the subordinate helpers in *P. dominulus* are actually full sisters of the dominant female. Therefore, these helpers are helping to protect and feed their nephews and nieces, possibly increasing the number that survive to reproduce, in which case the altruists gain indirect fitness [949]. On the other hand, about a third of all helpers in *P. dominulus* are neither sisters nor cousins of the female they assist. They cannot therefore derive indirect fitness benefits of any sort from their helpfulness. But, just as Reyer argued for pied kingfishers, perhaps these wasps are secondary helpers with a reasonable probability of reproducing on their own after the demise or displacement of the dominant female. Observations of 28 nests of *P. dominulus* yielded 13 records of takeovers, ten of which were accomplished by resident helpers [949].

In *P. dominulus,* subordinate females almost never reproduce in the presence of the dominant egg layer, but in other paper wasps and social insects, some subordinates do reasonably well. These cases raise another evolutionary question: Why do dominant females let subordinates reproduce at all? One possible answer is that the fitness cost of policing colonymate behavior is greater than the benefit of totally suppressing subordinate reproduction. In large groups, or in groups with serious challengers, the dominant breeder may find it especially difficult to control the behavior of all the other members of her group [229]. Or perhaps under some circumstances, such as high risk of parasitic infestation, females that share reproductive duties together create a genetically more diverse colony that is better able to resist parasites. Evidence in favor of this hypothesis comes from experiments in which bumblebee groups were manipulated to create colonies with high and low genetic variation; in these experiments, the colonies with greater genetic variation were less severely attacked by parasites (Figure 23) [696].

Or perhaps dominant females permit a certain amount of reproduction by subordinates in order to encourage their helpers to remain in place. By doing so, they might make the helpers' inclusive fitness higher than it would be if they were to depart and try to reproduce on their own. This hypothesis is derived from what has been called **concession theory,** which is based on Hamilton's rule. This theory focuses on differences in the degree of relatedness among group members and differences in the opportunities for subordinates to reproduce on their own in order to explain why groups differ in size and in the degree to which one or a few group members monopolize reproduction [958]. Concession theory yields the prediction, for example, that dominants should concede more reproductive chances to unrelated helpers than to close relatives. When unrelated females help a dominant queen paper wasp, they cannot derive

23 **Genetic variation reduces parasitism** within experimental colonies of bumblebees. Parasitism is measured in three different ways: prevalence, or the percentage of colonies with parasites; load, or the total number of parasites per colony; and richness, or the number of species of parasites per colony. After Liersch and Schmid-Hempel [696].

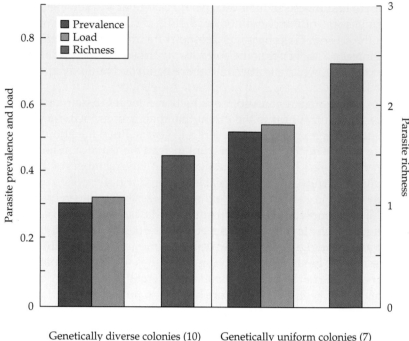

indirect fitness from their assistance, as the queen's relatives do, and so non-relatives should require some direct fitness payoffs for staying and helping. As predicted, when dominant females of the wasp *P. fuscatus* associate with unrelated helpers, the reproductive differences among dominants and subordinates are less than when the females in a colony are related to one another (Figure 24) [962]. Likewise, the skew in reproductive success is greater among large coalitions of male relatives in the lion than in smaller male bands, which are often composed of nonrelatives [878].

Concession theory also predicts that subordinates, whether related to dominants or not, should require fewer concessions to remain in a group if their

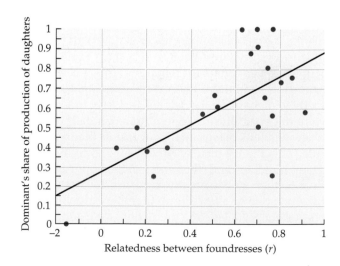

24   Who reproduces in paper wasp colonies? In *Polistes fuscatus* colonies, when foundress females are not closely related, they share reproduction more evenly, thereby reducing the degree to which a single female monopolizes reproduction at the nest. After Reeve et al. [962].

chances of reproducing personally elsewhere are very low. Recall that beta long-tailed manakin males permit the unrelated alpha male to do all the mating because solo betas have no chance of attracting females on their own. Under conditions of this sort, the fitness gained by cooperating need not be great in order for a subordinate to benefit from remaining with its group.

Although predictions from concession theory have been supported with data gathered from a wide range of insects and some vertebrates [958], the ability of the theory to account for most aspects of sociality is still under debate and will require additional tests [384].

The Evolution of Eusocial Behavior

Although the evolution of helping at the nest by reproductively capable female wasps unrelated to the queen is not yet completely understood, even more intriguing are the sterile castes of certain social insects (Figure 25). The anatomy and behavior of these caste members show extreme specialization for self-sacrifice. Thus, honey bee workers have a barbed sting designed to catch in the skin of vertebrate enemies, the better to deter them, even though it means the death of the hive defender (Figure 26). Soldier ants of some species have immense jaws, the better to pierce the bodies of predators, including fellow ants, that have invaded their colonies. Other ant species have grenade soldiers, whose suicidal mission is to rush at colony intruders while simultaneously constricting their abdominal muscles so violently that they burst a large abdominal gland, spilling a disabling glue over their enemies [764].

Species that include specialized nonreproductive castes are known as **eusocial** species. Eusocial insects caught Darwin's eye, and his solution for their evolution was closely related to the indirect fitness hypothesis already outlined. Dar-

25   Eusocial insects have sterile castes. (Left) In many ant species, several sterile castes exist, each designed for a particular servant role within the colony. Here a tiny worker ant perches on the head of a giant soldier female. (Right) In a nasute termite colony, the sterile soldiers have glands in their heads and can spray enemies with sticky repellent fluids. Left, photograph by Mark Moffett; right, photograph by the author.

26   Suicidal sacrifice by a worker bee. When a honey bee stings a vertebrate, she dies after leaving her stinger and the associated poison sac attached to the body of the victim. Photograph by Bernd Heinrich.

win noted that social insect colonies are extended families, so that when sterile members of the group help others survive to reproduce, the helpers (even if they die in the process) are helping to maintain family traits, including the ability of the reproducing members of the colony to generate some sterile helpers.

The next major advance on Darwin's explanation came 120 years later when W. D. Hamilton developed his now famous genetic cost–benefit analysis of worker altruism. Remember that according to Hamilton's rule, altruism can evolve when the altruist's loss in personal reproduction ($*c$) times the degree of relatedness of parent to offspring (r_c) is less than the added number of reproducing relatives that owe their existence to the altruist ($*b$) multiplied by the degree of relatedness between the altruist and the helped individuals (r_b). Hamilton realized that if r_b, the relatedness of the altruist to the relatives it helped, was particularly high, then the indirect fitness side of the equation would be increased. He also was the first to point out that r_b for sisters could indeed be unusually high in the Hymenoptera because of the haplodiploid system of sex determination in this group [484].

The potentially very close genetic similarity between sister bees, ants, and wasps arises because male Hymenoptera, which come from unfertilized eggs, are haploid (have only one set of chromosomes, not two). Therefore, all the haploid sperm a male ant, bee, or wasp makes are chromosomally (and thus genetically) identical. If a female ant, bee, or wasp mates with just one male, all the sperm she receives will have the same set of genes. When the female uses those sperm to fertilize some eggs, all her diploid daughters will carry the same set of paternal chromosomes and genes, which make up 50 percent of their total genotype. The other set of chromosomes carried by female hymenopterans comes from their mother. The mother's haploid eggs are not genetically uniform because she is diploid; gamete formation in animals with two sets of chromosomes involves the production of a cell with just one set drawn at random from those in the parent. An egg made by a female bee, ant, or wasp will on average share 50 percent of the alleles carried within her other eggs. Thus, when a queen bee's eggs unite with genetically identical sperm, the resulting offspring will share on average 75 percent of their alleles: 50 percent from their father and 0–50 percent from their mother (Figure 27).

Under the haplodiploid system of sex determination, hymenopteran sisters may therefore have a coefficient of relatedness of 0.75, higher than the 0.50 figure for a mother and her daughters and sons. As a consequence of this genetic fact, ($r_c \times *c$) should be less than ($r_b \times *b$) more often in the Hymenoptera than in

(A) Mother–offspring genetic relatedness

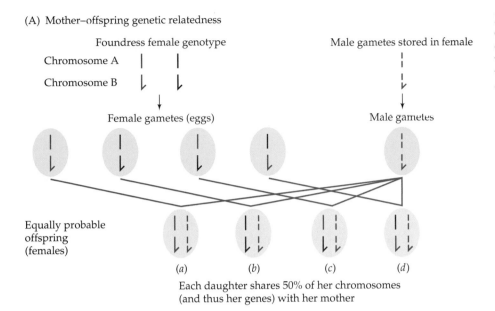

27 **Haplodiploidy and the evolution of sociality in the Hymenoptera.** The degree of genetic relatedness of a female wasp (A) to her offspring and (B) to her sisters. For the sake of simplicity, only two chromosomes are shown.

(B) Sister–sister genetic relatedness

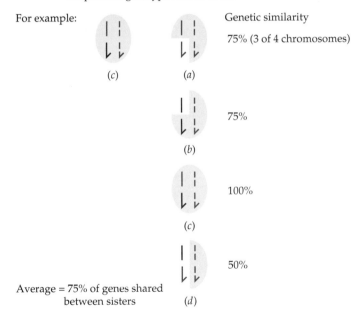

other groups, all other things being equal, which would facilitate the evolution of eusociality in these insects. If sisters really are especially closely related, indirect selection could more easily favor hymenopterans that, so to speak, put all their eggs (alleles) in a sister's basket rather than reproducing personally. Perhaps not coincidentally, the Hymenoptera have the greatest number of eusocial species of any insect family.

Testing the Haplodiploid Hypothesis

The haplodiploid hypothesis for eusociality generates testable predictions. First, if a female worker in a eusocial bee, ant, or wasp colony is to cash in on her potentially high degree of relatedness to her fellow females, she should bias her help toward reproductively competent sisters rather than toward male siblings. Although sister hymenopterans share up to 75 percent of their genes in common, a sister has only 25 percent of her haploid brother's genes (see Figure 27). Males do not receive any of the paternal genes that their sisters possess. The remaining half of the genome that sisters and brothers both receive from their mother ranges from 0 to 100 percent identical, averaging 50 percent; 50 percent of a half means that a sister shares one-quarter of the genes of her brothers on average ($r = 0.25$).

Bob Trivers and Hope Hare recognized that these brother–sister inequalities in r should lead to conflict between queens and workers over how workers allocate their assistance [1156]. If workers share three times as many genes on average with their sisters as with their brothers, the stable investment ratio from the workers' perspective would be 3:1 in favor of sisters. In contrast, their queen mother donates 50 percent of her genes to each offspring, male or female; she presumably gains no genetic advantage by investing more resources in the production of sons ($r = 0.5$) than in daughters ($r = 0.5$).

Various social conflicts have indeed been observed in ant colonies, including efforts by queens to consume haploid eggs that workers have laid—eggs that would have become sons of the workers had they not been destroyed [509]. In large eusocial colonies, however, if queen and workers come into conflict about the allocation of food to the queen's sons and daughters, the workers should hold the upper hand because the queen cannot possibly monitor the activities of thousands of workers. In these cases, workers could conceivably withhold food from brothers in order to nourish sisters instead. If workers attempt to maximize their own inclusive fitness, then the combined *weight* of all the female reproductives (a measure of the total resources devoted to the production of females) raised by the colony's workers should be three times as much as the combined weight of the male reproductives. When Trivers and Hare surveyed the literature, they found the expected 3:1 investment ratio [1156].

However, hymenopteran workers should bias their production of reproductives toward females only if their mother mates just once. If sperm from two or more haploid males are used by a queen to fertilize her eggs, the daughters with different fathers (i.e., half-sisters) will not be closely related at all. Only when females have the same father will they share 75 percent of their genes in common (see Figure 27). As it turns out, some queens in a species of *Formica* ant do mate with two or even three males, while others do not. Liselotte Sündstrom realized that this ant provided a wonderful opportunity to find out whether workers did indeed take r_b into account when allocating food to future reproductive sisters and brothers. They did. The daughters of single-mating mothers heavily biased their investment toward producing sister queens. But workers in colonies with multiply mated queens behaved quite differently. For them, brothers were as genetically valuable as sisters, and they did *not* bias the colony's production toward females [1110].

Ulrich Mueller has also shown that worker hymenopterans can alter their investment in colonymates according to their relatedness [824]. He experimentally manipulated colonies of a primitively eusocial bee, removing the foundress queen from some nests but leaving her alone in others. When a colony has its foundress queen, the usual asymmetry in relatedness persists between workers and their sisters ($r = 0.75$) and their brothers ($r = 0.25$). Under these conditions, a bias toward female progeny is expected under the indirect selection hypothesis. But in colonies from which the foundress has been removed, a daughter assumes reproductive leadership. Under these conditions, her sister-workers are now helping her produce nieces ($r = 0.375$) and nephews ($r = 0.375$), rather than additional siblings. Thus, the relatedness asymmetry disappears, and workers ought to treat male progeny more favorably in these colonies. In fact, workers in the experimental colonies did invest more in males (the combined weight of which equaled 63 percent of the total weight of all reproductives) than did workers in colonies that retained their foundress queen (in which males constituted 43 percent of the total weight).

Finally, if haplodiploidy makes it easier for eusocial systems to evolve, then sterile altruists should also appear in haplodiploid groups other than the Hymenoptera. The thrips are haplodiploid insects, and as predicted, a sterile self-sacrificing caste has evolved in some gall-forming thrips species [210]. Foundress female thrips induce the formation of a gall in plant tissue and then fill the protective hollow interior of the gall with their offspring, some of which form a soldier caste. These soldiers possess enlarged, spiny forelegs, the better to grasp and stab enemies that would enter the gall and eat their siblings (Figure 28). Direct measurements of the degree of relatedness between soldiers and the individuals they protect reveal an *r* value well over 0.5 for a species whose foundress females mate with only one male. Thus, members of this species exhibit the altruism that Hamilton predicted would be more likely to evolve in association with the haplodiploid system of sex determination.

The haplodiploid hypothesis is based on the premise that exceptionally close relatedness between helper and recipient promotes the evolution of eusociality. If this is true, then other mechanisms that result in extremely close genetic relatedness among members of social groups should also be associated with caste formation and extraordinary self-sacrificing behavior. Both inbreeding and clonal or asexual reproduction can result in very high coefficients of relatedness among family members. In those species of aphids in which mothers reproduce asexually, for example, daughters get carbon copies of their mother's genotype, so the *r* value for sisters is 1.0. A number of asexual aphids have converged anatomically and behaviorally with the gall-forming thrips [568, 1099]. These aphids also live in galls made by a foundress female, and they also have soldier castes. Aphid soldiers use their powerful, spiny forelegs (or thickened swordlike mouthparts) to dispatch enemies, such as syrphid fly larvae, when those predatory insects attempt to enter the gall to feast upon the soldiers' sisters (Figure 29). The soldier aphids are sterile and so never reproduce; indeed, many die in defense of their sisters. In experiments conducted by William Foster, an average of about 20 soldiers of *Pemphigus spyrothecae* fell in battle while dispatching one syrphid fly larva. In the absence of soldiers, however, one predatory larva could eat all of the 100 nonsoldier aphids that Foster brought together for his tests [396]. In additional experiments, all the aphids occupying a sample of galls were removed before being returned to their homes in reassembled groups that either had some soldiers or were without defenders. The soldierless galls were ten times as likely to be attacked by syrphids or other insect predators as those with a soldier caste [397]. In nature, soldier aphids do not die in vain, because they derive an indirect fitness payoff when the beneficiaries of their actions survive to reproduce.

Very Close Relatedness Is Not Essential for Eusociality to Evolve

Although the studies described above support the contention that haplodiploidy can facilitate the evolution of altruism and skewed sex ratios, other hypotheses on the evolution of eusociality do not require that sisters be more closely related to one another than to their brothers or their offspring [1273]. These other hypotheses receive indirect support from the finding that female siblings in many eusocial species are *not* especially closely related [33]. For one thing, the queens of some eusocial ants, bees, wasps, and thrips mate more than once, as is true for the promiscuous honey bee, as you may recall. A queen with several mates produces daughters that do not share high coefficients of relatedness. Direct measurements of *r* for female offspring of two species of

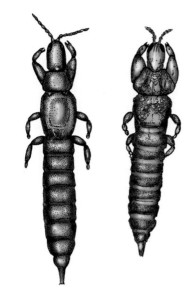

28   **A sterile thrips soldier** (right) next to a reproductive foundress female (left). Note the large forelegs of the soldier, which it uses to defend its gall-occupying relatives. Based on a photograph, courtesy of B. Kranz and Berni Crespi.

29 Altruism in aphids. Four species of aphids in which obligately sterile soldiers (left) with enlarged grasping legs and short, stabbing beaks protect their more delicate colonymates (right), which have the potential to reproduce when mature. The species were drawn at different scales by Christina Thalia Grant. After Stern and Foster [1099].

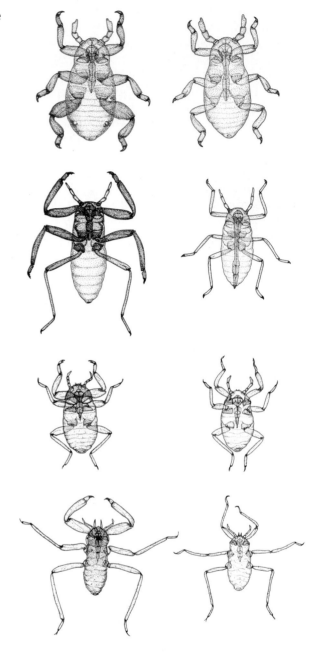

polyandrous eusocial wasps produced mean values no greater than 0.40 [992]. Likewise, in *Polistes* wasp colonies, the average *r* of nestmates almost never reaches the 0.75 maximum value, and often is less than 0.50 [1105]. These results indicate that many paper wasp queens either mate more than once or share reproduction with other females in their nests, something known to occur in many eusocial ants and wasps.

Therefore, the haplodiploid system of sex determination does not guarantee that workers in eusocial hymenopteran societies will be very closely related, nor do fairly low levels of *r* prevent the evolution of eusociality in these insects. Moreover, eusocial systems have evolved in other organisms that do not possess the haplodiploid method of sex determination. The termites, for example, are every bit as eusocial as honey bees and paper wasps, despite the fact that both males and females are diploid (see Figure 25, right).

30   A mammal with castes. Naked mole-rats live in large colonies made up of many workers who serve a queen and one or a few breeding males. Photograph by Raymond Mendez.

A final example comes from the existence of several diploid, eusocial vertebrates [179, 1048]. The best-studied of these is the bizarre-looking naked mole-rat, a little, hairless, sausage-shaped mammal (Figure 30) that lives in a complex maze of underground burrows that can house more than 200 individuals. The impressive size of their subterranean home stems from extraordinary cooperation among chain gangs of colony members, which work together to move tons of earth to the surface each year while burrowing in search of edible tubers under the African plains. Yet when it comes to reproducing, breeding is restricted to a single big "queen" and several "kings" that live in a centrally located nest chamber. Females other than the queen do not even ovulate. Instead, they serve as sterile helpers at the nest, consigned to specialized support roles for the queen and kings, as are most of the males in the colony [669].

Naked mole-rat workers are often very closely related to their fellow workers, thanks to inbreeding by their parents [963]. When two siblings mate, their offspring are likely to have coefficients of relatedness above 0.5 because their parents share many alleles in common as a result of their shared descent. Nonetheless, not every king and queen of a mole-rat colony are siblings or even cousins; some reproducing mole-rats evidently prefer to pair off with a nonrelative [218]. Moreover, mole-rat colonies produce some especially fat individuals that seem designed to leave their home colonies in order to found a new colony elsewhere, presumably with an unrelated individual of the opposite sex from another colony [145]. The fact that mole-rat workers in some colonies are not extraordinarily closely related tells us again that the evolution of eusociality is not dependent upon exceptionally high r values. In the naked mole-rat and many other species, is enough that helpers will usually care for siblings.

The Ecology of Eusociality

Hamilton's rule that altruism can evolve when $(r_c \times {}^*c)$ is less than $(r_b \times {}^*b)$ contains more elements than r_b. In fact, altruism can spread even when r_b is close to 0, provided that ${}^*c$ (the number of offspring the helper gave up to be an altruist) is also very low. In other words, if the ecology of the species is such that young dispersing adults have little chance of reproducing successfully, then nondispersers that remain at their natal site to help other relatives are likely to secure sufficient indirect fitness to make helping at the nest the adaptively superior option.

31 The costs of dispersal in eusocial insects may be high because of the low probability of building a new home as safe as the natal nest. Australian termite colonies live in huge, hard-packed mounds that are gradually built over years by millions of workers. The chance that a single termite from an established colony will successfully initiate a new equivalent colony is tiny. Photograph by the author.

For many social animals, especially the social insects, the chance that a dispersing individual will ever succeed in building something like the natal home is vanishingly small (Figure 31) [17]. Consider that a foundress female of a eusocial ambrosia beetle takes about half a year to gnaw her way just 5 centimeters into a eucalyptus tree [615]. Most foundresses die long before they get this far, which is only the first phase of building a tunnel complex in the tree. However, once a large burrow network is established, a colony can persist for decades, with helper daughters assured of a safe home in which to assist their mother in rearing reproductive males and new foundresses, a few of which may disperse successfully, raising the indirect fitness of their stay-at-home sisters.

One of the main functions of helpers is the defense of their mother's valuable nest against predators and potential nest usurpers of their own species. Defense of a fortress nest built in wood or plant tissue on which colony members can safely feed occurs in at least one ambrosia beetle and a few aphids and thrips, as well as many termite species. But defense of a fortress nest is not the only way in which sterile workers can increase the production of their relatives. David Queller and Joan Strassmann argue that in many of the eusocial ants, bees, and wasps, the sterile workers' most important service is gathering food for their larval relatives [947]. Whereas fortress defenders live amid a wealth of digestible plant material, the typical ant, bee, or wasp must roam far from the nest site in search of scarce food. In so doing, it runs a gauntlet of predators. Because forager mortality rates are high, a female that lived and foraged alone would often die before her brood achieved independence. If, however, a nesting female can enlist the aid of others, care will continue to be provided for her young even if she should die prematurely. Under these circumstances, helpers related to the primary reproductive female gain considerable indirect fitness by bringing the brood through to adulthood [383]. Applying these categories of social altruism to vertebrates, naked mole-rats can be considered fortress defenders whose workers help prevent usurpation of the network of protective underground tunnels, while avian helpers at the nest are life insurers, providing the extended care their helpless newborn siblings will require, should their parents die.

Thus, although certain genetic factors may give the evolution of altruism a boost, ecological factors that increase the effect of helping on the survival of relatives are equally important. However, our understanding of complex sociality is still incomplete. For example, there are seven species of African mole-rats, all of which are burrowing, parental animals. The costs of leaving a safe underground burrow would seem to be high, and the benefits of helping great, for all species, given the value of the burrow, the care required by helpless youngsters, and the advantages of communal burrow construction. Therefore, we might expect to observe eusocial life in all seven species. However, direct evidence of eusociality is available for only two species, the naked mole-rat and the Damaraland mole-rat (although some persons feel that the other mole-rats may also exhibit some elements of eusociality) [179]. Clear-cut examples of rodents with a worker caste are extremely rare [578], and in fact, joint occupation of a burrow by several adults has been reported for only a handful of subterranean rodents [670]. These facts pose uncomfortable questions for the argument that fortress nests promote the evolution of eusociality [578]. The use of subterranean tunnels and the associated high costs of dispersal must be only part of the ecological story behind the evolution of the eusocial lifestyle in mammals.

In general, it is easier to offer a tentative explanation for why a species *has* evolved a particular trait than for why a species has *not* evolved a particular trait. For example, researchers have accounted convincingly for the social life of the Florida scrub jay. But what about its close relative, the western scrub jay, a nonsocial species? Why haven't members of this species become social? There must be years when young western scrub jays have little chance of finding a suitable vacant territory. Why haven't they evolved the ability to remain as helpers at the nest under these conditions? Likewise, one wonders why a soldier caste has evolved in only some, not all, gall-forming aphid species. All these aphids have a valuable home to defend, but altruists are known for only 50 species [1100]. Perhaps concession theory will come to the rescue [958], but much more remains to be learned about the genetic and ecological bases of altruism and social living before we can close the book on this great evolutionary puzzle.

Summary

1. In animal societies, individuals often tolerate the close presence of other members of their species despite the reproductive interference, increased competition for limited resources, and heightened risk of disease that social living entails. Under some ecological circumstances, the advantages of sociality (often improved defense against predators) are great enough to outweigh the many and diverse costs of social living. The common view that social life is always evolutionarily superior to solitary life is incorrect.

2. Animals that live together may help one another in various ways. Helpful interactions have a variety of fitness consequences for the participants. Some cooperative acts may immediately elevate the personal reproductive success of both cooperators (mutualism). Still others may be performed at some cost that will be more than repaid when the recipient reciprocates in the next of a series of interactions (reciprocal altruism). Finally, some helpful actions

are examples of true altruism because they reduce the reproductive success of the helper while at the same time raising the reproductive output of another individual.

3. Mutualisms and reciprocal altruism can spread through a population through the action of direct (natural) selection. If, however, a helper really does permanently reduce its direct fitness while helping to raise the fitness of another, its altruism poses a major evolutionary puzzle. One solution to this puzzle may be that the direct fitness costs of certain kinds of altruism are outweighed by the indirect fitness gains generated when an individual increases the number of surviving nondescendant kin.

4. True altruism is found in nature, especially in species with helpers at the nest, which assist in rearing the broods of other individuals. In the eusocial insects, these helpers may be sterile. These individuals boost their inclusive fitness by aiding close relatives, often siblings.

5. Although the indirect fitness gained by helping is increased if the coefficient of relatedness between helper and beneficiary is high, eusociality can evolve even when the degree of relatedness is not great, provided that the ecology of the species is such that helpers can improve the production of relatives markedly, and especially if the odds are against successful personal reproduction for individuals dispersing from a safe natal nest.

Discussion Questions

1. Some vampire bats regurgitate blood meals to other hungry individuals [1237]. What is the minimum information that you would need in order to determine whether this helpful behavior is currently maintained by indirect selection? By direct selection?

2. Let's say that in calculating the inclusive fitness of a male in a coalition of lions, you measured his direct fitness by multiplying by 0.5 the number of offspring produced by the male, and then added as his indirect fitness the total number of all offspring produced by the other members of his coalition times the mean r of those offspring to the male in question. Your calculation of his inclusive fitness would be challenged on what grounds?

3. If a female of a social wasp species could help to produce more females with an r of 0.75, why would any females reproduce personally, given that reproducers are related to their offspring by just 0.5? Use Hamilton's rule to explain why some "worker" paper wasps produced early in the year leave their natal nests to wait for the next year's breeding season rather than becoming helpers at the nest of their mother.

4. In several ant species, two or more unrelated females may join forces to found a colony together after they have mated. The females may cooperate in digging the nest and producing the first generation of workers, but then they start fighting until only one is left alive [98]. How can it pay to join such an association? What prediction can you make about the survival rates and average productivity of colonies founded by a single female? Under what conditions is it accurate to call this social system a mutualism? Develop at least one cost–benefit hypothesis to account for the timing of the switch from cooperation to aggressive behavior. If the behavior of the two queens is the product of natural selection, not group selection, what prediction can you make about the interactions between them during the colony establishment phase prior to the fight-to-the-death phase?

5. This chapter examines the evolution of sterile workers strictly from the perspective of fitness gains and losses incurred by the workers themselves. But what if we view the issue from the standpoint of selection on queens [1212]? How might queens gain direct fitness benefits by "forcing" some of their offspring to

be nonreproductive helpers—even if this reduced the inclusive fitness of those helpers?

6. The figure below contains data on the relative reproductive success of alpha and beta males of the white-browed scrub wren in two different kinds of cooperative breeding associations [1227]. Is the observed pattern consistent with concession theory?

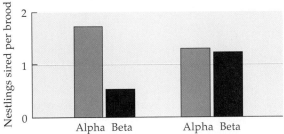

Suggested Reading

W. D. Hamilton's work [484] initiated a revolution in the understanding of social behavior; see also reviews by Richard Alexander [17], Mary Jane West-Eberhard [1212], and Steve Emlen [360]. I have relied heavily on Jerry Brown's *Helping and Communal Breeding in Birds* as a guide for understanding the kinds of selection that affect the evolution of social behavior [168]. Two papers highlight the differences between concession theory and alternative explanations for why subordinate members of social groups vary in how much they reproduce [229, 363]. Uli Reyer's study of the social pied kingfisher is superb [964], as are the papers by Jan Komdeur and colleagues on the Seychelles warbler [643, 645]. Excellent reviews of social behavior are available for ants [536] and other social insects [1246], lions [876], acorn woodpeckers [636], and naked mole-rats [1048].

15 The Evolution of Human Behavior

Humans are an animal species with an evolutionary history. Yes, we are unusual, but kittiwakes and hangingflies are unique creatures too. Having applied evolutionary thinking to kittiwakes and hangingflies, why not do the same for humans? Since our evolution has been shaped by natural selection, our behavioral attributes should tend to help us pass on our genes to the next generation. Admittedly, this proposition is resisted by many people, especially those who believe that our behavior is essentially "cultural" as opposed to "biological." And it is true that our cultural traditions influence our behavior. If, for example, I had been born and reared by tribesmen in Papua New Guinea, I would have considered it natural to be seen in public completely naked except for a long, thin, hollow gourd covering my penis. Were I to wander around Tempe, Arizona, in a similar outfit, I would surely attract unwanted attention. The pervasive effects of our cultures on our behavior obviously make it especially challenging to apply evolutionary theory to the human species. But natural selection theory has proved extremely useful in solving the Darwinian puzzles posed by kittiwakes and hangingflies. Let's not abandon

ogists of the logic of natural selection and the importance of thinking of behavior as a means by which individuals attempt to pass on their genes in competition with genetically different rivals (p. 14). This kind of Darwinian thinking is the foundation for the modern adaptationist approach.

"No one has ever identified the genes responsible for any human behavior." Because sociobiologists view behavior as the product of a genetic competition among individuals, many persons think incorrectly that the sociobiologist's goal is to discover genes "for" various human behaviors, such as altruism or aggression. If this goal really was the central mission of sociobiology, and if, as has also been claimed, sociobiologists really believed that our behavior is "genetically determined," then sociobiology would have died soon after it had been named. In reality, sociobiologists are well aware that all phenotypes develop as the result of gene–environment interactions (see Chapters 3 and 4). Moreover, the development of behavior is not the focus of sociobiological research, which deals with the possible fitness consequences of behavior, not the operating rules of developmental mechanisms.

Sociobiologists do assume that genes can affect behavioral development, because otherwise behavior could not evolve. This assumption is clearly valid for our species as well as for all others whose behavior genetics have been studied. Human genetic variation causes behavioral variation in populations today (see p. 55) and thus almost certainly did so in the past. As our knowledge of the role of genes in the development and operation of the human brain increases with the assistance of the Human Genome Project, we will identify more and more genes that can influence (but not determine) the development of human behavior.

In any event, we know even less about the genetics of Florida scrub jay helping at the nest and ground squirrel alarm calling than we do about the genetics of human behavior. Fortunately, this lack of proximate information has not prevented evolutionary biologists from asking, and often answering, ultimate questions about the evolution of scrub jay and ground squirrel behavior. As we shall see, one can test sociobiological hypotheses about the adaptive nature of certain human characteristics without any information on the genetic or developmental basis for these characteristics.

"But humans don't do things just because they want to raise their inclusive fitness." Some opponents of sociobiology have pointed out that, although humans desire many things, the wish to maximize our inclusive fitness is not what motivates our behavior [986]. If you had asked Picasso why he wished to produce attractive paintings, or Bill why he wanted to marry Jane, Picasso and Bill would not have explained that they wanted to increase their genetic success. But if a baby cuckoo could talk, it would not tell you that it rolled its host's eggs out of the nest "because I want to propagate as many copies of my genes as possible." Neither cuckoos nor humans need to be aware of the ultimate reasons for their activities in order to behave adaptively. Our brain's costly decision-making mechanisms were shaped by natural selection to enhance our fitness, not to provide us with the capacity to monitor the reproductive consequences of each and every action. It is enough that proximate mechanisms, like a well-developed sex drive, motivate individuals to do things, like copulate, that are correlated with fitness payoffs—the production of offspring. On the proximate level, we learn a language, we enjoy sweet foods, we fall in love, we have sex, and we derive satisfaction from our charitable actions because we possess physiological mechanisms that facilitate these goals. Because honey tastes good, we want to eat it, and when we do, we acquire useful calories that may contribute to our survival and reproductive success without our ever being aware that this is the evolved function of our fondness for sweets.

"But not all human behavior is biologically adaptive!" Critics of sociobiology often point out that certain cultural practices, such as blood donation, circumcision, prohibitions against eating perfectly edible foods, or a celibate priesthood, seem most unlikely to advance individual fitness. If some humans do things that reduce their fitness, these persons argue, then sociobiology cannot be correct. This argument assumes that natural selection theory requires that every aspect of every organism be currently adaptive [448], which it does not, as we learned in Chapter 7 (see p. 186). Nor is it necessary to think that every trait is an adaptation in order to use the adaptationist method to identify interesting puzzles, produce plausible hypotheses, and attempt to test alternative explanations. If, for example, a sociobiologist or adaptationist were to examine religiously motivated celibacy, he or she would be fully aware of the possibility that the trait was a *maladaptive* by-product of certain brain modules that controlled other generally adaptive abilities. Nonetheless, a sociobiologist might still try to produce a testable hypothesis on how acceptance of celibacy could paradoxically enable priests to leave more copies of their genes than if they were sexually active. Needless to say, this would be a challenge, but perhaps not an insuperable one for someone aware of indirect selection.

Even if twenty adaptationist hypotheses on celibacy were developed, there is no guarantee that any would withstand testing. This is as it should be. T. H. Huxley, the great defender of Darwinian theory, wrote, "There is a wonderful truth in [the] saying [that] next to being right in this world, the best of all things is to be clearly and definitely wrong, because you will come out somewhere" [562]. If our sociobiological hypotheses about the celibate priesthood were incorrect, effective tests would tell us so, and we would probably learn some interesting things along the way.

"Sociobiology is a politically reactionary doctrine that supports social injustice and inequality." As noted above, the original critics of sociobiology denounced it as politically dangerous because they feared it would provide scientific cover for immoral social policies of the sort advanced by racist and fascist demagogues in the past [21]. According to this view, the claim that such and such a trait is adaptive implies that it is both genetically determined and good, and therefore cannot and should not be changed. After all, if one claims that male dominance is adaptive, isn't this saying that the status quo in our society is desirable and that feminist claims fly in the face of what is genetically fixed and morally necessary?

Scientific findings can be employed in ways that often surprise, and even horrify, the investigator. Einstein's basic research on the relation between energy and matter contributed to the development of atomic weapons, much to his dismay; Darwin's theory of evolution has been misunderstood and misused by some persons to defend the principle that the rich are evolutionarily superior beings, as well as to promote unabashedly racist plans for the "improvement of the human species" by selective breeding.

We can hope that political perversions of evolutionary theory have been so discredited that they will not happen again. The critical point here, however, is that sociobiologists attempt to explain why social behavior exists, not to justify any given trait. This distinction is easily understood in cases involving other organisms. Biologists who study infanticide by male langurs or how a small marine copepod adaptively feeds on the eye of the Greenland shark are never accused of approving of infanticide or the blinding of sharks. To say that something is biologically or evolutionarily adaptive means only that it tends to elevate the inclusive fitness of individuals with the trait—nothing more.

Moreover, a hypothesis that a behavioral ability is adaptive does not mean that the characteristic is developmentally inflexible. All sociobiologists understand that development is an interactive process involving both genes and environment. Change the environment, and you will change the gene–environment interactions underlying a behavioral phenotype, with the result that the phenotype may change. A classic example in human biology involves language acquisition, a clearly adaptive ability that rests upon vast numbers of extraordinarily complex gene–environment interactions. Among our many genes are some that code for proteins that promote the development of certain specialized brain modules that facilitate language learning. These neural units evolved specifically in the context of language acquisition, rather than as a side effect of some sort of generalized intelligence, as we can see from the existence of two rare human phenotypes. On the one hand, certain profoundly retarded individuals chatter away, producing completely grammatical sentences with little inherent meaning. On the other hand, some English-speaking individuals with normal to above-normal intelligence have great trouble with the rules of grammar, often failing, for example, to add "-ed" to verbs when they wish to speak of past events [920].

Of course, the genes "for" language ability interact with all sorts of environmental inputs, which include the molecular building blocks needed for enzyme production and neural development. Alter the molecular or acoustical environment of a young person and you can change the nature of the phenotype, as demonstrated most obviously by the ability of babies to learn any of the 5000 or so languages distributed around the world. The proposition that we possess evolved proximate mechanisms underlying a capacity for language, or aggression, or sexual jealousy, or the urge to reproduce does not condemn us to one particular expression of these characteristics for all time. For example, thanks to an environment in which effective means of birth control exist, my wife and I chose to have only two children, even though as an evolutionary biologist I believe that in some sense we exist solely to propagate the genes within us.

Evolution and the Diversity of Human Cultures

Languages are not the only culturally variable aspect of human behavior. There are polyandrous, polygynous, and monogamous societies; cultures in which females make important political decisions and others in which they do not; human groups for which warfare is a constant fact of life and other groups that never fight. Depending on where you were born, you might delight in wearing an especially handsome penis sheath, you might take pride in memorizing the Koran, you might be forbidden to look at your mother-in-law, or you might be forced to marry someone you had never met. Given the rich variety of human traditions, how can anyone say that our behavior is the product of natural selection?

One answer is that selection may have endowed us with particular genes that promote the development of an adaptable nervous system that helps us acquire and adaptively use cultural trait X or Y. As already documented, the flexibility people exhibit in their charitable behavior or in food preparation with spices (p. 243) has adaptive value, and the same is true of the conditional strategies of blackbirds (p. 264), scorpionflies (p. 334), and pied kingfishers (p. 436). In all these species, individuals match their behavioral tactics to their social and environmental circumstances, and in so doing, often leave more copies of their genes than they would otherwise.

To reinforce this point, consider how young female white-fronted bee-eaters make adaptive reproductive decisions. This African bird nests in loose colonies in clay banks. Like male pied kingfishers, young female white-fronted bee-eaters can choose to breed, or to help a breeding pair at their nest burrow, or to sit out the breeding season altogether. If an unpaired, dominant, older male courts her, a young female almost always leaves her family and natal territory to nest in a different part of the colony, particularly if her mate has a group of helpers to assist in feeding the offspring they will produce. Her choice usually results in high direct fitness payoffs. But if young, subordinate males are the only potential mates available to her, the young female will usually refuse to set up housekeeping. Young males come with few or no helpers, and when they try to breed, they are often harassed by their fathers, who may force their sons to abandon their mates and return home to help rear their siblings.

A female that opts not to pair off under unfavorable conditions may choose to slip an egg into someone else's nest, or to become a helper at the nest in her natal territory—provided that the breeding pair there are her parents, to whom she is closely related. If one or both of her parents have died or moved away, she is unlikely to help rear the chicks there, which are at best half-siblings, and instead will simply wait, conserving her energy for a better time in which to reproduce [366].

Thus, despite their bird-sized brains, female bee-eaters base their reproductive decisions on sophisticated evaluations of their social environment (Figure 2). The behaviorally flexible bee-eaters generally manage to choose the option most likely to increase, rather than lower, their inclusive fitness. Some evolutionary biologists have proposed that the many and varied conditional strategies that underlie

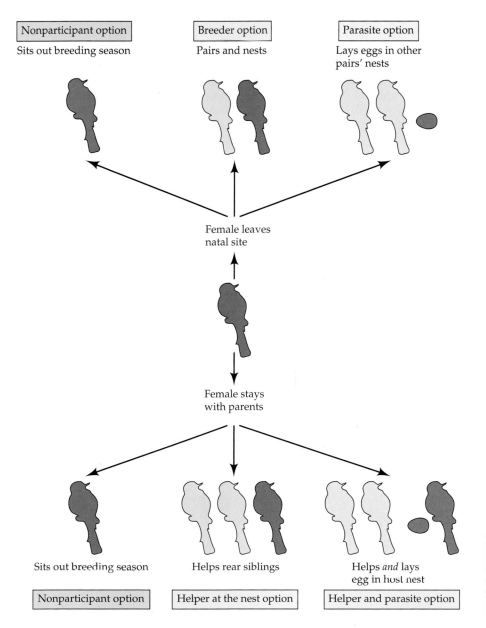

Nonparticipant option
Sits out breeding season

Breeder option
Pairs and nests

Parasite option
Lays eggs in other pairs' nests

Female leaves natal site

Female stays with parents

Sits out breeding season
Nonparticipant option

Helps rear siblings
Helper at the nest option

Helps *and* lays egg in host nest
Helper and parasite option

2 Complex social decisions are made by many animals. Female white-fronted bee-eaters choose adaptively among several behavioral options when they reach the age of reproduction. After Emlen et al. [366].

our behavior do for us what the young female bee-eater's conditional strategy does for her: namely, they provide us with a *limited*, not open-ended, set of options and the ability to make decisions that tend to *raise*, not decrease, our inclusive fitness under the cultural conditions we happen to encounter.

Sociobiology versus Arbitrary Culture Theory

The main alternative to the sociobiological approach to cultural diversity is the theory that human cultural traditions arise from accidents of history and the almost unlimited inventiveness of the human mind. According to this **arbitrary culture theory,** our cultural activities have nothing to do with fitness maximization, but instead reflect the more or less arbitrary process by which traditions originate and persist over time. The contrast between the two approaches can be illustrated by the differences between cultural anthropologists and sociobiologists in their analysis of adoption, which is common in some societies and much less so in others. At one time, for example, an amazing 30 percent of all children became adoptees in Oceania, the islands of the central Pacific Ocean. Because some persons in Oceania adopted children who were not their kin, the cultural anthropologist Marshall Sahlins concluded that evolutionary theory could not explain these people's behavior [1011].

But is adoption in Oceania really practiced in an arbitrary manner, with no positive effect on the inclusive fitness of individuals? When Joan Silk analyzed data on the relationships between moderately large samples of adopters and adoptees in 11 different cultures in Oceania [1057], she found that most adopters cared for children who were cousin equivalents or closer (minimum $r = 0.125$) (Figure 3). The highly nonrandom nature of adoption in these societies casts doubt on the arbitrary culture hypothesis while supporting an indirect fitness hypothesis for this form of human altruism.

Nevertheless, a minority of adopters in Oceania do take in the children of strangers, their genetic competitors, who may receive the same love and affection that parents typically supply to their own offspring. Are these exceptions to the typical adoption pattern in Oceania impossible to explain from a sociobiological perspective? No. Silk suggests that small families in some agricultural cultures might benefit from gaining adoptees, even if they were nonrelatives, because adopted persons can contribute to the family workforce, raising the economic productivity of the family unit and improving the survival chances of the adopters' genetic offspring. This direct fitness hypothesis produces the prediction that small families in Oceania should be more likely to adopt than large ones, a prediction that Silk showed was correct.

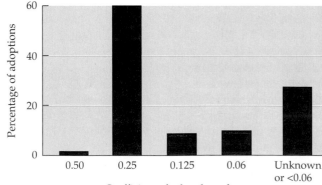

3 The indirect fitness hypothesis for adoption can be tested by examining the coefficient of relatedness between adoptive parents and their adopted children. In 11 island societies in Oceania, adopter and adoptee were usually close relatives, yielding indirect fitness gains for the adopters. After Silk [1057].

An alternative evolutionary hypothesis for adoption among nonrelatives recognizes that some decisions may be the maladaptive by-product of otherwise adaptive proximate mechanisms. Adoption of a nonrelative might, for example, be a consequence of motivational systems that cause adult humans to want to have children and raise a family. According to this hypothesis, although adults who adopt infant strangers may reduce their fitness, the urge to have and love children is *usually* adaptive. Because these psychological mechanisms tend to elevate fitness, they are maintained in human populations even though they *sometimes* induce people to behave maladaptively.

The maladaptive side effect hypothesis for adoption also generates testable predictions, one of which is that husbands and wives who have lost an only child, or who fail to produce children themselves, should be especially prone to adopt strangers. The proximate mechanism that produces the adaptive desire to be parents could cause some people to adopt substitutes for genetic offspring. Another prediction from this hypothesis is that adoption of nonrelatives should also sometimes occur in animal species not closely related to humans when adults have lost their offspring and fortuitously encounter a substitute (Figure 4) [594]. Cardinals have been known to feed goldfish [1205], and a white whale was seen trying to lift a floating log part way out of the water, as if the log were a distressed infant that needed help to reach the surface to breathe [147]. Adopting a goldfish or a log did nothing to promote the cardinal's or the white whale's genes. But these individuals had almost certainly lost their young recently and were employing generally adaptive parental behaviors in an unusual, nonadaptive fashion.

The risk to human beings of a maladaptive application of an adaptive psychological mechanism may be especially high in modern societies, environments that are very different from the ones in which our behavioral mechanisms evolved. In Western culture, babies are routinely made available to nonrelatives who do not know the parents of the adoptee, something that almost never occurred in the distant past. Under these novel conditions, our neural systems, which evolved long ago, can cause us to behave maladaptively and, in so doing, reveal something about the naturally selected features of our psyches.

4 Adoption occurs in nonhuman animals, often when adults have just lost an offspring but encounter a substitute. Here several emperor penguins compete for "possession" of a youngster. Photograph by Kim Westerskov.

The abundance of testable sociobiological hypotheses on adoption (and we have not discussed the possibility that adoption can be a form of charitable behavior that raises the social status of adopters) speaks to the productive nature of evolutionary theory. No one evolutionary hypothesis explains every case of adoption, just as no one hypothesis resolves every aspect of a complex anatomical or physiological phenomenon. However, an evolutionary approach seems far more likely to help us understand the complexity of adoptive behavior than the competing view that adoption is a purely arbitrary manifestation of human desire and intellect.

Adaptive Mating Decisions

Having briefly illustrated how evolutionary biologists can develop and test hypotheses on donating blood and adopting children, we now turn our attention to those behavioral traits whose relationship to fitness is most obvious. In analyzing the sexual tactics of men and women, we shall use the same theoretical framework that has proved so useful in understanding reproductive behavior in other animal species. As we saw in Chapter 11, interactions between males and females create selection pressure on both sexes, leading to the coevolution of their respective reproductive tactics. You cannot understand what males of a species are doing without paying attention to females, and vice versa.

In the human species, as in other mammals, females retain control of reproduction by virtue of their major physiological investments in egg production, nurturing of embryos, and feeding of infants after they are born. Although human males are also able and often willing to make large parental investments in offspring, their reproductive decisions nevertheless take place in a setting defined by female physiology and psychology [280, 422]. Women have eggs available for fertilization on only a few days each month, and then only if they are not pregnant or nursing babies. Prior to modern contraception, most women were pregnant or nursing for most of their reproductive lives, with the result that the ratio of females with fertilizable eggs to males with mature sperm has been heavily tilted toward males during our evolutionary history.

Male-biased operational sex ratios mean that males inevitably face stiff competition for females with fertilizable eggs. In turn, the abundance of sexually active males should enable females to choose sexual partners whose attributes are likely to promote their fitness. Since human males have the potential to provide material benefits to their partners and offspring, female choice can be expected to operate heavily on this criterion, with females favoring males able and willing to transfer wealth and other resources to their mates and offspring. Since these transfers are costly to males, men can be expected to exhibit mate choice themselves, making investments in women that are likely to provide fitness payoffs for them—namely, offspring that carry their genes, not those of other men. In turn, those other men may attempt to engage the wives of paternal males in extramarital affairs. The risk of being cuckolded creates intense selection pressure on paternal husbands, who can be expected to have evolved countermeasures to reduce the risk. Thus, an evolutionary view of human reproductive behavior leads us to anticipate both cooperation and conflict between men and women, with profound consequences for the evolution of the sexual psychology of both sexes [184].

Adaptive Mate Choice by Women

We can test this approach to our behavior in the customary fashion by using evolutionary theory to produce hypotheses with testable predictions. Since

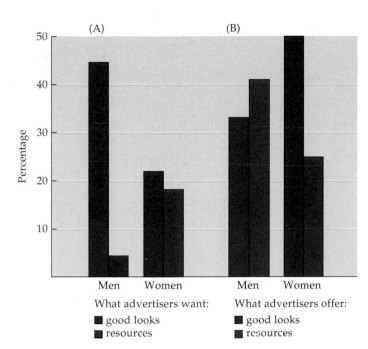

5 Sex differences in preferences and offers made to the opposite sex. (A) Women advertising in the Personals section of newspapers indicate that they are looking for someone with a good income or with good looks. Men overwhelmingly identify good looks as the key criterion. (B) Men are far more likely than women to offer wealth as an inducement to respond to the advertisement, whereas women are far more likely than men to assert that they are good-looking. After Waynforth and Dunbar [1190].

females are expected to set the rules for male mating competition, let's begin by analyzing female mate choice. If a woman's reproductive success is advanced by access to a partner's material assistance and parental care, then we can predict that women should find wealthy, paternal men more attractive than poor, uncaring men.

One of the nice things about studying behavior in humans is that you can ask them questions. When women in modern Western societies are surveyed about what they are looking for in a mate, they consistently rank "good earning power" highly, and so do women from around the world [183]. Moreover, women advertising for partners in newspapers are far more likely than men to specify that they are looking for someone who is rich (and men are much more likely to offer resources than are women) (Figure 5) [1190]. More indirectly, women placing personal ads in both Arizonan and Indian newspapers often say they are interested in someone older than they are (Figure 6) [611]. Since older men typically have access to greater income, a preference for older men should tend to result in access to more wealth [183].

It could be that women's interest in the earning power of potential mates is a purely rational response to the fact that males in almost every culture control their society's economy, making it difficult for a female to achieve material well-being on her own. If this nonevolutionary hypothesis explains why females favor wealthy males, then women who are themselves well off and not dependent on a partner's resources should place much less importance on male earning power. Contrary to this prediction, several surveys have shown that women with relatively high expected incomes actually put *more* emphasis, not less, on the financial status of prospective mates [1149, 1232].

Another way to evaluate the preferences of mate-seeking females is to use personal ads to calculate the "market value" of men of different ages, which can be done by dividing the number of individuals requesting a particular age class of partner in their advertisements by the number of individuals in that age class who are advertising their availability. This measure provides an indication of what age groups are in short supply. Men in their late thirties have the highest market value (Figure 7) [891], most likely because of their relatively high incomes coupled with the probability that men of this age will usually

6 Different cultures show similar mate preferences. Men advertise for younger women and women advertise for older men in both (A) the *Arizona Solo* and (B) the *Times* of New Delhi, India. The advertisers indicated their own age and the maximum and minimum ages they would accept in potential partners. After Kenrick and Keefe [613].

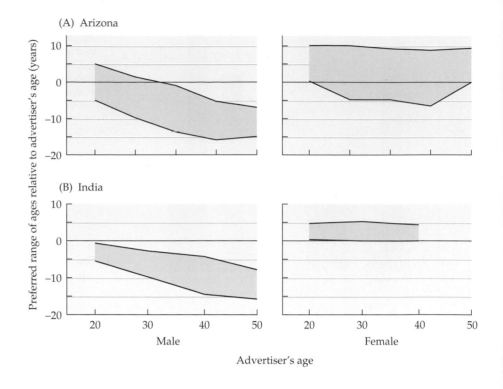

(A) Arizona

(B) India

Preferred range of ages relative to advertiser's age (years)

Male

Female

Advertiser's age

live long enough to invest a large amount in their wives' children, helping them reach adulthood. In other words, women evaluating potential mates are influenced by two criteria: current income and the total resources they and their children are likely to receive from a partner during that individual's remaining lifetime.

If women's mate preferences really have evolved as a means of maximizing reproductive success, then we might expect those preferences to be different when women are ovulating as opposed to when they are infertile. Copulating outside the fertile phase of the menstrual cycle contributes to the acquisition of a social partner or the maintenance of such a relationship, whereas copulating during the fertile period has the potential consequence of acquiring genes for one's offspring. Indeed, when conception is most likely, women rate highly masculine faces more attractive than at other times during the menstrual cycle [899]. If, as some suspect, certain features of a man's face reflect the competence of his immune system and his health during childhood, fertile women that copulate with men endowed with masculine faces may have healthier offspring than they would otherwise.

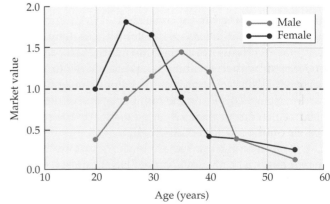

7 Age and the market value of men and women. Market value is measured by the number of advertisements requesting individuals of certain age classes divided by the number of individuals of those age classes announcing their availability in the Personals sections of newspapers. After Pawłowski and Dunbar [891].

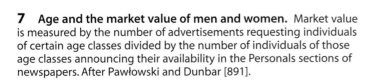

Market value

Age (years)

Male
Female

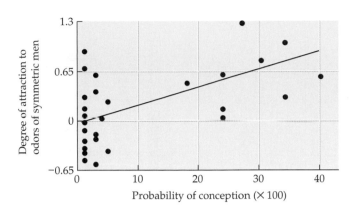

8 **Female mate preferences change during the menstrual cycle.** Women who are not on birth control pills are more likely to find the odors of symmetrical men attractive when the probability of conception is greater than 1 in 5 per copulation (that is, when women are in the ovulatory phase of the cycle). After Gangestad and Thornhill [416].

In addition to cyclical changes in their response to men's faces, the reaction of women to male body odors also changes over the menstrual cycle. During the ovulatory phase of the cycle, women prefer the odors of men with symmetrical bodies, a point established by measuring the body symmetry of a sample of men and then asking women to evaluate the relative attractiveness of the scents clinging to white T-shirts that the men had worn while sleeping (Figure 8) [416]. Once again, if body symmetry is a property of healthy men with "good genes" (see p. 108), then women's preferences may cause them to seek out sexual partners of high genetic quality during the very time when they are most likely to become pregnant.

Mating preferences are one thing, but mating behavior is another. Are women actually more likely to copulate with men who possess the attributes they prefer than with men who lack them? Among the Ache people of eastern Paraguay, good hunters with high social status are more likely to have extramarital affairs and produce illegitimate children than poor hunters, suggesting that females in this society find skillful providers sexually attractive [602]. Likewise, in Renaissance Portugal, noblemen were more likely to marry more than once, and more likely to produce illegitimate children, than men of lower social rank. These results are consistent with the prediction that females use possession of wealth as a cue when selecting a father for their children [121].

If we bear the imprint of past evolution on our psyches, then women in modern Western societies should also use resource control and its correlates, such as high social status, when deciding which men to accept as sexual partners. In order to study the relationship between male income and copulatory success in modern Quebec, Daniel Perusse secured data from a large sample of respondents on how often they copulated with each of their sexual partners in the preceding year. With this information, Perusse was able to estimate the *number of potential conceptions* (NPC) a male would have been responsible for, had he and his partner(s) abstained from birth control. Male mating success, as measured by NPC, was highly correlated with male income, especially for unmarried men (Figure 9). Perusse concluded that many single Canadian men attempt to mate often with more than one woman, but their ability to do so is heavily influenced by their wealth and status, thanks to the female preference for rich partners. Thus, the striving for high income and status exhibited by males in modern societies may be the product of past selection by choosy females, which occurred in environments in which potential conceptions had an excellent chance of being actual ones [902].

But do females who secure wealthy mates have higher fitness than females whose partners can provide fewer material benefits? This prediction deserves to be tested in societies like those of our ancestors—namely, cultures without modern birth control technology. Among the Ache, the children of men who

9 Income and copulatory success. Income is positively correlated with the number of potential conceptions (NPC) in the preceding year for unmarried Canadian men of various age groups, but especially for older men. After Perusse [902].

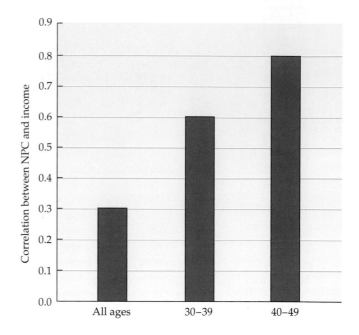

were good hunters were in fact more likely to survive to reproductive age than the children of less skillful hunters [602]. Likewise, several studies of traditional societies in Africa and Iran have revealed a positive correlation between a woman's reproductive success and her husband's wealth, as measured by land owned or number of domestic animals in the husband's herds [124, 566, 738]. Although traditional societies usually permit polygyny, women married to rich polygynists need not pay a fitness penalty (but see below and [502, 1104]) because one-half or one-third of a great deal can be more than all of a poor man's holdings. The same argument is used to explain why some females accept polygynous mates in birds and other animals.

The correlation between male wealth and female reproductive success applies not just to some traditional cultures, but also to pre-birth control societies more like our own. For example, James Boone found that during the 15th and 16th centuries, Portuguese women married to men of the highest nobility and presumed greatest wealth had more children than women whose husbands were untitled or in the military [121] (Table 1).

In most modern cultures, however, high family income is *not* positively related to the number of children produced (Figure 10). Indeed, poor couples often have more surviving children than rich ones. Although some persons have concluded that this finding invalidates an evolutionary analysis of human

TABLE 1 Reproductive performance of married men in Portuguese society in the 15th and 16th centuries

Rank	All offspring		Illegitimate offspring only	
	Number of men	Mean number of offspring	Number of men	Mean number of offspring
Royalty	96	4.75	50	0.52
Royal bureaucrats	168	4.62	43	0.36
Landed aristocrats	216	4.54	80	0.37
Untitled/military	553	2.33	127	0.23

Source: Boone [121]

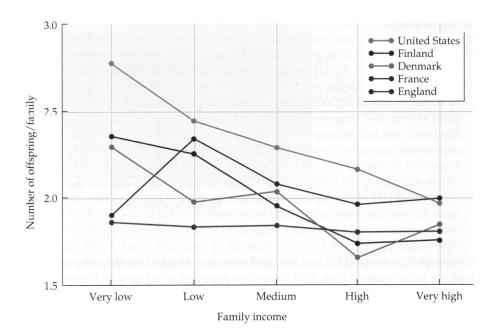

10 Fertility often declines as family income increases in modern industrial societies. After Population Series, No. 58, United Nations, 1976.

behavior [1171], sociobiologists have pointed out that the disconnect between wealth and reproductive success may be the product of certain evolutionarily novel features of modern society, including highly effective birth control and the existence of welfare for poor families. In addition, two other novel factors may be at work: first, the amount of wealth and material goods that individuals can acquire has never been greater, and second, the costs of rearing children have skyrocketed for wealthy people, who are in effect competing against others of their class to provide their children with high status [738]. In the past, fitness (as measured in grandchildren) may have been maximized in wealthy families that had relatively few offspring and so were able to give fewer children more resources, which would have provided them with an edge in the competition with others of their class for high socioeconomic status. Children able to achieve exceptional social success would have often produced many grandchildren for their wealthy parents, especially in cultures lacking birth control. (The drive to sequester resources may also have enabled small families to survive recurrent famines [122].) In the present, these previously adaptive psychological mechanisms may interact with a unique set of environmental circumstances to reduce, rather than increase, family size across generations, a proposition that deserves much additional investigation [126].

Adaptive Mate Choice by Men

Even today, a woman's lifetime reproductive success is limited by the number of pregnancies that she can sustain and the number of children that she can rear, which in turn is often affected by the material support she receives from her mate. In contrast, a man's reproductive success is dependent, first and foremost, upon access to the eggs of fertile women, and his fitness, therefore, can rise with the number of fecund women with whom he copulates and the frequency with which he does so. The point is that women and men are not expected to have the same interest in acquiring large numbers of fertile copulatory partners because the fitness payoffs for polygamy differ for the two sexes.

Evolutionary psychologists have therefore predicted that men should, on average, have a lower threshold for sexual activity than women, especially if they can secure attractive partners. Men in cultures around the world place great importance on "good looks" in women (see Figure 5), judging from their

13 Sex differences in mate selectivity. College men differ from college women in the minimum intelligence that they say they would require in a casual sexual partner. However, men and women have similar standards with respect to the minimum intelligence they say they would require in a marriage partner. After Kenrick et al. [614].

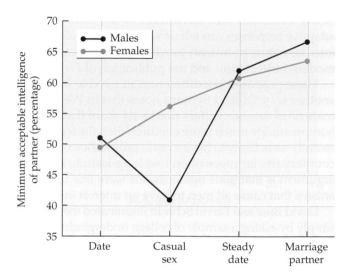

jects to evaluate the likelihood that they would be willing to have sex with a desirable potential mate after having known this person for periods ranging from 1 hour to 5 years, the differences between men and women were also dramatic (Figure 12B): "After knowing a potential mate for just 1 hour, men are slightly disinclined to consider having sex, but the disinclination is not strong. For most women, sex after just 1 hour is a virtual impossibility" [186].

The lower standards of males, at least when it comes to casual sex, are evident in the responses of college students when asked about their absolute *minimum* requirements for the intelligence of persons with whom they would be willing to have different kinds of sexual relationships. Men, but not women, are prepared to have sex with partners of far below average intelligence, provided that the encounter carries no emotional or material commitment (Figure 13) [614].

The enthusiasm of males for low-cost sexual opportunities has also been documented by two social psychologists who sent confederates, an attractive young man and an attractive young woman, on the following mission. They were to approach strangers of the opposite sex on a college campus, asking some of them, "Would you go to bed with me tonight?" Not one woman agreed to the proposition, but 75 percent of the men said "Yes." Remember that the male subjects had known the woman in question for about a minute [221].

Now, it is possible that all the women in this study who said "No" did so because they sensibly feared becoming pregnant or did not wish to risk injury in a sexual encounter with a male stranger. If so, then homosexual women should have no such inhibitions about casual sex, since sexual interactions between two women cannot result in pregnancy and are unlikely to lead to physical assault. But homosexual women are no more interested in having multiple partners than are heterosexual females [56].

Conflict between the Sexes

The reproductive interests of the sexes overlap, but they are not identical. Over evolutionary history, females of our species have probably gained fitness when their partners provided resources exclusively for them and their offspring (and not for other women and their children). In 19th-century Utah, for example, women married monogamously to relatively poor Mormon men actually had more surviving children on average (6.9) than women married to rich polygynous Mormons (5.5) [502]. Even though the polygynists' wives had lower fitness, the polygynists themselves did much better than monogamous men because they had children by several wives, not just one.

Throughout human history, the wives of many men have had to contend with the possibility that their husbands might use their wealth to attract and keep still more wives or extramarital partners. In today's societies as well, some married women are receptive to extramarital affairs, which may enable them to acquire additional material goods, or more protection or better genes for their offspring, from their extra-pair partners. Because both women and men can potentially elevate their fitness by mating with more than one partner, it is possible for a husband to reduce his wife's fitness, and vice versa.

One result of this fitness conflict between man and woman may have been the evolution of a capacity for sexual jealousy, which is an emotional state that helps individuals detect and interfere with the fitness-damaging sexual behavior of a mate. However, the nature of sexual jealousy should differ between the sexes, according to evolutionary psychologists, because the reproductive harm done by an unfaithful partner differs for men and women. A wife whose husband secures an additional mate (in a polygynous society) or goes on to divorce her in favor of a new partner usually loses some or all of her access to her husband's wealth, and thus the means to support herself and her children. Therefore, a woman's sexual jealousy, so the argument goes, should be focused on the possible loss of an attentive provider and helpmate that occurs when a man becomes emotionally involved with another woman. On the other hand, a husband whose wife mates with another man may eventually care for offspring fathered by that other male. A man's sexual jealousy should therefore revolve around the potential loss of paternity and parental investment arising from a wife's extramarital sexual activity rather than the loss of resources and emotional commitment [285].

If this view is correct, then if men and women were asked to imagine their responses to two scenarios, one in which a partner develops a deep friendship with another individual and one in which a partner engages in sexual intercourse with another individual, women should find the first scenario more disturbing than the second, whereas men should be more upset at the thought of a mate copulating with another man. Data from several cultures confirm these predictions [185, 285]. For example, in a study involving Swedish university students, who live in a fairly sexually permissive culture, 63 percent of the women found the prospect of emotional infidelity more troubling, whereas almost exactly the same percentage of the men deemed sexual infidelity more upsetting [1223].

One effect of male sexual jealousy and possessiveness is to reduce the possibility that another male will supply a man's partner with sperm. In this light, marriage, including the honeymoon, may be a cultural institution that serves the function of mate guarding (see p. 338). Everywhere men aspire to monopolize or restrict sexual access to their mates, although they do not necessarily succeed. Marriage rules institutionalize these ambitions. Although one sometimes hears of societies in which complete sexual freedom is the norm, the notion that such cultures actually exist appears to have been a (wistful?) misinterpretation on the part of outside observers. In *all* cultures studied to date, adultery committed by a woman, or even suspicion of it, is considered an offense against her husband and often precipitates violence [283]. A cuckolded husband may legally murder his wife or her lover in many societies [285].

The possibility that a woman may copulate with more than one man has also apparently shaped certain aspects of male anatomy and physiology, which seem designed to help men engage in sperm competition [1074]. Human testes, for example, weigh more than those of gorillas [492], and human ejaculates contain many more sperm as well. Gorillas are much larger animals than humans, but they have a mating system in which one male completely controls the sexual activities of several females. In such a system the risk that a fertilizable female will receive ejaculates from more than one male is almost nil. Therefore, delivery of

14 **A possible effect of sperm competition in humans.** The number of sperm in an ejaculate increases as a function of the time a man and woman have spent away from each other since their last copulation, perhaps to help the male swamp any rival ejaculates his partner may have received in his absence. After Baker and Bellis [59].

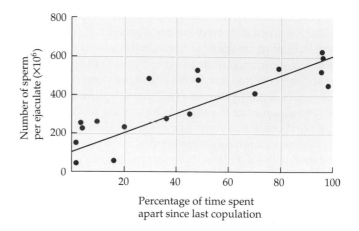

large quantities of sperm to a mate carries no special advantage for male gorillas. In contrast, the quantity of sperm transferred to a human female may affect a male's chances of fertilizing an egg, given the slight possibility that his partner will receive sperm from another male during the same ovulatory period.

In some other animals in which sperm competition is intense, males can adjust the number of sperm donated to a partner in relation to the risk that she has received sperm from a rival male [1064]. Do human males do the same (Figure 14)? If they do, then the more time a couple spend together between copulations, the fewer the sperm the man's ejaculate should contain, because the lower the probability that his sperm will have to swamp those of a rival. However, when a male is away from his partner for much of the time, the probability of sperm competition increases (Figure 15), favoring an added investment in sperm [59]. Robin Baker and Mark Bellis collected data from cooperative couples who supplied ejaculates captured in condoms for sperm counts. The results showed increased sperm delivery by men who had been out of sight of their partners for some time.

15 **Does mate guarding reduce the frequency of extra-pair matings?** In this study, the probability that a woman had copulated with more than one man was a function of how much time her primary partner had spent in her company. After Baker and Bellis [60].

An alternative hypothesis for sperm volume adjustment by men, if it indeed occurs, should also be considered. Martin Daly suggested that men who have been away from their partners may have had social interactions with other women, which the male psyche might interpret as potential opportunities for extra-pair copulations. Even looking at large numbers of women might do the trick, causing the brain to order the testes to produce extra quantities of sperm in case a potential EPC should become reality. If so, then a male stimulated by nonpartners in this way might eventually transfer the extra sperm to his regular partner when they were reunited.

Coercive Sex

One of the least attractive manifestations of sexual conflict in many species (Figure 16) is the occurrence of forced copulation. Although human rapists are often severely punished, rape occurs in every culture studied to date [1133]. Many persons have analyzed the phenomenon, including Susan Brownmiller in her influential book *Against Our Will* [175]. In her view, rapists act on behalf of all men to instill fear in all women, the better to intimidate and control them, thus keeping them "in their place."

16 Rape occurs in animals other than humans.
(Top) In the beetle *Tegrodera aloga,* a male (right) can court a female (left) decorously by repeatedly drawing her antennae into grooves on his head; copulation ensues only if the female responds to this courtship. (Bottom) Alternatively, a male (below) can force a female (above) to mate by running to her, grasping her, throwing her on her side, and then inserting his everted genitalia as the female struggles to break free. Photographs by the author.

Brownmiller's intimidation hypothesis implies that some males are willing to take the substantial risks associated with rape in order to provide a benefit for the rest of male society. This argument suffers from all the logical problems inherent in "for-the-good-of-the-group" hypotheses (with the added difficulty that groups composed of only one sex cannot be the focus of any realistic sort of group selection), but we can test it anyway. If the evolved function of the trait is to subjugate all women, then the rapist element in male society can be predicted to target older, dominant women (or young women who aspire to positions of power) to demonstrate the penalty that comes from stepping outside the traditional subordinate role. This prediction is not supported: most rape victims are young, *poor* women [1134].

An alternative evolutionary hypothesis proposed by Randy and Nancy Thornhill is that rape is an adaptive tactic in a conditional sexual strategy [1134]. According to the Thornhills, sexual selection has favored males with the capacity to commit rape *under some conditions* as a means of fertilizing eggs and leaving descendants. In this view, rape in humans is analogous to forced copulation in *Panorpa* scorpionflies (see p. 334), in which males unable to offer nuptial gifts use the low-gain, last-chance tactic of trying to force females to copulate with them. According to the rape-as-adaptive-tactic hypothesis, human males unable to attract willing sexual partners might use rape as a reproductive option of last resort.

The proposition that rape might serve an adaptive sexual function angered Brownmiller, who wrote, "It is reductive and reactionary to isolate rape from other kinds of violent antisocial behavior and to dignify it with adaptive significance" [176]. This view, however, confuses efforts to explain rape with attempts to excuse or justify the behavior. As noted earlier, when evolutionary biologists examine the adaptive value of a trait, whether it be rape in the human species, forced copulation in scorpionflies, or brood parasitism by cowbirds, their goal is to explain the evolutionary causes of the behavior and not to condone rape, or brood parasitism, or anything else.

The explanatory hypothesis that human rape is an evolved reproductive tactic controlled by a conditional strategy generates the prediction that some raped women will become pregnant, which they do, even in modern societies in which many women take birth control pills [1133]. In the past, in the absence of reliable birth control technology and abortion procedures, rapists would have had a still higher probability of fathering children through forced copulation. Furthermore, if rape is a reproductive tactic, then rapists should more often target women of high fertility, just as bank swallows and other birds identify fertile (egg-laying) females and try to force those individuals to copulate with them [84, 1086]. In contrast, Brownmiller's view that rape has nothing to do with reproduction, but is merely another form of violent antisocial behavior driven by a proximate desire of men to dominate women, suggests that the age distribution of rape victims should be the same as that of women murdered by male assailants. Crime data support the evolutionary hypothesis, not the position that rape has nothing to do with reproduction (Figure 17).

If rape is an evolved reproductive tactic of last resort, then it should occur more often in populations in which a high proportion of the fertile female population is monopolized by a relatively small part of the male population. Philip Starks and Caroline Blackie checked this prediction by looking at the correlation between rape and divorce rates in different populations. They reasoned that when divorce occurs, some men move on to acquire younger second (and even third) wives, depriving other men of the chance to marry these women and thereby increasing the overall variation in male reproductive success. The more serial monogamists in a population, the smaller the pool of fertile unmarried women relative to unmarried men, and the greater the potential benefits

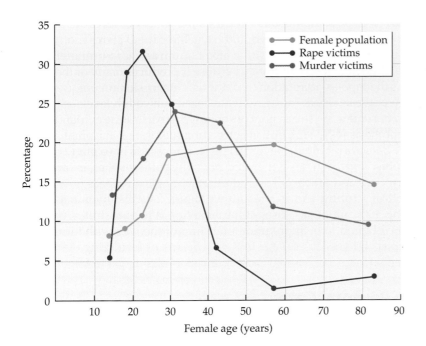

17 Testing explanations for rape. If rape were motivated purely by the intent to attack women violently (a proximate hypothesis), we would expect that the distribution of rape victims would match that of female murder victims. Instead, rape victims are especially likely to be young (fertile) women, a result consistent with ultimate hypotheses proposing that rape is linked to male reproductive tactics. Data on rape victims come from 1974–1975 police reports for 26 U.S. cities. After Thornhill and Thornhill [1134].

of rape as an alternative reproductive tactic for the excluded males. As Starks and Blackie predicted, the higher the divorce rate for a state in the United States, the higher the frequency of rape in 1965, 1975, 1985, and 1994 (Figure 18) [1093].

Although these findings suggest that rape may increase the fitness of some men, it is also possible that rape is not adaptive per se, but is instead a *maladaptive* by-product of the male sexual psyche, which causes quick sexual arousal, a desire for variety in sexual partners, and an interest in impersonal sex, all attributes that generate many adaptive consequences while also incidentally leading some men to rape some women [1133]. After all, many men engage in thoroughly nonreproductive sexual activities, including masturbation, homosexual rape, and rape of postmenopausal women and prepubertal girls, just as males of many nonhuman species also exhibit sexual activity that cannot possibly result in offspring, such as the copulatory mounting of weaned pups by male elephant seals [987]. Moreover, the only attempt to estimate the reproductive consequences of rape for men in a traditional society yielded the conclusion that the fitness cost to a rapist exceeded the benefit by a factor of about ten [1070].

← maladaptive

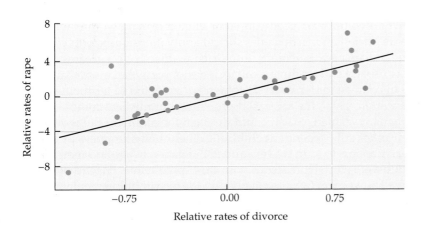

18 Rape as a conditional reproductive tactic. The correlation between rape and divorce rates in the United States. After Starks and Blackie [1093].

Even if coercive copulation usually reduces the fitness of its practitioners, the rape-as-by-product hypothesis would be tenable if the systems motivating male sexual behavior have a net positive effect on fitness. Unfortunately, this hypothesis is difficult to test definitively because it produces many of the same predictions as the rape-as-adaptation hypothesis. One prediction unique to the by-product hypothesis is that rapists will have unusually high levels of sexual activity with consenting as well as nonconsenting partners. Some evidence supports this hypothesis [674, 882], but as is true of many issues in human sociobiology, more data are required. Nevertheless, in this case as in so many others, the adaptationist approach has generated novel hypotheses that are entirely testable in principle and practice. The evolutionary angle on rape is now available for skeptical scrutiny, and as a result, we should eventually gain a better understanding of the ultimate causes of the behavior. When we do, we will not in any way be obliged to be more understanding of the illegal and immoral activity of rapists [1133].

Adaptive Parental Care

Ultimately, the sexual preferences and practices of men and women translate into the production of children and grandchildren. As we have noted, both men and women can provide parental care for their offspring, but there is great variation among individuals with respect to how much care they actually offer a given offspring [1258]. For example, infants that produce abnormally high-pitched cries have to cry for longer periods to elicit parental assistance than infants that cry in a more typical manner [413]. Moreover, occasionally an infant is not just ignored, but actually killed, by its parent. Here would seem to be the kind of behavior that must defeat evolutionary analysis.

But maybe not. Remember that an individual's impact on the next generation's gene pool is determined not by the number of babies conceived or born, but by the number of offspring that reach reproductive age. If babies with high-pitched cries have serious birth defects that make it unlikely that they will survive to reproduce, then parents that reduce or end their investment in these children may have more resources to invest in other offspring now or in the future. (And babies unable to cry normally often do have serious illnesses or congenital defects [413].) Likewise, if caring for a newborn or carrying a fetus to term threatens to reduce the lifetime reproductive success of a woman, then ending her investment in that offspring now can potentially increase, not reduce, her fitness [282]. An evolutionary prediction, therefore, is that single, pregnant women should be more likely to end their pregnancies than married women, who have a partner to help raise the offspring. Throughout most of human history, when single woman have attempted to rear a child unassisted, they have probably failed.

However, the evolutionary prediction that single pregnant women will seek abortions more often than married ones should apply more strongly to younger than to older women because young unmarried women have a greater chance of securing a husband and his fitness-enhancing support in the future than older women do. In contrast, older *married* women are expected to terminate pregnancies more often, given the increased probability of medical complications for older pregnant women, which could threaten the ability of the women to care for their other still-dependent children. When the willingness of British women to carry a fetus to term was examined in relation to age and marital status, the results were entirely in accord with these predictions (Figure 19) [730].

Men, as well as women, can potentially provide or withhold parental investment in their offspring. Because men run the risk of caring for a partner's child that was fathered by someone else, the sociobiological prediction is that married men should have evolved psychological mechanisms that protect them

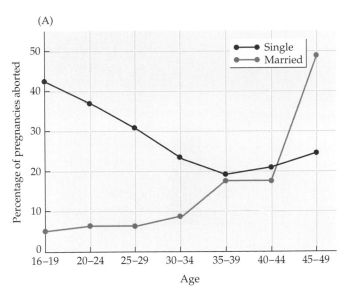

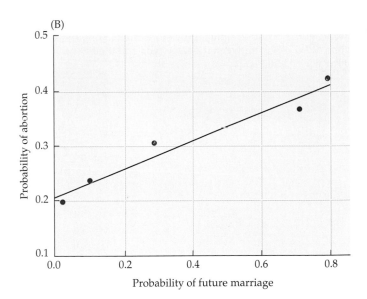

19 Do women employ abortion adaptively? (A) Single women in England and Wales are *less* likely to abort pregnancies as they grow older; married women are *more* likely to abort pregnancies as they age. These differences reflect the different costs and benefits of investing in an offspring when a woman has or does not have a husband's support. (B) The probability that a single woman will undergo an abortion is a function of the age-related probability that she will attract a husband in the future. After Lycett and Dunbar [730].

against that risk. As we have noted, males that learn of a wife's adultery are likely to react extremely negatively and often seek a divorce, with its consequent reduction or elimination of paternal care [1119]. Male concern about paternity is so obsessive that husbands of rape victims in many cultures may legally divorce their unfortunate wives [175].

Stepfathers are another category of males who are placed in the position of caring for offspring who are not their own. To test the prediction that stepfathers should favor their genetic offspring over stepchildren, Mark Flinn monitored some Trinidadian families in which a stepfather lived with children of his own as well as those his wife had by another man. In these families, men quarreled less often with their genetic offspring than with their stepchildren (Figure 20) [388]. Likewise, in modern American society, a man's genetic offspring by a current mate are far more likely to receive money for college than are his stepchildren by either a current or a previous mate (Figure 21) [30].

The same pattern applies to stepmothers, who are less likely to provide for their stepchildren than for their own genetic offspring. Households in which a mother cares for stepchildren, adopted children, or foster children spend less on food than households in which mothers reside with their genetic offspring [203]. Moreover, in blended families, the mother's own genetic offspring receive on average one more year of schooling than the stepchildren that reside with her. Note that by studying blended families, researchers have eliminated the possibility that stepchildren receive reduced educational opportunities because women who remarry are on average less capable of providing for the education of *any* children in their household [204].

Stepchildren not only tend to receive fewer resources from their stepparents, but are also at greater risk of being physically assaulted, a discovery made by Martin Daly and Margo Wilson when they tested the evolutionary hypothesis that our evolved psychological mechanisms encourage us to bias our parental care toward our genetic offspring [284]. Daly and Wilson tested this parental preference hypothesis by predicting that stepparents should contribute a dis-

20   Stepparenting and conflict. Conflicts between men and their stepchildren occurred significantly more often in one Trinidadian village than did conflicts between the same men and their genetic offspring. After Flinn [388].

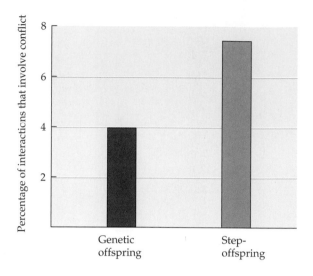

proportionately large fraction of the cases of criminal child abuse in Western societies. They found that in Hamilton, Ontario, children 4 years old or younger were *40 times more likely* to suffer abuse in families with a stepparent than in families with both genetic parents present! Note that for both categories of parents in this study, the absolute likelihood that a child would be abused was small (Figure 22), but the *relative* risk was far greater for children in households with a stepparent [281]. Daly and Wilson argued that the psychological systems that promote selective parental care generally lead to increased fitness, but may occasionally cause maladaptive side effects in a very small proportion of blended families.

Helping Children Marry

We have seen that adults care for children in a selective fashion, usually favoring genetic offspring over nonrelatives. But parents do not always treat their sons and their daughters identically when it comes to dishing out parental benefits. An example involves the material sacrifices parents make to help their offspring acquire spouses. In some societies, the groom and his family are called upon to donate resources such as cattle or money or labor—the bridewealth—

21   Parental favoritism. The odds that a man will give money to a child for college are much higher if the man is the genetic father of the potential recipient than if he is a stepparent to the child. The four categories of offspring examined in this study were genetic offspring living with their father, genetic offspring living with their father's previous mate, and the man's stepchildren living either with him or with a previous mate. The amount given to a genetic offspring living with a previous mate was used as a standard against which the other donations were measured. After Anderson et al. [30].

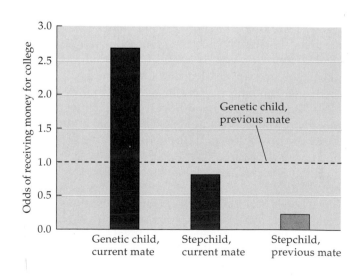

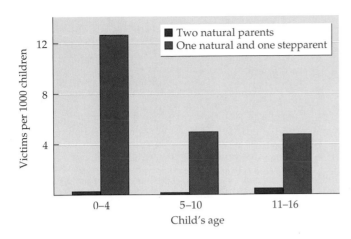

22 Child abuse and the relatedness of parents and offspring. Child abuse is far more likely to occur in households with a stepparent than in households with two genetic parents. After Daly and Wilson [281].

to the bride's family (Figure 23); in other groups, the family of the bride sends their daughter off to marry with a special donation—the dowry—to her new husband or his family.

If bridewealth or dowry payments were purely arbitrary traditions, the accidental products of cultural evolution, then we would predict that the two forms of payments should be equally represented among cultures worldwide. They are not [419]. (Before reading further, use sexual selection theory to predict whether bridewealth or dowry should be the more common cultural practice worldwide.)

23 Marriage requires bridewealth payments in many traditional African cultures, such as the Masai of Kenya and Tanzania. The father who stands here with his about-to-be-married daughter will receive cattle from her husband-to-be and his relatives. Photograph by Jason Lauré.

TABLE 2   The relationship between the mating systems of human cultures, bridewealth payments, and an inheritance system that favors sons

Mating system	Bridewealth payment		Sons favored	
	No	Yes	No	Yes
Monogamy	62%	38%	42%	58%
Limited polygyny	46%	54%	20%	80%
General polygyny	9%	91%	3%	97%

Note: The data are from Murdock's *Ethnographic Atlas* [831] for 112 monogamous cultures, 290 cultures that practice limited polygyny (less than 20 percent of men are polygynous) and 448 that practice general polygyny (more than 20 percent of men are polygynous). When sons are favored, they receive all or almost all of the parental inheritance.

Source: Hartung [495]

Because males typically compete for access to females, we can predict that bridewealth payments should be far more common than dowries. This is indeed the case. Bridewealth payments have been recorded in 66 percent of the 1267 societies described in the *Ethnographic Atlas* [831], whereas dowry is standard practice in just 3 percent of these societies. Bridewealth payments are found particularly frequently in polygynous cultures, occurring in more than 90 percent of those societies classified under the heading "general polygyny," in which more than 20 percent of married men have more than one wife (Table 2). When some males monopolize several females, marriageable females become an especially scarce and valuable commodity. One way to secure multiple wives, and so achieve exceptional reproductive success, is to provide a payment to their parents or other relatives. To do so often requires much wealth.

Rich men in polygynous societies can produce a great many children. Under these conditions, parents can potentially secure more descendants if they direct their wealth into the hands of a son, who can then use it to acquire a number of wives. Biased parental investment can even occur posthumously if wealthy parents pass on an inheritance to a son or sons, enabling their male offspring to become successful polygynists [125]. These sons can produce many more grandchildren for the deceased parent than a daughter, whose reproductive success can never exceed the number of embryos she can personally produce and nurture. John Hartung found the predicted correlation between the practice of polygyny in a society and the occurrence of inheritance rules that favor sons (see Table 2) [495].

Guy Cowlishaw and Ruth Mace confirmed that this pattern was real after controlling for the nonindependence of cultures—that is, after dealing with the statistical problem of how to treat information from several cultures that may have inherited similar practices as a result of sharing a recent common cultural ancestor. First, they constructed cladograms of cultures based on linguistic information; these diagrams identified which cultures were closely related and which were not. They then superimposed the mating system of each culture on the cladogram to identify those whose mating systems had changed to polygyny from a monogamous predecessor and vice versa. Independent cultural changes to polygyny were much more likely to be associated with inheritance rules favoring sons than were changes to monogamy [248].

Even in supposedly monogamous Western societies, rich men may have opportunities for unusual copulatory success (as demonstrated by Perusse's study of Canadian men) because their wealth makes them attractive to women. If parents in modern societies retain an ancestrally selected bias that causes them

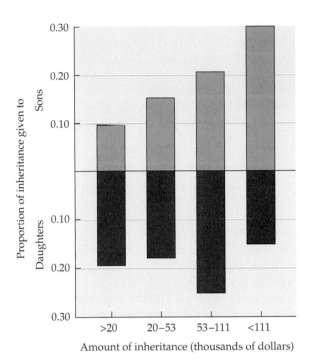

24 Inheritance decisions. Wealthy Canadian parents bias their legacies toward their sons, who are more likely than daughters to convert exceptional wealth into exceptional reproductive success. After Smith et al. [1073].

to favor those offspring with the highest reproductive potential, we can predict that even today, very wealthy parents should be inclined to give their sons more inheritance than their daughters—a prediction that has received support (Figure 24) [1073].

In contrast to the prevalence of biased parental investment in male offspring, societies in which parents give significantly more to their daughters are rare—as one might expect, given that females typically are in demand as marriage partners. However, Lee Cronk has found one tribal society, the Mukogodo of Kenya, in which parents often provide more food and more medical care (from a local Catholic mission clinic) to their daughters than to their sons [269]. The Mukogodo have only recently abandoned their traditional hunting–gathering lifestyle in favor of an economy built on sheep and goat herding. Their herds are small, and their standing with other pastoral tribes in the area is very low. A son of an impoverished Mukogodo family is unlikely to acquire a large enough flock to pay the bride price for a wife from any of the surrounding tribes. In contrast, a Mukogodo daughter has a good chance of marrying a member of a higher-status tribe, since polygyny is standard among these groups and women are in short supply. As a result of the greater ease of marriage for daughters than for sons, the average number of offspring of a Mukogodo daughter is nearly 4, whereas a son's direct fitness is only about 3. This inequality favors families that have more daughters than sons, an outcome that is achieved through greater investment in young girls than in young boys.

This case nicely illustrates the point that human behavior is adaptively flexible, not arbitrarily or infinitely variable. Whatever psychological mechanisms control parental solicitude toward offspring, these evolved systems permit parents to favor sons under some circumstances and daughters under others, while encouraging equal treatment under still others. The option chosen tends to enhance parental fitness under local conditions. The *differences* between the daughter-favoring Mukogodo and those nearby herding cultures in which sons receive preferential treatment are surely not genetic. Instead, these parental differences reflect our ability to use evolved conditional strategies to select among a limited set of options, choosing the one with the highest fitness payoff in a given environment, just as female white-fronted bee-eaters unconsciously select the best fitness-raising tactic available to them in a variable social environment.

Among the Mukogodo and most other tribal groups, men pay a price to acquire a bride. Why do a few contrary cultures sanction payments that help a woman secure a husband? One answer is that in monogamous societies in which males typically invest materially in their children, parents that help their daughter "buy" the right kind of man gain fitness as a result. Males in socially stratified monogamous cultures vary greatly in their status and access to wealth. In such cultures, women married to elite males should generally enjoy a reproductive advantage because a monogamous husband's wealth will not be divided among a bevy of wives, but instead will go to his only wife and her children. To the extent that wealth and high status translate into reproductive success, a woman's parents may gain by competing with the parents of other families for an "alpha" husband, even if this requires that they offer a material inducement to the right male or to his family. Steven Gaulin and James Boster found that dowry payments occur in substantially less than 0.5 percent of the nonstratified societies, whereas 9 percent of the stratified societies and 60 percent of the *monogamous* stratified cultures permit or encourage dowries [419]. This highly nonrandom set of associations supports the hypothesis that dowry practices are not a random cultural artifact, but part of an adaptive parental strategy.

Thus, we have yet another example of how sociobiologists, far from being flummoxed by cultural diversity, can make use of it to test evolutionary hypotheses about human behavior. The successes of sociobiology demonstrate that the Darwinian approach has much to tell us about ourselves. There is no guarantee, however, that what we will learn will be put to uses that most people would consider good. The brain mechanisms that push us to solve mysteries about ourselves and our world may lead to discoveries that will be used to worsen, rather than alleviate, the social and military crises that characterize modern life. But whatever our wishes, the fact that we are an evolved animal species is not going to change, and so we might as well understand the significance of this fact, if only to give ourselves insight into what Pogo meant when he declared, "We have met the enemy and it is us."

Summary

1. Human beings are an evolved animal species. Human sociobiology, a field of study that includes evolutionary psychology and evolutionary anthropology, employs natural selection theory to generate testable hypotheses about the possible adaptive value of our species' behavior.

2. The study of human sociobiology has been marked by intense controversy, in part because some persons have misunderstood the goals and foundations of the discipline. Contrary to the claims of some critics, sociobiologists are not motivated by any particular political agenda, nor are sociobiological hypotheses based on the premise that elements of human behavior are genetically determined or morally desirable. Sociobiological hypotheses are designed to explain, not justify, our behavior. Testing these hypotheses can help us understand why we have psychological mechanisms that motivate us to behave in certain ways.

3. Unlike arbitrary culture theory, which proposes that human behavior is the arbitrary product of cultural traditions constrained only by the limits of our imagination, sociobiological theory views human behavior as the product of conditional strategies involving limited options that tend to increase, rather than decrease, the fitness of individuals.

4. Many sociobiological hypotheses have been advanced for elements of human reproductive behavior, which has been shaped by cooperation and conflict between the sexes. On the one hand, both males and females can potentially gain fitness through parental care of their offspring. On the other hand, opportunities for conflict of interest also exist, since male reproductive potential is higher than that of females and since certainty of maternity is higher than certainty of paternity. Females run the risk of losing support for their offspring; males run the risk of investing their resources in offspring fathered by men other than themselves. These possibilities appear to have heavily influenced the evolution of the proximate psychological mechanisms that underlie much of human sexual and parental behavior.

5. To the extent that we are curious about the evolution of human behavior, a sociobiological approach offers avenues of exploration that other disciplines do not. Although there are problems in employing this approach, the same can be said about any scientific endeavor that touches on human concerns.

Discussion Questions

1. Marshall Sahlins has argued that sociobiology is contradicted because people in most cultures do not even have words to express fractions. Without fractions, a person cannot possibly calculate coefficients of relatedness, and without this information (Sahlins claims), people cannot determine how to behave in order to maximize their indirect fitness [1011]. Has Sahlins delivered a knockout blow to sociobiological theory?

2. Philip Kitcher states that "socially relevant science," such as sociobiology, demands "higher standards of evidence" because if a mistake is made (a hypothesis presented as confirmed when it is false), the societal consequences may be especially severe. For example, a hypothesis that men are more disposed to seek political power and high status in business and science than women is dangerous because it "threaten(s) to stifle the aspirations of millions" [628]. How would a sociobiologist respond to Kitcher's claim?

3. Develop an evolutionary analysis of divorce, complete with predictions about the different factors that might cause men and women to divorce their mates and the expected consequences for individuals who end their marriages.

4. In writing about the widespread occurrence of genocide, Stephen J. Gould reviews the adaptationist hypothesis that the capacity for large-scale murder evolved as a result of intense competition for resources or mates between small bands during our evolutionary history. Gould dismisses this hypothesis on the grounds that evolutionary theory has nothing novel to say about this aspect of human behavior. "An evolutionary speculation can only help if it teaches us something we don't know already—if, for example, we learned that genocide was biologically enjoined by certain genes, or even that a positive propensity, rather than a mere capacity, regulated our murderous potentiality. But the observational facts of human history speak against determination and only for potentiality. Each case of genocide can be matched with numerous incidents of social benevolence; each murderous clan can be paired with a pacific clan" [447]. Evaluate Gould's argument critically in the light of what you know about (1) the proximate–ultimate distinction, (2) conditional strategies, and (3) the differences between adaptationist and arbitrary culture hypotheses.

5. Discussions of rape are invariably emotionally charged. From an evolutionary perspective, why might women have an especially strong visceral response to the topic and an intense desire to punish rapists? Why might most men also wish to deter rape? Would an understanding of evolutionary theory have led to a revision in this legal ruling by the United States Supreme Court, which contained the claim that "rape is without doubt deserving of serious punishment; but in terms of moral depravity and of the injury to the person and to the pub-

lic, it does not compare with murder....[Rape] does not include...even the serious injury to another person..." (quoted in [593]).

6. Natalie Angier states that married men have the same probability of fertilizing an egg per copulation with their wives as rapists do when forcing copulation on a victim [39]. In the past, the probability that an offspring of a married man would survive to reproduce was almost certainly much higher than the probability that a rapist's child would reach reproductive age because married men assist their children, whereas rapists do not. Is Angier correct, therefore, in claiming that rape cannot be an adaptive tactic? (Remember that "adaptive" means "reproductively useful.")

7. Some persons believe that men in Western societies prefer sexual partners who are younger than they are because they have been taught this arbitrary cultural convention from an early age. What prediction follows from this hypothesis with respect to the dating preferences of teenage males for females of different ages? What conclusion follows from the data in the figure below [613]?

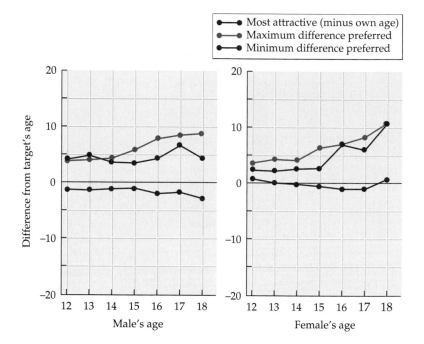

Suggested Reading

The debate on sociobiology began with E. O. Wilson; see the final chapter of *Sociobiology* [1247] and an attack on the book [22] with Wilson's reply [1248]. Representative recent books and articles that continue to criticize the evolutionary analysis of human behavior include [176, 335, 352, 423, 694, 988]. The misconceptions that many persons have about human sociobiology have been outlined by two of my colleagues at Arizona State, Doug Kenrick [610] and Owen Jones [593]. Most of the classic papers on the evolutionary analysis of human behavior can be found, along with updates and critiques, in *Human Nature*, edited by Laura Betzig [103]. Modern accounts of an evolutionary approach to human behavior include books by Richard Alexander [18, 19], Martin Daly and Margo Wilson [282, 284], and Robert Wright [1267], whose account of evolutionary psychology is extremely readable. The evolution of human sexual behavior and psychology has been explored in various ways by Donald Symons [1119], David Geary [422], and Geoffrey Miller [792]. Readers can also find many interesting articles on human behavior in the journal *Human Behavior and Evolution*.

Glossary

Action potential The neural signal; a self-regenerating change in membrane electrical charge that travels the length of a nerve cell and may elicit changes in the activity of a neighboring nerve cell or muscle.

Adaptation A characteristic that confers higher fitness on individuals than any other existing alternative exhibited by other individuals within the population; a trait that has spread, or will spread, or is being maintained in a population as a result of natural selection.

Adaptationist A behavioral biologist who develops and tests hypotheses on the possible adaptive value of particular traits. (An **adaptationist hypothesis** proposes that a particular trait improves the chances that individuals will propagate their special genes compared to the effects of alternative forms of the trait.)

Adaptive value The contribution that a trait or gene makes to fitness or individual genetic success.

Allele A form of a gene; different alleles typically code for distinctive variants of the same protein.

Altruism Helpful behavior that raises the recipient's direct fitness while lowering the donor's direct fitness.

Arbitrary culture theory The view that human behavior is the arbitrary product of whatever cultural traditions people are exposed to within a society; thus, our actions are not expected to be explicable in evolutionary terms.

Artificial selection *See* Selection.

Associated reproductive pattern A seasonal change in reproductive behavior that is associated with changes in the gonads and hormones in contrast to a disassociated reproductive pattern in which the onset of reproductive activity is not triggered by a sharp change in certain hormone concentrations.

Batesian mimic An edible animal that deceptively resembles another inedible species and gains protection against predators as a result.

Brood parasite An animal that exploits the parental care of individuals other than its parents.

By-product hypothesis An explanation for a maladaptive or nonadaptive attribute that is said to occur as a by-product of a proximate mechanism that has some other adaptive consequence for individuals.

Causal question In the scientific method, a question about the cause of a natural phenomenon.

Central pattern generator A group of cells in the central nervous system that can produce a particular pattern of signals that triggers a functional behavioral response.

Circadian rhythm A roughly 24-hour cycle of behavior that expresses itself independent of environmental changes.

Circannual rhythm A annual cycle of behavior that expresses itself independent of environmental changes.

Cladogram A diagram of the evolutionary relationships among species or groups of species; a phylogenetic tree.

Coefficient of relatedness The probability that an allele present in one individual will be present in a close relative; the proportion of the total genotype of one individual present in another as a result of shared ancestry.

Communication The cooperative transfer of information from a signaler to a receiver.

Comparative method A procedure for testing evolutionary hypotheses based on disciplined comparisons among species of known evolutionary relationships.

Concession theory The argument that social systems can form as a result of concessions made by dominant individuals to subordinates that have the effect of raising the fitness of these other individuals sufficiently to retain their membership in the group.

Conditional strategy *See* Strategy.

Convergent evolution The independent acquisition over time of similar characteristics in unrelated species subject to similar selection pressures.

Cooperation A mutually helpful interaction between two individuals.

Critical period A phase in an animal's life when certain experiences are particularly likely to have a potent developmental effect.

Cryptic female choice The ability of a female that has received sperm from more than one male to choose which male's sperm will fertilize her eggs.

Cumulative selection *See* Selection.

Darwinian puzzle A trait that appears to reduce the fitness of individuals that possess it.

Developmental homeostasis The capacity of developmental mechanisms within individuals to produce adaptive traits despite the potentially disruptive effects of mutant genes and suboptimal environmental conditions.

Dilution effect Safety in numbers that comes from swamping the consumption capacity of local predators.

Diploid Having two copies of each gene in one's genotype.

Direct fitness *See* Fitness.

Direct selection *See* Selection.

Display A stereotyped action used as a communication signal by individuals.

Dissociated reproductive pattern *See* Associated reproductive pattern

Divergent evolution The evolution of differences among closely related species because of differing selection pressures in their different environments.

Dominance hierarchy A social ranking within a group in which some individuals give way to others, often conceding useful resources to others without a fight.

Ethology The study of the proximate mechanisms and adaptive value of animal behavior.

Eusocial Of or relating to societies that contain specialized nonreproducing castes that work on behalf of the reproductive members of the society.

Evolutionarily stable strategy *See* Strategy.

Evolutionary cost-benefit approach A method of studying the adaptive value of alternative traits based on the recognition that phenotypes come with fitness costs and fitness benefits; an adaptation has a better cost–benefit ratio than alternative versions of the trait.

Evolutionary psychology The study of the adaptive value of psychological mechanisms, especially of human beings; a key component of sociobiology.

Explosive breeding assemblage A mating system in which a large proportion of the adult female population becomes receptive on one or a very few days of the year.

Extra-pair copulation A mating by a male or female with an individual other than his or her social partner in an apparently monogamous species.

Female defense polygyny *See* Polygyny.

Female distribution theory A general explanation for the diversity in mating systems that arises because females of different species are spatially distributed in different ways, which affects the fitness gains derived from different male mating tactics.

Female enforced monogamy *See* Monogamy

Female preference hypothesis An explanation for the formation of leks based on female preferences for mating with males whose attributes can be compared with those of the other members of a group.

Fertility insurance An increased likelihood of egg fertilization, which females can achieve by mating with several males.

Fitness A measure of the genes contributed to the next generation by an individual, often stated in terms of the number of offspring that survive to reproduce.

> **Direct fitness** The genes contributed to the next generation by an individual via personal reproduction in the bodies of its own offspring.
>
> **Indirect fitness** The genes contributed to the next generation by an individual indirectly by helping nondescendant kin, in effect creating relatives that would not have existed without the help of the individual.
>
> **Inclusive fitness** The sum of an individual's direct and indirect fitness.

Fitness benefit That aspect of a trait that tends to raise the fitness of individuals.

Fitness cost That aspect of a trait that tends to reduce the fitness of individuals.

Fixed action pattern An innate, highly stereotyped response that is triggered by a well-defined, simple stimulus; once the pattern is activated, the response is performed in its entirety.

Free-running cycle The cycle of activity of an individual that is expressed in a constant environment.

Frequency-dependent selection *See* Selection.

Game theory An optimality approach to the study of behavioral adaptation in which the fitness payoffs to an individual from a particular behavioral tactic are dependent on what the other members of the group are doing.

Genetic compatibility The ability of the genes present in some sperm to complement the genes present in some eggs, resulting in an increased likelihood that the zygote will develop into an especially competent individual.

Genomic imprinting The phenomenon in which the phenotypic effect of a gene depends upon the parent that supplied it.

Genotype The genetic constitution of an individual; may refer to the alleles of one gene possessed by the individual or to its complete set of genes.

Good genes Genes that advance the fitness of the bearer. The good genes hypothesis or theory argues that females exhibit mate choice in order to provide their offspring with a partner's genes that will advance their offspring's chances of survival or reproductive success.

Good parent theory An explanation for mate choice in which choosy individuals select partners on the basis of how well they will care for their offspring.

Group selection *See* Selection.

Hamilton's rule The theoretical point made by W. D. Hamilton that altruism can spread when $rb > c$ (with r being the coefficient of relatedness between the altruist and the individual helped, b being the fitness benefit received by the helped individual, and c being the cost of altruism in terms of the altruist's lost direct fitness).

Haploid Having only one copy of each gene in one's genotype, as is characteristic of the sperm and eggs of diploid organisms.

Healthy mate theory An explanation for female preferences for males healthy enough to produce and maintain elaborate ornaments; males of this sort are unlikely to pass communicable diseases or parasites to their mates.

Heritability The proportion of the total variance in the phenotypes in a population that arises because of genetic variance among individuals, or Vg divided by $Vg + Ve$, where Vg = phenotypic variance caused by genetic differences among individuals and Ve = phenotypic variance among individuals that is environmentally induced.

Home range An area that an animal occupies but does not defend, in contrast to a territory, which is defended.

Honest signal A signal that conveys accurate information to receivers about the signaler's phenotypic quality which may be relevant to its fighting ability or quality as a potential mate.

Hotshot A male whose attributes are especially appealing to sexually receptive females.

Hotspot A location whose properties attract sexually receptive females to the males able to hold the site.

Hypothesis A possible explanation for what causes something to occur.

Ideal free distribution The distribution of nonterritorial individuals in space when they are free to make decisions that maximize their individual fitness.

Illegitimate receiver An individual that uses information gained from the signals of another to advance its own fitness at a cost to the signaler.

Illegitimate signaler An individual that produces a signal that may deceive another into responding in ways that advance the signaler's fitness at a cost to the receiver.

Imprinting A form of learning in which individuals exposed to certain key stimuli, usually early in life, form an association with the stimulus and may later exhibit sexual behavior toward individuals with similar stimuli.

Inbreeding depression The tendency of the descendants of individuals that mate with close relatives to have lower fitness than non-inbred members of their species.

Inclusive fitness *See* Fitness.

Indirect fitness *See* Fitness.

Indirect reciprocity *See* Reciprocity.

Indirect selection *See* Selection.

Innate releasing mechanism A hypothetical neural mechanism responsible for controlling an innate response to a sign stimulus.

Instinct A behavior pattern that reliably develops in most individuals and that promotes a functional response to a releaser or sign stimulus the first time the stimulus elicits the response.

Interneuron A neuron that relays messages either from receptor neurons to the central nervous system (a sensory interneuron) or from the central nervous system to neurons commanding muscle cells (a motor interneuron).

Kin discrimination The capacity of an individual to react differently to others based on their degree of genetic relatedness to that individual.

Kin selection *See* Selection.

Learning A durable and usually adaptive change in an individual's behavior traceable to a specific experience in that individual's life.

Lek A traditional display site that females visit to select a mate from among the several to many males displaying on small, resource-free territories.

Lek paradox The Darwinian puzzle created by the persistence of extreme female mate preferences in lekking species even though female mate choice in the past could be expected to have eliminated genetic differences among males, thereby eliminating any current fitness benefit for choosy females.

Mate assistance When a male (usually) provides parental care for the offspring of his mate, contributing to the formation of monogamous bonds between male and female.

Mate guarding Actions taken by males (usually) to prevent a sexual partner from acquiring sperm from other males.

Mobbing behavior A behavior in which prey closely approach and attempt to harass a predator.

Monogamy A mating system in which one male mates with one female in a breeding season. In female enforced monogamy, males are unable to mate with more than one female because of actions taken by their partners.

More resources hypothesis The explanation for female choice based on the gains in material benefits that choosy individuals receive.

Mutualism A mutually beneficial relationship or cooperative interaction.

Natural selection *See* Selection.

Neuron A nerve cell.

Nondescendant kin Relatives other than offspring.

Nuclei Clusters of cell bodies of neurons within nervous systems.

Nuptial gift A food item or other useful present transferred by a male to a female just prior to or during copulation.

Operant conditioning A kind of learning based on trial and error, in which an action, or operant, becomes more frequently performed if it is rewarded.

Operational sex ratio The ratio of receptive males to receptive females during a given period.

Optimal foraging theory An explanation for the feeding decisions made by animals based on the assumption that individuals are attempting to maximize their net fitness gains by, for example, maximizing caloric intake or some other dietary gain that correlates with reproductive success.

Optimality theory A theory based on the assumption that the attributes of organisms are optimal, that is, better than others in terms of their cost–benefit ratios; the theory is used to generate hypotheses about the possible adaptive value of traits.

Parent-offspring conflict The clash of interests that occurs when parents can gain fitness by withholding parental care or resources from some offspring in order to invest in another round of reproduction at a time when existing offspring would benefit from receipt of the investment.

Parental investment Costly parental activities that increase the likelihood of survival for some existing offspring, but that reduce the parent's chances of producing offspring in the future.

Phenotype Any measurable trait of an individual that develops as a result of an interaction between the individual's genes and its environment.

Phenotype matching A proximate mechanism of kin discrimination in which an individual's behavior toward another is based on how similar they are in some way, such as odor or appearance.

Pheromone A volatile chemical released by an individual as a scent signal for another.

Photoperiod The number of hours of light in a 24-hour period.

Phylogeny An evolutionary genealogy of the relationships among a number of species or clusters of species.

Phylogenetic tree A diagram of the evolutionary relationships among species or clusters of species.

Polyandry A mating system in which a female has several partners in a breeding season.

Polygyny A mating system in which a male fertilizes the eggs of several females in a breeding season. This mating system takes many forms including **female defense polygyny** (in which males guard groups of mates), **lek polygyny** (in which males with unusually attractive traits or display territories attract several mates), **resource defense polygyny** (in which males with unusually large amounts of resources attract several mates), and **scramble competition polygyny** (in which males that are unusually good at finding mates become polygynous).

Polygyny threshold model An explanation for why females might enter into a polygynous mating relationship when the resources gained by joining an already mated male on his territory may exceed those to be gained by forming a partnership with an unmated male.

Polymorphism The coexistence of two or more distinctive forms or traits within a species.

Prediction In science, an expected result that should be observed if a particular hypothesis is true.

Prisoner's dilemma A game theory construct in which the fitness payoffs to individuals are set such that mutual cooperation between the players generates a lower return than defection, which occurs when one individual accepts assistance from the other but does not return the favor.

Proximate cause An immediate underlying cause based on the operation of internal mechanisms possessed by an individual.

Reciprocity Reciprocal altruism in which a helpful action is repaid at a later date by the recipient of assistance. (Indirect reciprocity involves reciprocal altruism in which a helpful action is repaid at a later date by observers of the helpful individual.)

Releaser A sign stimulus given by an individual as a social signal to another.

Reproductive success The number of surviving offspring produced by an individual; direct fitness.

Resource-holding power The ability of an individual to retain control of valuable resources, which is often a function of its size or experience.

Resource defense polygyny *See* Polygyny.

Runaway selection *See* Selection.

Satellite male A male that intercepts females drawn to a calling or resource-defending rival of the satellite.

Scientific conclusion In the scientific method, a hypothesis that has been tested and rejected or accepted.

Scientific method The process of testing hypotheses by examining the validity of the predictions derived from them.

Scramble competition polygyny *See* Polygyny.

Search image A perceptual screening mechanism used by predators to search visually for cryptic, edible prey.

Selection The effect of differences among individuals in their ability to transmit copies of their genes to the next generation.

    Artificial selection A process that is identical to natural selection, except that humans control the reproductive success of alternative phenotypes within the selected population.

    Chase away selection A form of sexual selection that occurs when males differ in their ability to exploit the sensory systems of females in ways that raise male and reduce female reproductive success, leading to counter-selection on females favoring those that are increasingly resistant to male courtship signals.

    Cumulative selection The effect of repeated bouts of natural selection, resulting in the accumulation of many small adaptive changes in an evolving population, which can add up to a large evolutionary change over time.

    Direct selection *See* Natural selection.

    Frequency-dependent selection A form of natural selection in which the fitness of a phenotype depends upon its relative frequency in the population.

    Group selection Selection that occurs when groups differ in their collective attributes and those differences affect the survival chances of those groups.

    Indirect selection A form of natural selection that occurs when individuals differ in their effects on the survival of nondescendant kin, creating differences in the indirect fitness of the individuals interacting with this category of kin.

    Kin selection A form of natural selection that occurs when individuals differ in ways that affect their parental care or helping behavior, and thus the survival of their own offspring or the survival of nondescendant kin.

    Natural selection The process that produces evolutionary change when individuals differ in heritable traits that are correlated with differences in their individual reproductive success; also known as direct selection.

    Runaway selection A form of sexual selection that occurs when female mating preferences for certain male attributes create a positive feedback loop favoring both males with these attributes and females that prefer them.

    Sexual selection A form of natural selection that occurs when individuals differ in their ability to compete with others for mates or to attract members of the opposite sex.

Selfish herd A group of individuals whose members use others as living shields against predators.

Sensory exploitation A situation in which a signaler is able to tap into a preexisting sensitivity or bias in the perceptual system of a receiver, thereby gaining an advantage in transmitting a message to that receiver.

Sex role reversal The occurrence of competition among females for access to males, which may choose selectively among potential mates.

Sexual dimorphism A difference between males and females of the same species.

Sexual selection *See* Selection.

Siblicide The killing of a sibling by a brother or sister.

Sign stimulus The effective component of an action or object that triggers a fixed action pattern.

Sociobiology A discipline that uses evolutionary theory as a foundation for the study of social behavior; often used to refer to adaptationist studies of human reproductive and social behavior.

Sperm competition The competition between males that determines whose sperm will fertilize a female's eggs when both males' sperm have been accepted by that female.

Stimulus filtering The capacity of neurons and neural networks to ignore stimuli that could potentially elicit a response from them.

Strategy A set of rules for behavior controlled by an inherited proximate mechanism such that the differences between individuals in their strategies are linked to differences in their genes.

Conditional strategy A set of rules that provides for different tactics under different environmental conditions; the inherited behavioral capacity to be flexible in response to certain cues or situations.

Evolutionarily stable strategy That set of rules of behavior that when adopted by a certain proportion of the population cannot be replaced by any alternative strategy.

Supernormal stimulus A sign stimulus that is more effective in eliciting a response than naturally occurring actions or objects.

Synapse The point of near contact between one nerve cell and another.

Tactic A behavior pattern that is specified by an evolved strategy.

Territory An area that an animal defends against intruders.

Test In the scientific method, actual results that permit one to evaluate a hypothesis by comparing the evidence against the predicted results.

Ultimate cause The evolutionary, historical reason why something is the way it is.

Bibliography

1 Able, K. P. 1993. Orientation cues used by migratory birds: A review of cue-conflict experiments. *Trends in Ecology and Evolution* 10:367–371.

2 Able, K. P. 1996. The debate over olfactory navigation by homing pigeons. Journal of Experimental Biology 199:121–124.

3 Acharya, L., and J. N. McNeil. 1998. Predation risk and mating behavior: The responses of moths to bat-like ultrasound. *Behavioral Ecology* 9:552–558.

4 Adkins-Regan, E. 1981. Hormone specificity, androgen metabolism, and social behavior. *American Zoologist* 21:257–271.

5 Airey, D. C., H. Castillo-Juarez, G. Casella, E. J. Pollak, and T. J. DeVoogd. 2000. Variation in the volume of zebra finch song control nuclei is heritable: Developmental and evolutionary implications. *Proceedings of the Royal Society of London B* 267:2099–2104.

6 Airey, D. C., and T. J. DeVoogd. 2000. Greater song complexity is associated with augmented song system anatomy in zebra finches. *NeuroReport* 11:2339–2344.

7 Alatalo, R. V., J. Höglund, and A. Lundberg. 1991. Lekking in the black grouse—a test of male viability. *Nature* 352:155–156.

8 Alatalo, R. V., J. Höglund, A. Lundberg, P. T. Rintamäki, and B. Silverin. 1996. Testosterone and male mating success on the black grouse leks. *Proceedings of the Royal Society of London B* 263:1697–1702.

9 Alatalo, R. V., and A. Lundberg. 1984. Polyterritorial polygyny in the pied flycatcher *Ficedula hypoleuca*—evidence for the deception hypothesis. *Annales Zoologici Fennici* 21:217–228.

10 Alcock, J. 1987. Leks and hilltopping in insects. *Journal of Natural History* 21:319–328.

11 Alcock, J. 1994. Postinsemination associations between males and females in insects: The mate-guarding hypothesis. *Annual Reviews of Entomology* 39:1–21.

12 Alcock, J. 1996. Provisional rejection of three alternative hypotheses on the maintenance of a size dichotomy in Dawson's burrowing bee (*Amegilla dawsoni*) (Apidae, Apinae, Anthophorini). *Behavioral Ecology and Sociobiology* 39:181–188.

13 Alcock, J. 2001. *The Triumph of Sociobiology*. Oxford University Press, New York.

14 Alcock, J., and W. J. Bailey. 1995. Acoustical communication and the mating system of the Australian whistling moth *Hecatesia exultans* (Noctuidae: Agaristidae). *Journal of Zoology* 237:337–352.

15 Alcock, J., and W. J. Bailey. 1997. Success in territorial defence by male tarantula hawk wasps *Hemipepsis ustulata*: The role of residency. *Ecological Entomology* 22:377–383.

16 Alcock, J., G. C. Eickwort, and K. R. Eickwort. 1977. The reproductive behavior of *Anthidium maculosum* and the evolutionary significance of multiple copulations by females. *Behavioral Ecology and Sociobiology* 2:385–396.

17 Alexander, R. D. 1974. The evolution of social behavior. *Annual Review of Ecology and Systematics* 5:325–383.

18 Alexander, R. D. 1979. *Darwinism and Human Affairs*. University of Washington Press, Seattle, WA.

19 Alexander, R. D. 1987. *The Biology of Moral Systems*. Aldine de Gruyter, Hawthorne, NY.

20 Alexander, R. D., J. L. Hoogland, R. D. Howard, K. M. Noonan, and P. W. Sherman. 1979. Sexual dimorphism and breeding systems in pinnipeds, ungulates, primates and humans. In *Evolutionary Biology and Human Social Behavior: An Anthropological Perspective*, N. A. Chagnon and W. Irons (eds.). Duxbury Press, North Scituate, MA.

21 Allen, E. et al. 1975. Against "sociobiology." *New York Review of Books*, November 13:82, 84–86.

22 Allen, E. et al. 1976. Sociobiology—another biological determinism. *BioScience* 26:183–186.

23 Allen, G. R., D. J. Kazmer, and R. F. Luck. 1994. Post-copulatory male behaviour, sperm precedence and multiple mating in a solitary parasitoid wasp. *Animal Behaviour* 48:635–644.

24 Allison, T., A. Puce, and G. McCarthy. 2000. Social perception from visual cues: The role of the STS region. *Trends in Cognitive Sciences* 4:267–278.

25 Allman, J. 1999. *Evolving Brains*. Scientific American Library, New York.

26 Allman, J., A. Rosin, R. Kumar, and A. Hasenstaub. 1998. Parenting and survival in anthropoid primates: Caretakers live longer. *Proceedings of the National Academy of Sciences* 95:6866–6869.

27 Ancel, A., H. Visser, Y. Handrich, D. Masman, and Y. L. Maho. 1997. Energy saving in huddling penguins. *Nature* 385:304–305.

107 Birkhead, T. R., and A. P. Møller. 1992. *Sperm Competition in Birds: Evolutionary Causes and Consequences.* Academic Press, London.

108 Birkhead, T. R., and A. P. Møller (eds.). 1998. *Sperm Competition and Sexual Selection.* Academic Press, San Diego, CA.

109 Bissoondath, C. J., and C. Wiklund. 1995. Protein content of spermatophores in relation to monandry/polyandry in butterflies. *Behavioral Ecology and Sociobiology* 37:365–372.

110 Bjorksten, T. A., K. Fowler, and A. Pomiankowski. 2000. What does sexual trait FA tell us about stress? *Trends in Ecology and Evolution* 15:163–166.

111 Black, A. H. 1971. The direct control of neural processes by reward and punishment. *American Scientist* 59:236–245.

112 Blackledge, T. A. 1998. Signal conflict in spider webs driven by predators and prey. *Proceedings of the Royal Society of London B* 265:1991–1996.

113 Blackledge, T. A. 1998. Stabilimentum variation and foraging success in *Argiope aurantia* and *Argiope trifasciata* (Araneae: Araneidae). *Journal of Zoology* 246:21–27.

114 Blackledge, T. A., and J. W. Wenzel. 1999. Do stabilimenta in orb webs attract prey or defend spiders? *Behavioral Ecology* 10:372–376.

115 Blest, A. D. 1957. The evolution of protective displays in the Saturnoidea and Sphingidae (Lepidoptera). *Behaviour* 11:257–309.

116 Blest, A. D. 1957. The function of eye-spot patterns in the Lepidoptera. *Behaviour* 11:257–309.

117 Boake, C. R. B., and A. Hoikkala. 1995. Courtship behaviour and mating success of wild-caught *Drosophila silvestris* males. *Animal Behaviour* 49:1303–1313.

118 Boggess, J. 1984. Infant killing and male reproductive strategies in langurs (*Presbytis entellus*). In *Infanticide: Comparative and Evolutionary Perspectives,* G. Hausfater and S. B. Hrdy (eds.). Aldine, Chicago.

119 Bolhuis, J. J., G. G. O. Zijlstra, A. M. den Boer-Visser, and E. A. Van der Zee. 2000. Localized neuronal activation in the zebra finch brain is related to the strength of song learning. *Proceedings of the National Academy of Sciences* 97:2282–2285.

120 Bolles, R. C. 1973. The comparative psychology of learning: The selective association principle and some problems with "general" laws of learning. In *Perspectives in Animal Behavior,* G. Bermant (ed.). Scott, Foresman & Company, Glenview, IL.

121 Boone, J. L. III. 1986. Parental investment and elite family structure in preindustrial states: A case study of late medieval-early modern Portuguese genealogies. *American Anthropologist* 88:859–878.

122 Boone, J. L. III, and K. L. Kessler. 1999. More status or more children? Social status, fertility reduction, and long-term fitness. *Evolution and Human Behavior* 20:257–277.

123 Boppré, M. 1986. Insects pharmacophagously using defensive plant-chemicals (pyrrolizidine alkaloids). *Naturwissenschaften* 73:17–26.

124 Borgerhoff Mulder, M. 1987. Resources and reproductive success in women with an example from the Kipsigis of Kenya. *Journal of Zoology* 213:489–505.

125 Borgerhoff Mulder, M. 1988. Reproductive consequences of sex-biased inheritance. In *Comparative Socioecology of Mammals and Man,* V. Standen and R. Foley (eds.). Blackwell, London.

126 Borgerhoff Mulder, M. 1998. The demographic transition: Are we any closer to an evolutionary explanation? *Trends in Ecology and Evolution* 13:266–269.

127 Borgia, G. 1985. Bower quality, number of decorations and mating success of male satin bowerbirds (*Ptilonorhynchus violaceus*). *Animal Behaviour* 33:266–271.

128 Borgia, G. 1986. Sexual selection in bowerbirds. *Scientific American* 254(June):92–100.

129 Borgia, G. 1995. Why do bowerbirds build bowers? *American Scientist* 83:542–547.

130 Borg-Karlson, A.-K. 1990. Chemical and ethological studies of pollination in the genus *Ophrys* (Orchidaceae). *Phytochemistry* 29:1359–1387.

131 Borries, C. 1997. Infanticide in seasonally breeding multimale groups of Hanuman langurs (*Presbytis entellus*) in Ramnagar (South Nepal). *Behavioral Ecology and Sociobiology* 41:139–150.

132 Borries, C., K. Launhardt, C. Epplen, J. T. Epplen, and P. Winkler. 1999. DNA analyses support the hypothesis that infanticide is adaptive in langur monkeys. *Proceedings of the Royal Society of London B* 266:901–904.

133 Bouchard, T. J., Jr. 1994. Genes, environment, and personality. *Science* 264:1700–1701.

134 Bouchard, T. J., Jr. 1997. IQ similarity in twins reared apart: Findings and responses to critics. In *Intelligence: Heredity and Environment,* R. Sternberg and E. Grigorenko (eds.). Cambridge University Press, New York.

135 Bouchard, T. J., Jr. 1998. Genetic and environmental influences on adult intelligence and special mental abilities. *Human Biology* 70:257–279.

136 Bouchard, T. J., Jr. , D. T. Lykken, M. McGue, N. L. Segal, and A. Tellegen. 1990. Sources of human psychological differences: The Minnesota study of twins reared apart. *Science* 250:223–228.

137 Bouchard, T. J., Jr. , and M. McGue. 1981. Familial studies of intelligence: A review. *Science* 212:1055–1059.

138 Boughman, J. W. 1998. Vocal learning by greater spear-nosed bats. *Proceedings of the Royal Society of London B* 265:227–233.

139 Braaten, R. F., and K. Reynolds. 1999. Auditory preference for conspecific song in isolation-reared zebra finches. *Animal Behaviour* 58:105–111.

140 Bradbury, J. W. 1977. Lek mating behavior in the hammer-headed bat. *Zeitschrift für Tierpsychologie* 45:225–255.

141 Bradbury, J. W. 1981. The evolution of leks. In *Natural Selection and Social Behavior,* R. D. Alexander and D. W. Tinkle (eds.). Chiron Press, New York.

142 Bradbury, J. W., and R. M. Gibson. 1983. Leks and mate choice. In *Mate Choice,* P. Bateson (ed.). Cambridge University Press, Cambridge.

143 Bradbury, J. W., and S. L. Vehrencamp. 1998. *Principles of Animal Communication.* Sinauer Associates, Sunderland, MA.

144 Bradbury, J. W., S. L. Vehrencamp, and R. M. Gibson. 1989. Dispersion of displaying male sage grouse. I. Patterns of temporal variation. *Behavioral Ecology and Sociobiology* 24:1–14.

145 Braude, S. 2000. Dispersal and new colony formation in wild naked mole-rats: Evidence against inbreeding as the system of mating. *Behavioral Ecology* 11:7–12.

146   Breininger, D. R., V. L. Larson, B. W. Duncan, R. B. Smith, D. M. Oddy, and M. F. Goodchild. 1995. Landscape patterns of Florida scrub jay habitat use and demographic success. *Conservation Biology* 9:1442–1453.

147   Bremmer, F. 1986. White whales on holiday. *Natural History* 95(Jan):40–49.

148   Brenowitz, E. A. 1991. Evolution of the vocal control system in the avian brain. *Seminars in the Neurosciences* 3:399–407.

149   Brenowitz, E. A. 1994. Flexibility and constraint in the evolution of animal communication. In *Flexibility and Constraint in Behavioral Systems*, R. J. Greenspan and C. P. Kyriacou (eds.). John Wiley & Sons, New York.

150   Brenowitz, E. A. 1997. Comparative approaches to the avian song system. *Journal of Neurobiology* 33:517–531.

151   Brenowitz, E. A., L. F. Baptista, K. Lent, and J. C. Wingfield. 1998. Seasonal plasticity of the song control system in wild Nuttall's white-crowned sparrows. *Journal of Neurobiology* 34:69–82.

152   Brenowitz, E. A., D. Margoliash, and K. W. Nordeen. 1997. An introduction to birdsong and the avian song system. *Journal of Neurobiology* 33:495–500.

153   Breven, K. A. 1981. Mate choice in the wood frog, *Rana sylvatica*. *Evolution* 35:707–722.

154   Bridges, R. S. 1998. The genetics of motherhood. *Nature Genetics* 20:108–109.

155   Briskie, J. V., P. R. Martin, and T. E. Martin. 1999. Nest predation and the evolution of nestling begging calls. *Proceedings of the Royal Society of London B* 266:2153–2159.

156   Briskie, J. V., C. Naugler, and S. M. Leech. 1994. Begging intensity of nestling birds varies with sibling relatedness. *Proceedings of the Royal Society of London B* 258:73–78.

157   Brockmann, H. J., T. Colson, and W. Potts. 1994. Sperm competition in horseshoe crabs (*Limulus polyphemus*). *Behavioral Ecology and Sociobiology* 35:153–160.

158   Brockmann, H. J., C. Nguyen, and W. Potts. 2000. Paternity in horseshoe crabs when spawning in multiple-male groups. *Animal Behaviour* 60:837–849.

159   Brockmann, H. J., and D. Penn. 1992. Male mating tactics in the horseshoe crab, *Limulus polyphemus*. *Animal Behaviour* 44:653–665.

160   Brotherton, P. N. M., and M. B. Manser. 1997. Female dispersion and the evolution of monogamy in the dik-dik. *Animal Behaviour* 54:1413–1424.

161   Brower, J. V. Z. 1958. Experimental studies of mimicry in some North American butterflies. 1. The monarch, *Danaus plexippus*, and viceroy, *Limenitis archippus*. *Evolution* 12:3–47.

162   Brower, L. P. 1996. Monarch butterfly orientation: Missing pieces of a magnificent puzzle. *Journal of Experimental Biology* 199:93–103.

163   Brower, L. P., and W. H. Calvert. 1984. Chemical defence in butterflies. In *The Biology of Butterflies*, R. I. Vane-Wright and P. R. Ackery (eds.). Academic Press, London.

164   Brown, C. R., and M. B. Brown. 1986. Ecto-parasitism as a cost of coloniality in cliff swallows (*Hirundo pyrrhonota*). *Ecology* 67:1206–1218.

165   Brown, C. R., and M. B. Brown. 1996. *Coloniality in the Cliff Swallow: The Effect of Group Size on Social Behavior*. University of Chicago Press, Chicago, IL.

166   Brown, C. R., and M. B. Brown. 2000. Heritable basis for choice of group size in a colonial bird. *Proceedings of the National Academy of Sciences* 97:14825–14830.

167   Brown, J. L. 1975. *The Evolution of Behavior*. W. W. Norton, New York.

168   Brown, J. L. 1987. *Helping and Communal Breeding in Birds: Ecology and Evolution*. Princeton University Press, Princeton, NJ.

169   Brown, J. L. 1997. A theory of mate choice based on heterozygosity. *Behavioral Ecology* 8:60–65.

170   Brown, J. L., and A. Eklund. 1994. Kin recognition and the major histocompatibility complex: An integrative review. *American Naturalist* 143:435–461.

171   Brown, J. L., and G. H. Orians. 1970. Spacing patterns in mobile animals. *Annual Review of Ecology and Systematics* 1:239–262.

172   Brown, J. L., and C. M. Vleck. 1998. Prolactin and helping in birds: Has natural selection strengthened helping behavior? *Behavioral Ecology* 9:541–545.

173   Brown, J. R., H. Ye, R. T. Bronson, P. Dikkes, and M. E. Greenberg. 1996. A defect in nurturing in mice lacking the immediate early gene *fosB*. *Cell* 86:297–309.

174   Brown, K. M. 1998. Proximate and ultimate causes of adoption in ring-billed gulls. *Animal Behaviour* 56:1529–1543.

175   Brownmiller, S. 1975. *Against Our Will: Men, Women and Rape*. Simon and Schuster, New York.

176   Brownmiller, S., and B. Merhof. 1992. A feminist response to rape as an adaptation in men. *Brain and Behavioral Sciences* 15:381–382.

177   Buchanan, K. L., and C. K. Catchpole. 2000. Song as an indicator of male parental effort in the sedge warbler. *Proceedings of the Royal Society of London B* 267:321–326.

178   Bumann, D., J. Krause, and D. Rubenstein. 1997. Mortality risk of spatial positions in animal groups: The danger of being in the front. *Behaviour* 134:1063–1074.

179   Burda, H., R. L. Honeycutt, S. Begall, O. Locker-Grutjen, and A. Scharff. 2000. Are naked and common mole-rats eusocial and if so, why? *Behavioral Ecology and Sociobiology* 47:293–303.

180   Burke, T. 1989. DNA fingerprinting and other methods for the study of mating success. *Trends in Ecology and Evolution* 4:139–144.

181   Burley, N. T., and R. Symanski. 1998. "A taste for the beautiful": Latent aesthetic mate preferences for white crests in two species of Australian grassfinches. *American Naturalist* 152:792–802.

182   Buskirk, R. E., C. Frolich, and K. G. Ross. 1984. The natural selection of sexual cannibalism. *American Naturalist* 123:612–625.

183   Buss, D. M. 1989. Sex differences in human mate preferences: Evolutionary hypothesis tested in 37 cultures. *Behavioral and Brain Sciences* 12:1–149.

184   Buss, D. M. 1999. *Evolutionary Psychology, The New Science of the Mind*. Allyn and Bacon, Boston, MA.

185   Buss, D. M., R. J. Larsen, D. Westen, and J. Semmelroth. 1992. Sex differences in jealousy: Evolution, physiology, and psychology. *Psychological Science* 3:251–255.

186   Buss, D. M., and D. P. Schmitt. 1993. Sexual strategies theory: An evolutionary perspective on human mating. *Psychological Review* 100:204–232.

500 Bibliography

187 Butchart, S. H. M., N. Seddon, and J. M. M. Ekstrom. 1999. Yelling for sex: Harem males compete for female access in bronze-winged jacanas. *Animal Behaviour* 57:637–646.

188 Cade, W. 1980. Alternative male reproductive strategies. *Florida Entomologist* 63:30–45.

189 Cade, W. 1981. Alternative male strategies: Genetic differences in crickets. *Science* 212:563–564.

190 Calvert, W. H., and L. P. Brower. 1986. The location of monarch butterfly (*Danaus plexippus* L.) overwintering colonies in Mexico in relation to topography and climate. *Journal of the Lepidopterists' Society* 40:164–187.

191 Camhi, J. M. 1984. *Neuroethology*. Sinauer Associates, Sunderland, MA.

192 Cantoni, D., and R. Brown. 1997. Paternal investment and reproductive success in the California mouse, *Peromyscus californicus*. *Animal Behaviour* 54:377–386.

193 Caraco, T., and L. L. Wolf. 1975. Ecological determinants of group sizes in foraging lions. *American Naturalist* 109:343–352.

194 Carew, T. J. 2000. *Behavioral Neurobiology: The Cellular Organization of Behavior*. Sinauer Associates, Sunderland, MA.

195 Carey, M., and V. Nolan, Jr. 1975. Polygyny in indigo buntings: A hypothesis tested. *Science* 190:1296–1297.

196 Carey, S. 1992. Becoming a face expert. *Philosophical Transactions of the Royal Society of London B* 335:95–103.

197 Carlsson, B.-G. 1991. Recruitment of mates and deceptive behavior by male Tengmalm's owls. *Behavioral Ecology and Sociobiology* 28:321–328.

198 Caro, T. M. 1986. The functions of stotting: A review of the hypotheses. *Animal Behaviour* 34:649–662.

199 Caro, T. M. 1986. The functions of stotting in Thomson's gazelles: Some tests of the predictions. *Animal Behaviour* 34:663–684.

200 Caro, T. M. 1995. Pursuit-deterrence revisited. *Trends in Ecology and Evolution* 10:500–503.

201 Carr, A. 1967. Adaptive aspects of the scheduled travel of *Chelonia*. In *Animal Orientation and Navigation*, R.M. Storm (ed.). Oregon State University Press, Corvallis, OR.

202 Carter, R. 1998. *Mapping the Brain*. Weidenfeld and Nicolson, London.

203 Case, A., I.-F. Lin, and S. McLanahan. 2000. How hungry is the selfish gene? *Economic Journal* 110:781–804.

204 Case, A., I.-F. Lin, and S. McLanahan. 2001. Educational attainment of siblings in step families. Evolution and Human Behavior. In Press.

205 Catania, K. C. 2000. Cortical organization in insectivora: The parallel evolution of the sensory periphery and the brain. *Brain Behavior and Evolution* 55:311–321.

206 Catania, K. C., and J. H. Kaas. 1996. The unusual nose and brain of the star-nosed mole. *BioScience* 46:578–586.

207 Catania, K. C., and J. H. Kaas. 1997. Somatosensory fovea in the star-nosed mole: Behavioral use of the star in relation to innervation patterns and cortical representation. *Journal of Comparative Neurology* 387:215–233.

208 Catchpole, C. K., and P. J. B. Slater. 1995. *Bird Song, Biological Themes and Variations*. Cambridge University Press, Cambridge.

209 Cezilly, F., and R. G. Nager. 1995. Comparative evidence for a positive association between divorce and extra-pair paternity in birds. *Proceedings of the Royal Society of London B* 262:7–12.

210 Chapman, T., and B. Crespi. 1998. High relatedness and inbreeding in two species of haplodiploid eusocial thrips (Insecta: Thysanoptera) revealed by microsatellite analysis. *Behavioral Ecology and Sociobiology* 43:301–306.

211 Chen, J.-S., and A. Amsel. 1980. Recall (versus recognition) of taste and immunization against aversive taste anticipations based on illness. *Science* 209:831–833.

212 Chilton, G., M. R. Lein, and L. F. Baptista. 1990. Mate choice by female white-crowned sparrows in a mixed-dialect population. *Behavioral Ecology and Sociobiology* 27:223–227.

213 Chivers, D. P., G. E. Brown, and R. J. F. Smith. 1996. The evolution of chemical alarm signals: Attracting predators benefits alarm signal givers. *American Naturalist* 148:649–659.

214 Chivers, D. P., M. H. Puttlitz, and A. R. Blaustein. 2000. Chemical alarm signaling by reticulate sculpins, *Cottus perplexus*. *Environmental Biology of Fishes* 57:347–352.

215 Christy, J. H. 1995. Mimicry, mate choice, and the sensory trap hypothesis. *American Naturalist* 146:171–181.

216 Church, S. C., A. T. D. Bennett, I. C. Cuthill, and J. C. Partridge. 1998. Ultraviolet cues affect the foraging behaviour of blue tits. *Proceedings of the Royal Society of London B* 265:1509–1514.

217 Cioffi, D., and R. Garner. 1998. The effect of response options on decisions and subsequent behavior: Sometimes inaction is better. *Personality and Social Psychology Bulletin* 24:463–472.

218 Ciszek, D. 2000. New colony formation in the "highly inbred" eusocial naked mole-rat: Outbreeding is preferred. *Behavioral Ecology* 11:1–6.

219 Clark, D. L., and B. G. Galef. 1995. Prenatal influences on reproductive life history strategies. *Trends in Ecology and Evolution* 10:151–153.

220 Clark, M. M., and B. G. Galef, Jr. 2000. Why some male Mongolian gerbils may help at the nest: Testosterone, asexuality and alloparenting. *Animal Behaviour* 59:801–806.

221 Clark, R. D., and E. Hatfield. 1989. Gender differences in receptivity to sexual offers. *Journal of Psychology and Human Sexuality* 2:39–55.

222 Clarke, F. M., and C. G. Faulkes. 1999. Kin discrimination and female mate choice in the naked mole-rat *Heterocephalus glaber*. *Proceedings of the Royal Society of London B* 266:1995–2002.

223 Clayton, D. F. 1997. Role of gene regulation in song circuit development and song learning. *Journal of Neurobiology* 33:549–571.

224 Clayton, N. S. 1998. Memory and the hippocampus in food-storing birds: A comparative approach. *Neuropharmacology* 37:441–452.

225 Clayton, N. S., and J. R. Krebs. 1994. Hippocampal growth and attrition in birds affected by experience. *Proceedings of the National Academy of Sciences* 91:7410–7414.

226 Clayton, N. S., J. C. Reboreda, and A. Kacelnik. 1997. Seasonal changes of hippocampus volume in parasitic cowbirds. *Behavioral Processes* 41:237–243.

227 Clotfelter, E. D. 1998. What cues do brown-headed cowbirds use to locate red-winged blackbird host nests? *Animal Behaviour* 55:1181–1189.

228 Clutton-Brock, T. H. 1991. *The Evolution of Parental Care*. Princeton University Press, Princeton, NJ.

229 Clutton-Brock, T. H. 1998. Reproductive skew, concessions and limited control. *Behavioral Ecology and Sociobiology* 13:288–292.

230 Clutton-Brock, T. H., and S. D. Albon. 1979. The roaring of red deer and the evolution of honest advertisement. *Behaviour* 69:145–170.

231 Clutton-Brock, T. H., S. D. Albon, R. M. Gibson, and F. E. Guinness. 1979. The logical stag: Adaptive aspects of fighting in red deer. *Animal Behaviour* 27:211–225.

232 Clutton-Brock, T. H., and P. Harvey. 1984. Comparative approaches to investigating adaptation. In *Behavioural Ecology: An Evolutionary Approach* (2nd edition), J. R. Krebs and N. B. Davies (eds.). Blackwell, Oxford.

233 Clutton-Brock, T. H., M. Hiraiwa-Hasegawa, and A. Robertson. 1989. Mate choice on fallow deer leks. *Nature* 340:463–465.

234 Clutton-Brock, T. H., M. J. O'Riain, P. N. M. Brotherton, D. Gaynor, R. Kansky, A. S. Griffin, and M. Manser. 1999. Selfish sentinels in cooperative mammals. *Science* 284:1640–1644.

235 Clutton-Brock, T. H., and G. A. Parker. 1992. Potential reproductive rates and the operation of sexual selection. *Quarterly Review of Biology* 67:437–456.

236 Clutton-Brock, T. H., and G. A. Parker. 1995. Sexual coercion in animal societies. *Animal Behaviour* 49:1345–1365.

237 Clutton-Brock, T. H., and A. C. J. Vincent. 1991. Sexual selection and the potential reproductive rates of males and females. *Nature* 351:58–60.

238 Collins, J. P., and J. E. Cheek. 1983. Effect of food and density on development of typical and cannibalistic salamander larvae in *Ambystoma tigrinum nebulosum*. *American Zoologist* 23:77–84.

239 Collins, S. A. 2000. Is female preference for male repertoires due to sensory bias? *Proceedings of the Royal Society of London B* 266:2309–2314.

240 Collins, S. A., C. Hubbard, and A. M. Houtman. 1994. Female mate choice in the zebra finch—the effect of male beak color and male song. *Behavioral Ecology and Sociobiology* 35:21–26.

241 Conover, M. R. 1994. Stimuli eliciting distress calls in adult passerines and response of predators and birds to their broadcast. *Behaviour* 131:19–37.

242 Cooney, R., and A. Cockburn. 1995. Territorial defence is the major function of female song in the superb fairy-wren, *Malurus cyaneus*. *Animal Behaviour* 49:1635–1647.

243 Cooper, W. E. 1995. Foraging mode, prey chemical discrimination, and phylogeny in lizards. *Animal Behaviour* 50:973–985.

244 Cooper, W. E. 2000. An adaptive difference in the relationship between foraging mode and responses to prey chemicals in two congeneric scincid lizards. *Ethology* 106:193–206.

245 Coss, R. G., and R. O. Goldthwaite. 1995. The persistence of old designs for perception. *Perspectives in Ethology* 11:83–148.

246 Court, G. S. 1996. The seal's own skin game. *Natural History* 105(8):36–41.

247 Cowlishaw, G., and R. I. M. Dunbar. 1991. Dominance rank and mating success in male primates. *Animal Behaviour* 41:1045–1056.

248 Cowlishaw, G., and R. Mace. 1996. Cross-cultural patterns of marriage and inheritance: A phylogenetic approach. *Ethology and Sociobiology* 17:97–98.

249 Cox, G. W. 1985. The evolution of avian migration systems between temperate and tropical regions of the New World. *American Naturalist* 126:452–474.

250 Coyne, J. 1998. Not black and white. *Nature* 396:35–36.

251 Craig, C. L., and G. D. Bernard. 1990. Insect attraction to ultraviolet-reflecting spider webs and web decorations. *Ecology* 71:616–623.

252 Craig, C. L., R. S. Weber, and G. D. Bernard. 1996. Evolution of predator-prey systems: Spider foraging plasticity in response to the visual ecology of prey. *American Naturalist* 147:205–229.

253 Craig, P. 1996. Intertidal territoriality and time-budget of the surgeonfish, *Acanthurus lineatus*, in American Samoa. *Environmental Biology of Fishes* 46:27–36.

254 Creel, S. 1997. Cooperative hunting and group size: Assumptions and currencies. *Animal Behaviour* 54:1319–1324.

255 Creel, S., and N. M. Creel. 1995. Communal hunting and pack size in African wild dogs, *Lycaon pictus*. *Animal Behaviour* 50:1325–1339.

256 Creel, S., N. M. Creel, M. G. L. Mills, and S. L. Monfort. 1997. Rank and reproduction in cooperatively breeding African wild dogs: Behavioral and endocrine correlates. *Behavioral Ecology* 8:298–306.

257 Crespi, B. J. 2000. The evolution of maladaptation. *Heredity* 84:623–629.

258 Cresswell, W. 1994. Flocking is an effective anti-predation strategy in redshanks, *Tringa totanus*. *Animal Behaviour* 47:433–442.

259 Crews, D. 1975. Psychobiology of reptilian reproduction. *Science* 189:1059–1065.

260 Crews, D. 1984. Gamete production, sex hormone secretion, and mating behavior uncoupled. *Hormones and Behavior* 18:22–28.

261 Crews, D. (ed.). 1987. *Psychobiology of Reproductive Behavior: An Evolutionary Perspective*. Prentice-Hall, Englewood Cliffs, NJ.

262 Crews, D. 1991. Trans-seasonal action of androgen in the control of spring courtship behavior in male red-sided garter snakes. *Proceedings of the National Academy of Sciences* 88:3545–3548.

263 Crews, D. 1992. Behavioral endocrinology and reproduction: An evolutionary perspective. *Oxford Reviews of Reproductive Biology* 14:303–370.

264 Crews, D. 1997. Species diversity and the evolution of behavioral controlling mechanisms. *Annals of the New York Academy of Sciences* 807:1–21.

265 Crews, D., and N. Greenberg. 1981. Function and causation of social signals in lizards. *American Zoologist* 21:273–294.

266 Crews, D., V. Hingorani, and R. J. Nelson. 1988. Role of the pineal gland in the control of annual reproductive behavioral and physiological cycles in the red-sided garter snake (*Thamnophis sirtalis parietalis*). *Journal of Biological Rhythms* 3:293–302.

267 Crews, D., and M. C. Moore. 1986. Evolution of mechanisms controlling mating behavior. *Science* 231:121–125.

268 Cristol, D. A., and P. V. Switzer. 1999. Avian prey-dropping behavior. II. American crows and walnuts. *Behavioral Ecology* 10:220–226.

269   Cronk, L. 1993. Parental favoritism toward daughters. *American Scientist* 81:272–279.

270   Cullen, E. 1957. Adaptations in the kittiwake to cliff nesting. *Ibis* 99:275–302.

271   Cumming, J. M. 1994. Sexual selection and the evolution of dance fly mating systems (Diptera: Empididae; Empidinae). *Canadian Entomologist* 126:907–920.

272   Cunningham, E. J. A., and A. F. Russell. 2000. Egg investment is influenced by male attractiveness in the mallard. *Nature* 404:74–77.

273   Cunningham, M. R., A. R. Roberts, C. H. Wu, A. P. Barbee, and P. B. Druen. 1995. Their ideas of beauty are, on the whole, the same as ours: Consistency and variability in the cross-cultural perception of female physical attractiveness. *Journal of Personality and Social Psychology* 68:261–279.

274   Curio, E. 1978. The adaptive significance of avian mobbing. I. Teleonomic hypotheses and predictions. *Zeitschrift für Tierpsychologie* 48:175–183.

275   Currie, C. R., J. A. Scott, R. C. Summerbell, and D. Malloch. 1999. Fungus-growing ants use antibiotic-producing bacteria to control garden parasites. *Nature* 398:701–704.

276   Dagg, A. I. 1998. Infanticide by male lions hypothesis: A fallacy influencing research into human behavior. *American Anthropologist* 100:940–950.

277   Dale, J., R. Montgomerie, D. Michaud, and P. Boag. 1999. Frequency and timing of extrapair fertilisation in the polyandrous red phalarope (*Phalaropus fulicarius*). *Behavioral Ecology and Sociobiology* 46:50–56.

278   Dale, S., H. Rinden, and T. Slagsvold. 1992. Competition for a mate restricts mate search of female pied flycatchers. *Behavioral Ecology and Sociobiology* 30:165–176.

279   Dale, S., and T. Slagsvold. 1994. Polygyny and deception in the pied flycatcher: Can females determine male mating status? *Animal Behaviour* 48:1207–1217.

280   Daly, M., and M. Wilson. 1983. *Sex, Evolution and Behavior* (2nd edition). Willard Grant Press, Boston.

281   Daly, M., and M. Wilson. 1985. Child abuse and other risks of not living with both parents. *Ethology and Sociobiology* 6:197–210.

282   Daly, M., and M. Wilson. 1988. *Homicide*. Aldine de Gruyter, Hawthorne, NY.

283   Daly, M., and M. Wilson. 1992. The man who mistook his wife for a chattel. In *The Adapted Mind*, J. H. Barkow, L. Cosmides, and J. Tooby (eds.). Oxford University Press, New York.

284   Daly, M., and M. Wilson. 1998. *The Truth about Cinderella*. Yale University Press, New Haven, CT.

285   Daly, M., M. Wilson, and S. J. Weghorst. 1982. Male sexual jealousy. *Ethology and Sociobiology* 3:11–27.

286   Darwin, C. 1859. *On the Origin of Species*. Murray, London.

287   Darwin, C. 1871. *The Descent of Man and Selection in Relation to Sex*. Murray, London.

288   Davies, N. B. 1977. Prey selection and social behaviour in wagtails (Aves: Motacillidae). *Journal of Animal Ecology* 46:37–57.

289   Davies, N. B. 1978. Territorial defence in the speckled wood butterfly (*Pararge aegeria*): The resident always wins. *Animal Behaviour* 26:138–147.

290   Davies, N. B. 1983. Polyandry, cloaca-pecking and sperm competition in dunnocks. *Nature* 302:334–336.

291   Davies, N. B. 1985. Cooperation and conflict among dunnocks, *Prunella modularis*, in a variable mating system. *Animal Behaviour* 33:628–648.

292   Davies, N. B. 1992. *Dunnock Behaviour and Social Evolution*. Oxford University Press, Oxford.

293   Davies, N. B. 2000. *Cuckoos, Cowbirds and Other Cheats*. T & A D Poyser, London.

294   Davies, N. B., and M. de L. Brooke. 1988. Cuckoos versus reed warblers: Adaptations and counteradaptations. *Animal Behaviour* 36:262–284.

295   Davies, N. B., and T. R. Halliday. 1978. Deep croaks and fighting assessment in toads *Bufo bufo*. *Nature* 275:683–685.

296   Davies, N. B., I. R. Hartley, B. J. Hatchwell, and N. E. Langmore. 1996. Female control of copulations to maximize male help: A comparison of polygynandrous alpine accentors, *Prunella collaris*, and dunnocks, *P. modularis*. *Animal Behaviour* 51:27–47.

297   Davies, N. B., and A. I. Houston. 1981. Owners and satellites: The economics of territory defense in the pied wagtail, *Motacilla alba*. *Journal of Animal Ecology* 50:157–180.

298   Davies, N. B., R. M. Kilner, and D. G. Noble. 1998. Nestling cuckoos *Cuculus canorus* exploit hosts with begging calls that mimic a brood. *Proceedings of the Royal Society of London B* 265:673–678.

299   Davies, N. B., and A. Lundberg. 1984. Food distribution and a variable mating system in the dunnock, *Prunella modularis*. *Journal of Animal Ecology* 53:895–912.

300   Davis-Walton, J., and P. W. Sherman. 1994. Sleep arrhythmia in the eusocial naked mole-rat. *Naturwissenschaften* 81:272–275.

301   Dawkins, R. 1977. *The Selfish Gene*. Oxford University Press, New York.

302   Dawkins, R. 1980. Good strategy or evolutionarily stable strategy? In *Sociobiology: Beyond Nature/Nurture?* G. W. Barlow and J. Silverberg (eds.). Westview Press, Boulder, CO.

303   Dawkins, R. 1982. *The Extended Phenotype*. W. H. Freeman, San Francisco.

304   Dawkins, R. 1986. *The Blind Watchmaker*. W. W. Norton, New York.

305   Dawkins, R., and J. Krebs. 1978. Animal signals: Information or manipulation? In *Behavioural Ecology: An Evolutionary Approach* (1st edition), J. R. Krebs and N. B. Davies (eds.). Blackwell, Oxford.

306   de Belle, J. S., A. J. Hilliker, and M. B. Sokolowski. 1989. Genetic localization of *foraging* (*for*): A major gene for larval behavior in *Drosophila melanogaster*. *Genetics* 123:157–163.

307   de Belle, J. S., and M. B. Sokolowski. 1987. Heredity of *rover/sitter*: Alternative foraging strategies of *Drosophila melanogaster* larvae. *Heredity* 59:73–83.

308   de Bono, M., and C. I. Bargmann. 1998. Natural variation in a neuropeptide Y receptor homolog modifies social behavior and food response in *C. elegans*. *Cell* 94:679–689.

309   de Renzi, E., and G. di Pellegrino. 1998. Prosopagnosia and alexia without object agnosia. *Cortex* 34:403–415.

310   DeCoursey, P. J., and J. Buggy. 1989. Circadian rhythmicity after neural transplant to hamster third ventricle: Specificity of suprachiasmatic nuclei. *Brain Research* 500:263–275.

311 DeHeer, C. J., M. A. D. Goodisman, and K. G. Ross. 1999. Queen dispersal strategies in the multiple-queen form of the fire ant *Solenopsis invicta*. *American Naturalist* 153:660–675.

312 Dennett, D. C. 1995. *Darwin's Dangerous Idea*. Simon & Schuster, New York.

313 Dethier, V. G. 1962. *To Know a Fly*. Holden-Day, San Francisco.

314 Dethier, V. G. 1976. *The Hungry Fly: A Physiological Study of the Behavior Associated with Feeding*. Harvard University Press, Cambridge, MA.

315 Deutsch, C. J., M. P. Haley, and B. J. Le Boeuf. 1990. Reproductive effort of male northern elephant seals: Estimates from mass loss. *Canadian Journal of Zoology* 68:2580–2593.

316 Deutsch, J. C. 1994. Uganda kob mating success does not increase on larger leks. *Behavioral Ecology and Sociobiology* 34:451–459.

317 DeVoogd, T. J. 1991. Endocrine modulation of the development and adult function of the avian song system. *Psychoneuroendocrinology* 16:41–66.

318 DeWolfe, B. B., L. F. Baptista, and L. Petrinovich. 1989. Song development and territory establishment in Nuttall's white-crowned sparrow. *Condor* 91:297–407.

319 DeWoody, J. A., D. E. Fletcher, M. Mackiewicz, S. D. Wilkins, and J. C. Avise. 2000. The genetic mating system of spotted sunfish (*Lepomis punctatus*): Mate numbers and the influence of male reproductive parasites. *Molecular Ecology* 9:2119–2128.

320 Dewsbury, D. A. 1982. Dominance rank, copulatory behavior, and differential reproduction. *Quarterly Review of Biology* 57:135–158.

321 Dhondt, A. A., and J. Schillemans. 1983. Reproductive success of the great tit in relation to its territorial status. *Animal Behaviour* 31:902–912.

322 Diamond, J. M. 1992. *The Third Chimpanzee*. HarperCollins Publishers, New York.

323 Diamond, J. M. 2000. Talk of cannibalism. *Nature* 407:25–26.

324 Dias, P. C. 1996. Sources and sinks in population biology. *Trends in Ecology and Evolution* 11:326–330.

325 Dickinson, J. L. 1995. Trade-offs between postcopulatory riding and mate location in the blue milkweed beetle. *Behavioral Ecology* 6:280–286.

326 Dickinson, J. L., and R. L. Rutowski. 1989. The function of the mating plug in the chalcedon checkerspot butterfly. *Animal Behaviour* 38:154–162.

327 Dickinson, J. L., and W. W. Weathers. 1999. Replacement males in the western bluebird: Opportunity for paternity, chick-feeding rules, and fitness consequences of male parental care. *Behavioral Ecology and Sociobiology* 45:201–209.

328 Dingle, H. 1996. *Migration: The Biology of Life on the Move*. Oxford University Press, New York.

329 Dixson, A. F. 1987. Observations on the evolution of genitalia and copulatory behavior in primates. *Journal of Zoology* 213:423–443.

330 Dodson, G. N. 1997. Resource defense mating system in antlered flies, *Phytalmia* spp. (Diptera: Tephritidae). *Annals of the Entomological Society of America* 90:496–504.

331 Dodson, G. N., and M. W. Beck. 1993. Pre-copulatory guarding of penultimate females by male crab spiders, *Misumenoides formosipes*. *Animal Behaviour* 46:951–959.

332 Double, M., and A. Cockburn. 2000. Pre-dawn infidelity: Females control extra-pair mating in superb fairy-wrens. *Proceedings of the Royal Society of London B* 267:465–470.

333 Doupe, A. J., and P. K. Kuhl. 1999. Birdsong and human speech: Common themes and mechanisms. *Annual Review of Neuroscience* 22:567–631.

334 Doupe, A. J., and M. M. Solis. 1997. Song- and order-selective neurons develop in the songbird anterior forebrain during vocal learning. *Journal of Neurobiology* 33:694–709.

335 Dover, G. A. 2000. *Dear Mr. Darwin: Letters on the Evolution of Life and Human Nature*. University of California Press, Berkeley, CA.

336 Downhower, J. F., and K. B. Armitage. 1971. The yellow-bellied marmot and the evolution of polygamy. *American Naturalist* 105:355–370.

337 Drea, C. M., M. L. Weldele, N. G. Forger, E. M. Coscia, L. G. Frank, P. Licht, and S. E. Glickman. 1998. Androgens and masculinization of genitalia in the spotted hyaena (*Crocuta crocuta*). 2. Effects of prenatal anti-androgens. *Journal of Reproduction and Fertility* 113:117–127.

338 Drews, C. 1996. Contests and patterns of injuries in free-ranging male baboons (*Papio cynocephalus*). *Behaviour* 133:443–474.

339 Dudley, R. 2000. Evolutionary origins of human alcoholism in primate frugivory. *Quarterly Review of Biology* 75:3–15.

340 Dufour, K. W., and P. J. Weatherhead. 1998. Bilateral symmetry as an indicator of male quality in red-winged blackbirds: Associations with measures of health, viability, and parental effort. *Behavioral Ecology* 9:220–231.

341 Eadie, J. M., and B. E. Lyon. 1998. Cooperation, conflict, and crèching behavior in goldeneye ducks. *American Naturalist* 151:397–408.

342 East, M. L., and H. Hofer. 1997. The peniform clitoris of female spotted hyaenas. *Trends in Ecology and Evolution* 12:401–402.

343 East, M. L., H. Hofer, and W. Wickler. 1993. The erect "penis" is a flag of submission in a female-dominated society: Greetings in Serengeti spotted hyenas. *Behavioral Ecology and Sociobiology* 33:355–370.

344 Ebensperger, L. A. 1998. Strategies and counterstrategies to infanticide in mammals. *Biological Reviews* 73:321–346.

345 Eberhard, W. G. 1980. The natural history and behavior of the bolas spider *Mastophora dizzydeani* sp. n. (Araneidae). *Psyche* 87:143–169.

346 Eberhard, W. G. 1990. Animal genitalia and female choice. *American Scientist* 78:134–141.

347 Eberhard, W. G. 1993. Evaluating models of sexual selection: Genitalia as a test case. *American Naturalist* 142:564–571.

348 Eberhard, W. G. 1996. *Female Control: Sexual Selection by Cryptic Female Choice*. Princeton University Press, Princeton, NJ.

349 Eberhard, W. G. 1998. Female roles in sperm competition. In *Sperm Competition and Sexual Selection*, T.H. Birkhead and A.P. Møller (eds.). Academic Press, New York.

350 Eens, M., R. Pinxten, and R. F. Verhegen. 1992. Song learning in captive European starlings, *Sturnus vulgaris*. *Animal Behaviour* 44:1131–1141.

351 Eggert, A.-K., and S. K. Sakaluk. 1995. Female-coerced monogamy in burying beetles. *Behavioral Ecology and Sociobiology* 37:147–154.

352 Ehrenreich, B. 2000. How "natural" is rape? *Time* 155:88.

353 Eisner, T., and J. Meinwald. 1995. The chemistry of sexual selection. *Proceedings of the National Academy of Sciences* 92:50–55.

354 Eisner, T., S. R. Smedley, D. K. Young, M. Eisner, B. Roach, and J. Meinwald. 1996. Chemical basis of courtship in a beetle (*Neopyrochroa flabellata*): Cantharidin as precopulatory "enticing" agent. *Proceedings of the National Academy of Sciences* 93:6494–6498.

355 Elgar, M. A. 1986. The establishment of foraging flocks in house sparrows: Risk of predation and daily temperature. *Behavioral Ecology and Sociobiology* 19:433–438.

356 Elliott, P. F. 1975. Longevity and the evolution of polygamy. *American Naturalist* 109:281–287.

357 Ellis, A. W., and A. W. Young. 1996. *Human Cognitive Neuropsychology.* Psychology Press, East Sussex, UK.

358 Ellis, L. 1995. Dominance and reproductive success among nonhuman animals: A cross- species comparison. *Ethology and Sociobiology* 16:257–333.

359 Emlen, S. T. 1978. Cooperative breeding. In *Behavioural Ecology: An Evolutionary Approach* (1st edition), J. R. Krebs and N. B. Davies (eds.). Blackwell, Oxford.

360 Emlen, S. T. 1991. Evolution of cooperative breeding in birds and mammals. In *Behavioural Ecology: An Evolutionary Approach* (3rd edition), J. R. Krebs and N. B. Davies (eds.). Blackwell Scientific, Oxford.

361 Emlen, S. T., N. J. Demong, and D. J. Emlen. 1989. Experimental induction of infanticide in female wattled jacanas. *Auk* 106:1–7.

362 Emlen, S. T., and L. W. Oring. 1977. Ecology, sexual selection and the evolution of mating systems. *Science* 197:215–223.

363 Emlen, S. T., H. K. Reeve, and L. Keller. 1998. Reproductive skew: Disentangling concessions from control. *Trends in Ecology and Evolution* 13:458–459.

364 Emlen, S. T., J. D. Rising, and W. L. Thompson. 1975. A behavioral and morphological study of sympatry in the indigo and lazuli buntings of the great plains. *Wilson Bulletin* 87:145–179.

365 Emlen, S. T., and P. H. Wrege. 1986. Forced copulations and intra-specific parasitism: Two costs of social living in the white-fronted bee-eater. *Ethology* 71:2–29.

366 Emlen, S. T., P. H. Wrege, and N. J. Demong. 1995. Making decisions in the family: An evolutionary perspective. *American Scientist* 83:148–157.

367 Emlen, S. T., P. H. Wrege, and M. S. Webster. 1998. Cuckoldry as a cost of polyandry in the sex-role-reversed wattled jacana, *Jacana jacana. Proceedings of the Royal Society of London B* 265:2359–2364.

368 Endler, J. A. 1991. Interactions between predators and prey. In *Behavioural Ecology: An Evolutionary Approach* (3rd edition), J. R. Krebs and N. B. Davies (eds.). Blackwell Scientific, Oxford.

369 Enquist, M., R. H. Rosenberg, and H. Temrin. 1998. The logic of Ménage á Trois. *Proceedings of the Royal Society of London B* 265:609–613.

370 Epstein, R., R. P. Lanza, and B. F. Skinner. 1980. Symbolic communication between two pigeons (*Columba livia domestica*). *Science* 207:543–545.

371 Evans, H. E. 1966. *Life on a Little Known Planet.* Dell, New York.

372 Evans, H. E. 1973. *Wasp Farm.* Anchor Press, Garden City, NY.

373 Ewer, R. F. 1973. *The Carnivores.* Cornell University Press, Ithaca, NY.

374 Faaborg, J., P. G. Parker, L. DeLay, T. J. de Vries, J. C. Bednarz, S. M. Paz, J. Naranjo, and T. A. Waite. 1995. Confirmation of cooperative polyandry in the Galapagos hawk (*Buteo galapagoensis*). *Behavioral Ecology and Sociobiology* 36:83–90.

375 Fahrbach, S. E., and G. E. Robinson. 1995. Behavioral development in the honey bee: Toward the study of learning under natural conditions. *Learning and Memory* 2:199–224.

376 Falls, J. B. 1988. Does song deter territorial intrusion in white-throated sparrows (*Zonotrichia albicollis*)? *Canadian Journal of Zoology* 66:206–211.

377 Farley, C. T., and C. R. Taylor. 1991. A mechanical trigger for the trot-gallop transition in horses. *Science* 253:306–308.

378 Farner, D. S. 1964. Time measurement in vertebrate photoperiodism. *American Naturalist* 95:375–386.

379 Farner, D. S., and R. A. Lewis. 1971. Photoperiodism and reproductive cycles in birds. *Photophysiology* 6:325–370.

380 Fenstemaker, S. B., L. Zup, L. G. Frank, S. E. Glickman, and N. G. Forger. 1999. A sex difference in the hypothalamus of the spotted hyena. *Nature Neuroscience* 2:943–945.

381 Ferguson, J. N., L. J. Young, E. F. Hearn, M. M. Matzuk, T. R. Insel, and J. T. Winslow. 2000. Social amnesia in mice lacking the oxytocin gene. *Nature Genetics* 25:284–288.

382 Fernald, R. D. 1993. Cichlids in love. *The Sciences* 33:27–31.

383 Field, J., G. Shreeves, S. Sumner, and M. Casiraghi. 2000. Insurance-based advantage to helpers in a tropical hover wasp. *Nature* 404:869–871.

384 Field, J., C. R. Solís, D. C. Queller, and J. E. Strassmann. 1998. Social and genetic structure of paper wasp cofoundress associations: Tests of reproductive skew models. *American Naturalist* 151:545–563.

385 Field, S. A., and M. A. Keller. 1993. Alternative mating tactics and female mimicry as post-copulatory mate-guarding behaviour in the parasitic wasp *Cotesia rubecula. Animal Behaviour* 46:1183–1189.

386 Fisher, J. 1954. Evolution and bird sociality. In *Evolution as a Process*, J. Huxley, A. C. Hardy and E. B. Ford (eds.). Allen & Unwin, London.

387 Fiske, P., P. T. Rintamäki, and E. Karvonen. 1998. Mating success in lekking males: A meta-analysis. *Behavioral Ecology* 9:328–338.

388 Flinn, M. V. 1988. Step-parent/step-offspring interactions in a Caribbean village. *Ethology and Sociobiology* 9:335–369.

389 Foltz, D. W., and P. L. Schwagmeyer. 1989. Sperm competition in the thirteen-lined ground squirrel: Differential fertilization success under field conditions. *American Naturalist* 133:257–265.

390 Ford, E. B. 1955. *Moths.* Collins, London.

391 Forster, L. M. 1992. The stereotyped behaviour of sexual cannibalism in *Latrodectus hasselti* Thorell

(Araneae: Theridiidae), the Australian redback spider. *Australian Journal of Zoology* 40:1–11.

392 Forsyth, A., and J. Alcock. 1990. Female mimicry and resource defense polygyny by males of a tropical rove beetle *Leistotrophus versicolor* (Coleoptera: Staphylinidae). *Behavioral Ecology and Sociobiology* 26:325–330.

393 Foster, M. S. 1977. Odd couples in manakins: A study of social organization and cooperative breeding in *Chiroxiphia linearis. American Naturalist* 111:845–853.

394 Foster, S. A. 1994. Inference of evolutionary pattern: Diversionary displays of three-spined sticklebacks. *Behavioral Ecology* 5:114–121.

395 Foster, S. A., and S. A. Cameron. 1996. Geographic variation in behavior: A phylogenetic framework for comparative studies. In *Phylogenies and the Comparative Method,* E. Martins (ed.). Oxford University Press, New York.

396 Foster, W. A. 1990. Experimental evidence for effective and altruistic colony defence against natural predators by soldiers of the gall-forming aphid *Pemphigus spyrothecae* (Hemiptera: Pemphigidae). *Behavioral Ecology and Sociobiology* 27:421–439.

397 Foster, W. A., and P. K. Rhoden. 1998. Soldiers effectively defend aphid colonies against predators in the field. *Animal Behaviour* 55:761–765.

398 Foulkes, N. S., G. Duval, and P. Sassone-Corsi. 1996. Adaptive inducibility of CREM as transcriptional memory of circadian rhythms. *Nature* 381:83–85.

399 Francis, R. C., K. K. Soma, and R. D. Fernald. 1993. Social regulation of the brain-pituitary-gonadal axis. *Proceedings of the National Academy of Sciences* 90:7794–7798.

400 Frank, L. G. 1986. Social organization of the spotted hyena *Crocuta crocuta.* II. Dominance and reproduction. *Animal Behaviour* 34:1510–1527.

401 Frank, L. G. 1997. Evolution of genital masculinization: Why do female hyaenas have such a large "penis?" *Trends in Ecology and Evolution* 12:58–62.

402 Frank, L. G., S. E. Glickman, and P. Licht. 1991. Fatal sibling aggression, precocial development, and androgens in neonatal spotted hyenas. *Science* 252:702–704.

403 Frank, L. G., H. E. Holekamp, and L. Smale. 1995. Dominance, demographics and reproductive success in female spotted hyenas: A long-term study. In *Serengeti II: Research, Management, and Conservation of an Ecosystem,* A. R. E. Sinclair and P. Arcese (eds.). University of Chicago Press, Chicago.

404 Frank, L. G., M. L. Weldele, and S. E. Glickman. 1995. Masculinization costs in hyaenas. *Nature* 377:584–585.

405 Freeman-Gallant, C. R. 1998. Fitness consequences of male parental care in savannah sparrows. *Behavioral Ecology* 9:486–492.

406 Fretwell, S. D., and H. K. Lucas, Jr. 1969. On territorial behavior and other factors influencing habitat distribution in birds. I. Theoretical development. *Acta Biotheoretica* 19:16–36.

407 Frey-Roos, F., P. A. Brodmann, and H.-U. Reyer. 1995. Relationships between food resources, foraging patterns, and reproductive success in the water pipit, *Anthus sp. spinoletta. Behavioral Ecology* 6:287–295.

408 Fuchs, S., and R. F. A. Moritz. 1999. Evolution of extreme polyandry in the honeybee *Apis mellifera* L. *Behavioral Ecology and Sociobiology* 45:269–275.

409 Fullard, J. H., J. W. Dawson, L. D. Otero, and A. Surlykke. 1997. Bat-deafness in day- flying moths (Lepidoptera, Notodontidae, Dioptinae). *Journal of Comparative Physiology A* 181:477–483.

410 Fullard, J. H., L. D. Otero, A. Orellana, and A. Surlykke. 2000. Auditory sensitivity and diel flight activity in Neotropical Lepidoptera. *Annals of the Entomological Society of America* 93:956–965.

411 Fullard, J. H., and J. E. Yack. 1993. The evolutionary biology of insect hearing. *Trends in Ecology and Evolution* 8:248–252.

412 Funk, D. H., and D. W. Tallamy. 2000. Courtship role reversal and deceptive signals in the long-tailed dance fly, *Rhamphomyia longicauda. Animal Behaviour* 59:411–421.

413 Furlow, F. B. 1997. Human neonatal cry quality as an honest signal of fitness. *Evolution and Human Behavior* 18:175–194.

414 Gagliardo, A., P. Ioalé, F. Odetti, and V. P. Bingman. 2001. The ontogeny of the homing pigeon navigational map: Evidence for a sensitive learning period. *Proceedings of the Royal Society of London B* 268:1–6.

415 Galef, B. G., Jr., and H. C. Kaner. 1980. Establishment and maintenance of preferences for maternal and artificial olfactory stimuli in juvenile rats. *Journal of Comparative and Physiological Psychology* 94:588–595.

416 Gangestad, S. W., and R. Thornhill. 1998. Menstrual cycle variation in women's preferences for the scent of symmetrical men. *Proceedings of the Royal Society of London B* 265:927–933.

417 Garcia, J., and F. R. Ervin. 1968. Gustatory-visceral and telereceptor-cutaneous conditioning: Adaptation in internal and external milieus. *Communications in Behavioral Biology (A)* 1:389–415.

418 Garcia, J., W. G. Hankins, and K. W. Rusiniak. 1974. Behavioral regulation of the milieu interne in man and rat. *Science* 185:824–831.

419 Gaulin, S. J. C., and J. S. Boster. 1990. Dowry as female competition. *American Anthropologist* 92:994–1005.

420 Gaulin, S. J. C., and R. W. FitzGerald. 1986. Sex differences in spatial ability: An evolutionary hypothesis and test. *American Naturalist* 127:74–88.

421 Gaulin, S. J. C., and R. W. FitzGerald. 1989. Sexual selection for spatial-learning ability. *Animal Behaviour* 37:322–331.

422 Geary, D. C. 1998. *Male, Female: The Evolution of Human Sex Differences.* American Psychological Association, Washington, DC.

423 Gee, H. 2000. Gee, Officer Krupke, it's my genes. Nature Science Update, 5 July 2000.

424 Gentner, T. Q., and S. H. Hulse. 2000. European starling preference and choice for variation in conspecific male song. *Animal Behaviour* 59:443–458.

425 Gentner, T. Q., S. H. Hulse, D. Duffy, and G. F. Ball. 2001. Response biases in auditory forebrain regions of female songbirds following exposure to sexually relevant variation in male song. *Journal of Neurobiology* 46:48–58.

426 Getting, P. A. 1983. Mechanisms of pattern generation underlying swimming in *Tritonia.* II. Network reconstruction. *Journal of Neurophysiology* 49:1017–1035.

427 Getting, P. A. 1989. A network oscillator underlying swimming in *Tritonia.* In *Neuronal and Cellular Oscillators,* J. W. Jacklet (ed.). Dekker, New York.

428 Gibson, R. M. 1996. A re-evaluation of hotspot settlement in lekking sage grouse. *Animal Behaviour* 52:993–1005.

429 Gil, D., J. Graves, N. Hazon, and A. Wells. 1999. Male attractiveness and differential testosterone investment in zebra finch eggs. *Science* 286:126–128.

430 Gil, D., J. A. Graves, and P. J. B. Slater. 1999. Seasonal patterns of singing in the willow warbler: Evidence against the fertility announcement hypothesis. *Animal Behaviour* 58:995–1000.

431 Gilardi, J. D., S. S. Duffey, C. A. Munn, and L. A. Tell. 1999. Biochemical functions of geophagy in parrots: Detoxification of dietary toxins and cytoprotective effects. *Journal of Chemical Ecology* 25:897–922.

432 Gill, F. B., and B. G. Murray. 1972. Song variation in sympatric blue-winged and golden-winged warblers. *Auk* 89:625–643.

433 Gill, F. B., and L. L. Wolf. 1975. Economics of feeding territoriality in the golden-winged sunbird. *Ecology* 56:333–345.

434 Gill, F. B., and L. L. Wolf. 1975. Foraging strategies and energetics of East African sunbirds at mistletoe flowers. *American Naturalist* 109:491–510.

435 Gill, F. B., and L. L. Wolf. 1978. Comparative foraging efficiencies of some montane sunbirds in Kenya. *Condor* 80:391–400.

436 Giray, T., Z.-Y. Huang, E. Guzman-Novoa, and G. E. Robinson. 1999. Physiological correlates of genetic variation for rate of behavioral development in the honeybee, *Apis mellifera*. *Behavioral Ecology and Sociobiology* 47:17–28.

437 Glander, K. E. 1981. Feeding patterns in mantled howler monkeys. In *Foraging Behavior: Ecological, Ethological and Psychological Approaches*, A. C. Kamil and T. D. Sargent (eds.). Garland Press, New York.

438 Glickman, S. E., L. G. Frank, P. Licht, T. Yalckinkaya, P. K. Siiteri, and J. Davidson. 1993. Sexual differentiation of the female spotted hyena: One of nature's experiments. *Annals of the New York Academy of Sciences* 662:135–159.

439 Golding, Y. C., and M. Edmunds. 2000. Behavioural mimicry of honeybees (*Apis mellifera*) by droneflies (Diptera : Syrphidae : *Eristalis* spp.). *Proceedings of the Royal Society of London B* 267:903–909.

440 Golla, W., H. Hofer, and M. L. East. 1999. Within-litter sibling aggression in spotted hyaenas: Effect of maternal nursing, sex and age. *Animal Behaviour* 58:715–726.

441 González, A., C. Rossini, M. Eisner, and T. Eisner. 1999. Sexually transmitted chemical defense in a moth. *Proceedings of the National Academy of Sciences* 96:5570–5574.

442 Goodall, J. 1988. *In the Shadow of Man*. Houghton Mifflin, Boston.

443 Goss-Custard, J. D., J. T. Cayford, and S. E. G. Lea. 1998. The changing trade-off between food finding and food stealing in juvenile oystercatchers. *Animal Behaviour* 55:745–760.

444 Gould, J. L. 1982. Why do honey bees have dialects? *Behavioral Ecology and Sociobiology* 10:53–56.

445 Gould, J. L. 1998. Sensory bases of navigation. *Current Biology* 8:R731–R738.

446 Gould, S. J. 1986. Evolution and the triumph of homology, or why history matters. *American Scientist* 74:60–69.

447 Gould, S. J. 1996. The diet of worms and the defenestration of Prague. *Natural History* 105(9):18–*ff*.

448 Gould, S. J., and R. C. Lewontin. 1979. The spandrels of San Marco and the Panglossian paradigm: A critique of the adaptationist programme. *Proceedings of the Royal Society of London B* 205:581–598.

449 Goymann, W., M. L. East, and H. Hofer. 2001. Androgens and the role of female 'hyperaggressiveness' in spotted hyenas (*Crocuta crocuta*). *Hormones and Behavior* 39:83–92.

450 Grant, B. S. 1999. Fine tuning the peppered moth paradigm. *Evolution* 53:980–984.

451 Grant, B. S., D. F. Owen, and C. A. Clarke. 1996. Parallel rise and fall of melanic peppered moths in America and Britain. *Journal of Heredity* 87:351–357.

452 Graves, J. A., and A. Whiten. 1980. Adoption of strange chicks by herring gulls, *Larus argentatus*. *Zeitschrift für Tierpsychologie* 54:267–278.

453 Gray, E. M. 1997. Do red-winged blackbirds benefit genetically from seeking copulations with extra-pair males? *Animal Behaviour* 53:605–623.

454 Gray, E. M. 1997. Female red-winged blackbirds accrue material benefits from copulating with extra-pair males. *Animal Behaviour* 53:625–639.

455 Greene, E. 1987. Individuals in an osprey colony discriminate between high and low quality information. *Nature* 329:239–241.

456 Greene, E., B. E. Lyon, V. R. Muehter, L. Ratcliffe, S. J. Oliver, and P. T. Boag. 2000. Disruptive sexual selection for plumage colouration in a passerine bird. *Nature* 407:1000–1003.

457 Greene, E., L. T. Orsak, and D. W. Whitman. 1987. A tephritid fly mimics the territorial displays of its jumping spider predators. *Science* 236:310–312.

458 Greenwood, P. J. 1980. Mating systems, philopatry, and dispersal in birds and mammals. *Animal Behaviour* 28:1140–1162.

459 Griffin, D. R. 1958. *Listening in the Dark*. Yale University Press, New Haven, CT.

460 Gross, M. R. 1982. Sneakers, satellites, and parentals: Polymorphic mating strategies in North American sunfishes. *Zeitschrift für Tierpsychologie* 60:1–26.

461 Gross, M. R. 1987. Evolution of diadromy in fishes. *American Fisheries Society Symposium* 1:14–25.

462 Gross, M. R. 1996. Alternative reproductive strategies and tactics: Diversity within species. *Trends in Ecology and Evolution* 11:92–98.

463 Gross, M. R., and A. M. MacMillan. 1981. Predation and the evolution of colonial nesting in bluegill sunfish (*Lepomis macrochirus*). *Behavioral Ecology and Sociobiology* 8:163–174.

464 Gross, M. R., and R. C. Sargent. 1985. The evolution of male and female parental care in fishes. *American Zoologist* 25:807–822.

465 Gubernick, D. J., and T. Teferi. 2000. Adaptive significance of male parental care in a monogamous mammal. *Proceedings of the Royal Society of London B* 267:147–150.

466 Gurney, M. E., and M. Konishi. 1980. Hormone-induced sexual differentiation of brain and behavior in zebra finches. *Science* 208:1380–1383.

467 Gwinner, E. 1996. Circannual clocks in avian reproduction and migration. *Ibis* 138:47–63.

468 Gwinner, E., and J. Dittami. 1990. Endogenous reproductive rhythms in a tropical bird. *Science* 249:906–908.

469 Gwinner, E., and W. Wiltschko. 1980. Circannual changes in migratory orientation of the garden warbler, *Sylvia borin*. *Behavioral Ecology and Sociobiology* 7:73–78.

470 Gwynne, D. T. 1981. Sexual difference theory: Mormon crickets show role reversal in mate choice. *Science* 213:779–780.

471 Gwynne, D. T., and L. W. Simmons. 1990. Experimental reversal of courtship roles in an insect. *Nature* 346:172–174.

472 Hack, M. A. 1998. The energetics of male mating strategies in field crickets. *Journal of Insect Behavior* 11:853–868.

473 Hafernik, J., and L. Saul-Gershenz. 2000. Beetle larvae cooperate to mimic bees. *Nature* 405:35–36.

474 Haftorn, S. 1999. Contexts and possible functions of alarm calling in the willow tit, *Parus montanus*: The principle of 'better safe than sorry.' *Behaviour* 137:437–449.

475 Hagstrum, J. T. 2000. Infrasound and the avian navigational map. *Journal of Experimental Biology* 203:1103–1111.

476 Hahn, T. P. 1995. Integration of photoperiodic and food cues to time changes in reproductive physiology by an opportunistic breeder, the red crossbill, *Loxia curvirostra* (Aves: Carduelinae). *Journal of Experimental Zoology* 272:213–226.

477 Hahn, T. P. 1998. Reproductive seasonality in an opportunistic breeder, the red crossbill, *Loxia curvirostra*. *Ecology* 79:2365–2375.

478 Hahn, T. P., J. C. Wingfield, R. Mullen, and P. J. Deviche. 1995. Endocrine bases of spatial and temporal opportunism in arctic-breeding birds. *American Zoologist* 35:259–273.

479 Haig, D. 1997. The social gene. In *Behavioural Ecology, An Evolutionary Approach* (4th edition), J. R. Krebs and N. B. Davies (eds.). Blackwell, Oxford.

480 Haley, M. P. 1994. Resource-holding power asymmetries, the prior residence effect, and reproductive payoffs in male northern elephant seal fights. *Behavioral Ecology and Sociobiology* 34:427–434.

481 Haley, M. P., C. J. Deutsch, and B. J. Le Boeuf. 1994. Size, dominance and copulatory success in male northern elephant seals, *Mirounga angustirostris*. *Animal Behaviour* 48:1249–1260.

482 Halgren, E., A. M. Dale, M. I. Sereno, R. B. H. Tootell, K. Marinkovic, and B. R. Rosen. 1999. Location of human face-selective cortex with respect to retinotopic areas. *Human Brain Mapping* 7:29–37.

483 Hall, J. C. 1994. The mating of the fly. *Science* 264:1702–1714.

484 Hamilton, W. D. 1964. The genetical theory of social behavior, I, II. *Journal of Theoretical Biology* 7:1–52.

485 Hamilton, W. D. 1971. Geometry for the selfish herd. *Journal of Theoretical Biology* 31:295–311.

486 Hamilton, W. D., and M. Zuk. 1982. Heritable true fitness and bright birds: A role for parasites? *Science* 218:384–387.

487 Hamilton, W. J., and G. H. Orians. 1965. Evolution of brood parasitism in altricial birds. *Condor* 67:361–382.

488 Hamner, W. M. 1964. Circadian control of photoperiodism in the house finch demonstrated by interrupted-night experiments. *Nature* 203:1400–1401.

489 Hanby, J. P., and J. D. Bygott. 1987. Emigration of subadult lions. *Animal Behaviour* 35:161–169.

490 Hansson, B., S. Bensch, and D. Hasselquist. 2000. Patterns of nest predation contribute to polygyny in the great reed warbler. *Ecology* 81:319–328.

491 Harbison, H., D. A. Nelson, and T. P. Hahn. 1999. Long-term persistence of song dialects in the mountain white-crowned sparrow. *Condor* 101:133–148.

492 Harcourt, A. H., P. H. Harvey, S. G. Larson, and R. V. Short. 1981. Testis weight, body weight and breeding system in primates. *Nature* 293:55–57.

493 Harlow, H. F., and M. K. Harlow. 1962. Social deprivation in monkeys. *Scientific American* 207:136–146.

494 Harlow, H. F., M. K. Harlow, and S. J. Suomi. 1971. From thought to therapy: Lessons from a primate laboratory. *American Scientist* 59:538–549.

495 Hartung, J. 1982. Polygyny and inheritance of wealth. *Current Anthropology* 23:1–12.

496 Harvey, P. H., and M. D. Pagel. 1991. *The Comparative Method in Evolutionary Biology*. Oxford University Press, London.

497 Haskell, D. G. 1999. The effect of predation on begging-call evolution in nestling wood warblers. *Animal Behaviour* 57:893–901.

498 Hasselquist, D. 1998. Polygyny in great reed warblers: A long-term study of factors contributing to fitness. *Ecology* 79:2376–2350.

499 Hausfater, G. 1975. Dominance and reproduction in baboons (*Papio cynocephalus*): A quantitative analysis. *Contributions in Primatology* 7:1–150.

500 Hayes, L. D. 2000. To nest communally or not to nest communally: A review of rodent communal nesting and nursing. *Animal Behaviour* 59:677–688.

501 Healy, S. D., E. Gwinner, and J. R. Krebs. 1996. Hippocampal volume in migratory and non-migratory warblers: Effects of age and experience. *Behavioural Brain Research* 81:62–68.

502 Heath, K. M., and C. Hadley. 1998. Dichotomous male reproductive strategies in a polygynous human society: Mating versus parental effort. *Current Anthropology* 39:369–374.

503 Hebblethwaite, M. L., and W. M. Shields. 1990. Social influences on barn swallow foraging in the Adirondacks: A test of competing hypotheses. *Animal Behaviour* 39:97–104.

504 Heinrich, B. 1979. *Bumblebee Economics*. Harvard University Press, Cambridge, MA.

505 Heinrich, B. 1984. *In a Patch of Fireweed*. Harvard University Press, Cambridge, MA.

506 Heinrich, B. 1988. Winter foraging at carcasses by three sympatric corvids, with emphasis on recruitment by the raven, *Corvus corax*. *Behavioral Ecology and Sociobiology* 23:141–156.

507 Heinrich, B. 1989. *Ravens in Winter*. Summit Books, New York.

508 Heinze, J., and B. Hölldobler. 1993. Fighting for a harem of queens: Physiology of reproduction in *Cardiocondyla* male ants. *Proceedings of the National Academy of Sciences* 90:8412–8414.

509 Heinze, J., B. Hölldobler, and C. Peeters. 1994. Conflict and cooperation in ant societies. *Naturwissenschaften* 81:489–497.

510 Heinze, J., B. Hölldobler, and K. Yamauchi. 1998. Male competition in *Cardiocondyla* ants. *Behavioral Ecology and Sociobiology* 42:239–246.

511 Helbig, A. J. 1991. Inheritance of migratory direction in a bird species: A cross-breeding experiment with SE- and SW-migrating blackcaps (*Sylvia atricapilla*). *Behavioral Ecology and Sociobiology* 29:9–12.

512 Heller, K.-G., P. Fleischman, and A. Klutz-Rider. 2000. Carotenoids in the spermatophore of bushcrickets (Orthoptera: Ephippigerinae). *Proceedings of the Royal Society of London B* 267:1905–1908.

513 Hennessy, D. F., and D. H. Owings. 1978. Snake species discrimination and the role of olfactory cues in the snake-directed behavior of the California ground squirrel. *Behaviour* 65:115–124.

514 Herberstein, M. E. 2000. Foraging behaviour in orb-web spiders (Araneidae): Do web decorations increase prey capture success in *Argiope keyserlingi* Karsch, 1878? *Australian Journal of Zoology* 48:217–223.

515 Herberstein, M. E., C. L. Craig, J. A. Coddington, and M. A. Elgar. 2001. The functional significance of silk decorations of orb-web spiders: A critical review of the empirical evidence. *Biological Reviews* 76. In press.

516 Hessler, N. A., and A. J. Doupe. 1999. Singing-related neural activity in a dorsal forebrain-basal ganglia circuit of adult zebra finches. *Journal of Neuroscience* 19:10461–10481.

517 Hettema, J. M., M. C. Neale, and K. S. Kendler. 1995. Physical similarity and the equal-environment assumption in twin studies of psychiatric disorders. *Behavior Genetics* 25:327–335.

518 Hill, G. E. 1990. Female house finches prefer colourful males: Sexual selection for a condition-dependent trait. *Animal Behaviour* 40:563–572.

519 Hill, G. E. 1991. Plumage coloration is a sexually selected indicator of male quality. *Nature* 350:337–339.

520 Hill, G. E. 1995. Ornamental traits as indicators of environmental health. *BioScience* 45:25–31.

521 Hill, G. E., and W. R. Brawner, III. 1998. Melanin-based plumage coloration in the house finch is unaffected by coccidid infection. *Proceedings of the Royal Society of London B* 265:1105–1109.

522 Hitchcock, C. L., and D. F. Sherry. 1990. Long-term memory for cache sites in the black-capped chickadee. *Animal Behaviour* 40:701–712.

523 Hoelzel, A. R., B. J. Le Boeuf, J. Reiter, and C. Campagna. 1999. Alpha-male paternity in elephant seals. *Behavioral Ecology and Sociobiology* 46:298–306.

524 Hofmann, H. A., M. E. Benson, and R. D. Fernald. 1999. Social status regulates growth rate: Consequences for life-history strategies. *Proceedings of the National Academy of Sciences* 96:14171–14176.

525 Hogg, J. T. 1984. Mating in bighorn sheep: Multiple creative male strategies. *Science* 225:526–529.

526 Höglund, J., and R. V. Alatolo. 1995. *Leks*. Princeton University Press, Princeton, NJ.

527 Höglund, J., and A. Lundberg. 1987. Sexual selection in a monomorphic lek-breeding bird: Correlates of male mating success in the great snipe *Gallinago media*. *Behavioral Ecology and Sociobiology* 21:211–216.

528 Höglund, J., and J. G. M. Robertson. 1990. Spacing of leks in relation to female home ranges, habitat require-ments and male attractiveness in the great snipe (*Gallinago media*). *Behavioral Ecology and Sociobiology* 26:173–180.

529 Högstedt, G. 1983. Adaptation unto death: Function of fear screams. *American Naturalist* 121:562–570.

530 Holden, C. 1980. Identical twins reared apart. *Science* 207:1323–1328.

531 Holekamp, K. E. 1984. Natal dispersal in Belding's ground squirrels (*Spermophilus beldingi*). *Behavioral Ecology and Sociobiology* 16:21–30.

532 Holekamp, K. E., and P. W. Sherman. 1989. Why male ground squirrels disperse. *American Scientist* 77:232–239.

533 Holland, B., and W. R. Rice. 1998. Chase-away sexual selection: Antagonistic seduction versus resistance. *Evolution* 52:1–7.

534 Holland, B., and W. R. Rice. 1999. Experimental removal of sexual selection reverses intersexual antagonistic coevolution and removes a reproductive load. *Proceedings of the National Academy of Sciences* 96:5083–5088.

535 Hölldobler, B. 1971. Communication between ants and their guests. *Scientific American* 224(Mar):86–95.

536 Hölldobler, B., and E. O. Wilson. 1990. *The Ants*. Harvard University Press, Cambridge, MA.

537 Hölldobler, B., and E. O. Wilson. 1994. *Journey to the Ants*. Harvard University Press, Cambridge, MA.

538 Holley, A. J. F. 1984. Adoption, parent-chick recognition, and maladaptation in the herring gull *Larus argentatus*. *Zeitschrift für Tierpsychologie* 64:9–14.

539 Holloway, C. C., and D. F. Clayton. 2001. Estrogen synthesis in the male brain triggers development of the avian song control pathway in vitro. *Nature Neuroscience* 4:170–175.

540 Holmes, R. T., P. P. Marra, and T. W. Sherry. 1996. Habitat-specific demography of breeding black-throated blue warblers (*Dendroica caerulescens*): Implications for population dynamics. *Journal of Animal Ecology* 65:183–195.

541 Holmes, W. G. 1984. Predation risk and foraging behavior of the hoary marmot in Alaska. *Behavioral Ecology and Sociobiology* 15:293–302.

542 Holmes, W. G. 1986. Identification of paternal half-siblings by captive Belding's ground squirrels. *Animal Behaviour* 34:321–327.

543 Holmes, W. G. 1995. The ontogeny of littermate preferences in juvenile golden-mantled ground squirrels: Effects of rearing and relatedness. *Animal Behaviour* 50:309–322.

544 Holmes, W. G., and P. W. Sherman. 1982. The ontogeny of kin recognition in two species of ground squirrels. *American Zoologist* 22:491–517.

545 Holmes, W. G., and P. W. Sherman. 1983. Kin recognition in animals. *American Scientist* 71:46–55.

546 Hoogland, J. L. 1998. Why do female Gunnison's prairie dogs copulate with more than one male? *Animal Behaviour* 55:351–359.

547 Hoogland, J. L., and P. W. Sherman. 1976. Advantages and disadvantages of bank swallow (*Riparia riparia*) coloniality. *Ecological Monographs* 46:33–58.

548 Hopkins, C. D. 1998. Design features for electric communication. *Journal of Experimental Biology* 202:1217–1228.

549 Hori, M. 1993. Frequency-dependent natural selection in the handedness of scale-eating cichlid fish. *Science* 260:216–219.

550 Horn, A. G., M. L. Leonard, and D. M. Weary. 1995. Oxygen consumption during crowing by roosters: Talk is cheap. *Animal Behaviour* 50:1171–1175.

551 Hotker, H. 2000. Intraspecific variation in size and density of Avocet colonies: Effects of nest-distances on hatching and breeding success. *Journal of Avian Biology* 31:387–398.

552 Howlett, R. J., and M. E. N. Majerus. 1987. The understanding of industrial melanism in the peppered moth (*Biston betularia*) (Lepidoptera: Geometridae). *Biological Journal of the Linnean Society* 30:31–44.

553 Hrdy, S. B. 1977. Infanticide as a primate reproductive strategy. *American Scientist* 65:40–49.

554 Hrdy, S. B. 1977. *The Langurs of Abu.* Harvard University Press, Cambridge, MA.

555 Hrdy, S. B. 1999. *Mother Nature: A History of Mothers, Infants, and Natural Selection.* Pantheon, New York.

556 Huang, Z.-Y., and G. E. Robinson. 1992. Honeybee colony integration: Worker-worker interactions mediate hormonally regulated plasticity in division of labor. *Proceedings of the National Academy of Sciences* 89:11726–11729.

557 Hume, D. K., and R. D. Montgomerie. 2001. Facial attractiveness signals different aspects of "quality" in women and men. *Evolution and Human Behavior* 22:93–112.

558 Hunter, F. M., M. Petrie, M. Otronen, T. Birkhead, and A. P. Møller. 1993. Why do females copulate repeatedly with one male? *Trends in Ecology and Evolution* 8:21–26.

559 Hunter, M. L., and J. R. Krebs. 1979. Geographic variation in the song of the great tit (*Parus major*) in relation to ecological factors. *Journal of Animal Ecology* 48:759–785.

560 Hurst, L. D., and G. T. McVean. 1998. Do we understand the evolution of genomic imprinting? *Current Opinion in Genetics & Development* 8:701–708.

561 Hurst, L. D., and J. R. Peck. 1996. Recent advances in understanding of the evolution and maintenance of sex. *Trends in Ecology and Evolution* 11:46–52.

562 Huxley, T. H. 1910. *Lectures and Lay Sermons.* E. P. Dutton, New York.

563 Ichikawa, N. 1995. Male counterstrategy against infanticide of the female giant water bug *Lethocerus deyrollei* (Hemiptera: Belostomatidae). *Journal of Insect Behavior* 8:181–188.

564 Immelmann, K. 1969. Song development in the zebra finch and other estrildid finches. In *Bird Vocalizations,* R. A. Hinde (ed.). Cambridge University Press, Cambridge.

565 Ioalé, P., M. Nozzolini, and F. Papi. 1990. Homing pigeons do extract directional information from olfactory stimuli. *Behavioral Ecology and Sociobiology* 26:301–305.

566 Irons, W. 1979. Cultural and biological success. In *Evolutionary Biology and Human Social Behavior: An Anthropological Perspective,* N. A. Chagnon and W. Irons (eds.). Duxbury Press, North Scituate, MA.

567 Isbell, L. A., D. L. Cheney, and R. M. Seyfarth. 1993. Are immigrant vervet monkeys, *Cercopithecus aethiops,* at greater risk of mortality than residents? *Animal Behaviour* 45:729–734.

568 Itô, Y. 1989. The evolutionary biology of sterile soldiers in aphids. *Trends in Ecology and Evolution* 4:69–73.

569 Iwasa, Y. 1998. The conflict theory of genomic imprinting: How much can be explained? *Current Topics in Developmental Biology* 40:255–293.

570 Jackson, R. R., and R. S. Wilcox. 1998. Spider-eating spiders. *American Scientist* 86:350–357.

571 Jacobs, L. F. 1996. Sexual selection and the brain. *Trends in Ecology and Evolution* 11:82–86.

572 Jacobs, L. F., S. J. C. Gaulin, D. F. Sherry, and G. E. Hoffman. 1990. Evolution of spatial cognition: Sex-specific patterns of spatial behavior predict hippocampal size. *Proceedings of the National Academy of Sciences* 87:6349–6352.

573 Jamieson, I. G. 1989. Behavioral heterochrony and the evolution of birds' helping at the nest: An unselected consequence of communal breeding? *American Naturalist* 133:394–406.

574 Jamieson, I. G. 1991. The unselected hypothesis for the evolution of helping behavior: Too much or too little emphasis on natural selection? *American Naturalist* 138:271–282.

575 Jamieson, I. G., J. S. Quinn, P. A. Rose, and B. N. White. 1994. Shared paternity among non-relatives is a result of an egalitarian mating system in a communally breeding bird, the pukeko. *Proceedings of the Royal Society of London B* 257:271–277.

576 Janetos, A. C. 1980. Strategies of female mate choice: A theoretical analysis. *Behavioral Ecology and Sociobiology* 7:107–112.

577 Jarvis, E. D., and C. V. Mello. 2000. Molecular mapping of brain areas involved in parrot vocal communication. *Journal of Comparative Neurology* 419:1–31.

578 Jarvis, J. U. M., and N. C. Bennett. 1993. Eusociality has evolved independently in two genera of bathygerid mole-rats—but occurs in no other subterranean mammal. *Behavioral Ecology and Sociobiology* 33:253–260.

579 Jaycox, E. R., and S. G. Parise. 1980. Homesite selection by Italian honey bee swarms, *Apis mellifera ligustica* (Hymenoptera: Apidae). *Journal of the Kansas Entomological Society* 53:171–178.

580 Jaycox, E. R., and S. G. Parise. 1981. Homesite selection by swarms of black-bodied honey bees, *Apis mellifera caucasia* and *A. m. carnica. Journal of the Kansas Entomological Society* 54:697–703.

581 Jennions, M. D., and D. W. Macdonald. 1994. Cooperative breeding in mammals. *Trends in Ecology and Evolution* 9:89–93.

582 Jia, F. Y., M. D. Greenfield, and R. D. Collins. 2000. Genetic variance of sexually selected traits in waxmoths: Maintenance by genotype x environment interaction. *Evolution* 54:953–967.

583 Jiménez, J. A., K. A. Hughes, G. Alaks, L. Graham, and R. C. Lacy. 1994. An experimental study of inbreeding depression in a natural habitat. *Science* 266:271–273.

584 Jin, H., and D. F. Clayton. 1997. Localized changes in immediate-early gene regulation during sensory and motor learning in zebra finches. *Neuron* 19:1049–1059.

585 Johns, T. 1990. *With Bitter Herbs They Shall Eat It: Chemical Ecology and the Origins of Human Diet and Medicine.* University of Arizona Press, Tuscon, AZ.

Knowing where and getting there: A human navigation network. *Science* 280:921–934.

741   Maguire, E. A., D. G. Gadian, I. S. Johnsrude, C. D. Good, J. Ashburner, R. S. J. Frackowiak, and C. D. Frith. 2000. Navigation-related structural change in the hippocampi of taxi drivers. *Proceedings of the National Academy of Sciences* 97:4398–4403.

742   Mallach, T. J., and P. L. Leberg. 1999. Use of dredged material substrates by nesting terns and black skimmers. *Journal of Wildlife Management* 63:137–146.

743   Mangel, M., and C. Clark. 1988. *Dynamic Modelling in Behavioral Ecology*. Princeton University Press, Princeton, NJ.

744   Marden, J. H. 1995. How insects learned to fly. *The Sciences* 35(6):26–30.

745   Marden, J. H., and M. G. Kramer. 1994. Surface-skimming stoneflies: A possible intermediate stage in insect flight evolution. *Science* 266:427–430.

746   Marden, J. H., and M. G. Kramer. 1995. Locomotor performance of insects with rudimentary wings. *Nature* 377:332–334.

747   Marden, J. H., B. C. O'Donnell, M. A. Thomas, and J. Y. Bye. 2000. Surface-skimming stoneflies and mayflies: The taxonomic and mechanical diversity of two-dimensional aerodynamic locomotion. *Physiological and Biochemical Zoology* 73:751-764.

748   Marden, J. H., and R. A. Rollins. 1994. Assessment of energy reserves by damselflies engaged in aerial contests for mating territories. *Animal Behaviour* 48:1023–1030.

749   Marden, J. H., and J. K. Waage. 1990. Escalated damselfly territorial contests and energetic wars of attrition. *Animal Behaviour* 39:954–959.

750   Marden, J. H., M. R. Wolf, and K. E. Weber. 1997. Aerial performance of *Drosophila melanogaster* from populations selected for upwind flight ability. *Journal of Experimental Biology* 200:2747–2755.

751   Maret, T. J., and J. P. Collins. 1994. Individual responses to population size structure: The role of size variation in controlling expression of a trophic polyphenism. *Oecologia* 100:279–285.

752   Margulis, S. W., and J. Altmann. 1997. Behavioural risk factors in the reproduction of inbred and outbred old-field mice. *Animal Behaviour* 54:397–408.

753   Margulis, S. W., W. Saltzman, and D. H. Abbott. 1995. Behavioural and hormonal changes in female naked mole-rats (*Heterocephalus glaber*) following removal of the breeding female from a colony. *Hormones and Behavior* 29:227–247.

754   Marler, C. A., and M. C. Moore. 1989. Time and energy costs of aggression in testosterone-implanted free-living male mountain spiny lizards (*Sceloporus jarrovi*). *Physiological Zoology* 62:1334–1350.

755   Marler, C. A., and M. C. Moore. 1991. Supplementary feeding compensates for testosterone-induced costs of aggression in male mountain spiny lizards, *Sceloporus jarrovi*. *Animal Behaviour* 42:209–219.

756   Marler, P. 1955. Characteristics of some animal calls. *Nature* 176:6–8.

757   Marler, P. 1970. Birdsong and speech development: Could there be parallels? *American Scientist* 58:669–673.

758   Marler, P., and M. Tamura. 1964. Culturally transmitted patterns of vocal behavior in sparrows. *Science* 146:1483–1486.

759   Marra, P. P. 2000. The role of behavioral dominance in structuring patterns of habitat occupancy in a migrant bird during the nonbreeding season. *Behavioral Ecology* 11:299–308.

760   Marra, P. P., K. A. Hobson, and R. T. Holmes. 1998. Linking winter and summer events in a migratory bird by using stable-carbon isotopes. *Science* 282:1884–1886.

761   Marshall, J. T., Jr. 1964. Voice communication and relationships among brown towhees. *Condor* 66:345–356.

762   Martín, J., and P. López. 2000. Chemoreception, symmetry and mate choice in lizards. *Proceedings of the Royal Society of London B* 267:1265–1269.

763   Marzluff, J. M., B. Heinrich, and C. S. Marzluff. 1996. Raven roosts are mobile information centres. *Animal Behaviour* 51:89–103.

764   Maschwitz, U., and E. Maschwitz. 1974. Platzende Arbeiterinnen: Eine neue Art der Feindabwehr bei sozialen Hautflüglern. *Oecologia* 14:289–294.

765   Masonjones, H. D., and S. M. Lewis. 2000. Differences in potential reproductive rates of male and female seahorses related to courtship roles. *Animal Behaviour* 59:11–20.

766   Massaro, D. W., and D. G. Stork. 1998. Speech recognition and sensory integration. *American Scientist* 86:236–244.

767   Mateo, J. M., and W. G. Holmes. 1997. Development of alarm-call responses in Belding's ground squirrels: The role of dams. *Animal Behaviour* 54:509–524.

768   Mateo, J. M., and R. E. Johnston. 2000. Kin recognition and the "armpit effect": Evidence of self-reference phenotype matching. *Proceedings of the Royal Society of London B* 267:695–700.

769   Mather, M. H., and B. D. Roitberg. 1987. A sheep in wolf's clothing: Tephritid flies mimic spider predators. *Science* 236:308–310.

770   Matsumoto-Oda, A. 1999. Female choice in the opportunistic mating of wild chimpanzees (*Pan troglodytes schweinfurthii*) at Mahale. *Behavioral Ecology and Sociobiology* 46:258–266.

771   Maxwell, M. R. 1998. Lifetime mating opportunities and male mating behaviour in sexually cannibalistic praying mantids. *Animal Behaviour* 55:1011–1028.

772   May, M. 1991. Aerial defense tactics of flying insects. *American Scientist* 79:316–329.

773   Maynard Smith, J. 1974. The theory of games and the evolution of animal conflicts. *Journal of Theoretical Biology* 47:209–221.

774   Mayr, E. 1961. Cause and effect in biology. *Science* 134:1501–1506.

775   Mayr, E. 1963. *Animal Species and Evolution*. Harvard University Press, Cambridge, MA.

776   McCann, T. S. 1981. Aggression and sexual activity of male southern elephant seals, *Mirounga leonina*. *Journal of Zoology* 195:295–310.

777   McClearn, G. E., B. Johansson, S. Berg, N. L. Pedersen, F. Ahern, S. Petrill, and R. Plomin. 1997. Substantial genetic influence on cognitive abilities in twins 80 or more years old. *Science* 276:1560–1563.

778   McCracken, G. F. 1984. Communal nursing in Mexican free-tailed bat maternity colonies. *Science* 223:1090–1091.

779   McCracken, G. F. 1993. Locational memory and female-pup reunions in Mexican free- tailed bat maternity colonies. *Animal Behaviour* 45:811–813.

780 McCracken, G. F., and J. W. Bradbury. 1981. Social organization and kinship in the polygynous bat *Phyllostomus hastatus*. *Behavioral Ecology and Sociobiology* 8:11–34.

781 McCracken, G. F., and M. K. Gustin. 1991. Nursing behavior in Mexican free-tailed bat maternity colonies. *Ethology* 89:305–321.

782 McDonald, D. B., and W. K. Potts. 1994. Cooperative display and relatedness among males in a lek-mating bird. *Science* 266:1030–1032.

783 McIver, J. D., and G. Stonedahl. 1993. Myrmecomorphy: Morphological and behavioral mimicry of ants. *Annual Review of Entomology* 38:351–379.

784 McNab, B. K., and J. F. Eisenberg. 1989. Brain size and its relation to the rate of metabolism in mammals. *American Naturalist* 133:157–167.

785 McNicol, D., Jr., and D. Crews. 1979. Estrogen/progesterone synergy in the control of female sexual receptivity in the lizard *Anolis carolinensis*. *General and Comparative Endocrinology* 38:68–74.

786 Mech, L. D. 1970. *The Wolf: The Ecology and Behavior of an Endangered Species*. Doubleday (Natural History Press), Garden City, NY.

787 Medvin, M. B., and M. D. Beecher. 1986. Parent-offspring recognition in the barn swallow (*Hirundo rustica*). *Animal Behaviour* 34:1627–1639.

788 Medvin, M. B., P. K. Stoddard, and M. D. Beecher. 1993. Signals for parent-offspring recognition: A comparative analysis of the begging calls of cliff swallows and barn swallows. *Animal Behaviour* 45:841–850.

789 Meire, P. M., and A. Ervynck. 1986. Are oystercatchers (*Haemoptopus ostralegus*) selecting the most profitable mussels (*Mytilus edulis*)? *Animal Behaviour* 34:1427–1435.

790 Mello, C. V., and S. Ribeiro. 1998. ZENK protein regulation by song in the brain of songbirds. *Journal of Comparative Neurology* 383:426–438.

791 Miller, D. E., and J. T. Emlen, Jr. 1975. Individual chick recognition and family integrity in the ring-billed gull. *Behaviour* 52:124–144.

792 Miller, G. F. 2000. *The Mating Mind*. Doubleday, New York.

793 Miller, L. A. 1971. Physiological responses of green lacewings (*Chrysopa*, Neuroptera) to ultrasound. *Journal of Insect Physiology* 17:491–506.

794 Miller, L. A. 1983. How insects detect and avoid bats. In *Neuroethology and Behavioral Physiology*, F. Huber and H. Markl (eds.). Springer-Verlag, Berlin.

795 Miller, L. A., and J. Olesen. 1979. Avoidance behavior in green lacewings. I. Behavior of free flying green lacewings to hunting bats and ultrasound. *Journal of Comparative Physiology* 131:113–120.

796 Mittwoch, U. 1996. Sex-determining mechanisms in animals. *Trends in Ecology and Evolution* 11:63–67.

797 Mock, D. W. 1984. Siblicidal aggression and resource monopolization in birds. *Science* 225:731–733.

798 Mock, D. W., H. Drummond, and C. H. Stinson. 1990. Avian siblicide. *American Scientist* 78:438–449.

799 Mock, D. W., and G. A. Parker. 1997. *The Evolution of Sibling Rivalry*. Oxford University Press, Oxford.

800 Mock, D. W., and G. A. Parker. 1998. Siblicide, family conflict and the evolutionary limits of selfishness. *Animal Behaviour* 56:1–10.

801 Mock, D. W., and B. J. Ploger. 1987. Parental manipulation of optimal hatch asynchrony in cattle egrets: An experimental study. *Animal Behaviour* 35:150–160.

802 Moffat, S. D., E. Hampson, and M. Hatzipantelis. 1998. Navigation in a "virtual" maze: Sex differences and correlation with psychometric measures of spatial ability in humans. *Human Behavior and Evolution* 19:73–87.

803 Moiseff, A., G. S. Pollack, and R. R. Hoy. 1978. Steering responses of flying crickets to sound and ultrasound: Mate attraction and predator avoidance. *Proceedings of the National Academy of Sciences* 75:4052–4056.

804 Møller, A. P. 1988. Female choice selects for male sexual tail ornaments in the monogamous swallow. *Nature* 332:640–642.

805 Møller, A. P. 1991. Density-dependent extra-pair copulations in the swallow *Hirundo rustica*. *Ethology* 87:316–329.

806 Møller, A. P. 1991. Why mated songbirds sing so much: Mate guarding and male announcement of mate fertility status. *American Naturalist* 138:994–1014.

807 Møller, A. P. 1992. Female swallow preference for symmetrical male sexual ornaments. *Nature* 357:238–240.

808 Møller, A. P. 1999. Asymmetry as a predictor of growth, fecundity and survival. *Ecology Letters* 2:149–156.

809 Møller, A. P., P. Y. Henry, and J. Erritzoe. 2000. The evolution of song repertoires and immune defences in birds. *Proceedings of the Royal Society of London B* 267:165–169.

810 Møller, A. P., and J. P. Swaddle. 1997. *Asymmetry, Developmental Stability, and Evolution*. Oxford University Press, New York.

811 Money, J., and A. A. Ehrhardt. 1972. *Man and Woman, Boy and Girl*. Johns Hopkins University Press, Baltimore, MD.

812 Montgomerie, R. D., B. E. Lyon, and K. Holder. 2001. Dirty ptarmigan: Behavioral modification of conspicuous male plumage. *Behavioral Ecology* In Press.

813 Montgomerie, R. D., and P. J. Weatherhead. 1988. Risks and rewards of nest defence by parent birds. *Quarterly Review of Biology* 63:167–187.

814 Montgomery, H. E. et al. 1998. Human gene for physical performance. *Nature* 393:221–222.

815 Moore, J., and R. Ali. 1984. Are dispersal and inbreeding avoidance related? *Animal Behaviour* 32:94–112.

816 Moore, M. C., and B. Kranz. 1983. Evidence for androgen independence of male mounting behavior in white-crowned sparrows (*Zonotrichia leucophrys gambelii*). *Hormones and Behavior* 17:414–423.

817 Mooring, M. S., and B. L. Hart. 1995. Differential grooming rate and tick load of territorial male and female impala, *Aepyceros melampus*. *Behavioral Ecology* 6:94–101.

818 Moreno, J., J. P. Veiga, P. J. Cordero, and E. Mínguez. 1999. Effects of paternal care on reproductive success in the polygynous spotless starling *Sturnus unicolor*. *Behavioral Ecology and Sociobiology* 47:47–53.

819 Morley, R., and A. Lucas. 1997. Nutrition and cognitive development. *British Medical Bulletin* 53:123–124.

820 Morton, E. S. 1975. Ecological sources of selection on avian sounds. *American Naturalist* 109:17–34.

821 Moss, C. 1988. *Elephant Memories*. William Morrow, New York.

822 Mountjoy, D. J., and R. E. Lemon. 1996. Female choice for complex song in the European starling: A field experiment. *Behavioral Ecology and Sociobiology* 38:65–72.

823   Mousseau, T. A., and C. W. Fox. 1998. The adaptive significance of maternal effects. *Trends in Ecology and Evolution* 13:403–407.

824   Mueller, U. G. 1991. Haplodiploidy and the evolution of facultative sex ratios in a primitively eusocial bee. *Science* 254:442–444.

825   Mumme, R. L. 1992. Delayed dispersal and cooperative breeding in the Seychelles warbler. *Trends in Ecology and Evolution* 7:330–331.

826   Mumme, R. L. 1992. Do helpers increase reproductive success? An experimental analysis in the Florida scrub jay. *Behavioral Ecology and Sociobiology* 31:319–328.

827   Mumme, R. L., W. D. Koenig, and F. A. Pitelka. 1983. Reproductive competition in the communal acorn woodpecker: Sisters destroy each other's eggs. *Nature* 306:583–584.

828   Mumme, R. L., W. D. Koenig, and F. L. W. Ratnieks. 1989. Helping behaviour, reproductive value, and the future component of indirect fitness. *Animal Behaviour* 38:331–343.

829   Munn, C. A. 1986. Birds that "cry wolf." *Nature* 319:143–145.

830   Munn, C. A. 1991. Macaw biology and ecotourism, or "when a bird in the bush is worth two in the hand." In *New World Parrots in Crisis,* S. R. Beisinger and N. F. R. Snyder (eds.). Smithsonian Institution Press, Washington, D.C.

831   Murdock, G. P. 1967. *Ethnographic Atlas.* Pittsburgh University Press, Pittsburgh, PA.

832   Nams, V. O. 1997. Density-dependent predation by skunks using olfactory search images. *Oecologia* 110:440–448.

833   Nealen, P. M., and D. J. Perkel. 2000. Sexual dimorphism in the song system of the Carolina wren *Thryothorus ludovicianus. Journal of Comparative Neurology* 418:346–360.

834   Nelson, D. A. 1999. Ecological influences on vocal development in the white-crowned sparrow. *Animal Behaviour* 58:21–36.

835   Nelson, D. A. 2000. A preference for own-subspecies' song guides vocal learning in a song bird. *Proceedings of the National Academy of Sciences* 97:13348–13353.

836   Nelson, R. J. 2000. *An Introduction to Behavioral Endocrinology* (2nd edition). Sinauer Associates, Sunderland, MA.

837   Nevitt, G. 1999. Foraging by seabirds on an olfactory landscape. *American Scientist* 87:46–53.

838   Newcomer, S. D., J. A. Zeh, and D. W. Zeh. 1999. Genetic benefits enhance the reproductive success of polyandrous females. *Proceedings of the National Academy of Sciences* 96:10236–10241.

839   Newton, P. N. 1986. Infanticide in an undisturbed forest population of hanuman langurs, *Presbytis entellus. Animal Behaviour* 34:785–789.

840   Nieh, J. C. 1998. The role of a scent beacon in the communication of food location by the stingless bee, *Melipona panamica. Behavioral Ecology and Sociobiology* 43:47–58.

841   Nieh, J. C. 1999. Stingless-bee communication. *American Scientist* 87:428–435.

842   Noë, R., and A. A. Sluijter. 1990. Reproductive tactics of male savanna baboons. *Behaviour* 113:117–170.

843   Nolan, P. M., G. E. Hill, and A. M. Stoehr. 1998. Sex, size, and plumage redness predict house finch survival during an epidemic. *Proceedings of the Royal Society of London B* 265:961–965.

844   Nolen, T. G., and R. R. Hoy. 1984. Phonotaxis in flying crickets: Neural correlates. *Science* 226:992–994.

845   Noonan, K. C. 1983. Female choice in the cichlid fish *Cichlasoma nigrofasciatum. Animal Behaviour* 31:1005–1010.

846   Nordby, J. C., S. E. Campbell, and M. D. Beecher. 1999. Ecological correlates of song learning in song sparrows. *Behavioral Ecology* 10:287–297.

847   Nottebohm, F., and A. P. Arnold. 1976. Sexual dimorphism in vocal control areas of songbird brain. *Science* 194:211–213.

848   Nowak, M., and K. Sigmund. 1993. A strategy of win-stay, lose-shift that outperforms tit-for-tat in the Prisoner's Dilemma game. *Nature* 364:56–58.

849   Nowak, M. A., and K. Sigmund. 1994. The alternating prisoner's dilemma. *Journal of Theoretical Biology* 168:219–226.

850   Nowicki, S., D. Hasselquist, S. Bensch, and S. Peters. 2000. Nestling growth and song repertoire size in great reed warblers: Evidence for song learning as an indicator mechanism in mate choice. *Proceedings of the Royal Society of London B* 267:2419–2424.

851   Nowicki, S., S. Peters, and J. Podos. 1998. Song learning, early nutrition, and sexual selection in birds. *American Zoologist* 38:179–190.

852   Nowicki, S., W. A. Searcy, and M. Hughes. 1998. The territory defense function of song in song sparrows: A test with the speaker occupation design. *Behaviour* 135:615–628.

853   Ober, C., T. Hyslop, S. Eilas, L. R. Wietkamp, and W. W. Hauck. 1998. Human leucocyte antigen matching and fetal loss: A result of a 10-year prospective study. *Human Reproduction* 13:33–38.

854   O'Connell, S. M., and G. Cowlishaw. 1994. Infanticide avoidance, sperm competition and mate choice: The function of copulation calls in female baboons. *Animal Behaviour* 48:687–694.

855   Olsén, H. K., M. Grahn, J. Lohm, and A. Lángefors. 1998. MHC and kin discrimination in juvenile Arctic charr, *Salvelinus alpinus* (L.). *Animal Behaviour* 56:319–327.

856   Olson, D. J., A. C. Kamil, R. P Balda, and P. J. Nims. 1995. Performance of four seed-caching corvid species in operant tests of nonspatial and spatial memory. *Journal of Comparative Psychology* 109:173–181.

857   O'Neill, K. M. 1983. Territoriality, body size, and spacing in males of the bee wolf *Philanthus basilaris* (Hymenoptera; Sphecidae). *Behaviour* 86:295–321.

858   O'Neill, W. E., and N. Suga. 1979. Target range-sensitive neurons in the auditory cortex of the mustache bat. *Science* 203:69–72.

859   Orians, G. H. 1962. Natural selection and ecological theory. *American Naturalist* 96:257–264.

860   Orians, G. H. 1969. On the evolution of mating systems in birds and mammals. *American Naturalist* 103:589–603.

861   Orians, G. H., and G. Christman. 1968. A comparative study of the behavior of red-winged, tricolored, and yellow-headed blackbirds. *University of California Publications in Zoology* 84:1–83.

862 Oring, L. W. 1985. Avian polyandry. *Current Ornithology* 3:309–351.

863 Oring, L. W., M. A. Colwell, and J. M. Reed. 1991. Lifetime reproductive success in the spotted sandpiper (*Actitis macularia*): Sex differences and variance components. *Behavioral Ecology and Sociobiology* 28:425–432.

864 Oring, L. W., R. C. Fleischer, J. M. Reed, and K. E. Marsden. 1992. Cuckoldry through stored sperm in the sequentially polyandrous spotted sandpiper. *Nature* 359:631–633.

865 Oring, L. W., and M. L. Knudson. 1973. Monogamy and polyandry in the spotted sandpiper. *The Living Bird* 11:59–73.

866 Orr, M. R. 1992. Parasitic flies (Diptera: Phoridae) influence foraging rhythms and caste division of labor in the leaf-cutter ant, *Atta cephalotes* (Hymenoptera: Formicidae). *Behavioral Ecology and Sociobiology* 30:395–402.

867 Osborne, K. A., A. Robichon, E. Burgess, S. Butland, R. A. Shaw, A. Coulthard, H. S. Pereira, R. J. Greenspan, and M. B. Sokolowski. 1997. Natural behavior polymorphism due to a cGMP-dependent protein kinase of *Drosophila*. *Science* 277:834–836.

868 Osorno, J. L., and H. Drummond. 1995. The function of hatching asynchrony in the blue-footed booby. *Behavioral Ecology and Sociobiology* 37:265–274.

869 Östlund, S., and I. Ahnesjö. 1998. Female fifteen-spined sticklebacks prefer better fathers. *Animal Behaviour* 56:1177–1183.

870 Otter, K., L. Ratcliffe, D. Michaud, and P. T. Boag. 1998. Do female black-capped chickadees prefer high-ranking males as extra-pair partners? *Behavioral Ecology and Sociobiology* 43:25–36.

871 Ottosson, U., J. Báckman, and H. G. Smith. 1997. Begging affects parental effort in the pied flycatcher, *Ficedula hypoleuca*. *Behavioral Ecology and Sociobiology* 41:381–384.

872 Owens, D. D., and M. J. Owens. 1979. Communal denning and clan associations in brown hyenas (*Hyaena brunnea*, Thunberg) in the central Kalahari Desert. *African Journal of Ecology* 17:35–44.

873 Owens, I. P. F., and R. V. Short. 1995. Hormonal basis of sexual dimorphism in birds: Implications for sexual selection theory. *Trends in Ecology and Evolution* 10:44–47.

874 Owings, D. H., and R. G. Coss. 1977. Snake mobbing by California ground squirrels: Adaptive variation and ontogeny. *Behaviour* 62:50–69.

875 Packer, C. 1977. Reciprocal altruism in *Papio anubis*. *Nature* 265:441–443.

876 Packer, C. 1986. The ecology of sociality in felids. In *Ecological Aspects of Social Evolution*, D. I. Rubenstein and R. W. Wrangham (eds.). Princeton University Press, Princeton, NJ.

877 Packer, C. 1994. *Into Africa*. University of Chicago Press, Chicago.

878 Packer, C., D. A. Gilbert, A. E. Pusey, and S. J. O'Brien. 1991. A molecular genetic analysis of kinship and cooperation in African lions. *Nature* 351:562–565.

879 Packer, C., D. Scheel, and A. E. Pusey. 1990. Why lions form groups: Food is not enough. *American Naturalist* 136:1–19.

880 Page, T. L. 1985. Clocks and circadian rhythms. In *Comprehensive Insect Physiology, Biochemistry, and Pharmacology*, G. A. Kerkut and L. I. Gilbert (eds.). Pergamon Press, New York.

881 Palmer, A. R. 2000. Quasireplication and the contract of error: Lessons from sex ratios, heritabilities and fluctuating asymmetry. *Annual Review of Ecology and Systematics* 31:441–480.

882 Palmer, C. T. 1991. Human rape: Adaptation or by-product? *Journal of Sex Research* 28:365–386.

883 Palombit, R. A., R. M. Seyfarth, and D. L. Cheney. 1997. The adaptive value of "friendships" to female baboons: Experimental and observational evidence. *Animal Behaviour* 54:599–614.

884 Panhuis, T. M., and G. S. Wilkinson. 1999. Exaggerated male eye span influences contest outcome in stalk-eyed flies (Diopsidae). *Behavioral Ecology and Sociobiology* 46:221–227.

885 Papaj, D. R., and R. H. Messing. 1998. Asymmetries in physiological state as a possible cause of resident advantage in contests. *Behaviour* 135:1013–1030.

886 Papi, F. 1975. La navigazione dei colombi viaggiatori. *Le Scienze* 78:66–75.

887 Papi, F. 1986. Pigeon navigation: Solved problems and open questions. *Monitore Zoologici Italiana* 20:471–517.

888 Parker, G. A. 1970. Sperm competition and its evolutionary consequences in the insects. *Biological Reviews* 45:526–567.

889 Parker, G. A., R. R. Baker, and V. G. F. Smith. 1972. The origin and evolution of gamete dimorphism and the male-female phenomenon. *Journal of Theoretical Biology* 36:529–553.

890 Pavey, C. R., and A. K. Smyth. 1998. Effects of avian mobbing on roost use and diet of powerful owls, *Ninox strenua*. *Animal Behaviour* 55:313–318.

891 Pawłowski, B., and R. I. M. Dunbar. 1999. Impact of market value on human mate choice decisions. *Proceedings of the Royal Society of London B* 266:281–285.

892 Payne, R. B., and L. L. Payne. 1998. Brood parasitism by cowbirds: Risks and effects on reproductive success and survival in indigo buntings. *Behavioral Ecology* 9:64–73.

893 Peakall, R. 1990. Responses of male *Zaspilothynnus trilobatus* Turner wasps to females and the sexually deceptive orchid it pollinates. *Functional Ecology* 4:159–167.

894 Pengelly, E. T., and S. J. Asmundson. 1974. Circannual rhythmicity in hibernating animals. In *Circannual Clocks*, E.T. Pengelley (ed.). Academic Press, New York.

895 Penn, D. J., and W. K. Potts. 1998. How do major histocompatibility complex genes influence odor and mating preferences? *Advances in Immunology* 69:411–436.

896 Penn, D. J., and W. K. Potts. 1998. MHC-disassortative mating preferences reversed by cross-fostering. *Proceedings of the Royal Society of London B* 265:1299–1306.

897 Penn, D. J., and W. K. Potts. 1999. The evolution of mating preferences and major histocompatibility complex genes. *American Naturalist* 153:145–164.

898 Pennisi, E. 2000. Fruit fly genome yields data and a validation. *Science* 287:1374.

899 Penton-Voak, I. S., and D. I. Perrett. 2000. Female preference for male faces changes cyclically: Further evidence. *Evolution and Human Behavior* 21:39–48.

900 Perrigo, G., W. C. Bryant, and F. S. vom Saal. 1990. A unique neural timing system prevents male mice from harming their own offspring. *Animal Behaviour* 39:535–539.

901 Perrill, S. A., H. C. Gerhardt, and R. Daniel. 1978. Sexual parasitism in the green tree frog (*Hyla cinerea*). *Science* 200:1179–1180.

902 Perusse, D. 1993. Cultural and reproductive success in industrial societies: Testing the relationship at the proximate and ultimate levels. *Behavioral and Brain Sciences* 16:267–283.

903 Petit, L. J. 1991. Adaptive tolerance of cowbird parasitism by prothonotary warblers: A consequence of site limitation? *Animal Behaviour* 41:425–432.

904 Petrie, M. 1992. Peacocks with low mating success are more likely to suffer predation. *Animal Behaviour* 44:585–586.

905 Petrie, M. 1994. Improved growth and survival of offspring of peacocks with more elaborate trains. *Nature* 371:585–586.

906 Petrie, M., and T. Halliday. 1994. Experimental and natural changes in the peacock's (*Pavo cristatus*) train can affect mating success. *Behavioral Ecology and Sociobiology* 35:213–217.

907 Petrie, M., T. Halliday, and C. Sanders. 1991. Peahens prefer peacocks with elaborate trains. *Animal Behaviour* 41:323–332.

908 Petrie, M., and B. Kempenaers. 1998. Extra-pair paternity in birds: Explaining variation between species and populations. *Trends in Ecology and Evolution* 13:52–57.

909 Pfennig, D. W. 1992. Polyphenism in spadefoot toad tadpoles as a locally adjusted evolutionarily stable strategy. *Evolution* 46:1408–1420.

910 Pfennig, D. W., and J. P. Collins. 1993. Kinship affects morphogenesis in cannibalistic salamanders. *Nature* 362:836–838.

911 Pfennig, D. W., G. J. Gamboa, H. K. Reeve, J. S. Reeve, and I. D. Ferguson. 1983. The mechanism of nestmate discrimination in social wasps (*Polistes*, Hymenoptera: Vespidae). *Behavioral Ecology and Sociobiology* 13:299–305.

912 Pfennig, D. W., and P. W. Sherman. 1995. Kin recognition. *Scientific American* 272(June):68–73.

913 Pfennig, D. W., P. W. Sherman, and J. P. Collins. 1994. Kin recognition and cannibalism in polyphenic salamanders. *Behavioral Ecology* 5:225–232.

914 Picciotto, M. R. 1999. Knock-out mouse models used to study neurobiological systems. *Critical Reviews in Neurobiology* 13:103–149.

915 Pierce, G. J., and J. G. Ollason. 1987. Eight reasons why optimal foraging theory is a complete waste of time. *Oikos* 49:111–118.

916 Pierotti, R., and E. C. Murphy. 1987. Intergenerational conflicts in gulls. *Animal Behaviour* 35:435–444.

917 Pietrewicz, A. T., and A. C. Kamil. 1977. Visual detection of cryptic prey by blue jays (*Cyanocitta cristata*). *Science* 195:580–582.

918 Pietrewicz, A. T., and A. C. Kamil. 1979. Search image formation in the blue jay *Cyanocitta cristata*. *Science* 204:1332–1333.

919 Pietsch, T. W., and D. B. Grobecker. 1978. The compleat angler: Aggressive mimicry in the antennariid anglerfish. *Science* 201:369–370.

920 Pinker, S. 1994. *The Language Instinct*. W. Morrow & Co., New York.

921 Pinxten, R., and M. Eeens. 1998. Male starlings sing most in the late morning, following egg-laying: A strategy to protect their paternity? *Behaviour* 135:1197–1211.

922 Piper, W. H., D. C. Evers, M. W. Meyer, K. B. Tischler, J. D. Kaplan, and R. C. Fleischer. 1997. Genetic monogamy in the common loon (*Gavia immer*). *Behavioral Ecology and Sociobiology* 41:25–32.

923 Pitcher, T. 1979. He who hesitates lives: Is stotting anti-ambush behavior? *American Naturalist* 113:453–456.

924 Pitnick, S., T. A. Markow, and G. S. Spicer. 1995. Delayed male maturity is a cost of producing large sperm in *Drosophila*. *Proceedings of the National Academy of Sciences* 92:10614–10618.

925 Pizzari, T., and T. R. Birkhead. 2000. Female feral fowl eject sperm of subdominant males. *Nature* 405:787–789.

926 Plomin, R., and I. Craig. 1997. Human behavioural genetics of cognitive abilities and disabilities. *BioEssays* 19:1117–1124.

927 Plomin, R., J. C. De Fries, G. E. McClearn, and M. Rutter. 1997. *Behavioral Genetics* (3rd edition). W. H. Freeman, New York.

928 Plomin, R., D. W. Fulker, R. Corley, and J. C. DeFries. 1997. Nature, nurture, and cognitive development from 1 to 16 years: A parent-offspring adoption study. *Psychological Science* 8:442–447.

929 Polak, M., L. L. Wolf, W. T. Starmer, and J. S. F. Barker. 2001. Function of the mating plug in *Drosophila hibisci* Bock. *Behavioral Ecology and Sociobiology* 49: 196–205.

930 Pollak, G., D. Marsh, R. Bodenhamer, and A. Souther. 1977. Echo-detecting characteristics of neurons in inferior colliculus of unanesthetized bats. *Science* 196:675–677.

931 Pomiankowski, A., and A. P. Møller. 1995. A resolution of the lek paradox. *Proceedings of the Royal Society of London B* 260:21–29.

932 Porter, R. H., V. J. Tepper, and D. M. White. 1981. Experiential influences on the development of huddling preferences and "sibling" recognition in spiny mice. *Developmental Psychobiology* 14:375–382.

933 Powell, A. N., and C. L. Collier. 2000. Habitat use and reproductive success of western snowy plovers at new nesting areas created for California least terns. *Journal of Wildlife Management* 64:24–33.

934 Powell, G. V. N. 1974. Experimental analysis of the social value of flocking by starlings (*Sturnus vulgaris*) in relation to predation and foraging. *Animal Behaviour* 22:501–505.

935 Power, H. W. 1998. Quality control and the important questions in avian conspecific brood parasitism. In *Parasitic Birds and Their Hosts: Studies in Coevolution*, S. I. Rothstein and S. K. Robinson (eds.). Oxford University Press, New York.

936 Preston-Mafham, R., and K. Preston-Mafham. 1993. *The Encyclopedia of Land Invertebrate Behaviour*. MIT Press, Cambridge, MA.

937 Prete, F. R. 1995. Designing behavior: A case study. *Perspectives in Ethology* 11:255–277.

938 Proctor, H. C. 1991. Courtship in the water mite *Neumania papillator*: Males capitalize on female adaptations for predation. *Animal Behaviour* 42:589–598.

939 Proctor, H. C. 1992. Sensory exploitation and the evolution of male mating behaviour: A cladistic test. *Animal Behaviour* 44:745–752.

940 Puce, A., T. Allison, and G. McCarthy. 1999. Electrophysiological studies of human face perception. III: Effects of top-down processing of face-specific potentials. *Cerebral Cortex* 9:445–458.

941 Purseglove, J. W., E. G. Brown, C. L. Green, and S. R. J. Robbins. 1981. *Spices*. Longman, London.

942 Pusey, A. E., and C. Packer. 1987. The evolution of sex-biased dispersal in lions. *Behaviour* 101:275–310.

943 Pusey, A. E., and C. Packer. 1994. Infanticide in lions. In *Infanticide and Parental Care,* S. Parmigiani and F. S. vom Saal (eds.). Harwood Academic Press, Chur, Switzerland.

944 Pusey, A. E., and M. Wolf. 1996. Inbreeding avoidance in animals. *Trends in Ecology and Evolution* 11:201–206.

945 Queller, D. C. 1996. The measurement and meaning of inclusive fitness. *Animal Behaviour* 51:229–232.

946 Queller, D. C. 1997. Why do females care more than males? *Proceedings of the Royal Society of London B* 264:1555–1557.

947 Queller, D. C., and J. E. Strassmann. 1998. Kin selection and social insects. *BioScience* 48:165–175.

948 Queller, D. C., J. E. Strassmann, and C. R. Hughes. 1993. Microsatellites and kinship. *Trends in Ecology and Evolution* 8:285–288.

949 Queller, D. C., F. Zacchi, R. Cervo, S. Turillazzi, M. T. Henshaw, L. A. Santorelli, and J. E. Strassmann. 2000. Unrelated helpers in a social insect. *Nature* 405:784–787.

950 Quinn, J. S., G. E. Woolfenden, J. W. Fitzpatrick, and B. N. White. 1999. Multi-locus DNA fingerprinting supports genetic monogamy in Florida scrub-jays. *Behavioral Ecology and Sociobiology* 45:1–10.

951 Quinn, V. S., and D. K. Hews. 2000. Signals and behavioural responses are not coupled in males: Aggression affected by replacement of an evolutionarily lost colour signal. *Proceedings of the Royal Society of London B* 267:755–758.

952 Rachlow, J. L., E. V. Berkeley, and J. Berger. 1998. Correlates of male mating strategies in white rhinos (*Ceratotherium simum*). *Journal of Mammalogy* 79:1317–1324.

953 Ralls, K., K. Brugger, and J. Ballou. 1979. Inbreeding and juvenile mortality in small populations of ungulates. *Science* 206:1101–1103.

954 Raouf, S. A., P. G. Parker, E. D. Ketterson, J. V. Nolan, and C. Ziegenfus. 1997. Testosterone affects reproductive success by influencing extra-pair fertilizations in male dark-eyed juncos (Aves: *Junco hyemalis*). *Proceedings of the Royal Society of London B* 264:1599–1603.

955 Ratnieks, F. L. W., and L. Keller. 1998. Queen control of egg fertilization in the honey bee. *Behavioral Ecology and Sociobiology* 44:57–62.

956 Rattenborg, N. C., S. L. Lima, and C. J. Amlaner. 1999. Facultative control of avian unihemispheric sleep under risk of predation. *Behavioural Brain Research* 105:163–172.

957 Reboreda, J. C., N. S. Clayton, and A. Kacelnik. 1996. Species and sex differences in hippocampus size in parasitic and non-parasitic cowbirds. *NeuroReport* 7:505–508.

958 Reeve, H. K., and L. Keller. 2001. Tests of reproductive-skew models in social insects. *Annual Review of Ecology and Systematics* 46:347–385.

959 Reeve, H. K., J. M. Peters, P. Nonacs, and P. T. Starks. 1998. Dispersal of first "workers" in social wasps: Causes and implications of an alternative reproductive strategy. *Proceedings of the National Academy of Sciences* 95:13737–12742.

960 Reeve, H. K., and P. W. Sherman. 1993. Adaptation and the goals of evolutionary research. *Quarterly Review of Biology* 68:1–32.

961 Reeve, H. K., and P. W. Sherman. 2000. Optimality and phylogeny: A critique of current thought. In *Adaptationism and Optimality,* E. Orzack and E. Sober (eds.). Cambridge University Press, Cambridge.

962 Reeve, H. K., P. T. Starks, J. M. Peters, and P. Nonacs. 2000. Genetic support for the evolutionary theory of reproductive transactions in social wasps. *Proceedings of the Royal Society of London B* 267:75–79.

963 Reeve, H. K., D. F. Westneat, W. A. Noon, P. W. Sherman, and C. F. Aquadro. 1990. DNA "fingerprinting" reveals high levels of inbreeding in colonies of the eusocial naked mole-rat. *Proceedings of the National Academy of Sciences* 87:2496–2500.

964 Reyer, H.-U. 1984. Investment and relatedness: A cost/benefit analysis of breeding and helping in the pied kingfisher. *Animal Behaviour* 32:1163–1178.

965 Reynolds, J. D. 1996. Animal breeding systems. *Trends in Ecology and Evolution* 11:68–72.

966 Reynolds, J. D., and M. R. Gross. 1990. Costs and benefits of female mate choice: Is there a lek paradox? *American Naturalist* 136:230–243.

967 Rhodes, G., F. Proffitt, J. M. Grady, and A. Sumich. 1998. Facial symmetry and the perception of beauty. *Psychonomic Bulletin and Review* 5:659–669.

968 Richardson, H., and N. A. M. Verbeek. 1986. Diet selection and optimization by northwestern crows feeding on Japanese littleneck clams. *Ecology* 67:1219–1226.

969 Riedel, S., A. Hinz, and R. Schwarz. 2000. Attitude towards blood donation in Germany: Results of a representative survey. *Infusion Therapy and Transfusion Medicine* 27:196–199.

970 Rintamäki, P. T., R. V. Alatalo, J. Höglund, and A. Lundberg. 1995. Male territoriality and female choice on black grouse leks. *Animal Behaviour* 49:759–767.

971 Rintamäki, P. T., A. Lundberg, R. V. Alatalo, and J. Höglund. 1998. Assortative mating and female clutch investment in black grouse. *Animal Behaviour* 56:1399–1403.

972 Robert, D., J. Amoroso, and R. R. Hoy. 1992. The evolutionary convergence of hearing in a parasitoid fly and its cricket host. *Science* 258:1135–1137.

973 Robinson, G. E. 1998. From society to genes with the honey bee. *American Scientist* 86:456–462.

974 Robinson, G. E. 1999. Integrative animal behavior and sociogenomics. *Trends in Ecology and Evolution* 14:202–205.

975 Rodrigues, M. 1996. Song activity in the chiffchaff: Territorial defence or mate guarding? *Animal Behaviour* 51:709–716.

976 Roeder, K. D. 1963. *Nerve Cells and Insect Behavior*. Harvard University Press, Cambridge, MA.

977 Roeder, K. D. 1970. Episodes in insect brains. *American Scientist* 58:378–389.

978 Roeder, K. D., and A. E. Treat. 1961. The detection and evasion of bats by moths. *American Scientist* 49:135–148.

979 Roff, D. A. 1986. The evolution of wing dimorphism in insects. *Evolution* 40:1009–1020.

980 Rohwer, S., J. C. Herron, and M. Daly. 1999. Stepparental behavior as mating effort in birds and other animals. *Evolution and Human Behavior* 20:367–390.

981 Rohwer, S., and C. D. Spaw. 1988. Evolutionary lag versus bill-size constraints: A comparative study of the acceptance of cowbird eggs by old hosts. *Evolutionary Ecology* 2:27–36.

982 Rolland, C., E. Danchin, and M. de Fraipont. 1998. The evolution of coloniality in birds in relation to food, habitat, predation, and life-history traits: A comparative analysis. *American Naturalist* 151:514–529.

983 Romer, H., and W. J. Bailey. 1998. Strategies for hearing in noise: Peripheral control over auditory sensitivity in the bushcricket *Sciarasaga quadrata* (Austrosaginae: Tettigoniidae). *Journal of Experimental Biology* 201:1023–1033.

984 Romey, W. L. 1995. Position preferences within groups: Do whirligigs select positions which balance feeding opportunities with predator avoidance? *Behavioral Ecology and Sociobiology* 37:195–200.

985 Roper, T. J. 1999. Olfaction in birds. *Advances in the Study of Behavior* 28:247–332.

986 Rose, M. 1998. *Darwin's Spectre*. Princeton University Press, Princeton, NJ.

987 Rose, N. A., C. J. Deutsch, and B. J. Le Boeuf. 1991. Sexual behavior of male northern elephant seals: III. The mounting of weaned pups. *Behaviour* 119:171–192.

988 Rose, S. 1997. *Lifelines, Biology Beyond Determinism*. Oxford University Press, Oxford.

989 Rosengaus, R. B., A. B. Maxmen, L. E. Coates, and J. F. A. Traniello. 1998. Disease resistance: a benefit of sociality in the dampwood termite *Zootermopsis angusticollis* (Isoptera: Termopsidae). *Behavioral Ecology and Sociobiology* 44:125–134.

990 Rosenqvist, G. 1990. Male mate choice and female-female competition for mates in the pipefish *Nerophis ophidion*. *Animal Behaviour* 39:1110–1116.

991 Rosenqvist, G., and K. Johansson. 1995. Male avoidance of parasitized females explained by direct benefits in a pipefish. *Animal Behaviour* 49:1039–1045.

992 Ross, K. G. 1986. Kin selection and the problem of sperm utilization in social insects. *Nature* 323:798–800.

993 Rowe, L. 1994. The costs of mating and mate choice in water striders. *Animal Behaviour* 48:1049–1056.

994 Rowe, L., G. Arnqvist, A. Sih, and J. Krupa. 1994. Sexual conflict and the evolutionary ecology of mating patterns: Water striders as a model system. *Trends in Ecology and Evolution* 9:289–293.

995 Rowley, I., and G. Chapman. 1986. Cross-fostering, imprinting, and learning in two sympatric species of cockatoos. *Behaviour* 96:1–16.

996 Ruffieux, L., J. M. Elouard, and M. Sartori. 1998. Flightlessness in mayflies and its relevance to hypotheses on the origin of insect flight. *Proceedings of the Royal Society of London B* 265:2135–2140.

997 Rutowski, R. L. 1998. Mating strategies in butterflies. *Scientific American* 279:64–69.

998 Ryan, M. J. 1985. *The Túngara Frog*. University of Chicago Press, Chicago.

999 Ryan, M. J. 1998. Sexual selection, receiver biases, and the evolution of sex differences. *Science* 281:1999–2003.

1000 Ryan, M. J., J. H. Fox, W. Wilczynski, and A. S. Rand. 1990. Sexual selection for sensory exploitation in the frog *Physalaemus postulosus*. *Nature* 343:66–67.

1001 Ryan, M. J., and A. Keddy-Hector. 1992. Directional patterns of female mate choice and the role of sensory biases. *American Naturalist* 139: S4–S35.

1002 Ryan, M. J., and A. S. Rand. 1990. The sensory basis of sexual selection for complex calls in the túngara frog, *Physalaemus pustulosus* (sexual selection for sensory exploitation). *Evolution* 44:305–314.

1003 Ryan, M. J., M. D. Tuttle, and L. K. Taft. 1981. The costs and benefits of frog chorusing behavior. *Behavioral Ecology and Sociobiology* 8:273–278.

1004 Ryan, M. J., and W. E. Wagner, Jr. 1987. Asymmetries in mating behavior between species: Female swordtails prefer heterospecific males. *Science* 236:595–597.

1005 Rydale, J., H. Roininen, and K. W. Philip. 2000. Persistence of bat defence reactions in high Arctic moths (Lepidoptera). *Proceedings of the Royal Society of London B* 267:553–557.

1006 Sacks, O. W. 1985. *The Man Who Mistook His Wife for a Hat and Other Clinical Tales*. Summit Books, New York.

1007 Sadowski, J. A., A. J. Moore, and E. D. Brodie, III. 1999. The evolution of empty nuptial gifts in a dance fly, *Empis snoddyi* (Diptera: Empididae): Bigger isn't always better. *Behavioral Ecology and Sociobiology* 45:161–166.

1008 Sæther, S. A., P. Fiske, and J. A. Kalas. 1990. Pushy males and choosy females: Courtship disruption and mate choice in the lekking great snipe. *Proceedings of the Royal Society of London B* 266:1227–1234.

1009 Sætre, G.-P., S. Dale, and T. Slagsvold. 1994. Female pied flycatchers prefer brightly coloured males. *Animal Behaviour* 48:1407–1416.

1010 Saetre, G.-P., T. Fossnes, and T. Slagsvold. 1995. Food provisioning in the pied flycatcher: Do females gain direct benefits from choosing bright-coloured males? *Journal of Animal Ecology* 64:21–30.

1011 Sahlins, M. 1976. *The Use and Abuse of Biology*. University of Michigan Press, Ann Arbor.

1012 Saino, N., P. Ninni, S. Calza, R. Martinelli, F. de Bernardi, and A. P. Møller. 2000. Better red than dead: Carotenoid-based mouth coloration reveals infection in barn swallow nestlings. *Proceedings of the Royal Society of London B* 267:57–61.

1013 Sandberg, R., and F. R. Moore. 1996. Migratory orientation of red-eyed vireos, *Vireo olivaceus*, in relation to energetic condition and ecological context. *Behavioral Ecology and Sociobiology* 39:1–10.

1014 Sargent, R. C. 1989. Allopaternal care in the fathead minnow, *Pimephales promelas*: Stepfathers discriminate against their adopted eggs. *Behavioral Ecology and Sociobiology* 25:379–386.

1015 Sargent, T. D. 1976. *Legion of Night—The Underwing Moths*. University of Massachusetts Press, Amherst, MA.

1016 Sato, T. 1994. Active accumulation of spawning substrate: A determinant of extreme polygyny in a shell-brooding cichlid fish. *Animal Behaviour* 48:669–678.

1017 Schaller, G. B. 1964. *The Year of the Gorilla*. University of Chicago Press, Chicago.

1018 Schaller, G. B. 1972. *The Serengeti Lion*. University of Chicago Press, Chicago.

1019 Scheel, D. 1993. Watching for lions in the grass: The usefulness of scanning and its effects during hunts. *Animal Behaviour* 46:695–704.

1020 Scheel, D., and C. Packer. 1991. Group hunting behaviour of lions: A search for cooperation. *Animal Behaviour* 41:711–722.

1021 Schieb, J. E., S. W. Gangestad, and R. Thornhill. 1999. Facial attractiveness, symmetry and cues of good genes. *Proceedings of the Royal Society of London B* 266:1913–1917.

1022 Schluter, D. 1994. Experimental evidence that competition promotes divergence in adaptive radiation. *Science* 266:798–801.

1023 Schluter, D., and J. D. McPhail. 1993. Character displacement and replicate adaptive radiation. *Trends in Ecology and Evolution* 8:197–200.

1024 Schmidt-Koenig, K., and J. U. Ganzhorn. 1991. On the problem of bird navigation. *Perspectives in Ethology* 9:261–283.

1025 Schoech, S. J. 1998. Physiology of helping in Florida scrub-jays. *American Scientist* 86:70–77.

1026 Schwabl, H. 1983. Auspragung und Bedeutung des Teilzugverhaltnes einer sudwestdeutschen Population der Amsel *Turdus merula*. *Journal für Ornithologie* 124:101–116.

1027 Schwabl, H., D. W. Mock, and J. A. Gieg. 1997. A hormonal mechanism for parental favouritism. *Nature* 386:231.

1028 Schwagmeyer, P. L. 1994. Competitive mate searching in the 13–lined ground squirrel: Potential roles of spatial memory? *Ethology* 98:265–276.

1029 Schwagmeyer, P. L. 1995. Searching today for tomorrow's mates. *Animal Behaviour* 50:759–767.

1030 Sealy, S. G. 1995. Burial of cowbird eggs by parasitized yellow warblers: An empirical and experimental study. *Animal Behaviour* 49:877–889.

1031 Searcy, W. A. 1992. Measuring response of female birds to male songs. In *Playback and Studies of Animal Communication*, P. K. McGregor (ed.). Plenum Press, New York.

1032 Searcy, W. A., and E. A. Brenowitz. 1988. Sexual differences in species recognition of avian song. *Nature* 332:152–154.

1033 Seeley, T. D. 1977. Measurement of nest cavity volume by the honey bee (*Apis mellifera*). *Behavioral Ecology and Sociobiology* 2:201–227.

1034 Seeley, T. D. 1985. *Honeybee Ecology: A Study of Adaptation in Social Life*. Princeton University Press, Princeton, NJ.

1035 Seeley, T. D. 1995. *The Wisdom of the Hive*. Harvard University Press, Cambridge, MA.

1036 Seeley, T. D., and S. C. Buhrman. 1999. Group decision making in swarms of honey bees. *Behavioral Ecology and Sociobiology* 45:19–32.

1037 Seeley, T. D., and P. K. Visscher. 1988. Assessing the benefits of cooperation in honeybee foraging: Search costs, forage quality, and competitive ability. *Behavioral Ecology and Sociobiology* 22:229–237.

1038 Segal, N. L. 2000. *Entwined Lives: Twins and What They Tell Us about Human Behavior*. Plume, New York.

1039 Segal, N. L. 2000. Virtual twins: New findings on within-family environmental influences on intelligence. *Journal of Educational Psychology* 92:442–448.

1040 Sharma, J., A. Angelucci, and M. Sur. 2000. Induction of visual orientation modules in auditory cortex. *Nature* 404:841–847.

1041 Shellman-Reeve, J. S. 1997. Advantages of biparental care in the wood-dwelling termite, *Zootermopsis nevadensis*. *Animal Behaviour* 54:163–170.

1042 Shellman-Reeve, J. S. 1999. Courting strategies and conflicts in a monogamous, biparental termite. *Proceedings of the Royal Society of London B* 266:137–144.

1043 Sherman, P. M. 1994. The orb-web: An energetic and behavioural estimator of a spider's dynamic foraging and reproductive strategies. *Animal Behaviour* 48:19–34.

1044 Sherman, P. W. 1977. Nepotism and the evolution of alarm calls. *Science* 197:1246–1253.

1045 Sherman, P. W. 1981. Kinship, demography and Belding's ground squirrel nepotism. *Behavioral Ecology and Sociobiology* 8:251–259.

1046 Sherman, P. W. 1988. The levels of analysis. *Animal Behaviour* 36:616–618.

1047 Sherman, P. W., and W. G. Holmes. 1985. Kin recognition: Issues and evidence. In *Experimental Behavioral Ecology and Sociobiology*, B. Hölldobler and M. Lindauer (eds.). G. Fischer Verlag, Stuttgart.

1048 Sherman, P. W., J. U. M. Jarvis, and R. D. Alexander (eds.). 1991. *The Biology of the Naked Mole-Rat*. Princeton University Press, Princeton, NJ.

1049 Sherry, D. F. 1984. Food storage by black-capped chickadees: Memory of the location and contents of caches. *Animal Behaviour* 32:451–464.

1050 Sherry, D. F., M. R. L. Forbes, M. Kjurgel, and G. O. Ivy. 1993. Females have a larger hippocampus than males in the brood-parasitic brown-headed cowbird. *Proceedings of the National Academy of Sciences* 90:7839–7843.

1051 Shubin, N., C. Tabin, and S. Carroll. 1997. Fossils, genes and the evolution of animal limbs. *Nature* 388:639–648.

1052 Shuster, S. M. 1989. Male alternative reproductive strategies in a marine isopod crustacean (*Paracerceis sculpta*): The use of genetic markers to measure differences in the fertilization success among α-, β-, and γ-males. *Evolution* 43:1683–1689.

1053 Shuster, S. M. 1992. The reproductive behaviour of α-, β-, and γ-male morphs in *Paracerceis sculpta*, a marine isopod crustacean. *Behaviour* 121:231–258.

1054 Shuster, S. M., and C. A. Sassaman. 1997. Genetic interaction between male mating strategy and sex ratio in a marine isopod. *Nature* 338:373–377.

1055 Shuster, S. M., and M. J. Wade. 1991. Equal mating success among male reproductive strategies in a marine isopod. *Nature* 350:608–610.

1056 Shutler, D., and P. J. Weatherhead. 1991. Owner and floater red-winged blackbirds: Determinants of status. *Behavioral Ecology and Sociobiology* 28:235–242.

1057 Silk, J. B. 1980. Adoption and kinship in Oceania. *American Anthropologist* 82:799–820.

1058 Silk, J. B., R. M. Seyfarth, and D. L. Cheney. 1999. The structure of social relationships among female savanna baboons in Moremi Reserve, Botswana. *Behaviour* 136:679–703.

1059 Silva, A. J., R. Paylor, J. M. Wehner, and S. Tonegawa. 1992. Impaired spatial learning in alpha-calcium-calmodulin kinase II mutant mice. *Science* 257:206–211.

1060 Silverthorne, Z. A., and V. L. Quinsey. 2000. Sexual partner age preferences of homosexual and heterosex-

522   Bibliography

ual men and women. *Archives of Sexual Behavior* 29:67–76.

1061   Simmons, L. W., and M. G. Ritchie. 1996. Symmetry in the songs of crickets. *Proceedings of the Royal Society of London B* 263:305–311.

1062   Simmons, L. W., and M. T. Siva-Jothy. 1998. Sperm competition in insects: Mechanisms and the potential for selection. In *Sperm Competition and Sexual Selection*, T. R. Birkhead and A. P. Møller (eds.). Academic Press, San Diego, CA.

1063   Simmons, P., and D. Young. 1999. *Nerve Cells and Animal Behaviour* (2nd edition). Cambridge University Press, Cambridge.

1064   Simmons, R. E., and L. Scheepers. 1996. Winning by a neck: Sexual selection in the evolution of giraffe. *American Naturalist* 148:771–786.

1065   Singer, P. 1981. *The Expanding Circle: Ethics and Sociobiology*. Farrar, Straus, and Giroux, New York.

1066   Singh, D. 1993. Adaptive significance of female physical attractiveness: Role of the waist-to-hip ratio. *Journal of Personality and Social Psychology* 65:293–307.

1067   Skinner, B. F. 1966. Operant behavior. In *Operant Behavior*, W. Honig (ed.). Appleton-Century-Crofts, New York.

1068   Slagsvold, T. 1998. On the origin and rarity of interspecific nest parasitism in birds. *American Naturalist* 152:264–272.

1069   Smale, L., K. E. Holekamp, and P. A. White. 1999. Siblicide revisited in the spotted hyaena: Does it conform to obligate or facultative models? *Animal Behaviour* 58:545–551.

1070   Smith, E. A., M. Borgerhoff Mulder, and K. Hill. 2001. Controversies in the evolutionary social sciences: A guide for the perplexed. *Trends in Ecology and Evolution* 16:128–135.

1071   Smith, H. G. 1991. Nestling American robins compete with siblings by begging. *Behavioral Ecology and Sociobiology* 29:307–312.

1072   Smith, H. G., and R. D. Montgomerie. 1991. Sexual selection and the tail ornaments of North American barn swallows. *Behavioral Ecology and Sociobiology* 28:195–201.

1073   Smith, M. S., B. J. Kish, and C. B. Crawford. 1987. Inheritance of wealth as human kin investment. *Ethology and Sociobiology* 8:171–182.

1074   Smith, R. L. 1984. Human sperm competition. In *Sperm Competition and the Evolution of Animal Mating Systems*, R. L. Smith (ed.). Academic Press, New York.

1075   Smith, R. L. 1997. Evolution of paternal care in giant water bugs (Heteroptera: Belostomatidae). In *Social Competition and Cooperation among Insects and Arachnids*, II. *Evolution of Sociality*, J. C. Choe and B. J. Crespi (eds.). Cambridge University Press, Cambridge.

1076   Smith, S. M. 1978. The "underworld" in a territorial species: Adaptive strategy for floaters. *American Naturalist* 112:571–582.

1077   Smuts, B. B., and R. W. Smuts. 1993. Male aggression and sexual coercion of females in nonhuman primates and other mammals: Evidence and theoretical implications. *Advances in the Study of Behavior* 22:1–63.

1078   Sneddon, L. U., F. A. Huntingford, A. C. Taylor, and J. F. Orr. 2000. Weapon strength and competitive success in the fights of shore crabs (*Carcinus maenas*). *Journal of Zoology* 250:397–403.

1079   Snow, D. W. 1956. Courtship ritual: The dance of the manakins. *Animal Kingdom* 59:86–91.

1080   Sober, E., and D. S. Wilson. 1998. *Unto Others: The Evolution and Psychology of Unselfish Behavior*. Harvard University Press, Cambridge, MA.

1081   Sokolowski, M. B. 1998. Genes for normal behavioral variation: Recent clues from flies and worms. *Neuron* 21:463–466.

1082   Sokolowski, M. B., H. S. Pereira, and K. A. Hughes. 1997. Evolution of foraging behavior in *Drosophila* by density-dependent selection. *Proceedings of the National Academy of Sciences* 94:7373–7377.

1083   Soler, M., J. J. Soler, J. G. Martinez, and A. P. Møller. 1995. Magpie host manipulation by great spotted cuckoos: Evidence for an avian Mafia? *Evolution* 49:770–775.

1084   Soma, K. K., A. D. Tramontin, and J. C. Wingfield. 2000. Oestrogen regulates male aggression in the non-breeding season. *Proceedings of the Royal Society of London B* 267:1089–1092.

1085   Sommer, V. 1994. Infanticide among the langurs of Jodhpur: Testing the sexual selection hypothesis with a long-term record. In *Infanticide and Parental Care*, S. Parmigiani and F.S. vom Saal (eds.). Harwood Academic Press, Chur, Switzerland.

1086   Sorenson, L. G. 1994. Forced extra-pair copulation in the white-cheeked pintail: Male tactics and female responses. *Condor* 96:400–410.

1087   Soukup, S. S., and C. F. Thompson. 1998. Social mating system and reproductive success in house wrens. *Behavioral Ecology* 9:43–48.

1088   Spitler-Nabors, K. J., and M. C. Baker. 1987. Sexual display response of female white-crowned sparrows to normal, isolate and modified conspecific songs. *Animal Behaviour* 35:380–386.

1089   Stachowicz, J. J., and M. E. Hay. 1999. Reducing predation through chemically mediated camouflage: Indirect effects of plant defenses on herbivores. *Ecology* 80:495–509.

1090   Stachowicz, J. J., and M. E. Hay. 2000. Geographic variation in camouflaged specialization by a decorator crab. *American Naturalist* 156:59–71.

1091   Stander, P. E. 1992. Cooperative hunting in lions: The role of the individual. *Behavioral Ecology and Sociobiology* 29:445–454.

1092   Stanford, C. B. 1995. Chimpanzee hunting behavior. *American Scientist* 83:256–261.

1093   Starks, P., and C. Blackie. 2000. The relationship between serial monogamy and rape in the United States (1960–1995). *Proceedings of the Royal Society of London B* 267:1259–1263.

1094   Starks, P. T., D. J. Fischer, R. E. Watson, G. L. Melikian, and S. D. Nath. 1998. Context-dependent nestmate-discrimination in the paper wasp, *Polistes dominulus*: A critical test of the optimal acceptance threshold model. *Animal Behaviour* 56:449–458.

1095   Stehle, J. H., N. S. Foulkes, C. A. Molina, V. Simmonneaux, P. Pévet, and P. Sassone-Corsi. 1993. Adrenergic signals direct rhythmic expression of transcriptional repressor CREM in the pineal gland. *Nature* 365:314–320.

1096   Stein, Z., M. Susser, G. Saenger, and F. Marolla. 1972. Nutrition and mental performance. *Science* 178:708–713.

1097 Stenmark, G., T. Slagsvold, and J. T. Lifjeld. 1988. Polygyny in the pied flycatcher, *Ficedula hypoleuca*: A test of the deception hypothesis. *Animal Behaviour* 36:1646–1657.

1098 Sterck, E. H. M., D. P. Watts, and C. P. van Schaik. 1997. The evolution of female social relationships in nonhuman primates. *Behavioral Ecology and Sociobiology* 41:291–310.

1099 Stern, D. L., and W. A. Foster. 1996. The evolution of soldiers in aphids. *Biological Reviews* 71:27–80.

1100 Stern, D. L., and W. A. Foster. 1997. The evolution of sociality in aphids: A clone's-eye-view. In *Social Competition and Cooperation in Insects and Arachnids: II. Evolution of Sociality*, J. Choe and B. Crespi (eds.). Princeton University Press, Princeton, NJ.

1101 Stewart, K. J. 1987. Spotted hyaenas: The importance of being dominant. *Trends in Ecology and Evolution* 2:88–89.

1102 Stone, G. N. 1995. Female foraging responses to sexual harassment in the solitary bee *Anthophora plumipes*. *Animal Behaviour* 50:405–412.

1103 Stoutamire, W. P. 1974. Australian terrestrial orchids, thynnid wasps and pseudocopulation. *American Orchid Society Bulletin* 43:13–18.

1104 Strassmann, B. I. 1997. Polygyny as a risk factor for child mortality among the Dogon. *Current Anthropology* 38:688–695.

1105 Strassmann, J. E., C. R. Hughes, D. C. Queller, S. Turillazzi, R. Cervo, S. K. Davis, and K. F. Goodnight. 1989. Genetic relatedness in primitively eusocial wasps. *Nature* 342:268–269.

1106 Strum, S. C. 1987. *Almost Human*. W. W. Norton, New York.

1107 Stumpner, A., and R. Lakes-Harlan. 1996. Auditory interneurons in a hearing fly (*Therobia leonidei*, Ormiini, Tachinidae, Diptera). *Journal of Comparative Physiology A* 178:227–233.

1108 Stutt, A. D., and P. Willmer. 1998. Territorial defence in speckled wood butterflies: Do the hottest males always win? *Animal Behaviour* 55:1341–1347.

1109 Sullivan, J. P., O. Jassim, S. E. Fahrbach, and G. E. Robinson. 2000. Juvenile hormone paces behavioral development in the adult worker honey bee. *Hormones and Behavior* 37:1–14.

1110 Sündstrom, L. 1994. Sex ratio bias, relatedness asymmetry and queen mating frequency in ants. *Nature* 367:266–268.

1111 Sur, M., A. Angelucci, and J. Sharma. 1999. Rewiring cortex: The role of patterned activity in development and plasticity of neocortical circuits. *Journal of Neurobiology* 41:33–43.

1112 Surlykke, A., and J. H. Fullard. 1989. Hearing of the Australian whistling moth, *Hecatesia thyridion*. *Naturwissenschaften* 76:132–134.

1113 Susser, M., and Z. Stein. 1994. Timing in prenatal nutrition: A reprise of the Dutch famine study. *Nutrition Reviews* 52:84–94.

1114 Svensson, B. G. 1997. Swarming behavior, sexual dimorphism, and female reproductive status in the sex role-reversed dance fly species *Rhamphomyia marginata*. *Journal of Insect Behavior* 10:783–804.

1115 Swaddle, J. P. 1999. Limits to length asymmetry detection in starlings: Implications for biological signalling. *Proceedings of the Royal Society of London B* 266:1299–1303.

1116 Swaddle, J. P., and I. C. Cuthill. 1994. Preference for symmetrical males by female zebra finches. *Nature* 367:165–166.

1117 Swan, L. W. 1970. Goose of the Himalayas. *Natural History* 79(Dec):68–75.

1118 Sweeney, B. W., and R. L. Vannote. 1982. Population synchrony in mayflies: A predator satiation hypothesis. *Evolution* 36:810–821.

1119 Symons, D. 1979. *The Evolution of Human Sexuality*. Oxford University Press, Oxford, New York.

1120 Székely, T., J. D. Reynolds, and J. Figuerola. 2000. Sexual size dimorphism in shorebirds, gulls, and alcids: The influence of sexual and natural selection. *Evolution* 54:1404–1413.

1121 Taborsky, M. 1994. Sneakers, satellites, and helpers: Parasitic and cooperative behavior in fish reproduction. *Advances in the Study of Behavior* 23:1–100.

1122 Taborsky, M., and A. Grantnerm. 1998. Behavioural time-energy budgets of cooperatively breeding *Neolamprologus pulcher* (Pisces: Cichlidae). *Animal Behaviour* 56:1375–1382.

1123 Tella, J. L., M. G. Forero, J. A. Donázar, J. J. Negro, and F. Hiraldo. 1997. Non-adaptive adoptions of nestlings in the colonial lesser kestrel: Proximate causes and fitness consequences. *Behavioral Ecology and Sociobiology* 40:253–260.

1124 Temrin, H., and A. Arak. 1989. Polyterritoriality and deception in passerine birds. *Trends in Ecology and Evolution* 4:106–108.

1125 Temrin, H., and B. S. Tullberg. 1995. A phylogenetic analysis of the evolution of avian mating systems in relation to altricial and precocial young. *Behavioral Ecology* 6:296–307.

1126 Thomas, M. A., K. A. Walsh, M. R. Wolf, B. A. McPheron, and J. H. Marden. 2000. Molecular phylogenetic analysis of evolutionary trends in stonefly wing structure and locomotor behavior. *Proceedings of the National Academy of Sciences* 97:13178–13183.

1127 Thompson, C. W., N. Hillgarth, M. Leu, and H. E. McClure. 1997. High parasite load in house finches (*Carpodacus mexicanus*) is correlated with reduced expression of a sexually selected trait. *American Naturalist* 149:270–294.

1128 Thornhill, R. 1975. Scorpion-flies as kleptoparasites of web-building spiders. *Nature* 258:709–711.

1129 Thornhill, R. 1976. Sexual selection and nuptial feeding behavior in *Bittacus apicalis* (Insecta: Mecoptera). *American Naturalist* 119:529–548.

1130 Thornhill, R. 1981. *Panorpa* (Mecoptera: Panorpidae) scorpionflies: Systems for understanding resource-defense polygyny and alternative male reproductive efforts. *Annual Review of Ecology and Systematics* 12:355–386.

1131 Thornhill, R. 1992. Female preference for the pheromone of males with low fluctuating asymmetry in the Japanese scorpionfly (*Panorpa japonica*: Mecoptera). *Behavioral Ecology* 3:277–283.

1132 Thornhill, R., and S. W. Gangestad. 1999. Facial attractiveness. *Trends in Cognitive Sciences* 3:452–460.

1133 Thornhill, R., and C. T. Palmer. 2000. *A Natural History of Rape: The Biological Bases of Sexual Coercion*. MIT Press, Cambridge, MA.

1134 Thornhill, R., and N. W. Thornhill. 1983. Human rape: An evolutionary analysis. *Ethology and Sociobiology* 4:137–173.

1135 Tinbergen, L. 1960. The natural control of insects in pinewoods. 1. Factors influencing the intensity of predation by songbirds. *Archives Neerlandaises de Zoologie* 13:265–343.

1136 Tinbergen, N. 1951. *The Study of Instinct*. Oxford University Press, New York.

1137 Tinbergen, N. 1958. *Curious Naturalists*. Doubleday, Garden City, NY.

1138 Tinbergen, N. 1960. *The Herring Gull's World*. Doubleday, Garden City, NY.

1139 Tinbergen, N. 1963. On the aims and methods of ethology. *Zeitschrift für Tierpsychologie* 20:410–433.

1140 Tinbergen, N. 1963. The shell menace. *Natural History* 72(Aug):28–35.

1141 Tinbergen, N., and A. C. Perdeck. 1950. On the stimulus situations releasing the begging response in the newly hatched herring gull (*Larus argentatus* Pont.). *Behaviour* 3:1–39.

1142 Tobias, J. 1997. Asymmetric territorial contests in the European robin: The role of settlement costs. *Animal Behaviour* 54:9–21.

1143 Tobias, M. C., and G. E. Hill. 1998. A test of sensory bias for long tails in the house finch. *Animal Behaviour* 56:71–78.

1144 Toh, K. L., C. R. Jones, Y. He, E. J. Eide, W. A. Hinz, D. M. Virshup, L. J. Ptácek, and Y.-H. Fu. 2001. An h*Per2* phosphorylation site mutation in familial advanced sleep phase syndrome. *Science* 291:1040–1043.

1145 Tokarz, R. R., and D. Crews. 1981. Effects of prostaglandins on sexual receptivity in the female lizard, *Anolis carolinensis*. *Endocrinology* 109:451–457.

1146 Toma, D. P., G. Bloch, D. Moore, and G. E. Robinson. 2000. Changes in *period* mRNA levels in the brain and division of labor in honey bee colonies. *Proceedings of the National Academy of Sciences* 97:6914–6919.

1147 Tomkins, J. L., and L. W. Simmons. 1998. Female choice and manipulations of forceps size and symmetry in the earwig *Forficula auricularia* L. *Animal Behaviour* 56:347–356.

1148 Tovée, M. J., D. S. Maisey, J. L. Emery, and P. L. Cornelissen. 1999. Visual cues to female physical attractiveness. *Proceedings of the Royal Society of London B* 266:211–218.

1149 Townsend, J. M. 1989. Mate selection criteria: A pilot study. *Ethology and Sociobiology* 10:241–253.

1150 Tramontin, A. D., V. N. Hartman, and E. A. Brenowitz. 2000. Breeding conditions induce rapid and sequential growth in adult avian song control circuits: A model of seasonal plasticity in the brain. *Journal of Neuroscience* 20:854–861.

1151 Tramontin, A. D., J. C. Wingfield, and E. A. Brenowitz. 1999. Contributions of social cues and photoperiod to seasonal plasticity in the adult avian song control system. *Journal of Neuroscience* 19:476–483.

1152 Trivers, R. L. 1971. The evolution of reciprocal altruism. *Quarterly Review of Biology* 46:35–57.

1153 Trivers, R. L. 1972. Parental investment and sexual selection. In *Sexual Selection and the Descent of Man*, B. Campbell (ed.). Aldine, Chicago.

1154 Trivers, R. L. 1974. Parent-offspring conflict. *American Zoologist* 14:249–264.

1155 Trivers, R. L. 1985. *Social Evolution*. Benjamin/Cummings, Menlo Park, CA.

1156 Trivers, R. L., and H. Hare. 1976. Haplodiploidy and the evolution of the social insects. *Science* 191:249–263.

1157 Truman, J., and L. Riddiford. 1970. Neuroendocrine control of ecdysis in silkmoths. *Science* 167:1624–1626.

1158 Trumbo, S. T., and A.-K. Eggert. 1994. Beyond monogamy: Territory quality influences sexual advertisement in male burying beetles. *Animal Behaviour* 48:1043–1047.

1159 Tsubaki, Y., R. E. Hooper, and M. T. Siva-Jothy. 1997. Differences in adult and reproductive lifespan in the two male forms of *Mnais pruinosa costalis* Selys (Odonata: Calopterygidae). *Research in Population Ecology* 39:149–155.

1160 Tuttle, E. M., S. Pruett-Jones, and M. S. Webster. 1996. Cloacal protuberances and extreme sperm production in Australian fairy-wrens. *Proceedings of the Royal Society of London B* 263:1359–1364.

1161 Tyler, W. A. 1995. The adaptive significance of colonial nesting in a coral-reef fish. *Animal Behaviour* 49:949–966.

1162 Uetz, G. W., and E. I. Smith. 1999. Asymmetry in a visual signaling character and sexual selection in a wolf spider. *Behavioral Ecology and Sociobiology* 45:87–94.

1163 Urquhart, F. A. 1960. *The Monarch Butterfly*. University of Toronto Press, Toronto.

1164 Uy, J. A. C., and G. Borgia. 2000. Sexual selection drives rapid divergence in bowerbird display traits. *Evolution* 54:273–278.

1165 Vahed, K. 1998. The function of nuptial feeding in insects: Review of empirical studies. *Biological Reviews* 73:43–78.

1166 van Schaik, C. P., and P. M. Kappeler. 1997. Infanticide risk and the evolution of male-female associations in primates. *Proceedings of the Royal Society of London B* 264:1687–1694.

1167 van Staaden, M. J., and H. Romer. 1998. Evolutionary transition from stretch to hearing organs in ancient grasshoppers. *Nature* 394:773–778.

1168 Vetter, R. S. 1980. Defensive behavior of the black widow spider *Latrodectus hesperus* (Araneae: Theridiidae). *Behavioral Ecology and Sociobiology* 7:187–193.

1169 Vincent, A.C. J., I. Ahnesjö, A. Berglund, and G. Rosenqvist. 1992. Pipefishes and seahorses: Are they sex role reversed? *Trends in Ecology and Evolution* 7:237–241.

1170 Vincent, A. C. J., and L. M. Sadler. 1995. Faithful pair bonds in wild seahorses, *Hippocampus whitei*. *Animal Behaviour* 50:1557–1569.

1171 Vining, D. R., Jr. 1986. Social versus reproductive success: The central theoretical problem of human sociobiology. *Behavioral and Brain Sciences* 9:167–187.

1172 Vleck, C. M., and J. L. Brown. 1999. Testosterone and social and reproductive behaviour in *Aphelocoma* jays. *Animal Behaviour* 58:943–951.

1173 vom Saal, F. S., W. M. Grant, C. W. McMullen, and K. S. Laves. 1983. High fetal estrogen concentrations: Correlation with increased adult sexual activity and decreased aggression in male mice. *Science* 220:1306–1309.

1174 von Frisch, K. 1956. *The Dancing Bees*. Harcourt Brace Jovanovich, New York.

1175 von Frisch, K. 1967. *The Dance Language and Orientation of Bees*. Harvard University Press, Cambridge, MA.

1176 Vos, D. R. 1995. The role of sexual imprinting for sex recognition in zebra finches: A difference between males and females. *Animal Behaviour* 50:645–653.

1177 Waage, J. K. 1979. Dual function of the damselfly penis: Sperm removal and transfer. *Science* 203:916–918.

1178 Waage, J. K. 1997. Parental investment—minding the kids or keeping control? In *Feminism and Evolutionary Biology: Boundaries, Interactions, and Frontiers*, P.A. Gowaty (ed.). Chapman and Hall, New York.

1179 Wagner, R. H. 1992. Mate guarding by monogamous female razorbills. *Animal Behaviour* 44:533–538.

1180 Wagner, W. E., Jr. 1992. Deceptive or honest signalling of fighting ability? A test of alternative hypotheses for the function of changes in call dominant frequency by male cricket frogs. *Animal Behaviour* 44:449–462.

1181 Walcott, C. 1972. Bird navigation. *Natural History* 81(June):32–43.

1182 Walcott, C. 1996. Pigeon homing: Observations, experiments and confusions. *Journal of Experimental Biology* 199:21–27.

1183 Ward, P., and A. Zahavi. 1973. The importance of certain assemblages of birds as "information-centres" for food finding. *Ibis* 115:517–534.

1184 Watanabe, T. 1999. Prey attraction as a possible function of the silk decoration of uloborid spider *Octonoba sybotides*. *Behavioral Ecology* 10:607–611.

1185 Waterman, J. M. 1998. Mating tactics of male Cape ground squirrels, *Xerus inauris*: Consequences of year-round breeding. *Animal Behaviour* 56:459–466.

1186 Watson, P. J. 1998. Multi-male mating and female choice increase offspring growth in the spider *Neriene litigiosa* (Linyphiidae). *Animal Behaviour* 55:387–403.

1187 Watson, P. J., G. Arnqvist, and R. R. Stallmann. 1998. Sexual conflict and the energetic costs of mating and mate choice in water striders. *American Naturalist* 151:46–58.

1188 Watt, P. J., and R. Chapman. 1998. Whirligig beetle aggregations: What are the costs and the benefits? *Behavioral Ecology and Sociobiology* 42:179–184.

1189 Watt, P. J., S. F. Nottingham, and S. Young. 1997. Toad tadpole aggregation behaviour: Evidence for a predator avoidance function. *Animal Behaviour* 54:865–872.

1190 Waynforth, D., and R. I. M. Dunbar. 1995. Conditional mate choice strategies in humans—evidence from lonely hearts advertisements. *Behaviour* 132:755–779.

1191 Weatherhead, P. J., R. Montgomerie, H. L. Gibbs, and P. T. Boag. 1994. The cost of extra-pair fertilizations to female red-winged blackbirds. *Proceedings of the Royal Society of London B* 258:315–320.

1192 Weathers, W. W., R. S. Seymour, and R. V. Baudinette. 1993. Energetics of mound-tending behaviour in the malleefowl, *Leipoa ocellata* (Megapodiidae). *Animal Behaviour* 45:333–341.

1193 Weathers, W. W., and K. A. Sullivan. 1989. Juvenile foraging proficiency, parental effort, and avian reproductive success. *Ecological Monographs* 59:223–246.

1194 Webster, M. S. 1994. Female-defence polygyny in a Neotropical bird, the Montezuma oropendula. *Animal Behaviour* 48:779–794.

1195 Wedekind, C., and S. Füri. 1997. Body odour preferences in men and women: Do they aim for specific MHC combinations or simply heterozygosity? *Proceedings of the Royal Society of London B* 264:1471–1479.

1196 Wedekind, C., and M. Milinski. 1996. Human cooperation in the simultaneous and the alternating Prisoner's Dilemma: Pavlov versus Generous Tit-for-Tat. *Proceedings of the National Academy of Sciences* 93:2686–2689.

1197 Wedekind, C., and M. Milinski. 2000. Cooperation through image scoring in humans. *Science* 288:850–852.

1198 Wedell, N., and T. Tregenza. 1999. Successful fathers sire successful sons. *Evolution* 53:620–625.

1199 Wehner, R., M. Lehrer, and W. R. Harvey (eds.). 1966. Navigation: Migration and homing. *Journal of Experimental Biology* 199:1–261.

1200 Wehner, R., and S. Wehner. 1990. Insect navigation: Use of maps or Ariadne's thread? *Ethology, Ecology and Evolution* 2:27–48.

1201 Weidenmuller, A., and T. D. Seeley. 1999. Imprecision in waggle dances of the honeybee (*Apis mellifera*) for nearby food sources: Error or adaptation? *Behavioral Ecology and Sociobiology* 46:190–199.

1202 Weidinger, K. 2000. The breeding performance of blackcap *Sylvia atricapilla* in two types of forest habitat. *Ardea* 88:225–233.

1203 Weimerskirch, H., M. Salamolard, F. Sarrazin, and P. Jouventin. 1993. Foraging strategy of wandering albatrosses through the breeding season: A study using satellite telemetry. *Auk* 110:325–342.

1204 Welch, A. M., R. D. Semlitsch, and H. C. Gerhardt. 1998. Call duration as an indicator of genetic quality in male gray tree frogs. *Science* 280:1928–1930.

1205 Welty, J. 1975. *The Life of Birds* (2nd edition). Saunders, Philadelphia, PA.

1206 Wenner, A. M., and P. Wells. 1990. *Anatomy of a Controversy*. Columbia University Press, New York.

1207 Werren, J. H., M. R. Gross, and R. Shine. 1980. Paternity and the evolution of male parental care. *Journal of Theoretical Biology* 82:619–631.

1208 West, M. J., and A. P. King. 1985. Social guidance of vocal learning by female cowbirds: Validating its functional significance. *Zeitschrift für Tierpsychologie* 70:225–235.

1209 West, M. J., and A. P. King. 1990. Mozart's starling. *American Scientist* 78:106–114.

1210 Westcott, D. 1994. Leks of leks: A role for hotspots in lek evolution? *Proceedings of the Royal Society of London B* 258:281–286.

1211 Westcott, D., and J. N. M. Smith. 1997. Lek size variation and its consequences in the ochre-bellied flycatcher, *Mionectes oleagineus*. *Behavioral Ecology* 8:396–403.

1212 West-Eberhard, M. J. 1975. The evolution of social behavior by kin selection. *Quarterly Review of Biology* 50:1–33.

1213 West-Eberhard, M. J. 1979. Sexual selection, social competition, and evolution. *Proceedings of the American Philosophical Society* 123:222–234.

1214 Westneat, D. F. 1992. Do female red-winged blackbirds engage in a mixed mating strategy? *Ethology* 92:7–28.

1215   Westneat, D. F. 1994. To guard mates or go forage: Conflicting demands affect the paternity of male red-winged blackbirds. *American Naturalist* 144:343–354.

1216   Westneat, D. F., and R. C. Sargent. 1996. Sex and parenting: The effects of sexual conflict and parentage on parental strategies. *Trends in Ecology and Evolution* 11:87–91.

1217   Westneat, D. F., P. W. Sherman, and M. L. Morton. 1990. The ecology and evolution of extra-pair copulations in birds. *Current Ornithology* 7:330–369.

1218   Wetsman, A., and F. Marlowe. 1999. How universal are preferences for female waist-to-hip ratios? Evidence from the Hazda. *Evolution and Human Behavior* 20:219–228.

1219   Whaling, C. S., M. M. Solis, A. J. Doupe, J. A. Soha, and P. Marler. 1997. Acoustic and neural basis for innate recognition of song. *Proceedings of the National Academy of Sciences* 94:12694–12698.

1220   Wheelwright, N. T., C. B. Schultz, and P. J. Hodum. 1992. Polygyny and male parental care in savannah sparrows: Effects on female fitness. *Behavioral Ecology and Sociobiology* 31:279–289.

1221   Whitham, T. G. 1979. Habitat selection by *Pemphigus* aphids in response to resource limitation and competition. *Ecology* 59:1164–1176.

1222   Whitham, T. G. 1979. Territorial defense in a gall aphid. *Nature* 279:324–325.

1223   Whitham, T. G. 1980. The theory of habitat selection examined and extended using *Pemphigus* aphids. *American Naturalist* 115:449–466.

1224   Whitham, T. G. 1986. Costs and benefits of territoriality: Behavioral and reproductive release by competing aphids. *Ecology* 67:139–147.

1225   Whiting, M. J. 1999. When to be neighbourly: Differential agonistic responses in the lizard *Platysaurus broadleyi*. *Behavioral Ecology and Sociobiology* 46:210–214.

1226   Whittingham, L. A., and P. O. Dunn. 1998. Male parental effort and paternity in a variable mating system. *Animal Behaviour* 55:629–640.

1227   Whittingham, L. A., P. O. Dunn, and R. D. Macgrath. 1997. Relatedness, polyandry and extra-group paternity in the cooperatively-breeding white-browed scrub-wren (*Sericornis frontalis*). *Behavioral Ecology and Sociobiology* 40:261–270.

1228   Wickler, W. 1968. *Mimicry in Plants and Animals*. World University Library, London.

1229   Wickler, W., and U. Seibt. 1981. Monogamy in Crustacea and man. *Zeitschrift für Tierpsychologie* 57:215–234.

1230   Widemo, F. 1998. Alternative reproductive strategies in the ruff, *Philomachus pugnax*: A mixed ESS? *Animal Behaviour* 56:329–336.

1231   Widemo, F., and I. P. F. Owens. 1995. Lek size, male mating skew and the evolution of lekking. *Nature* 373:148–151.

1232   Wiederman, M. W., and E. R. Allgeier. 1992. Gender differences in mate selection criteria: Sociobiological or socioeconomic explanation? *Ethology and Sociobiology* 13:115–124.

1233   Wiederman, M. W., and E. Kendall. 1999. Evolution, sex, and jealousy: Investigation with a sample from Sweden. *Evolution and Human Behavior* 20:121–128.

1234   Wikelski, M., and S. Baurle. 1996. Pre-copulatory ejaculation solves time constraints during copulations in marine iguanas. *Proceedings of the Royal Society of London B* 263:439–444.

1235   Wikelski, M., M. Hau, and J. C. Wingfield. 1999. Social instability increases plasma testosterone in a year-round territorial neotropical bird. *Proceedings of the Royal Society of London B* 266:551–556.

1236   Wiklund, C. G., and M. Andersson. 1994. Natural selection of colony size in a passerine bird. *Journal of Animal Ecology* 63:765–774.

1237   Wilkinson, G. S. 1984. Reciprocal food sharing in the vampire bat. *Nature* 308:181–184.

1238   Wilkinson, G. S., and G. N. Dodson. 1997. Function and evolution of antlers and eye stalks in flies. In *The Evolution of Mating Systems in Insects and Arachnids*, J. C. Choe and B. J. Crespi (eds.). Cambridge University Press, Cambridge.

1239   Williams, G. C. 1966. *Adaptation and Natural Selection*. Princeton University Press, Princeton, N J.

1240   Williams, G. C. 1975. *Sex and Evolution*. Princeton University Press, Princeton, NJ.

1241   Williams, T. C., and J. M. Williams. 1978. An oceanic mass migration of land birds. *Scientific American* 239 (Oct.):166–176.

1242   Williams, T. J., M. E. Pepitone, S. E. Christensen, B. M. Cooke, A. D. Huberman, N. J. Breedlove, T. J. Breedlove, C. L. Jordan, and S. M. Breedlove. 2000. Finger-length ratios and sexual orientation. *Nature* 404:455–456.

1243   Willows, A. O. D. 1971. Giant brain cells in mollusks. *Scientific American* 224 (Feb):68–75.

1244   Wilmer, J. W., P. J. Allen, P. P. Pomeroy, S. D. Twiss, and W. Amos. 1999. Where have all the fathers gone? An extensive microsatellite analysis of paternity in the grey seal (*Halichoerus grypus*). *Molecular Ecology* 8:1417–1429.

1245   Wilson, D. S. 1997. Altruism and organism: Disentangling the themes of multilevel selection theory. *American Naturalist* 150:S122–S134.

1246   Wilson, E. O. 1971. *The Insect Societies*. Harvard University Press, Cambridge, MA.

1247   Wilson, E. O. 1975. *Sociobiology, The New Synthesis*. Harvard University Press, Cambridge, MA.

1248   Wilson, E. O. 1976. Academic vigilantism and the political significance of sociobiology. *BioScience* 26:187–190.

1249   Wilson, E. O. 1980. Caste and division of labor in leaf-cutter ants (Hymenoptera: Formicidae: *Atta*), II. The ergonomic optimization of leaf cutting. *Behavioral Ecology and Sociobiology* 7:157–165.

1250   Wilson, E. O. 1983. Caste and division of labor in leaf-cutter ants (Hymenoptera: Formicidae: *Atta*), III. Ergonomic resiliency in foraging by *A. cephalotes*. *Behavioral Ecology and Sociobiology* 14:47–54.

1251   Wilson, M. I., M. Daly, and S. Gordon. 1998. The evolved psychological apparatus of decision-making is one source of environmental problems. In *Behavioral Ecology and Conservation Biology*, T. M. Caro (ed.). Oxford University Press, New York.

1252   Wilson, N., S. C. Tubman, P. E. Eady, and G. W. Robertson. 1997. Female genotype affects male success in sperm competition. *Proceedings of the Royal Society of London B* 264:1491–1495.

1253   Wiltschko, W., P. Weindler, and R. Wiltschko. 1998. Interaction of magnetic and celestial cues in the migratory orientation of passerines. *Journal of Avian Biology* 29:606–617.

1254 Winfree, R. 1999. Cuckoos, cowbirds and the persistence of brood parasitism. *Trends in Ecology and Evolution* 14:338–343.

1255 Wingfield, J. C., J. Jacobs, and N. Hillgarth. 1997. Ecological constraints and the evolution of hormone-behavior interrelationships. *Annals of the New York Academy of Sciences* 807:22–41.

1256 Wingfield, J. C., and M. C. Moore. 1987. Hormonal, social and environmental factors in the reproductive biology of free-living male birds. In *Psychobiology of Reproductive Behavior: An Evolutionary Perspective*, D. Crews (ed.). Prentice-Hall, Englewood Cliffs, NJ.

1257 Wingfield, J. C., and M. Ramenofsky. 1997. Corticosterone and facultative dispersal in response to unpredictable events. *Ardea* 85:155–166.

1258 Winterhalder, B., and E. A. Smith. 2000. Analyzing adaptive strategies: Human behavioral ecology at twenty-five. *Evolutionary Anthropology* 9:51–72.

1259 Wise, K. K., M. R. Conover, and F. F. Knowlton. 1999. Response of coyotes to avian distress calls: Testing the startle-predator and predator-attraction hypotheses. *Behaviour* 136:935–949.

1260 Withers, G. S., S. E. Fahrbach, and G. E. Robinson. 1995. Effects of experience and juvenile hormone on the organization of the mushroom bodies of honey bees. *Journal of Neurobiology* 26:130–144.

1261 Wittenberger, J. F. 1981. *Animal Social Behavior*. Duxbury Press, Boston.

1262 Wolanski, E., E. Gereta, M. Borner, and S. Mduma. 1999. Water, migration and the Serengeti ecosystem. *American Scientist* 87:526–533.

1263 Woodroffe, R., and A. Vincent. 1994. Mother's little helpers: Patterns of male care in mammals. *Trends in Ecology and Evolution* 9:294–297.

1264 Woodward, J., and D. Goodstein. 1996. Conduct, misconduct and the structure of science. *American Scientist* 84:479–490.

1265 Woolfenden, G. E., and J. W. Fitzpatrick. 1984. *The Florida Scrub Jay: Demography of a Cooperative-Breeding Bird*. Princeton University Press, Princeton, NJ.

1266 Woyciechowski, M., L. Kabat, and E. Król. 1994. The function of the mating sign in honey bees, *Apis mellifera* L.: New evidence. *Animal Behaviour* 47:733–735.

1267 Wright, R. 1994. *The Moral Animal: Evolutionary Psychology and Everday Life*. Pantheon, New York.

1268 Wynne-Edwards, V. C. 1962. *Animal Dispersion in Relation to Social Behaviour*. Oliver & Boyd, Edinburgh.

1269 Yack, J. E. 1992. A multiterminal stretch receptor, chordotonal organ, and hair plate at the wing-hinge of *Manduca sexta*: Unravelling the mystery of the noctuid moth ear B cell. *Journal of Comparative Neurology* 324:500–508.

1270 Yack, J. E., and J. H. Fullard. 1990. The mechanoreceptive origin of insect tympanal organs: A comparative study of similar nerves in tympanate and atympanate moths. *Journal of Comparative Neurology* 300:523–534.

1271 Yack, J. E., and J. H. Fullard. 2000. Ultrasonic hearing in nocturnal butterflies. *Nature* 403:265–266.

1272 Yager, D. D., and M. L. May. 1990. Ultrasound-triggered, flight-gated evasive maneuvers in the flying praying mantis, *Parasphendale agrionina*. II. Tethered flight. *Journal of Experimental Biology* 152:41–58.

1273 Yanega, D. 1996. Sex ratio and sex allocation in sweat bees (Hymenoptera: Halictidae). *Journal of the Kansas Entomological Society* 69:98–115.

1274 Young, L. J., R. Nilsen, K. G. Waymire, G. M. MacGregor, and T. R. Insel. 1999. Increased affiliative response to vasopressin in mice expressing the V-1a receptor from a monogamous vole. *Nature* 400:766–768.

1275 Young, M. W. 2000. Marking time for a kingdom. *Science* 288:451–453.

1276 Yu, D. W., and G. H. Shepard, Jr. 1998. Is beauty in the eye of the beholder? *Nature* 396:321–322.

1277 Zach, R. 1979. Shell-dropping: Decision-making and optimal foraging in northwestern crows. *Behaviour* 68:106–117.

1278 Zahavi, A. 1975. Mate selection—A selection for a handicap. *Journal of Theoretical Biology* 53:205–214.

1279 Zeh, D. W., and R. L. Smith. 1985. Paternal investment in terrestrial arthropods. *American Zoologist* 25:785–805.

1280 Zeh, D. W., J. A. Zeh, and E. Bermingham. 1997. Polyandrous, sperm-storing females: Carriers of male genotypes through episodes of adverse selection. *Proceedings of the Royal Society of London B* 264:119–125.

1281 Zeh, J. A. 1997. Polyandry and enhanced reproductive success in the harlequin beetle-riding pseudoscorpion. *Behavioral Ecology and Sociobiology* 40:111–118.

1282 Zeh, J. A., S. D. Newcomer, and D. W. Zeh. 1998. Polyandrous females discriminate against previous mates. *Proceedings of the National Academy of Sciences* 95:13273–13736.

1283 Zeh, J. A., and D. W. Zeh. 1996. The evolution of polyandry I: Intragenomic conflict and genetic incompatibility. *Proceedings of the Royal Society of London B* 263:1711–1717.

1284 Zeh, J. A., and D. W. Zeh. 1997. The evolution of polyandry II: Post-copulatory defenses against genetic incompatibility. *Proceedings of the Royal Society of London B* 264:69–75.

1285 Zeh, J. A., and D. W. Zeh. 2001. Reproductive mode and the genetic benefits of polyandry. *Animal Behaviour*, in press.

1286 Zera, A. J., J. Potts, and K. Kobus. 1998. The physiology of life-history trade-offs: Experimental analysis of a hormonally induced life-history trade-off in *Gryllus assimilis*. *American Naturalist* 152:7–23.

1287 Zielinski, W. J., F. S. vom Saal, and J. G. Vandenbergh. 1992. The effect of intrauterine position on the survival, reproduction and home range of female house mice (*Mus musculus*). *Behavioral Ecology and Sociobiology* 30:185–192.

1288 Zucker, I. 1983. Motivation, biological clocks and temporal organization of behavior. In *Handbook of Behavioral Neurobiology: Motivation*, E. Satinoff and P. Teitelbaum (eds.). Plenum Press, New York.

1289 Zuk, M., and T. S. Johnson. 1999. Social environment and immunity in male red jungle fowl. *Behavioral Ecology* 11:146–153.

1290 Zuk, M., R. Thornhill, J. D. Ligon, K. Johnson, S. N. Austad, S. Ligon, N. Thornhill, and C. Costin. 1990. The role of male ornaments and courtship behavior in female choice of red jungle fowl. *American Naturalist* 136:459–473.

Illustration Credits

Unless cited below, full bibliographic information for all copyrighted illustrations, tables, or photographs can be found in the Bibliography. We are grateful to the following publishers for their permission to adapt or reprint copyrighted material.

Figures 2.10 © 1991; 2.14 © 1981; 2.19 © 1996; 4.4A © 1966; 4.10 © 1995; 4.12 © 1998; 4.24 © 1989; 6.12 © 1974; 6.18 © 1990; 8.4 © 1995; 8.15 © 1990; 8.23 © 1986; 8.24 © 1986; 8.25 © 1998; 8.30 © 1997; 9.25 © 1978; 9.26 © 1990; 10.10 © 1991; 10.11 © 1992; 10.21 © 1999; 11.8 © 2000; 11.34 © 1994; 12.5 © 2000; 12.31 © 1995; 12.32 © 1996; 13.8 © 1993; 13.9 © 1998; 13.18 © 1999; 14.1A © 1998; 15.14 © 1989 by Academic Press Limited, London.

Figures 2.6 © 1976; 3.9 © 1992; 5.33 © 1998; 5.35 © 1969; 6.8 © 2000; 6.13 © 1990; 9.21 © 1998; 10.15 © 1991; 11.23 © 1979; 13.11 © 1992 by the American Association for the Advancement of Science.

Figure 5.24 © 1996 by the American Institute of Biological Sciences.

Figures 5.34A © 1993; 6.16 © 1990 by the American Ornithological Union.

Figures 2.12 © 1979; 4.22 © 1990; 9.17 © 1977; 9.27 © 1997; 14.3 © 1994; 15.13 © 1990 by Blackwell Science.

Figures 11.21 © 1992; 15.5 © 1995; Discussion Question 2.2 © 1986 by Brill Academic Publishers.

Figure 3.10 © 1996 by Cell Press.

Discussion Question 1.2 from "The Far Side" by Gary Larson. Reproduced by permission of Chronicle Features, San Francisco.

Figures 3.15 © 1997; 5.38 © 1998; 9.15 © 1996 by the Company of Biologists, Limited.

Figures 8.20 © 2000; 11.33 © 1992 by CSIRO (*Australian Journal of Zoology*).

Figures 10.32 © 1997; 13.6 © 1997; 14.29 © 1996; 15.6 © 1992; 15.9 © 1993 by Cambridge University Press.

Figures 7.16 © 1999; 8.19 © 1990; 8.22 © 1986; 9.6 © 1986; 12.25 © 1999; 14.5 © 1986 by the Ecological Society of America.

Figures 4.17 © 1998; 5.30 © 2000; 11.44 © 1994; 15.21 © 1999; 15.24 © 1987; Discussion Question 4.5 © 1998; Discussion Question 12.5 © 1998 by Elsevier Science.

Figure 5.34B © 1990 by *Ethology, Ecology and Evolution*.

Figure 8.3 © 2000 by Karger AG Basel.

Figures 8.8 © 1998; 9.22 © 1996 by Kluwer Academic/Plenum Publishers.

Figures 2.11 © 1988; 2.17 © 1987; 3.3 © 1992; 3.11 © 2000; 3.12 © 1998; 6.9 © 1987; 6.10 © 1996; 8.16 © 1987; 10.2 © 1995; 10.8 © 1997; 10.9 © 1997; 10.29 © 1955; 10.30 © 1978; 11.24 © 1983; 11.28 © 2000; 11.37 © 1998; 11.41 © 1994; 12.11 © 1992; 12.35 © 1995; 13.23 © 1994; 14.9 © 2000; 14.22 © 1997 by Macmillan Magazines Limited.

Figure 5.31 © by MIT Press.

Figures 3.17 © 1997; 4.7 © 1993; 4.25 © 1993; 11.29 © 1996; 11.43 © 1999 by the National Academy of Sciences, USA.

Figures 5.18 © 1998; 6.25 © 1997 by the New York Academy of Sciences.

Figures 1.5 © 1951; 5.32 © 1999; 10.13 © 1998; 14.19 © 1998 by Oxford University Press.

Figure 10.18 © 1987 by Prentice-Hall, Inc., Englewood Cliffs, NJ.

Figure 4.34 © 1998 by the Psychonomic Society.

Figures 4.14B © 2000; 5.29 © 1998; 6.26 © 2000; 7.20 © 2000; 9.12 © 1998; 10.14 © 2000; 10.16 © 1995; 10.23 © 1998; 10.24 © 1998; 10.25 © 1994; 11.3 © 1997; 11.17 © 1996; 11.27 © 1999; 11.35 © 1998; 11.40 © 2000; 12.6 © 2000; 12.7 © 1997; 13.4 © 1998; 13.10 © 1998; 13.22 © 2000; 14.23 © 1998; 14.24 © 2000; 15.7 © 1999; 15.8 © 1998; 15.11 © 1999; 15.19 © 1999; Discussion Question 6.6 © 2000 by the Royal Society of London.

Figures 5.15 © 1991; 6.7 © 1986; 8.11 © 1999; 11.36 © 1990; 15.2 © 1995 by Sigma Xi (*American Scientist*).

Figures 4.19 © 2000; 7.26 © 1982; 8.27 © 1992; 11.42 © 1998; 12.12 © 1999; Discussion Question 2.1 © 1965 by the Society for the Study of Evolution (*Evolution*).

Figures 3.4 © 1991; 4.3 © 1982; 5.40 © 1990; 7.25 © 1986; 7.27 © 1998; 7.29 © 1995; 8.2 © 1997; 8.12 © 1988; 8.13 © 1988; 8.14 © 1999; 9.3 © 1999; 9.13 © 1996; 10.22 © 1997; 11.38 © 1994; 12.8 © 1987; 12.9 © 1999; 12.33 © 1993; 13.20 © 1999; Discussion Question 14.6 © 1997 by Springer Verlag.

Figures 6.15 © 1964; 6.24 © 1999; 8.33 © 1998; 9.7 © 1998; 10.17B © 1998; 13.12 © 1998; 14.4 © 1998 by the University of Chicago Press.

Figure 9.10 © 1976 by the University of Massachusetts Press, Amherst, MA.

Figures 2.5 © 1997; 2.8 © 1997; 3.6 © 1997; 4.30 © 1999; 5.25 © 1997; 10.5 © 1992; 10.6 © 1978 by Wiley-Liss, Inc., a subsidiary of John Wiley & Sons, Inc.

Photography Credits

Copyrights to the photographs in this textbook are held by the photographers listed in the figure captions and by the following agencies.

Bruce Coleman Inc.: Figures 1.12 and 7.22.

The Biological Photo Service: Figure 15.1.

Corbis Images: Figures 7.1 and 9.1B.

Minden Pictures: Figures 8.34 and 14.25A; Chapter 1 and Chapter 9 openers.

Oxford Scientific Films: Figures 4.9, 5.7, 9.22A, and 13.2A.

Photo Researchers, Inc.: Frontispiece; Chapter 3, Chapter 5, Chapter 6, Chapter 8, Chapter 10, and Chapter 15 openers; Figures 1.7, 1.10, 7.10A, 7.10B, 8.32, 10.23, 10.26, 10.31C, 11.13, and 14.12.

Planet Earth Pictures: Figure 8.28.

TimePix: Figures 4.8, 4.31, and 4.32.

VIREO: Figures 9.23 and 12.29.

Index